ECOLOGICAL COMMUNITIES

Ecological Communities:
Conceptual Issues and the Evidence

Edited by

DONALD R. STRONG, JR.

DANIEL SIMBERLOFF

LAWRENCE G. ABELE

ANNE B. THISTLE

PRINCETON UNIVERSITY PRESS
PRINCETON, NEW JERSEY

Library of Congress Cataloging in Publication Data will be found on the last
printed page of this book

ISBN 0-691-08340-1
0-691-08341-X (pbk.)

This book has been composed in Linotron Times Roman with Helvetica
display
Clothbound editions of Princeton University Press books are printed on
acid-free paper, and binding materials are chosen for strength and durability.
Paperbacks, although satisfactory for personal collections, are not usually
suitable for library rebinding

Printed in the United States of America by Princeton University Press
Princeton, New Jersey

This book is dedicated to
RICHARD P. SEIFERT
whose critical experiments
with natural communities
are a model for
contemporary ecology.

Preface

Ecological communities are groups of species living closely enough together for the potential of interaction. Various levels of ecological effect might result from interactions that do occur, from slight, intermittent influences to intense, enduring ones that are so profound as to lead to regulation of the interactant populations. We organized a symposium, held on March 11, 12, and 13, 1981, at Wakulla Springs, Florida, to emphasize the tremendous range of possibilities for organization and working of different communities; this book is the result. Species might compete with, prey upon, parasitize, or infect one another, or even aid one another mutualistically. However, all possible interactions do not occur, and those that do need not be intense or persistent and need not be cybernetic, nor need they have enduring evolutionary effect. One possibility that we expect to obtain fairly commonly is the community with so few strong interactions that organization arises primarily from mutually independent autecological processes rather than from synecological ones. Such communities would not be holistic entities, but rather just collections of relatively autonomous populations in the same place at the same time. Independent coexisting populations would likely be exposed to some of the same exogenous forces, such as the weather, but equally likely their reactions to externalities would be sufficiently different to make knowledge of each species' autecology necessary to an understanding of the group as a whole. Thus, good population ecology is a basis for good community ecology.

The contemporary questions in community ecology concern the existence, importance, looseness, transience, and contingency of interactions. Which interactions actually occur? How do interactions vary among species, environments, and locales? What fraction of ecological interactions between species have any enduring influence upon morphology, behavior, or reproductive phenotype? How do different kinds of interactions among a group of coexisting species—predation, disease, competition, etc.—meld for some net influence? The most profound issue of contemporary ecology, indeed the issue that may most distinguish ecological phenomena from those of finer levels of biological organization, stochasticity, makes deductive answers to all of these questions doubly difficult to find. Are interactions among species substantially changed by externally imposed fluctuations in exogenous factors such as the weather or casually associated predators? What influence does the inexorable wander of climate have on interactions with subtle effects, which may take many generations to work their influence on species?

The key to resolving these conceptual issues is evidence. We have emphasized evidence as a topic of this symposium because community ecology does not have a strong tradition of using the diversity of evidence that has been so powerful in other disciplines. Sciences that have progressed rapidly (physics, chemistry, molecular biology) have made great use of sorts of evidence that ecology has not, of vigorous hypothesis testing and experimentation. Vigorous hypothesis testing maximizes the potential of falsifying ideas. In contrast, much of community ecology has often been content with generalized mathematical theory and passively (rather than experimentally) collected observations. Most important, mainstream community ecology has accepted confirmatory evidence almost exclusively. One conventional approach to community ecology is to seek corroboration in broad patterns among species for the generalized theory of competitive relations among species. Confirmation of the orthodox "neo-Malthusian" view, that interspecific competition for limited, depletable resources is the prime factor in community organization and evolution, has been inferred from observational data of many sorts, e.g. from relative abundances of species, their taxonomic mixes in communities, their relative morphologies, etc. (Hutchinson, 1978).

Certainly, rapidly progressing sciences have used confirmatory evidence too, but of a fundamentally different sort than is traditional in community ecology. In physics, E. Rutherford's experiments showing heterogeneous scattering of alpha particles by metal foil corroborated a planetary, non-homogeneous model of the atom. In molecular genetics, experiments showing semi-conservative reduplication of DNA corroborated the double-helix model of the genetic material. The fundamental difference is that these classic corroborations were of surprising, "risky" theoretical predictions, whereas those in conventional community ecology are not. "Riskiness" of a theoretical prediction is a characteristic argued by Karl Popper to be a key to vigorous science, and we agree. The risk of a theoretical prediction increases with the number of distinct, possible, qualitatively contrary outcomes for an experiment. *A priori*, alpha particles could have been scattered homogeneously, to imply a very different sort of atomic model from that implied by Rutherford's experiment. Before the experiments, replication of DNA could have been found to be at either extreme, conservative or semi-conservative, or some fraction in between. The important feature of the experiment was that it had tremendous potential to falsify the double-helix model.

In community ecology, we have learned that there is little general risk for the orthodox competition theory in broad interspecific patterns in nature; the gamut of possible patterns are all seen as supporting the theory. Classic examples of this approach are the inference that competition, albeit in different forms, causes the different shapes of observed relative abundance curves of species in large samples. In this volume, Colwell and Winkler correctly point out that correlations of dispersal success with taxonomy can result in morphological patterns among species that do not reflect the effects of interspecific competition on character displacement. This is a result adumbrated by Grant and Abbott (1980). Thus, interspecific patterns alone are no risk for competition theory, and conversely, interspecific patterns are only the softest corroboration of the theory. Without hard evidence on the existence of interspecific competition, its intensity, endurance, and ability to overcome the influences of other factors, and most importantly, without hard evidence on the ability of competition to cause morphological differences between species, the first link in the necessary chain of evidence for establishing competition-caused character displacement as a general property of communities is missing. Equally important, hard evidence is needed on how other factors, such as dispersal, can modify any previous interspecific patterns among species. Although hard evidence on these crucial questions has not yet been produced, we hope that the acute scrutiny that this particular facet of community ecology receives in this volume will stimulate critical studies.

Of course, that much evidence in community ecology to date is but "soft corroboration," with little potential to falsify theory, does not mean that the theory is incorrect, only that the evidence is flimsy. The difficulty of experiments on communities, deriving from the large number of simultaneously relevant variables, the sometimes vast scale, and often long periods of time that must be accommodated for meaningful data, means that passively obtained observational data will continue to play a role in our science. So, one of our major goals for this symposium is to include discussions of the most critical possible means of dealing with non-experimental data, of tests that can potentially falsify theoretical predictions, even when manipulation and experimentation are impossible.

However, we believe that the most powerful means of critically testing propositions in community ecology is just that which has proven most powerful in other fields, experimentation. However, doing experiments in ecology is not just like doing them in other fields. Community ecology's uniqueness lies in the extraordinary diversity, variation, and heterogeneity of its objects of study, and in their historical legacy accumulated through evolution. Individual organisms, populations, species, and communities all are many orders of magnitude more variable internally than are the objects studied in physics, chemistry, or molecular biology, and the art in our science must reflect the difficulty of dealing with this variability. Differences among individuals such as age, sex, and nutritional condition are extremely important to how populations function, and likewise up through communities and ecosystems. Slight differences among species in trophic habit, ecological amplitude, and life history greatly affect the likelihood, intensity, and persistence of interactions among species. Most challenging, the stochasticity of the external environmental further varies population and community behavior. The same species in different environmental circumstances may have quite distinct ecology and influences upon other species. These contingencies mean that powerful experiments in community ecology must take into account autecology and natural history. Knowledge of autecology and natural history concerns species particulars and idiosyncracies, and only with this knowledge can experiments in community ecology reflect the actual influences of interactions among species.

L. G. Abele
D. S. Simberloff
D. R. Strong
A. B. Thistle

Tallahassee, Florida
May 1982

Contents

Marine Community Paradigms

Morphology, Species Combinations, and Coexistence

Food Web Design

Community Changes in Time and Space

Introduction

1.

An Overview: Real and Apparent Patterns in Community Structure

ROBERT M. MAY

Biology Department, Princeton University, Princeton, New Jersey 08544

"Two noticeable characteristics of papers recently published are the widespread interest in field quantitative methods in the study of population density, rates of spread, fluctuation, reproduction, feeding, or mortality; and an increasing awareness of evolutionary dynamic problems such as selection and competition. At the same time there is a general adoption and a tightening up of the statistical treatment of ecological data, which, though entirely sound and necessary, would become a partly bad feature if it tended to exclude the equally valuable type of observations on the pattern of nature, and the habitats and distribution of animals, that ecologists and naturalists can contribute." (Elton, 1947)

INTRODUCTION

In 1944, the British Ecological Society held a symposium on "The Ecology of Closely Allied Species," at which Lack, Elton, Varley, and others used various lines of evidence to argue that competition is a major factor in structuring plant and animal communities. Others argued to the contrary, with Diver contending that "the mathematical and experimental approaches had been dangerously oversimplified and omitted consideration of many factors [including] sources of energy and their relative availability, predator attack, mobility, population structure and growth, individual growth rate and bulk, relation of life cycle to annual cycle, range of tolerance, means of dispersal, and the like" (Anon., 1944). He concluded "there was little direct evidence that cohabitation or separation of related species was determined by space and food, since other factors usually kept populations below the point at which serious pressure was developed." Broadly similar themes dominated the celebrated Cold Spring Harbor Symposium in 1957, with some arguing that density-dependent effects arising from biological interactions are of predominant importance in setting population levels, while others argued the importance of the density-independent regulatory effects caused by the weather and other environmental factors. The Brookhaven Symposium of 1969 on "Diversity and Stability in Ecological Systems" again drew together many of the contemporary leaders of the subject; I think it gave a less polarized and more synthetic account of the issues, although (as the title itself suggests) there may have been too much of a tendency to view communities as orderly, patterned "systems." The present volume stems from a conference held at Wakulla Springs in 1981, and the same themes still interweave, albeit now greatly enriched by a rapidly expanding body of field observations, carefully planned experimental manipulations in the field and laboratory, and more rigorous techniques of statistical evaluation of the data. Whether these themes are drawing toward their resolution, or whether we are still in the opening passages of the work, is for the reader of this volume to decide.

An eager and naive pattern-seeker might note that these landmark meetings are regularly spaced, with a 12-year period, and might even go on to speculate on the underlying cause of this cycle (12 years is roughly the time from entering graduate school to the tenure decision?). This is silly. The "cycle" does, however, serve to illustrate one central concern of the Wakulla Springs Conference: given some apparent pattern in the organization of an ecological community, does it really derive from biological interactions among and within species? Or is it the sort of coincidence one often finds when the data are few? Or may the investigator have produced it, unconsciously, by making observations or designing experiments to conform to a preconceived notion? Or may the pattern simply be a statistical property of the system—a true pattern, but having no biological significance? Such questions, involving the disentangling of real from apparent patterns, occur in many other areas of science (see, for example, the debate between Arp and Bahcall (1973) on whether there are, or are not, significant spatial associations between astronomical objects with large but different red shifts); the questions are rarely easy to answer.

The apparent 12-year cycle is spoiled by, *inter alia*, the meeting on the "Ecology and Evolution of Communities" held in 1973 as a memorial to Robert MacArthur (Cody and Diamond, 1975a). Unlike the more wide-ranging meetings mentioned above, this one was mainly concerned with those areas, and that style, where MacArthur's own contributions had been so stimulating: field observations and theoretical models aimed at understanding how communities are structured by biological interactions, particularly competition. In some ways, the 1981 Wakulla Springs meeting represents a healthy reaction against too enthusiastic and uncritical an acceptance of some of the "pattern-seeking" field and theoretical studies of the two past decades. One technique, set out in some recent papers in this volume and elsewhere, is to construct "neutral models" or "null hypotheses," which aim to elucidate those apparent patterns that might be exhibited when comparisons are made among a given set of communities, assuming that some, or all, classes of biological interactions are absent. Insofar as comparative studies of real communities yield patterns beyond those found in such neutral models, one can be more confident that community structure is indeed being forged by particular kinds of interactions among species. More specifically, neutral models are being employed in an effort to see whether biologically based patterns must necessarily be attributed to competition, or whether predation or other effects (including, I would emphasize, pathogens or parasites) could equally well be the cause.

Although the phrases null/neutral hypothesis/model are characteristic of the relatively recent literature, the essential ideas have been applied in various contexts in community ecology over the years. (Indeed, any assignment of a significance level to a regression line is explicitly a statement about the rejection of a null hypothesis.) Unfortunately, an appropriate null hypothesis is not always easy to construct; in particular, a proper neutral model may require data that simply are not available. It is just as easy—and just as foolish—to construct an inappropriate or misleading neutral model as it is glibly to deduce evidence for competition from data that are susceptible to other interpretations.

In what follows, I briefly outline a miscellany of examples in which (whether or not the contemporary phraseology is employed) neutral models have been used in the hope of elucidating theoretical or empirical aspects of community structure; some of these examples are covered in more detail elsewhere in this volume, although most are not. Most of these vignettes are complex, and do not admit of Manichean division into white-hatted and black-hatted people. To my mind, no simple moral emerges from these tales, other than the broad injunction that alternative explanations in general, and appropriate neutral models in particular, should always be kept in view when experiments are designed or data analyzed.

I must emphasize, most strongly, that the contents of this introductory chapter do not accurately reflect the amounts of time spent on various topics at the conference itself. One of the main aims of the conference was to focus on analytically designed field studies that test theoretical ideas. Many of the papers did just this, presenting interesting and previously unpublished data (as, for example, in the chapters by Lawton, by Strong, by Rey, and by Grant and Schluter). Although such case studies predominated at the conference, and form the bulk of this consequent book, many participants' clearest memories will be of the disagreements—good-humored but nonetheless sharp—over theoretical and methodological issues; my chapter dwells exclusively on these issues.

NUMBER OF SPECIES PER GENUS

At the British Ecological Society meeting in 1944, Elton presented an analysis of 55 animal communities (including some parasite ones) and 27 plant communities, each from a relatively small geographical area. He showed that, in these communities, the average number of species per genus (the "S/G ratio") was markedly smaller than that for faunal lists from any large region, and attributed this difference to "existing or historical effects of competition between species of the same genus, resulting in a strong tendency for the species of any genus to be distributed as ecotypes in different habitats, or if not, to be unable to coexist permanently on the same area of the same habitat" (Elton, 1946).

C. B. Williams (1964), however, subsequently pointed out that it is a property of the statistical distribution of species among genera that, as the number of species and genera in a sample decrease (as they will when one goes from a larger region to a smaller), the ratio S/G will decrease. More recently, Grant (1966b) and Moreau (1966) have sought to find evidence for competition in the smaller S/G ratios observed for birds on islands or in restricted habitats, while Simberloff (1970) has given an incisive analysis (including extensive numerical simulations) to show that just such decreases in S/G with decreasing S are mathematical properties of the S-G distribution.

This cautionary tale is fairly straightforward. Although the observed S/G pattern appears to be just what one would expect if communities are structured by competition, closer examination shows the pattern to be primarily a statistical artifact, a mathematical property of the way the average S/G ratio varies with S. Insofar as the observed S/G patterns do differ slightly from mathematical expectation, the S/G ratios in restricted habitats appear to be relatively high rather than relatively low (the data points tend to lie slightly above the line derived from the null hypothesis); a more full discussion is given by Strong (1980) and by Simberloff in this volume. Notice that, as Simberloff has repeatedly emphasized, the explanation of the S/G pattern by a null hypothesis does not mean that competition is necessarily unimportant in determining which species co-occur in the communities studied by Elton, Grant, Moreau, and others; rather, it means this particular line of inquiry simply sheds little light.

STABILITY AND COMPLEXITY

The idea that complex ecosystems, with many species and a rich web of interactions, should be more stable than simple ones is an intuitively appealing one; it may seem that a community is better able to cope with disturbance if there are many alternative pathways along which energy and nutrients may flow. Elton (1958) advanced a set of six arguments in support of this notion that complexity begets stability. One of the six was a theoretical argument and consisted of the observation that mathematical models of simple prey-predator associations exhibit instability (Elton had in mind the neutrally stable Lotka-Volterra model for one prey and one predator, and the unstable Nicholson-Bailey model for host-parasitoid interactions). Whatever the status of the other five arguments (May, 1973, pp. 37–40, 173), this theoretical observation is meaningless until one has determined the stability properties of the analogous models with many predators and many prey. Such multispecies models turn out, in general, to be less stable the more species are present. That is, increasing dynamical stability is not a general mathematical consequence of increasing complexity; rather, the contrary is true.

I think this example belongs in a broad discussion of the uses of null hypotheses, because it provides an illuminating instance where an attractive idea was long accepted (and still is in many Introductory Biology texts) on the basis of logically incomplete arguments. Real communities, of course, are not random selections from the universe of general mathematical models, and the current task is to try to understand the special structural features that complex ecosystems may possess to help them reconcile stability with complexity. Are apparently complex tropical ecosystems actually constituted of many loosely coupled subsystems (Gilbert, 1977; Root, 1973)? Do dynamical considerations constrain the length of trophic chains (Pimm and Lawton, 1977; DeAngelis et al., 1978; Lawton and Pimm, 1978)? Is "donor control" (DeAngelis, 1975), or the character of predators' functional responses (Nunney, 1980), or some other feature, crucial in distinguishing real food webs from those that may seem possible in general? Or may it be that complex ecosystems really are typically more fragile, being found only in environments where disturbances are typically less severe or more localized than is the case for simpler ecosystems (Wolda, 1978; May, 1979)?

The example is also interesting for the light it sheds on the generation of hypotheses, null and otherwise. It is lunacy to imagine that the dynamical behavior of real communities bears anything but the vaguest metaphorical relation to the linearized stability properties of the conventional "community matrix" (Levins, 1975; May, 1973). But analyses of abstract community matrices have led to the generation of new ideas and the framing of testable hypotheses, such as those about the patterns of connectance in real food webs (Yodzis, 1980; Rejmánek and Starý, 1979), about the lengths of trophic chains, about the structuring of communities in terms of subunits or guilds, and so on. Some of this work is developed more fully in the chapters by Pimm, Auerbach, and Lawton.

PATTERNS IN THE RELATIVE ABUNDANCE OF SPECIES

An interesting attempt to employ neutral hypotheses in the exploration of patterns of numbers and relative abundance of species in communities is by Caswell (1976). He begins by attempting to determine the likely distribution of relative abundance of individuals among S species, assuming *no* biological interactions among them. Comparing these "neutral" distributions of species relative abundance with real distributions for birds, fish, insects, and plants in tropical and in temperate zones, Caswell finds real communities to be less diverse (both in the sense of fewer species and in the sense of greater dominance by a few common species) than would be the case in the absence of interspecific interactions. The discrepancy is greatest in the tropics, where biotic effects are thought to be most pronounced. Caswell concludes that the diversity of natural communities may be maintained in spite of, rather than because of, such biological interactions between species. Although it is possible to cavil, on technical grounds, at the appropriateness of the neutral hypothesis used by Caswell to generate interaction-free distributions of relative abundance, his attempt to bring these methods to bear on questions of relative diversity is most original, and deserves more attention than it has received.

One pervasive pattern that has been widely remarked for mature (as opposed to early successional or disturbed) communities of plants, moths, birds, diatoms, and other taxa, is that the distribution of relative abundance is not only lognormal, but is "canonically" lognormal (Preston, 1962; MacArthur and Wilson, 1967). This canonical lognormal is a particular member of the one-dimensionally infinite family of lognormal distributions, corresponding to a particular relationship between the number of species, S, and the variance of the distribution, σ^2 (this relationship is often conveniently parameterized as "$\gamma = 1$"; for a full discussion, see May, 1975). Another associated pattern, first explicitly remarked by Hutchinson (1953), is that the conventional parameter a (which is essentially an inverse measure of the standard deviation of the distribution of species relative abundance, $a = 0.71/\sigma$) seems always to have a value $a \sim 0.2$. Such generally observed patterns cry out for explanation. I have sought to explain both these phenomena as being no more than likely mathematical properties of any lognormal distribution of species, provided the collection is large enough ($S \gg 1$; May, 1975). While I think it remains true that this "neutral hypothesis" accounts for the roughly constant magnitude of the inverse variance, $a \sim 0.2$, a more careful analysis of the available data by Sugihara (1980) suggests that real distributions of species relative abundance conform to the "canonical" relationship between S and σ more closely than seems explainable by mathematical properties of the distribution alone. Sugihara has, indeed, gone on to advance a possible biological mechanism underlying the structure of mature communities: the sequential division of niche space by closely related species. This mechanism does lead to expectations of distributions of species relative abundance that are remarkably close to the observed data (Sugihara, 1980).

This is an instructive story. Earlier work took it for granted that the canonical pattern derived in some way from the biological structuring of the community. My subsequent attempt to explain the observed canonical distribution on "neutral" grounds, as being a mathematical property of the essentially randomly determined relative abundance of individuals among a large number of species, now appears to have been too glib; the canonical relationship is obeyed too closely to be explained in this general way. But, as Sugihara emphasizes, the fact that his "sequential niche breakage" model accounts for the observed distributions does not mean it is necessarily correct. It could yet be that a carefully framed neutral model could account for the observed patterns, thus bringing this story full circle for the second time.

SPECIES-AREA RELATIONS

Many authors have studied the empirical relation between the number of species on an island, S, and the area of the island, A. These studies of birds, plants, insects, and other taxa embrace both archipelagoes and other collections of real islands, and assemblies of virtual islands such as ponds, woodlots, or nature reserves. A log-log plot of S against A usually shows a linear relation, of the form

$$S = (\text{constant})\, A^z, \tag{1}$$

with z having a value around 0.2–0.3. As pointed out by Preston (1962) and by MacArthur and Wilson (1967), by assuming a canonical lognormal distribution of N individuals among S species, and adding the assumption that N is linearly proportional to island area, A, one can derive equation (1) with $z \simeq 0.25$ for $S \gg 1$ (see May, 1975).

Connor and McCoy (1979) have suggested these observations may be explained by a null model. They observe that z is evaluated as the slope of a regression line, and thus can be expressed as the product of a correlation coefficient (r) and the ratio of the standard deviations of the dependent and independent variables (S_y/S_x). As a null hypothesis, they suggest that both r and the ratio S_y/S_x vary independently randomly between 0 and 1. This hypothesis leads to a relation of the form of equation (1) between S and A, with the regression coefficient z being, on average, the expected value of the product of two numbers each distributed uniformly on the interval 0 to 1, whence $\langle z \rangle = 0.25$. Thus Connor and McCoy's neutral hypothesis seems to be consistent with the empirical facts, apparently undercutting any need to invoke biological interactions in understanding equation (1).

Sugihara (1981) has, however, noted that the neutral model of Connor and McCoy gives not merely the average value of z, but also the full distribution of z-values that we would expect to see when many such studies are tabulated. Sugihara shows that the z-values actually found in analyses of real data cluster more closely around 0.2–0.3 than does the distribution of values to be expected from the neutral model. Thus, on the basis of this more sensitive test, Connor and McCoy's null hypothesis may be rejected as an explanation for the observed S-A relationships. To put it another way, although the mean value of z predicted by the null model is consistent with the data, the variance in the null model's distribution of z-values is significantly greater than that observed in nature.

There is a useful message here. The null hypothesis cannot be rejected if we look only at the predictions it makes about the average values of a pertinent parameter. But if we probe deeper, looking at the variance or at the full distribution of the relevant parameter values, the actual data can be seen to be too tightly clustered to be consistent with the null hypothesis. This general theme will be heard again, more diffusely, below.

ASSEMBLY RULES AND INDIRECT EVIDENCE FOR COMPETITION OR CONVERGENCE

Many studies seek to show the convergence in the structure of communities in environmentally similar but geographically separated parts of the world. Other broadly related studies aim to find empirical rules governing the incidence of particular species on real or virtual islands of a given size, and governing the assembly of such island communities.

Recently, several people have endeavored to test this work against null models, in order to determine whether the apparent patterns are significantly different from what would be observed if the communities consisted of random collections of species, unstructured by competition or other biological interactions. There is, however, a profound difficulty in the construction of some of these neutral models, which usually are obtained by reshuffling of the observed data. This procedure is open to the objection that, if the real communities have been highly influenced by competition, one cannot construct a truly neutral model by pooling and redistributing the data. The present book contains a good representation of the range of views that are held on this subject, particularly in the chapters by Simberloff, Grant, Gilpin and Diamond, and Colwell and Winkler.

An illustrative example not covered elsewhere in this book is provided by the exchange between Fuentes (1976, 1980) and Crowder (1980). Fuentes (1976) studied the structure of lizard communities in physiognomically similar sites in North and South America, and showed that such communities were more similar to each other than to lizard communities in nearby sites on an altitudinal and vegetational gradient. Crowder (1980) suggested this evidence could not be taken to indicate community-level convergence, because the patterns were not significantly distinguishable from a neutral model that he constructed. But Fuentes (1980) objected, in my view with justification, that "to use the same species and ecospecies numbers I used [in order to construct the so-called neutral model] is to assume my results," rather than to subject them to an independent test.

Gilpin and Diamond (in this volume) present an illustrative example in which a null model that is constructed by drawing at random (subject to specified constraints) from the pooled data can, in certain limits, lead to nonsense. Suppose one has N islands, on each of which either species A or species B is present, but never both and never neither. If N is large, one would be inclined, intuitively, to regard this "checkerboard" pattern as evidence for competition between species A and B. But it is, of course, desirable to test the pattern against some neutral model, to add rigor to this conclusion. The way neutral models have usually been generated in these contexts is randomly to reshuffle the pool of species, subject to various restrictions. Suppose the set of constraints is that: (i) the number of islands remains equal to the actual number; (ii) the number of species remains equal to the actual number; (iii) each species is present on exactly as many islands as in actuality; and (iv) each island has as many species as does the real island. Although this procedure may appear reasonable, the constraints (iii) and (iv) can have the effect of convolving a lot of biological interactions into the supposedly null hypothesis if the data are in fact strongly structured by competitive or other biological interactions. In the limiting case of the "checkerboard" pattern, these constraints are so severe as to guarantee that the null model is identical with the observed pattern: each island in the hypothetical, "neutrally constructed" archipelago also must have one and only one of species A and B, and in the observed proportions. Clearly one should not reject the hypothesis that competition forged this pattern, but rather should reject, as inappropriate, the construction of the null model.

In this context, the chapter by Colwell and Winkler is illuminating. Using a computer program called GOD, they first generate assemblies of species, whose phylogenetic lineages obey specified rules. Subsets of this "mainland biota" then colonize archipelagoes, as described by a program called WALLACE; in this colonization, competitive interactions may or may not be important, depending on how WALLACE's rules are specified. Colwell and Winkler can now take the "field data," thus generated, and can see to what extent the emergent community patterns stand out against various neutral models. The key difference between this exercise and actual field studies is, of course, that—because GOD and WALLACE are immanent in Colwell and Winkler—they know whether or not their communities really are structured by competitive effects. The gist of their findings is that the conventional neutral models, constructed by reshuffling of the data, are often indistinguishable from the "observed" field data, even when competition has in fact been a strong force in WALLACE's colonization of the archipelago.

As Simberloff correctly emphasizes in his chapter, in such circumstances we should not give up, but rather should search for some more appropriate way of framing a neutral model. A constructive suggestion advanced by Gilpin and Diamond is to abandon the deterministic kind of model based on averages, and work with the full statistical distribution given by null models (*cf.* Sugihara, 1981). In the "checkerboard" case, the null model would have species A present on any island with probability p_A (where p_A is the proportion of the actual N islands on which A is found) and species B present with probability p_B. This neutral model would thus have a statistical distribution of presences and absences, with some islands having both species, some having one only, and some having none; for large N, the strict "checkerboard" (with one and only one species, A or B, per island) would then obviously be too patterned to be accounted for on neutral grounds. More generally, Colwell and Winkler's studies strongly suggest that null tests based on the full statistical distribution of the ensemble of possible models, or at least based on measures of the variance of species' distribution, are more likely to confirm the existence of underlying biological patterns than are simpler tests based on observed average values (of species per island and so on).

Other chapters offer additional new ideas and new studies aimed at elucidating aspects of the way particular communities are structured. James and Boecklen, for instance, use an innovative statistical analysis to explore the extent to which changes in the population densities of individual species of birds (as observed in a forest site in Maryland over 7 years) are correlated with the morphological relationships among the bird species. In a remarkably exhaustive computer investigation, Schoener compares the actual distribution of size differences among n co-occurring species of hawks ($n = 2, 3, 4, \ldots$) with the "null" distribution obtained by considering every conceivable combination of 2, 3, 4, ... species drawn from the global pool. Clearly, the above discussion of indirect evidence for competition does no more than hint at themes and studies that are developed elsewhere in the book.

If, however, the views I have expressed in this section are accepted, it appears that neutral models can be used to confirm the presence of convergence, or of assembly and incidence rules, but that their use in rejecting such patterns will continue to be contentious. The asymmetry arises because neutral models that give patterns indistinguishable from the real data can always be called into question if they have been constructed by pooling and rearranging the data itself. Given the awkward ambiguities and frustrations inherent in this situation, it is not surprising that feelings sometimes run high!

OTHER COMMUNITY PATTERNS: EFFECTS OF SIZE AND SCALE

It will often be that systematic patterns, in both populations and communities, are apparent if one examines the system on a sufficiently large scale or over a sufficiently long time, but that such patterns do not show up if the study is restricted in space or time. Yet the exigencies of researchers' lives are unfortunately often such that studies must be spatially and/or temporally localized.

One among many possible examples is to be found in studies of coral fish communities. The elegant and careful experimental studies conducted by Sale (1977, this volume) show little evidence of MacArthurian niche structure in these communities. Sale's study sites are, however, about the size of the average seminar room, and it has been argued that such structure is apparent when one looks at the communities over a much larger spatial scale (even stretching, in the studies of Anderson et al., 1981, from the outer Barrier Reef to the Queensland coastline). Likewise, many different views are held about whether intertidal communities are structured by competition or by predation, or whether indeed there is any particular structure (some, but not all, of the views are represented in this volume); again, the answer may depend partly on the scale and detail of the study.

Not only broad questions of scale, but also geographical and climatic details, can be important in forming an appropriate neutral model. Here, one cautionary tale must suffice. Stiles (1977) presented data suggesting that the peak flowering times of the different species within a particular assembly of plants appear to be roughly uniformly spaced, a fact that he argued to be consistent with the notion that the plants have specialized into temporal niches in their competition for pollinators. Poole and Rathcke (1979), however, showed that the distribution of peak flowering times was not significantly different from that generated by the null hypothesis that each species flowers at some random time of the year. But, as pointed out by Cole (1981), there are two pronounced flowering seasons in the region where Stiles's studies were done. Once this fact is taken into account, the flowering peaks are indeed more uniformly spaced than an appropriate neutral model suggests (and the overlap between the peak flowering periods of temporally adjacent species is lower than neutrally explainable). Thus Cole's analysis revives Stiles's conclusions, but now with an added rigor stemming from Poole and Rathcke's constructive criticisms.

PRACTICAL PROBLEMS AND TENTATIVE SCIENCE

Despite recent advances, both in the acquisition of data and in its analysis, I doubt that any multispecies community is sufficiently well understood for us to make confident predictions about its response to particular disturbances, especially those caused by man. Many important practical problems need further ecological studies, of a carefully planned kind, before anything other than crude and tentative generalizations can be made about dynamical behavior in response to perturbation. Unfortunately, in many of these practical situations, decisions must be made today; fishing and whaling quotas will be set for next year, and as habitats are destroyed at an accelerating rate, reserves must be set aside now. The choice in many circumstances is not between perfect and imperfect advice to managers, but between crudely imperfect advice and none at all.

Examples abound. For the multispecies fisheries of the North Sea, the North Pacific, and the Gulf of Thailand, advice based on tentative generalizations and oversimplified models is the best that ecologists can offer. Similarly, the Convention of the Southern Ocean (which aims to enunciate a set of scientific principles as a basis for managing the complex layering of trophic levels from krill to baleen and sperm whales) appeals to broad generalizations; see May *et al.* (1979). In a similar way, plans for the establishment and management of conservation areas and refuges in many different parts of the world rest on guesstimates and principles that are not yet—and may never be—established on an unarguable factual foundation (Soulé and Wilcox, 1980; Jewell, 1981). An instructive example, from a past age, of such practical action based on plausibility rather than certitude is Snow's suspicion that cholera was transmitted by water contaminated with sewage, and his suggestion that the Broad Street pump was the focus of the cholera outbreak in London in 1848. The epidemic stopped soon after the handle was, at Snow's urging, removed from the pump (Winslow, 1943).

In short, in assessing the contributions that ecological theory can make to management decisions, it must be kept in mind that practical decisions are often, of necessity, made in haste and in the absence of full information. This, needless to say, is never an excuse for bad science or overconfident claims based on uncertain knowledge, but it does, it my view, often justify accepting rough and tentative generalizations or patterns that have not been rigorously established.

STUDYING COMMUNITIES

One opportunity that is much neglected by managers of natural resources is the chance to make decisions in such a way as to maximize the flow of information about the system, as a foundation for future management choices. Thus, given that fish and whale quotas are set on the basis of frankly crude models, different models and recommendations could be made in different geographical areas, so that the management regime assumes some of the aspects of a controlled experiment. The same ideas can be (and to a limited extent are) employed both in establishing reserves and in managing locally abundant populations of endangered or protected animals (see the papers in the volume edited by Jewell, 1981). Snow's suggestion about the pump was a falsifiable hypothesis as well as a public health recommendation.

Both in practical management problems and in academic studies, any attempt to elucidate patterns of community structure must deal with the question of how to delimit the community. Much academic research restricts itself to a particular taxonomic group—birds, or lizards, or insects—instead of first consciously deciding which group of species comprises a coherent and irreducible community. And harvesting studies typically ignore all species that are not exploited, often in ways that are detrimental to the future well-being of the community, harvested and unharvested species alike. Hairston stresses this point later in the book. Recent studies of the structure of communities of seed-eating rodents and ants in desert environments in the American Southwest (Brown and Davidson, 1977; Brown, this volume), and of lizards and birds in the West Indies (Wright, 1979, 1981), are showing the way to a better tradition for the future.

The problem of identifying a coherent community can be exacerbated by a blinkered vision that focuses upon one particular kind of biological interaction to the exclusion of others. Thus too narrow a concern for competitive interactions may lead to important predatory species' being neglected in what purports to be a community study (as stressed, for example, by Faeth and Simberloff, 1981a), and, conversely, too much emphasis on prey-predator relations can cause competing species to be overlooked (particularly if they are taxonomically different from the main species being studied). Mounting my own current hobbyhorse, I note that few indeed are the community studies that take account of the influence of pathogens and parasites on population dynamics and community structure. Yet parasites—broadly defined to include viruses, bacteria, protozoans, fungi and helminths—arguably play major roles in shaping many communities, even on a biogeographical scale (Anderson and May, 1979; May and Anderson, 1979).

CONCLUDING REMARKS

Ecology is a difficult science, partly because evolution has only given us one world, and it is not easy to perform controlled experiments. There nevertheless exist a variety of techniques whereby the evolution and ecology of communities can be elucidated in an unambiguous way; these include systematically compiled comparative studies, manipulative experiments in the field and laboratory, and (as repeatedly evidenced in this book) the testing of putative patterns against appropriate "neutral models." The philosophical status of these investigative techniques has been lucidly discussed by several people recently (Hull, 1974; Ruse, 1977, 1979; McIntosh, 1980; Simberloff, 1980; Strong, 1980; Wimsatt, 1980), especially in relation to the more ineluctable procedures found in the physical sciences. As observed by Southwood (1980) and others, evolutionary biology and ecology are characterized by mixtures of probability and pattern (Monod's "chance and necessity"). Even when trends and patterns can be confidently identified, predictions will usually need to be cast as probabilistic statements; this characteristic is often disconcertingly at variance with the crisp determinacy of most predictions in physics.

The complications inherent in most studies of ecological communities are unfortunately such that it can be hard to keep a balanced view of all the relevant factors and contending hypotheses. Although this danger is present in all the sciences, the unconscious temptation to superimpose one's prejudices upon the data is more easily yielded to by virtue of these complexities.

If any simple lesson can be said to emerge from the examples discussed above, or more generally from the papers assembled in this volume, it is that no single method—theoretical or experimental—can be guaranteed to give useful results about community patterns. Past advances have come about in many different ways, from many different styles of investigation. Without going to Feyerabend's (1975) extreme of "anything goes," I believe that the creative tensions among different schools of researchers are a continuing source of new insights and new approaches; it is paradoxical that some of those who are most sensitively aware of the need to keep sight of alternative explanations for observed patterns in community structure seem, at the same time, ocassionally to accept that there is only one True Way to do science. I believe that, both in our pursuit of an understanding of the structure of ecological communities and in the scientific methods we employ to this end, we should be guided by Whitehead's precept (as cited by Birch, 1979): "seek simplicity, and distrust it."

Acknowledgements

Among those whose comments have shaped this manuscript are H. L. Caswell (who told me of the 1944 British Ecological Society meeting), B. J. Cole, J. M. Diamond, M. E. Gilpin, H. S. Horn, J. H. Lawton, D. S. Simberloff, T.R.E. Southwood, D. R. Strong, and G. Sugihara. This work was supported in part by NSF grant DEB81-02783.

Experimental Tests

2.

Inferences and Experimental Results in Guild Structure

NELSON G. HAIRSTON, SR.

Department of Zoology, The University of North Carolina, Chapel Hill,
North Carolina 27514

"We cannot go out and describe the world in any old way we please and then sit back and demand that an explanatory and predictive theory be built on that description. The description may be dynamically insufficient. Such is the agony of community ecology. We do not really know what a sufficient description of a community is because we do not know what the laws of transformation are like. . . ." (R. C. Lewontin, 1974, p. 8)

INTRODUCTION

In this paper, I report two studies. The results show the information that may be required for both a dynamically sufficient description of these guilds and an understanding of the laws of transformation that determine their composition. In neither example is the interspecific interaction that determines the laws of transformation obvious from the original description of the community, and I cannot claim that those laws have been specified yet. I do hope to show the scope that they must eventually cover. If guilds are to be considered ecologically meaningful groups of species, some such specification must be sought.

Both studies are of guilds of salamanders in the southern Appalachians. In one case, the guild is defined as those species of salamanders that spend most of their lives on the deciduous forest floor. Studies of stomach contents (Hairston, 1949; Whitaker and Rubin, 1971; Powders and Tietjen, 1974; Burton, 1976; Sarah Stenhouse, pers. comm.) demonstrate that all of the species share the resource of food: insects and other small invertebrates. They thus conform to the usually accepted definition of a guild. The second guild consists of the coexisting species of the genus *Desmognathus*. Guild membership is based on overlapping food (Hairston, 1949; Krzysik, 1979), similarity of life histories, and the assumption of similarity of requirements implicit in the close taxonomic relationship.

SALAMANDERS OF THE DECIDUOUS FOREST FLOOR

The composition of this guild varies slightly from location to location in the southern Appalachians. In the two areas where I carried out my experiments, 10 species were found in both areas, and an 11th species was present in one of them. Four of the 10 species have life histories and/or ecological distributions that preclude them from being considered part of the guild. *Pseudotriton ruber* and *Gyrinophilus porphyriticus* spend 32 months or more as larvae, and the adults are uncommon, although both were seen some distance from the streams. *Desmognathus quadramaculatus* is mostly aquatic (Hairston, 1949, 1973, 1980b; Organ, 1961), and in this study was only seen on the plots immediately adjacent to running water. *Desmognathus monticola*, while less aquatic than *quadramaculatus*, is still a stream-bank salamander, 85.4% of the specimens seen having been on the same plots to which *quadramaculatus* was confined.

The remaining species spend all or most of their lives on the forest floor, and in no case could the abundance of any of them be related to the distance of the plots from streams or seepage areas. These species are *Plethodon jordani*, *P. glutinosus*, *P. serratus*, *Eurycea bislineata*, *Desmognathus ochrophaeus*, and *D. wrighti*. *Desmognathus imitator* was also present in the Great Smoky Mountains, but only part of the specimens are separable from *D. ochrophaeus* in the field, and it has been lumped with that species in the following account. The compositions of the guilds and the relative abundances of the species are shown in Table 2.1.

The study was carried out in two areas where the altitudinal distributions of *Plethodon jordani* and *P. glutinosus* are quite different. In the Great Smoky Mountains, *P. jordani* is confined to higher elevations and *P. glutinosus* to lower ones. Their distributions overlap in a zone that is 70–120 m wide vertically, the altitude depending on the direction in which

Table 2.1. The membership and abundances of the guild of salamanders of the deciduous forest floor in the southern Appalachians. In the Great Smoky Mountains, *Desmognathus ochrophaeus* includes *D. imitator*. Numbers are total seen on 90 searches of control plots in each area.

Species	Balsam Mountains	Great Smoky Mountains
Plethodon jordani	3182	2473
Plethodon glutinosus	392	450
Plethodon serratus	89	61
Eurycea bislineata	18	72
Desmognathus ochrophaeus	15	297
Desmognathus wrighti	25	7

the slope faces. In the Balsam Mountains, the two species are found to-gether over an altitudinal range of 1220 m (Hairston, 1980a). This differ-ence prompted me to carry out paired experiments in the two areas to test the hypothesis that competition was enough stronger in the Smokies to result in competitive exclusion there. The experiments involved setting up a series of marked plots in each of the two areas. *Plethodon jordani* was removed from one pair of replicates at each location; *P. glutinosus* was removed from another pair of replicate plots.

Although the original purpose of the experiments was to test for com-petitive release of the two species, the abundances of all salamanders were monitored throughout the study, which was continued for 5 years because of the long generation times of the two most abundant species. Thus, the experiments also tested the effects of each of those two species on the other guild members.

The methods, which involved complete searches of the plots at night, are described in detail elsewhere (Hairston, 1980a). The plots are equi-lateral octagons 24.384 m (80 ft) in diameter, covering approximately 0.04 ha. Each plot was searched 6 times per year, twice each in May–June, July–August, and September–October. As is shown in Table 2.1, the abundances of the different species were very unequal. A result of the in-equality was that successful removal of about 56–64% of the *P. jordani* reduced the total salamander biomass by 40–50%. Similarly, removal of *P. glutinosus*, the largest member of the guild, reduced the total sala-mander biomass by 10%. It must be admitted that the foregoing estimates are based on the assumption that all species are, on the average, equally represented in night censuses, since all of my data are in number per plot search. The assumption may not be valid. For example, in preliminary studies I removed all *P. jordani* from two plots 7 times in 20 nights. The data conform reasonably well to a constant loss rate of 0.25 per removal, and I conclude that on an average night about one-fourth of the *P. jordani* are active and available for removal, the remainder being under-ground. I have no comparable data for any of the other species, but it is obvious that major deviations from the estimate based on *P. jordani* could cause deviations in the estimates of the proportion of total biomass removed. There are limits to the error, however, because *P. jordani* con-stitutes such a large proportion of the total number of salamanders seen (73% in the Great Smoky Mountains). If all of the other species have only one-tenth of their individuals active on one night, then only 52% of the total population are *jordani*. If half of their numbers are active at one time, *jordani* constitutes 84% of the total. These percentages are uncor-rected for biomass, but they give a realistic picture of the potential error in the estimates, as *jordani* is the second largest member of the guild.

Results

The most striking result of the removal experiments was the response of *P. glutinosus* to the removal of *P. jordani*. In the Great Smoky Mountains, *P. glutinosus* was significantly more abundant where *jordani* was removed than on the control plots, starting in the third year of the experiment and continuing through the fifth year. In the Balsams, the same effect was found, but only at the end of the fourth year and during the fifth year (Hairston, 1980a). When *P. glutinosus* was removed, *P. jordani* showed an increase in reproduction, reflected in a significant increase in the proportion of one- and two-year-olds, but the difference in abundance between experimental and control plots was not significant statistically. Again, the effect was greater in the Smokies.

None of the remaining members of the guild was affected by the removal of either of the two most abundant members. This pattern is well exemplified by the congeneric species *Plethodon serratus* (Figure 2.1). Sixteen statistical tests compared mean densities of each of the 4 species on control plots with their densities on *jordani*-removal plots and on *glutin-*

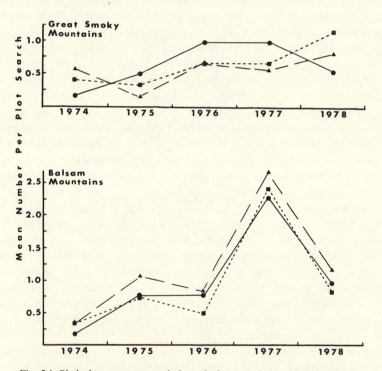

Fig. 2.1. *Plethodon serratus* populations during experimental removal of *P. jordani* (dashed lines connecting triangles) and *P. glutinosus* (dotted lines connecting squares), compared with controls (solid lines connecting circles).

osus-removal plots in both Smokies and Balsams (Hairston, 1981). In one case, that of *Desmognathus ochrophaeus* on plots in the Balsam Mountains from which *P. glutinosus* had been removed, the mean density on treatment plots was significantly greater than on controls. Inasmuch as "significance" is expected by chance once in 20 comparisons, no important meaning can be attributed to this isolated example. Furthermore, it was possible to test the same differences year by year for 71 comparisons. In two of these, there was a significant difference, but both of them were for 1974, the first year of the experiment. For any meaning to be attributed to the result, significant differences would have to be found at the end of the experiment.

We have here a guild that can only be demonstrated to contain 2 species. The remaining 5 species were not close enough ecologically to those 2 to be affected by reductions in their density, even though as much as half of the total biomass was removed, and even though the diets of all 7 species overlap broadly. There does not appear to be an acceptable *a priori* way in which to define this guild so as to include those species that interact ecologically, and so as to exclude those that do not. Restricting the guild to a single genus (*Plethodon*) would not solve the problem, because *P. serratus* is not affected by the removal of either congener. I conclude that we need to know the exact ecological relationships to a degree that is unlikely to be attained for this or for most other guilds, or that this and many other claimed examples of guilds are products of our imaginations.

SALAMANDERS OF THE GENUS *DESMOGNATHUS*

In the southern Appalachians, the members of this genus occur together in combinations of 3, 4, or 5 species. They are usually thought of as stream-bank salamanders, but individually the species vary from being mostly aquatic (*D. quadramaculatus*) through the "typical" *monticola*, *fuscus*, and *santeetlch* (see Tilley, 1981) to mostly terrestrial *ochrophaeus* and *imitator* and fully terrestrial *wrighti* and *aeneus*. All species lay their eggs in or near water, and the female remains with them until they hatch 2 or 3 months later. Except for *wrighti* and *aeneus*, all have larval stages, which vary between species in duration from a few weeks to 2 years.

For a number of years, I have had classes record the distance from surface water for each metamorphosed salamander. Data covering the years 1979 and 1980 are shown in Table 2.2. The results are typical for intermediate elevations (see Hairston, 1949, 1973, 1980b; Organ, 1961). The 4 species overlap broadly, but each has a characteristic spatial distribution that is different from all of the others. Until recently, such data were interpreted as representing the result of interspecific competition, natural

Table 2.2. Ecological distribution of four species of *Desmognathus*, elevation 686 m, Coweeta Hydrologic Laboratory, North Carolina. Class data, 1979 and 1980. The proportional distribution of each species is given.

		Distance from Stream (m)				
Species	N	0	<.3	.3–1.5	1.5–6	>6
quadramaculatus	48	.812	.104	.020	.040	.020
monticola	179	.290	.318	.251	.133	.005
ochrophaeus	97	.175	.257	.340	.185	.041
aeneus	16			.187	.312	.500

selection favoring the adoption of a sufficiently different distribution for each species to reduce competition to the level that permitted coexistence.

It is generally agreed (Dunn, 1926; Piatt, 1935; Wake, 1966) that the evolution of the genus has proceeded from a largely aquatic ancestor to increasingly terrestrial species, and that the series we see provides a reasonably accurate model of what happened over evolutionary time. Newly acquired data (Hairston, 1980b) and new interpretations (Tilley, 1968; Huheey and Brandon, 1973; Hairston, 1980b) have raised doubts about the conventional interpretation of the evolutionary history of the *Desmognathus* species, although that interpretation continues to be used (Krzysik, 1979). The decrease in size with an increasingly terrestrial habitat is inconsistent with competition as the driving force of natural selection, because competition requires increasing efficiency, and increased evaporative water loss with increased surface-volume ratio accompanying smaller size would seem not to promote increased efficiency.

There are also considerations arising from the fact that some species are missing from certain areas. In the Black Mountains of North Carolina, for example, *D. monticola* does not occur above an elevation of about 1200 m and *D. wrighti* virtually disappears below that elevation. If competition is important in structuring the present community, I would expect different ecological distributions of *D. quadramaculatus* and *D. ochrophaeus* in the presence and absence of the other 2 species. Both should converge on the stream-bank habitat of *monticola* at high elevations and tend to avoid it at low elevations. They show no such tendencies (Hairston, 1980b), even though the absence of *wrighti* at low elevations should make the forest a favorable place for *ochrophaeus*.

If, on the other hand, the sizes of the species are the result of competition forcing them to be different, I expect *D. quadramaculatus* to be larger at the low elevations, where it occurs with *monticola*, than it is at high elevations, where *monticola* is absent. As a matter of fact, *quadramaculatus* is larger at high elevations, although not significantly so for the data in hand.

Tilley (1968) and I (Hairston, 1980b) have advanced the hypothesis that predation, rather than competition, has been the most important force of natural selection in driving the evolutionary history of the species of *Desmognathus*. It is argued that predation on salamanders is much heavier in the streams than it is in the forest, and that, at each speciation, the smaller of the two was under selective pressure to become more terrestrial, thus reducing the effect of predation by large aquatic salamanders and fish.

This hypothesis readily explains the size-habitat relationship that causes difficulty for the hypothesis that competition has been the driving force in selection in the genus. Predation in the streams is also the most plausible reason for Organ's (1961) finding that survival of juveniles is positively related to how terrestrial the habitat of the species is. Predictions from the absence of *D. monticola* and *D. wrighti* do not apply to the predation hypothesis.

I have begun experiments designed to permit a choice between these two hypotheses. On a series of streams, which are used as blocks, *D. ochrophaeus* is being removed from one series of plots. If competition is the important interspecific interaction, *D. monticola* should be favored in comparison with its performance on control plots. With predation important, no such prediction is permissible, and *monticola* and perhaps *quadramaculatus* might decline in abundance because of the loss of an important prey item. On another group of plots, *monticola* is being removed. This removal should result in a benefit to *ochrophaeus* under either hypothesis, but *quadramaculatus* should gain only if competition is important. If the predation hypothesis is correct, it might decrease.

Tilley (1980) attempted the removal of *D. monticola* from a rockface habitat. He removed 128 specimens on 7 visits over a $3\frac{1}{4}$-year period. He observed no effect on the *ochrophaeus* population, but did observe an increase in the proportion of young *monticola*. The absolute number of adults was not affected by the removal, and although the lack of control areas makes complete interpretation hazardous, 2 removals per year do not appear to have been enough to have an effect on those *monticola* most likely to prey on *ochrophaeus*. Tilley does not report any impact on *quadramaculatus*, although that species was present. His study shows that, in order to reduce the population of a species of *Desmognathus*, a much more intense effort will be required.

DISCUSSION

The 2 guilds that I have described demonstrate that intuitively reasonable and frequently used descriptions of nature may be dynamically insufficient for community ecology, to continue borrowing Lewontin's term. The insufficiency seems to be related to the implicit assumptions that are the basis for the forms of the descriptions. The choice of the salamanders of the forest floor involves the assumption that consuming overlapping resources (food species) is sufficient to bind them together in an ecologically important way. If food is not limiting the populations of these salamanders, the assumption is false, and the basis used for the delimitation of the guild is nonexistent. But food is so important generally that the assumption seems intuitively reasonable, and it is fair to ask what better basis could be found for delimiting the guild. It can be maintained that the nature of the species' life histories would form a better basis for describing the community, thus separating the 4 species that return to the streams to breed. It has already been noted, however, that most of their feeding time is spent on the forest floor, and to abandon food *a priori* would require knowledge that the populations of the 4 species are regulated during the time when they are at the streams. Even if this knowledge were available, there would remain the problem posed by *Plethodon serratus*, a completely terrestrial congener of the 2 competing species. It is true that it is most active in spring and fall, while the larger species have summer as their period of activity, and it seems likely that the eggs of *serratus* are laid in a more superficial situation. To make use of all such facts in delimiting a guild would, if it is necessary, make the delimitation pointless. After all, the simultaneous consideration of ecologically related species is the heart of community ecology, and to admit a completely reductionist requirement would remove any hope that we have a scientifically valid field.

To consider the co-occurring species of *Desmognathus* to be a guild involves a different set of assumptions from the ones already discussed. It has been assumed that taxonomic affinity implies ecological impact, and furthermore that the impact is interspecific competition. The nature of the resource for which competition is assumed does not yet need to be specified. It is further assumed that the ways in which the species differ are the result of natural selection operating through the impact of the competitive relationships. I believe that I have shown that these assumptions are neither necessary nor sufficient to explain the observations (Hairston, 1980b), even though I advocated their acceptance in the past (Hairston, 1949, 1973). The experiments may or may not demonstrate ecologically important interactions, but given two plausible explanations, some means of choosing between them is necessary, and experiments are to be preferred to *a posteriori* arguments.

There are three kinds of direct ecological interaction between species: competition, predation, and mutualism. Up to now, theoretical ecologists have been principally concerned with how communities can be constructed from these interactions, principally competition. It has been common to cite data confirming the proposed mechanisms, and to conclude that the correct inferences have been drawn about the way in which nature works. This approach is always subject to the weakness that the explanation is sufficient but not necessary. It is my contention that community ecology will never escape from its agony until we begin the self-conscious application of a rigorous scientific method. Its success in rocky intertidal communities is encouraging, but in many areas, great ingenuity will be required to identify the implicit assumptions and to devise adequate experimental or other truly *a priori* tests of the explanations that are offered for what we see in nature. The task will be particularly difficult because we have only unfounded ideas about what to look for. The greatest need is for legitimate means of identification of interacting groups of species. The identification should be based on characteristics that are determined by the nature and strength of the interactions, and we need characteristics that discriminate between the effects of different interactions as well as those that separate interacting from non-interacting groups. Three characteristics that I have shown to be insufficient alone or in combination are co-occurrence, taxonomic affinity, and utilization of a common resource. Until we can discover the necessary characteristics, I believe that separate experiments testing hypotheses applicable to individual "communities" provide our only assurance of progress.

3.

Exorcising the Ghost of Competition Past: Phytophagous Insects

DONALD R. STRONG, JR.

Department of Biological Science, Florida State University, Tallahassee, Florida 32306

SUMMARY

Orthodox community theory is dominated by the assumptions of density-dependence and interspecific competition. Population growth inexorably depletes resources and shortages govern interactions both within and between species in this theory. Here, I summarize research with phytophagous insects of the tropical monocot *Heliconia*, which coexist in a manner quite opposite to this orthodox theory. Resources are not depleted by these insects. They do not feed from or occupy host plants in a density-dependent manner, and species do not compete among themselves. Although analyses of the forces that do affect these populations are not complete, preliminary observations indicate that natural enemies such as parasitoids greatly affect populations of these insects and may hold them at their very low densities. Other factors, such as host phenology, seasonality, and environmental stochasticity may also substantially affect these communities.

Insects on *Heliconia* are not unique among phytophagous insects. Many insects apparently coexist normally without the neo-Malthusian forces of interspecific competition. Non-competitive coexistence may be the usual situation for insects on plants, which comprise over 25% of the diversity of macroscopic organisms now in existence. Even for some organisms other than insects, several authors indicate that, upon scrutiny, the assumption of inevitable interspecific competition is also often untenable. Without inevitable interspecific competition the ghost of competition past need not be conjured up to explain species differences, niches, and "resource partitioning" in present communities.

Mainstream community theory is based upon Gause's extrapolation of Malthusian population theory to interspecific relations (Hutchinson, 1978). The centerpiece of this neo-Malthusian theory is deterministic density-dependence based upon resources. Population growth rate changes primarily as a function of density; populations inexorably grow until density is so high that food or accommodations limit more growth. Gause extended Malthusian theory to the interspecific realm with the notion of ecological niche. "It is admitted that as a result of competition two similar species scarcely ever occupy similar niches, but displace each other in such a manner that each takes possession of certain peculiar kinds of food and modes of life in which it has an advantage over its competitor" (Gause, 1934, p. 19). The definition and implications of niche were modified and amended from Elton (1927, p. 63). Elton used niche at that time only to denote where an animal lived and what is fed upon. Inferrences about competition and displacement were Gause's.

After Gause, much community theory continued on the neo-Malthusian track. Lack (1947) and Hutchinson (1959) interpreted the structure of natural communities in terms of competition. Orthodox interpretation of niche theory is based upon Gause's "admission" and infers that interspecific competition in the past caused niches to diverge, with the differences between species that we see today in communities as the result (Schoener, 1947b). The assumption that former competition is the general cause of species differences in communities has been satirically termed "the ghost of competition past" (Connell, 1980).

My purpose is to focus critical attention upon the major assumptions of neo-Malthusian community ecology for phytophagous insects in nature. Phytophages have had relatively little influence upon this theory, even though they make up over 25% of extant animal species (Southwood, 1978). One of my main points is that interspecific competition is not common in communities of phytophagous insects, probably because other natural factors frequently intervene to hold densities so low that populations do not usually deplete crucial resources. Natural communities of insects might be contrasted with those in mathematical theories or in very simple laboratory environments, where one has eliminated the normal influences of seasons, host plant phenology, spatial patchiness, predation, parasitism, and the effects of the larger composite community that interacts intermittently and relatively diffusely with the species in question, and where disturbances from the weather and other natural forces that cause environmental stochasticity have been eliminated. In these simplified models, populations can be forced to behave in a density-dependent manner, and communities may be forced into neo-Malthusian behavior. It is just this complex of additional factors listed above that are frequently quite important in real communities and that are ignored by neo-Malthusian theory in favor of competition. Suites of the above factors tend to reduce both density-dependence and the importance of competition in natural communities. How the complex of real factors actually

affects communities in nature can only be discovered by careful description and experiment. Thus, the ultimate community questions are empirical. How do communities really behave?

ROLLED-LEAF HISPINES ON *HELICONIA*

Hispine beetles that live in the scrolls formed by young *Heliconia* leaves
are good material for community studies. Adults of as many as eight species can intermingle in the scrolls at a single site, and as many as five
species can simultaneously occupy one scroll (Strong, 1977a, b). The scroll
is the food as well as the shelter. Adults spend their entire long lives in
or moving among scrolls, feeding and mating. Scrolls unfurl within a few
days, so associations are repeatedly reshuffled during the life of a beetle.

I have studied interactions, associations, and the use of resources by
these hispines in Central America (Strong, 1982a, b). Interspecific and intraspecific relations of adults are distinctly harmonious. Beetles have no
aggressive tendencies toward their own or other species, and live in intimate contact. No evidence for interference competition could be found in
patterns of species association in scrolls, in comparisons from many sites
from Trinidad through Costa Rica. The presence and abundance of a species in a scroll has no discernible influence upon other species, and
density has no influence upon interspecific association in scrolls. My experiments produced patterns of neutral interspecific association quite
similar to those found in nature, and reinforce the idea that hispine species
mix independently of one another in scrolls. Finally, the number of *Heliconia* species that a hispine species uses at a site is apparently not affected
by other hispine species.

These earlier studies concentrated mainly upon potential interference
competition in hispine communities. Here I will concentrate upon potential exploitation competition. For hispines, interference might result from
aggression that led to beetles' leaving scrolls or beetles' avoiding scrolls
as a function of their occupants. Fouling or marking of scrolls could cause
interference, just as could aggression. The most straightforward form of
exploitation for these insects would be depletion of the food in scrolls by
feeding beetles. Of course, resource depletion might precipitate interference. If accommodations inside scrolls were effectively used up by occupants, exploitation competition might cause interference competition. My
experimental studies with potential exploitation among hispines have
been done mainly at Finca La Selva in Costa Rica, with the insects of
Heliconia imbricata. The two most abundant species of rolled-leaf hispines
on *H. imbricata* at La Selva are *Chelobasis perplexa* and *Cephaloleia
consanguinea*.

The most striking fact germane to the question of exploitation competition is that densities of hispines are usually very low relative to the amount

of accommodation and food in scrolls. Leaves can be as large as a square meter, and the interiors of scrolls grow to many times the volume of beetles ever found inside, and as the studies described above indicate, beetles do not stay in or leave scrolls in a density-dependent manner. Food in scrolls is abundant relative to that eaten by hispines and other phytophagous insects. Figure 3.1 shows the fractions of leaf area eaten by the 13 most abundant phytophage species of *Heliconia imbricata* during the first

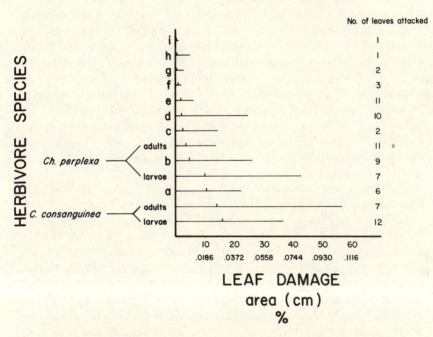

TOTAL DAMAGE
1.53% of leaf area

LEAF DAMAGE
area (cm)
%

Fig.3.1. Insect feeding from *Heliconia imbricata* over the first 60 days of life of 12 leaves, at Finca La Selva, Costa Rica, during June, July, and August 1976. Leaves were randomly chosen along the River Road Trail. All leaves were within an abandoned cacao plantation and shaded by canopy trees. Length of each horizontal line indicates the maximum feeding by that insect on any one of the 12 leaves. Vertical marks on the lines indicate the average feeding for the 12 leaves. Numbers to the right of lines indicate the number of leaves that each species fed upon. "a" = larvae of leaf miner sp. *a* (Coleoptera, Buprestidae), "b" = Orthoptera, "c" = larvae of leaf miner sp. *c* (Coleoptera, Buprestidae), "d" = larvae of Lepidoptera, "e" = adults of *Hemispherota* sp. (Coleoptera, Cassidinae), "f" = larvae of unidentified leaf minner, "g" = larvae of *Cheirispa dorsata* (Coleoptera, Hispinae), "h" = larvae of leaf miner sp. *h* (Coleoptera, Buprestidae), "i" = adults of *Cheirispa dorsata*.

60 days of life of 12 leaves. Each day the leaves were inspected, and the amount of new feeding by each insect was recorded. In this way an accurate cumulative tally of each species' feeding was built up. Larvae and adults of *Cephaloleia consanguinea*, a monophagous rolled-leaf hispine, caused most feeding damage to the leaves. Another species of rolled-leaf hispine that was abundant during the study was *Chelobasis perplexa*. Its larvae ranked fourth, and its adults ranked sixth, in damage caused to the leaves.

The fractions of *H. imbricata* leaf eaten by these herbivores are very small. In total, all insects ate only 1.53% of the leaf area. Even though leaves normally last between 12 and 18 months, most feeding is done within the first 30 days of a leaf's life. Most leaf damage, on the average, was done by the larvae of *C. consanguinea*, which took approximately 0.03% of the leaf area. The maximum feeding damage on any single leaf was caused by *C. consanguinea* adults, which ate between 0.09 and 0.11% of one of the leaves. *Cephaloleia consanguinea* larvae found their way onto all 12 rolled leaves in this study; adults of this insect fed upon only 7. Note that other insect species attacked fewer than the total 12 leaves. I paralleled this study of feeding upon *H. imbricata* with studies of *H. latispatha*, *H. tortuosa*, *H. pogonantha*, and the related plant *Ishnosiphon* sp. (Marantaceae). All of these plants are attacked by rolled-leaf hispines and other insects, which include both host-specific and more polyphagous species. The amounts eaten from *H. imbricata* were greater than those from any of these other species. No insect species came close to depleting the area of leaf that it fed upon. These results are typical of the very low rates of herbivory on *Heliconia* in nature.

One possibility is that, although phytophages use only a very small fraction of *Heliconia* leaf, less than the total amount of a leaf or fewer than all leaves are desirable to the insects. In a previous study, I had found that 20% to 40% of scrolls were unoccupied by hispine adults (Strong, 1982a). Are the unoccupied leaves chemically or physically less desirable than occupied leaves? One indication that leaves do not vary in inherent desirability comes from feeding experiments comparing the desirability of scrolls of *H. imbricata* that contained no hispines with that of scrolls that contained relatively high densities of hispines. For these two classes of scrolls I compared feeding by hispines in experiments done in Petri plates. Scrolls of the two classes were brought back to the field station. From each scroll, a circle of tissue 10 cm in diameter was cut from the area on which the beetles feed. The circles were cut from the section of the leaf that had most recently been exposed by the natural unfurling of the scroll. Each circle was cut in half, and each half placed in a separate Petri plate. I recorded all previous damage, then placed two *C. consanguinea* adults in one of the Petri plates and two *Ch. perplexa* in the other. Twenty-four hours later the beetles were removed and released, and the area that had been eaten was measured. The results showed that neither beetle species ate discernibly different amounts from the two classes of scrolls (empty and occupied) (Figure 3.2). My suspicion of inherent heterogeneity among leaves was not confirmed.

Thus, empty and occupied scrolls are apparently not differentially desirable to hispines on the basis of inherent physiological or phytochemical

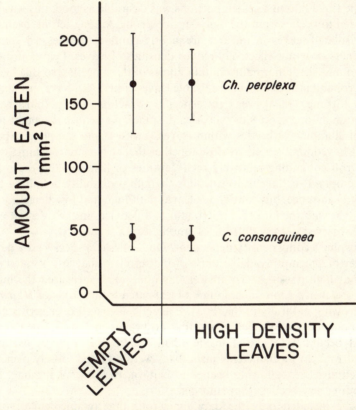

Fig. 3.2. Amounts eaten in laboratory experiments from *H. imbricata* scrolls found to be empty and scrolls found to contain relatively high densities of hispines in nature. Experiments were performed in Petri plates on leaf tissue that had been cut from fresh leaves, at the Finca La Selva field station. Dots indicate means, and bars extend one standard deviation above and below each mean. Each of the four treatments consists of 12 replicated observations. Each observation is the amount of feeding done in 24 hours by one pair of adults of the same species in a Petri plate. For *Ch. perplexa* the t statistic = 0.004, for *C. consanguinea* t = 0.155, when the amounts eaten from tissue excised from empty scrolls are compared to the amounts eaten from high-density scrolls.

properties. Of course, other sources of heterogeneity could reduce desirable scrolls to fewer than the total number. But, in my 9 years of experience with these insects and plants, I have found no pattern that indicates fewer than all *Heliconia* scrolls to be physiologically desirable and available to rolled-leaf hispines.

It is possible that food is depleted within desirable sections of leaves without depletion of a large fraction of the total leaf area. This possibility is extremely unlikely because only small fractions of leaf area are eaten even within parts of the scroll. The above tests (Figure 3.2) indicate that the leaf adjacent to eaten portions was desirable as food; this was the tissue fed to beetles from the high-density scrolls. Among scrolls, even the most isolated leaves in an area frequently contain hispines, so spatial patchiness does not effectively reduce densities of leaves. Leaves in both the sun and shade normally contain adults of these beetles, so this factor of microhabitat does not reduce usable leaves either. As I have speculated before (Strong 1982a), weak intraspecific attraction, and perhaps weak attraction among species of rolled-leaf hispines, may cause the slight pattern of clumping of beetles within leaves. Twice in my sampling experience, I have found scrolls so close together that they were touching, with one scroll containing relatively high densities of hispines and the other scroll empty. It is difficult to calculate the null probability of finding this sort of occurrence, but contiguous leaves with high and low densities are consistent with the idea that microhabitat differences in the forest do not cause observed density differences among leaves.

Is hispine feeding density-dependent? Do beetles of a species crowd one another or preempt available parts of the scroll so that less is eaten per beetle at high densities? To answer this question, I measured the influences of density upon the amounts of leaf eaten from scrolls of *H. imbricata* growing naturally in the forest. Plastic bags placed over very young scrolls created virgin rolled leaves that no beetles had occupied. Experimental beetles were placed inside when the tips of virgin scrolls had opened to 2 cm in diameter. Plastic bags were replaced loosely over experimental leaves, allowing beetles to exit via the base of the bag, but preventing new beetles from entering.

I have included only the data for scrolls that retained all of their allotment of experimental beetles for the full 24-hour duration of the experiment. Because beetles tend to move frequently among scrolls independently of density or previous occupants (Strong, 1982a), and many beetles left scrolls before the experiment was over, more trials were run at each density than produced usable data. If I had not allowed beetles to leave the bagged leaves, the feeding data would have been much more variable, because tying the bases of the plastic bags tightly would have produced data composed of at least two types of scrolls, those containing only beetles that fed without attempting to exit and those containing would-be emigrants, which were less interested in feeding. I found no obvious qualitative difference between scrolls that retained all of their experimental beetles and others. Both were healthy, and tissue from abandoned virgin scrolls was eaten normally by beetles in the lab.

The results of the intraspecific experiments are shown in Figure 3.3. *Cephaloleia consanguinea* is the smaller of the two hispine species in

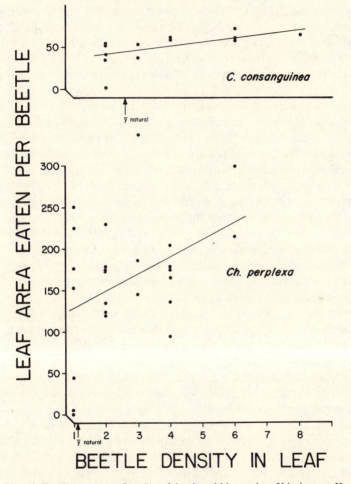

Fig. 3.3. Feeding rates as a function of density within species of hispines on *H. imbricata*. Data are from scrolls in nature. The average natural density of the two experimental beetle species in *H. imbricata* scrolls during July 1976, when the experiments were done, is indicated by arrows on the abscissa. Lines indicate linear regressions through the points. For *Ch. perplexa*, area eaten = 20.77 density + 109.11. For *C. consanguinea*, area = 5.89 density + 28.39.

these experiments and ate less than *Ch. perplexa*. Both hispine species responded to density in an inversely density-dependent fashion; both species ate more from scrolls at high than at low density, per individual. This result is indicated by correlation coefficients that were significantly positive (*C. consanguinea*, $r = 0.65$, $n = 13$, $p < 0.05$; *Ch. perplexa*, $r = 0.41$, $n = 25$, $p < 0.05$). The inverse density-dependence appears even though the feeding response was highly variable within densities. Most variability appears at low densities, as would be expected, where fewer beetles were responsible for a datum than at higher densities. At higher densities the fraction of feeding done by single individuals is a small part of the datum, so extreme individuals affect the datum less. Suggestions of non-linearity can be found in the relationships for both species in Figure 3.3. However, with the relatively few points and great scatter in the data, complex curve fitting is unsatisfactory.

The data show no evidence of crowding, interference, or resource depletion that would be manifested through density-dependence. Most scrolls in nature have densities much lower than the highest densities in these experiments. The average density of beetles in leaves during the time of experiments is shown on the abscissa of Figure 3.3. Average densities at other times were never as high as the median densities in these experiments (Strong, 1982a). The reason for inverse density-dependence, and not just a lack of any relationship to density, is still a mystery similar to the unknown reason for the slight clumping of these beetles among leaves that has been found in previous studies (Strong, 1982a). My current idea is that social interactions cause beetles to remain longer in high- than low-density scrolls and that remaining beetles feed intermittently.

Interspecific effects upon feeding rates were tested by placing beetles of different species together in the same virgin scrolls (Figure 3.4). Loose plastic bags over the scrolls excluded other beetles and allowed emigrants to leave. As in the intraspecific comparisons, I include only data from scrolls that retained all experimental beetles for the full 24 hours of the experiment. In Figure 3.4, I compare the amount eaten when the species were alone at the particular densities (from Figure 3.3) with the amount eaten when the species were physically together in the same leaf. Means for the species grown alone are the sums of the means for each species at the appropriate density from the data of Figure 3.3. The estimated sample variance for the species alone is calculated as by Steele and Torre (1960, eq. 5.13, p. 81). By the test of Cochran and Cox, I did an approximate t test of feeding, comparing beetle species along with those together in scrolls (Steele and Torre, 1960, p. 81). The approximate t value for the low species-density (1 *Ch. perplexa*, 2 *C. consanguinea*) = 0.11 (ns); that for the high species-density combination (2 *Ch. perplexa*, 4 *C. consanguinea*) = 0.59 (ns). Thus, just as with the intraspecific comparisons, there

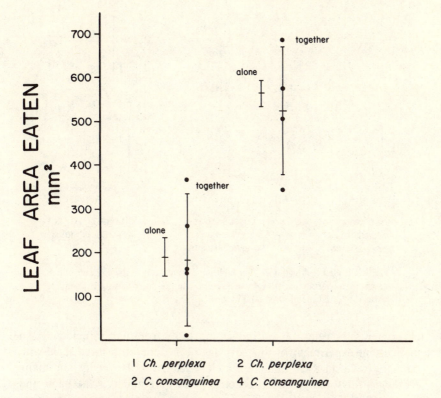

Fig. 3.4. Feeding rates of hispines for species alone in scrolls compared to rates for two species together. Means and standard deviations for species alone are shown as the center and ends, respectively, of the bar. For species alone, means and standard deviations were calculated by pooling data for the two species at the particular density. See text.

is no indication that densities of hispines roughly equal to those that occur in nature deplete resources sufficiently to cause competition.These experiments indicate little or no interspecific competition for food and corroborate the earlier finding of no interspecific competition for occupancy of scrolls (Strong, 1982a). Seifert (this volume) reaches a similar conclusion about hispine interactions with other arthropod species on *Heliconia*.

PARASITOIDS AND HOST PLANT CHEMISTRY

Parasitoids and predators are likely very important in keeping the densities of *Heliconia* phytophages so low (Figure 3.5) (Morrison and Strong, 1980; Strong, 1982a). Of course, critical tests will come only from

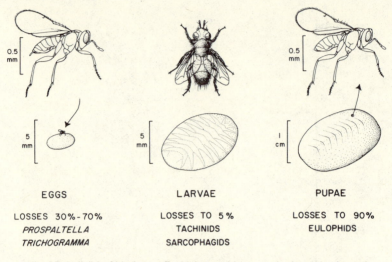

EGGS	LARVAE	PUPAE
LOSSES 30%-70%	LOSSES TO 5%	LOSSES TO 90%
PROSPALTELLA	TACHINIDS	EULOPHIDS
TRICHOGRAMMA	SARCOPHAGIDS	

Fig. 3.5. Rolled-leaf hispines suffer high parasitism rates in the egg and pupal stages mainly from parasitic hymenoptera. Larvae are parasitized at much lower rates, and mainly by tachinid flies.

experiments that elevate and reduce mortality from natural enemies. An inadvertent experiment of this sort has been done on a large scale with bananas in Costa Rica. Banana is a close relative of *Heliconia*, and many *Heliconia* phytophages have become pests of bananas (Auerbach and Strong, 1981). During the 1950's and 60's, agriculturalists sprayed bananas heavily with organo-synthetic insecticides, often with a result opposite to that expected. Parasitoids disappeared, phytophage populations grew to very high densities, and defoliation followed. Parasitoids and other natural enemies of banana pests recolonized plantations when wholesale aerial application of insecticides was stopped about 1970, and then populations of phytophages fell to quite tolerable levels. Now, most banana growers no longer apply large amounts of insecticide (Ostmark, 1974; Clyde Stevens, pers. comm.). I have found most of the known leaf-eating species of banana pests, as well as very high rates of parasitism upon these insects, in unsprayed Costa Rican plantations since aerial spraying was stopped. The few outbreaks of phytophages that my colleagues and I have been able to find on bananas since spraying was stopped were on plantations with new owners, who were ignorant of the history of insecticide use on bananas. Every banana outbreak that I have seen followed heavy application of insecticides from aircraft. I have never found higher than 5% parasitism rates of any life history stage of any insect in sprayed banana plantations during an outbreak. Phytophage eggs and pupae in

unsprayed plantations normally have parasitism rates in excess of 50%
(Strong, unpublished).

The idea that parasitoids prevent high densities of folivorous insects,
and thus prevent defoliation, is not new (Howard and Fiske, 1911). Parasitoid control of phytophages has had great application in agriculture
(DeBach, 1974). In recent ecological theory, parasitoids have been advocated as a major cause of low phytophage populations by Hairston,
Smith, and Slobodkin (1960, = HSS). This part of the HSS theory has
been distinctly unpopular. The major enduring objection to the HSS
hypothesis has been that the empirical evidence adduced by its authors,
that green vegetation is commonly abundant and thus not in shortage to
folivores, is misleading. The objection can be paraphrased thus: foliage
contains noxious phytochemicals, so what is green is not necessarily
edible; folivores may be in shortage of food even with an abundance of
vegetation (Janzen, 1975a; Murdoch, 1966). However, vegetation does not
have to be generally edible to many species of herbivores for the HSS
hypothesis to be valid (Slobodkin et al., 1967). Even if phytochemistry and
other factors restrict the number of potential defoliators to a small fraction of the local herbivore species, a few species always remain that are
adapted to the particular chemistry and can defoliate the plant. Parasitoids can certainly prevent adapted phytophages from defoliating plant
species that they specialize on. I agree with HSS, and do not argue that
phytochemistry has no effect on any insect, but that it usually does not
regulate phytophages adapted to a particular plant species.

The serendipitous "experiment" by banana growers with pesticides provides ample evidence that phytochemistry does not prevent defoliation by
specialist phytophages in the absence of parasitoids. The potential complication of insecticides' "fertilizing" the plants, or changing their chemistry in some odd way to increase vulnerability to herbivores, is not a
factor in this system. Bananas grow quite rapidly, and the outbreaks last
many months after a single spraying. Rolled-leaf hispines do not occur
on bananas, but many lepidopteran pest species have come to this crop
from Heliconia. The lepidopterans are very sparse on Heliconia, just as
they normally are on unsprayed bananas, and they suffer high rates of
parasitism on Heliconia just as on bananas.

The lack of herbivore outbreaks and the chronically low densities of
both hispines and Lepidoptera on Heliconia probably do not result from
phytochemical protection, because Heliconia lacks many of the potently
noxious phytochemicals that protect other plants (Gage and Strong,
1981). An intense effort has uncovered none of the alkaloids, saponins,
cardenolides or bufadienolides, cyanogenic glycosides, tannins (con-
densed or hydrolysable), or insect gut enzyme inhibitors that have been

found in other host plants. Moreover, extracts of *Heliconia* leaf do not inhibit feeding or growth in laboratory feeding experiments.

NEED THE GHOST OF COMPETITION PAST BE EXORCISED?

I have emphasized the influence of natural enemies in maintaining populations of insects on *Heliconia* far below densities that would cause neo-Malthusian forces to operate among species. For phytophagous insects in general, other factors are known to complement the influences of natural enemies (Lawton and Strong, 1981), and I would argue that hispine communities are, *a priori*, no exception. Host plant phenology, seasons, and environmental stochasticity in the form of weather variations may also contribute to population dynamics and densities of these insects.

The orthodox neo-Malthusian explanation of any absence of competition in communities is well described by Connell's (1980) wonderful rubric "the ghost of competition past"; species differences that reduce competition are assumed to be fossils of competition that selected for species differences. Advocates of this theory infer that "overdispersed niches," "niche compensation," and "resource partitioning" are evidence of ancient competition, but that such mechanisms can never be investigated experimentally because the molding forces of competition have been obviated by the evolutionary changes that were caused by the competition (Schoener, 1974b, p. 24). Like Connell (1980) and Krebs (1978), I feel that this theory is most unsatisfying because it is virtually impossible to falsify. What sort of communities would have occurred in the absence of the postulated competition? Species differences occur for a plethora of reasons other than competition. Certainly, only peculiar circular logic underlies interpreting the absence of competition as proof that it once existed.

Communities of phytophagous insects in general offer a model alternative to the assumptions of theory that invokes the ghost of competition past. Neo-Malthusian forces do not often come into play among insects eating plants, and there is little reason to conjure spirits from the past because we can see today good reasons for community structure. Actually, the lack of significant interspecific competition is not restricted to phytophagous insects. Birch (1979) has argued that the non-competitive coexistence of organisms sharing resources occurs fairly widely in ecological nature, especially among marine invertebrates. A similar conclusion is reached by Connell (1975): "the evidence... from controlled field experiments on invertebrates and plants, suggests that many species seldom reach population densities great enough to compete for resources, because either physical extremes or predation eliminates or suppresses them in their young stages." On another tack, introduced species would be quite prone to overt interspecific competition according to orthodox competition theory, because they would not have had time to evolve differences to reduce competition. However, this implication of the theory is not borne out by evidence. Simberloff (1981) analyzed many case histories of diverse species introduced by man into new geographical regions and concluded that interspecific competition is usually absent or so weak that no effect can be detected upon other species. This result militates against invoking competition as a general evolutionary antecedent to longstanding communities. Den Boer (1980) finds a similar lack of support for inexorable neo-Malthusian forces in communities of carabid beetles, as do Wise (this volume) for spiders and Lawton (this volume) for insects of bracken fern. Andrews and Petney (1981) have cleverly worked through the implications of the competition's-ghost theory for the niches of reptile ticks, to find that "there is no evidence to suggest that such competition in the past, or competition at present, maintains the parapatric boundaries found in these species. . . ." Thus, the ghost need not be exorcised for many organisms because there is little reason to suspect haunting in the first place.

4.

The Role of Competition in Spider Communities: Insights from Field Experiments with a Model Organism

DAVID H. WISE

Department of Biological Sciences, University of Maryland Baltimore County (UMBC), Catonsville, Maryland 21228

Considerable ecological theory postulates a central role for interspecific competition as a cause of patterns in resource utilization, distribution, and relative abundance of animal species. Tests for competition frequently have been indirect, relying upon inferences from patterns that are consistent with competition theory but that may have alternative explanations. Continuing calls for more direct tests (*e.g.* Reynoldson and Bellamy, 1971; Connell, 1975; Colwell and Fuentes, 1975; Pianka, 1976; Wiens, 1977a) reflect an increasing awareness of the value of carefully controlled field experiments. However, field experiments have not been widely used to test for competition in animal communities, probably for two major reasons. First, most species and their resources are not amenable to straightforward manipulation. Second, many community ecologists have avoided detailed studies of a few species in preference to a holistic approach that attempts to explain broad patterns by comparing different communities.

Reliance upon field experiments requires substantial acceptance of a reductionist approach to community ecology. The question of whether we can understand multispecies interactions from studies of one or a few species fuels the controversy between "population-reductionist" and "community-holistic" practitioners of ecology (*cf*. Simberloff, 1980; Levins and Lewontin, 1980). Generality of the reductionist-empiricist approach can be increased by selecting appropriate model systems for experimentation. Ecologists justifiably envy the reductionist successes of molecular biology and physiology, and often regret that the inherent variability of ecological systems precludes finding an ecological white rat or an *E. coli* of evolutionary ecology. Although nearly everyone finds joy in the indeterminacy of natural variation, this noise that is sometimes "music to the ecologist" (Simberloff, 1980) also often grates on the ecological ear. The need to generalize requires identification of appropriate model systems with more flesh than sets of differential equations. Such an ecological model will never be a single, or even a few, species, but rather must be a type of organism represented by many similar, yet different, species. A powerful model will be amenable to field experiments, will be widely distributed, and will have life-history characteristics that permit numerous associated behavioral, physiological, and non-experimental ecological studies. Also, the most useful ecological models will incorporate a diversity of adaptive lines, so that inferences can be generated about how biotic interactions and physical factors shape the evolution of communities.

I will examine the evidence, indirect and direct, for interspecific competition among individuals of a model terrestrial predator, the spider. My goals are (1) to address the question of competition using an ecological model that is well suited for field experiments and (2) to launch a general discussion of field experimentation as a means of testing and building ecological theory.

COMPETITION AMONG SPIDERS

Although occasionally afflicted with an irrational fascination with vertebrates, most ecologists, after detached deliberation, would name the spider as a typical terrestrial carnivore. Spiders are ubiquitous in terrestrial ecosystems and capture a major fraction of the energy in consumer species that escapes the decomposers (Menhinick, 1967; Van Hook, 1971; Moulder and Reichle, 1972; Turnbull, 1973). Species diversity is high, as is the variety of spider guilds, which range from wanderers to sit-and-wait ambushers to kleptoparasites, which steal the prey of others. Spider abundance and diversity facilitate comparative evolutionary studies (*e.g.* Enders, 1975, 1976; Anderson, 1978; Olive, 1980). The accessibility of most life-history stages makes spiders suitable for detailed behavioral and physiological studies, and for holistic investigations of community organization. Apart from these substantial virtues, many spiders, particularly web-spinners, are well suited for manipulations of resources and densities of conspecifics and hetereospecifics under relatively unaltered natural conditions.

Reynoldson and Bellamy (1971) proposed five criteria that, when satisfied together, establish interspecific competition beyond "reasonable doubt": (1) indirect evidence for resource limitation, (2) indirect evidence for intraspecific competition, (3) evidence for interspecific competition from the comparative distribution and/or relative abundances of the possibly competing species, (4) direct evidence of resource limitation and intraspecific competition from controlled manipulations of resources and population density, and (5) appropriate responses by a species to the experimental addition or removal of potentially competing species. Satisfaction of the first three criteria does not confirm the existence of competition, nor does the absence of patterns that meet these criteria prove that competition is absent. Nevertheless, indirect evidence makes field experiments more interpretable and helps unite reductionist and holistic approaches. I will summarize the current indirect and direct evidence concerning interspecific competition in spider communities.

Several ecologists have attributed differences in distribution or abundance of spiders to interspecific competition. Luczak (1963, 1966) hypothesized that differential shifts in the dominance relationships of both web-builders and wandering spiders may have resulted from changing intensities of competitive interactions in the communities she studied. Vollrath (1976) found that a kleptoparasitic species apparently shifts its diurnal activity pattern in response to decreased abundance of a congeneric species. Different phenologies have been interpreted as adaptations evolved to reduce interspecific competition (*e.g.* Luczak, 1959; Breymeyer, 1966). Uetz (1977) and Turner and Polis (1979) hypothesized that closely related wandering and raptorial spider species of similar size avoid competition by spatial and temporal niche separation. The contiguously allotopic distribution of two wolf spider species (Greenstone, 1980) supports this interpretation of the organization of wandering spider communities. Enders (1974) suggested that two closely related orb-weavers, *Argiope aurantia* and *A. trifasciata*, respond to competition by placing webs at different heights in old-field vegetation. Uetz *et al.* (1978) have suggested that differences in both mesh size and placement of the web facilitate coexistence of these two species. Taub (1977) and Brown (1981) have also interpreted niche differences between these *Argiope* species in the context of interspecific competition. Uetz *et al.* (1978) also argue that reduced overlap in habitat utilization between congeneric forest orb-weavers, and between non-congeners with similar web characteristics, permits coexistence by reducing or eliminating interspecific competition.

Not all ecologists have concluded from their holistic studies of spider communities that competition is a major interaction among these predators. Gertsch and Riechert (1976), Post and Riechert (1977), Turner and Polis (1979), and Maelfait *et al.* (1980) conclude from their niche analyses that interspecific competition is unimportant for most spiders. High or low niche overlap can be evidence for or against competition, depending upon assumptions made about the relative importance of exploitation and interference competition and the relevance of the evidence to evolutionary versus ecological time scales. Such indirect evidence cannot be decisive, but several observed patterns in abundance, local distribution, and niche overlap are at least consistent with the theory that interspecific competition exerts a significant influence on the organization of spider communities.

Indirect evidence indicates that spider populations are resource limited, which satisfies another of Reynoldson and Bellamy's criteria. The relatively low metabolic rates of spiders, particularly under food deprivation, suggest they have evolved under situations of frequent prey scarcity (Miyashita, 1968a; Anderson, 1970). Extensive yearly variation in size at maturity and fecundity also suggests that natural variation in prey availability can significantly limit spider growth and reproduction (Kajak, 1967; Wise, 1983). Temporal changes in population density and fecundity of the linyphiid *Erigone arctica* correlate with changes in abundance and activity of a collembolan, its major prey (Wingerden, 1975, 1978). Comparing the growth and reproduction of spiders in the laboratory with their performance in nature also indicates that prey often is a limited resource for spiders (Miyashita, 1968b; Kessler, 1973; Anderson, 1974; Kajak, 1978). However, this evidence suffers not only from being indirect but also from failing to indicate whether spiders compete for food. Since resource scarcity could limit growth and reproduction without behaving as a density-dependent factor, Reynoldson and Bellamy proposed experimental studies of competition within a species as their fourth criterion for establishing interspecific competition. Fortunately, both resources and spider densities, especially those of web-building species, can be manipulated under field conditions to test directly for resource limitation and intraspecific competition.

Field experiments have proven that web sites and prey levels may be limiting for web-building spiders. Colebourn (1974) modified the natural vegetation and also introduced artificial web substrate to establish that a shortage of suitable web sites limits the abundance of an orb-weaver in heather habitat. In a series of elegant studies, Riechert (1977, 1979, 1981) has proven the importance of intraspecific agonistic interactions between female funnel-web spiders as the mechanism of interference competition for a limited number of high-quality web sites. Schaefer (1978) demonstrated that a limited number of suitable sites is a major component of density-dependent population regulation of a sheet-web weaver. By manipulating both spider densities and prey levels of another sheet-web weaver, the filmy dome spider, I uncovered exploitation competition for prey among adult females (Wise, 1975).

Four of the five criteria of Reynoldson and Bellamy for establishing interspecific competition have been satisfied for the ecological model, *i.e.* spiders in general. However, clear experimental evidence of major competitive interactions between closely related spider species is lacking. Removal and density-manipulation experiments have not established interspecific competition to be a significant interaction in spider communities. Schaefer (1975) found no competitive release among wolf spiders after reducing one species on open plots and adding different combinations of species to enclosed areas. Schaefer (1978) also found that increasing the density of a sheet-web weaver in its natural habitat had no effect upon the population of a related species in which he had experimentally demonstrated intraspecific competition for web sites. Results of a field experiment, in which I supplemented the prey of adult females, showed that natural prey levels can limit the egg production of two syntopic species of forest orb-weavers, the basilica spider and the labyrinth spider (Wise, 1979). The following year I manipulated spider densities and uncovered no evidence of interspecific competition (Wise, 1981a). Neither species affected the web placement, web-site tenacity, or egg production of the other.

Other recent or ongoing experimental studies of interspecific competition among web-building spiders have also uncovered no evidence for major interactions between species. Indirect evidence from patterns of vertical stratification, habitat utilization, and prey differences has suggested the importance of interspecific competition between the orb-weavers *Argiope trifasciata* and *A. aurantia* (Enders, 1974; Taub, 1977; Uetz *et al.*, 1978; Brown, 1981). However, two years of field experiments did not produce evidence of significant competitive interactions between these species (Horton and Wise, 1983). Two replicates of mixed and single-species populations at two densities were established in open, 12 m × 12 m plots by removal and addition of juvenile spiders early in the summers of 1979 and 1980. The manipulations uncovered intraspecific competition, which was weak and not consistently present from one season to the other. Clear evidence of interspecific competition was lacking. Much of the variation in habitat utilization and dietary overlap resulted from changes in the weather and the vegetation, not from competition with the other species. Riechert and Cady (1983) conducted a removal experiment with a community of web-builders inhabiting rock outcrops in the mountains of Tennessee. Four species were studied, each from a different family. Although the species build quite different webs, they showed high overlap along several major niche parameters. Eggs and spiders of three species, in different combinations, were removed from different cliff areas. There was no evidence of competitive release of the remaining species as measured by niche expansions or increases in demographic parameters such as survival or egg production.

The only clearly documented example of significant interspecific competition among spiders is prey-stealing by kleptoparasites such as *Argyrodes*, which are adapted for living in the webs of larger spiders (Vollrath, 1979; Rypstra, 1981). However, the competitive interaction is almost entirely one-way and represents highly specialized foraging behavior. In fact, apparently many *Argyrodes* species rarely spin their own webs.

Interspecific competition may not be a major interaction among spiders, our model terrestrial carnivore. Of course interspecific competition between non-araneid predators may be important over ecological time; the spider has been proposed as one, not the sole, ecological model of a terrestrial predator. Of what value is the spider as a model if it does not represent a broader class of organisms? Two points are relevant: (1) non-mathematical models in ecology, *i.e.* representative species systems, will never play roles identical to those of models in the more reductionist biological sciences; (2) spiders by themselves constitute a rich ecological model that encompasses a broad range of organisms. But can a diverse assemblage function as a useful model? Returning to the question of interspecific competition in spider communities will illustrate how such a broadly defined model can function. Despite indirect evidence of interspecific competition, and despite direct experimental evidence of resource limitation and intraspecific competition in some species, experimental proof of interspecific competition among spiders is lacking. Failure to find direct evidence brings into question the relevance of competition theory to understanding spider communities. However, one might argue that the theory is weakened only when experimental evidence of interspecific competition is absent among particular species for which all four of Reynoldson and Bellamy's other criteria are satisfied. Such a view is too extreme because it defines the model organism too narrowly. To be useful, a body of general theory should work for organisms as different, yet as similar, as spiders in general.

Testing the relevance of competition theory with a particular, broadly viewed model system is valuable because additional studies with that model will yield explanations of the absence or presence of competition in terms of other interactions. Such meshing of diverse theories and concepts is necessary if ecology is to develop useful general principles. Future field experiments may uncover widespread competitive interactions between closely related spiders, but the immediate problem is to explain the absence of experimental evidence for significant interspecific competition in light of evidence, both indirect and direct, of resource limitation among spiders.

Traditional theory leads to the conclusion that similar spider species have evolved niche differences to avoid competition. Schaefer (1978) presents convincing evidence that the two species he studied do not compete because they have different web-site requirements. Niche theory, however, does not provide the best explanation for the absence of competition between the basilica and labyrinth spiders. The prey captured by the two species is remarkably similar, especially in view of their dissimilar webs (Wise and Barata, 1983). The absence of competition between these species in the field experiment discussed previously (Wise, 1981a) is most directly explained by the fact that intraspecific competition was weak or absent.

Variation in prey abundance might alleviate competition in some years. Evidence for food limitation of basilica and labyrinth spider fecundity comes from a 1977 prey supplementation experiment (Wise, 1979). The experiment that uncovered only minor intraspecific competition was done in the same habitat but during the following year. During this study, in which food was not experimentally supplemented, the rate of egg production by each species was equal to that of females that had received extra prey the preceding year. Thus, I concluded that spiders were not competing in 1978 because prey levels were not limiting fecundity that year (Wise, 1981a). However, in 1978 food limitation was not directly tested, nor was an adequate experimental test of competition conducted the previous year. Results of two 1980 experiments (Wise, 1983) with the labyrinth spider indicate that exploitation competition is absent even in years when a shortage of prey limits growth and fecundity. In one study I tested for intraspecific competition by establishing populations of adult female labyrinth spiders at two densities on open experimental units. Initial densities varied 10-fold between the two density treatments, representing a major portion of the crowding spectrum of non-experimental populations. Although no direct information on food limitation of mature females was collected in 1980, prey probably were relatively scarce, since fecundity was lower than in the year (1977) when the field experiment demonstrated food limitation. Despite evidence of agonistic interactions over webs, and despite indirect evidence of food limitation, survival was not lower in the high-density treatment, nor was exploitation competition affecting fecundity at the high density. A similar experiment in 1980 with immature labyrinth spiders, but which incorporated prey supplementation, also uncovered no negative effects of density upon survival or growth, despite experimental evidence that scarcity of prey was limiting growth rate. Previous field experiments also showed food supply to be a density-independent factor limiting the growth of juvenile filmy dome spiders (Wise, 1975).

Exploitation competition is possibly weak or absent in these situations because spiders may capture a small fraction of the prey that enters the air space surrounding their webs; if so, one or a few additional webs in a portion of this volume might not noticeably decrease rates of prey capture. The presence of additional webs might actually facilitate prey capture (Rypstra, 1979; Uetz, in press), thereby counteracting opposing negative effects of exploitation competition. Such a positive "knockdown" effect has been documented for tropical colonial spiders, but without actual proof of its occurring among non-social temperate species, it must remain a highly speculative explanation for species such as the labyrinth spider. Some spiders may space themselves at distances that minimize competition for prey in environments where exploitation competition would occur at high spider densities. Riechert (1974, 1978) has proposed that behaviorally based spacing is widespread among spiders, and she predicts that interspecific competition among spiders should be rare because of intraspecific territoriality. She has shown that female *Agelenopsis aperta* maintain territories that are large enough to minimize exploitation competition for prey and that these territories do not change size in response to within-generation fluctuations in prey abundance (Riechert, 1979, 1981).

In addition to more behavioral studies like those pioneered by Riechert, further research is needed on the role of other mechanisms that may reduce spider densities to levels where exploitation competition is weak or absent. Dispersal of young instars, with subsequent mortality from physical factors, natural enemies, or starvation, may reduce densities of many species to levels where competition is unimportant. Abiotic and biotic mortality may also be key factors in the dynamics of populations of older juveniles and adults, keeping densities low enough to reduce or remove competition.

Two patterns are emerging from experimental studies of spider populations and communities. One is the lack of significant interspecific competition among spiders. The other is the failure of traditional niche theory to predict consistently the presence of competition or account convincingly for its absence in many spider communities. More experimental studies of intraspecific and intertrophic interactions may contribute importantly to understanding why spider species do not compete with each other.

FIELD EXPERIMENTS AS TESTS OF COMPETITION

Failure of field experiments to uncover competition over ecological time scales does not disprove its impact over evolutionary time (Schoener, 1974b; Thomson, 1980). However, establishing whether interspecific competition has been a significant selective force is a formidable challenge (*cf*. Connell, 1980; Strong, 1980; and the discussion on neutral models in this volume). I will restrict my discussion to the role of field experiments in assessing the current importance of interactions such as competition. Field experiments so far have not produced evidence of significant interspecific competition among spiders. Statistically non-significant outcomes are often termed "negative results," an unfortunate term that implies the absence of findings. Early pedagogical warnings against committing type II errors, coupled with the realization that a null hypothesis can never be proven true, may make some ecologists uneasy about statistically non-significant results. Hence arises a tendency sometimes to reject negative results prematurely as resulting from improper experimental design. Improper design, of course, may be the explanation, but statistically significant results may also be artifacts of an improperly planned and executed experiment. An experiment should be judged on the relevance of its design to interpreting natural phenomena. Its findings should be interpreted in the context of the biology of the organisms, not solely according to the degree of statistical significance read from F tables. For example, even statistically significant findings can provide negative evidence if the relative magnitude of the response to the treatment is minor (*e.g.* Wise, 1981a).

Negative results could have more impact than those that confirm a generally held hypothesis, since frequent failure to find evidence for a phenomenon should force reformulation of theory. If a theory is worth testing, any results of adequately controlled experiments are relevant. Why test the theory if only positive results are interesting? A natural and valid disappointment over negative results may accompany the failure to produce an example that confirms theory. Any disappointment, however, should be with the theory, not with the experiment.

Only additional field experiments will determine whether the absence of interspecific competition is prevalent in spider communities. More such studies should be performed, but a reluctance to attempt field experiments does exist. Objections to experimental studies of competition derive primarily from perceived problems of generalizing from specific field experiments to broad explanations of community structure. These problems may be due to: (1) difficulty in manipulating single variables without changing other variables in an undetermined fashion; (2) spatial heterogeneity of ecological communities; (3) temporal variation in the intensity of competitive interactions; (4) the limited number of pair-wise tests of species interactions that can be accomplished via controlled removal experiments; and (5) indirect effects of changes in species abundances within complex networks of species interactions, which make it difficult to interpret the results of simple species manipulations. Cody (1974a) has argued that the limitations of field experiments make the "natural experiment" more attractive. The natural experiment involves comparing populations in different circumstances, among which only the variable of interest (*e.g.* density of a potential competitor) is assumed to vary. However, is it reasonable to assume that limiting factors differ between areas in such a way as to affect only one species, but not others? Another problem with a natural experiment is the unavoidable tendency to conclude that the absence of pattern results from nature's failure to perform the experiment, rather than from absence of the phenomenon under investigation. Basically, natural experiments do not appropriately test null hypotheses because they are *a posteriori* tests. Natural experiments are major sources of hypotheses and constitute valid indirect evidence. Although in some communities the natural experiment may be the only alternative, it cannot completely substitute for a carefully planned and controlled manipulative study.

Establishing appropriate controls in nature is a major difficulty with manipulative field experiments and forms the substance of the first three objections to experimentation listed above. The challenge of providing adequate controls arises at many levels. First is the need to perform the manipulation with minimum disturbance and maximum fidelity to natural conditions, preferably by manipulation of densities, resource levels, predators, etc. in open plots without use of artificial barriers. Also, levels of variables should reflect natural variation. In some cases artificial barriers and similar modifications are necessary. Often a complete control cannot be designed for the manipulation procedure itself, as is possible with laboratory systems. One must rely on what is known about the natural history of the system to assess whether the manipulative techniques may have altered the system in a way that seriously weakens the ability to generalize to non-manipulated nature.

A second difficulty in establishing controls derives from the inherent spatial heterogeneity of natural systems. Classical mathematical population theory and the paradigms of laboratory experimentation in population ecology have largely ignored heterogeneity, or have modeled it as a simple combination of basically homogeneous systems. An experimentalist necessarily imposes this outlook upon heterogeneous natural systems by selecting experimental areas that are as homogeneous as possible, assigning treatments to plots at random, and assigning individuals randomly to treatments whenever feasible. Despite the best efforts of the investigator, though, spatial variation in nature prevents field experiments from being equivalent to the laboratory model, in which all variables are tightly controlled. Field experiments always will be hybrids between carefully controlled laboratory studies and the uncontrolled natural experiment. However, controlled manipulations under field conditions can provide valuable tests of theory as long as the domain of generalization reflects the degree of replication. Hurlbert (in press) lucidly discusses problems of replication and inference in field experiments.

Replication of experiments over years is critical to understanding the significance of different interactions in animal communities. Wiens (1977a) has stressed the need to repeat competition studies for several years. Dunham's (1980) set of experiments with lizards, which uncovered competition but not in every year, is one good example of the need for temporal replication. Few such long-term studies have been conducted. Reluctance to perform long-term field manipulations is understandable: they are tedious, risky, and not particularly glamorous. However, temporally repeated experiments are crucial to testing theory, especially given increasing doubts over the equilibrium status of natural communities.

The last two objections to field experiments that I listed are not limited to manipulative studies. These points relate to the broader question of whether hypotheses about complex interactions can be tested by detailed studies of small components (*i.e.* one or a few species) of the larger system.

One point regarding field experiments is germane. Manipulations of one or a few variables (such as population density or resource levels) may produce a response, yet the underlying mechanisms may be unknown (Levins, 1975; Levine, 1976; Holt, 1977; Davidson, 1980; Levins and Lewontin, 1980). However, knowing how, or whether, the system responds to the perturbation is valuable. Knowledge of mechanisms depends upon what else is known about the system and upon future experimental and non-manipulative research.

Clearly field experiments are not the only valid approach to studying ecological interactions. Not all species, resources, or limiting factors can be manipulated, even in a partially controlled fashion. And an experimental approach is not always the most efficient use of resources. Testing and development of hypotheses in ecology require a mixture of approaches and outlooks (Southwood, 1980). However, the general hypotheses that will ultimately prove most useful are those that can be tested through field experimentation with powerful ecological models. Spiders possess the attributes of such a model, and in addition, they are well suited for field experiments. Web-building species in particular offer several advantages for manipulative studies, in part because of their semi-sessile habits: (1) densities can be manipulated by introduction of individuals into unenclosed plots (Wise, 1975, 1979, 1981a; Horton and Wise, 1983); (2) two major resources, web sites and prey, can be supplemented in a straightforward manner (Colebourn, 1974; Wise, 1975, 1979, 1983; Schaefer, 1978); (3) predation can be studied either by exclusion of vertebrate predators (Askenmo *et al.*, 1977) or by addition of invertebrate predators, such as other spiders, to replicated populations (Wise, 1982); (4) the accessibility of most life stages, even the egg sacs of several species, makes it possible to monitor directly the effects of experimental treatments upon population density, prey capture, individual growth, fecundity, agonistic interactions, survival, and parasitism; (5) the small size, large numbers, and annual generation time of many spiders enhance the feasibility of spatial and temporal replication of experiments. So far field experiments have demonstrated the absence of significant interspecific competition in spider communities. Future research will determine the extent to which the pattern is general, and which interactions and environmental variables determine whether or not interspecific competition occurs.

Testing and development of ecological theory require both field experiments and non-manipulative studies with several carefully selected and defined systems of model organisms. This approach has not characterized most ecological research. A tradition exists in ecology of testing general theories, often derived from simplified mathematical models, by conducting one type of study with a diverse array of organisms and communities. Many systems have been studied by only one or a few investigators.

Because ecologists strive to be as unique as the phenomena they study, it is perhaps not surprising that they have confirmed very few satisfying generalizations about animal communities. Increasing awareness of the heterogeneity of nature suggests that a more fruitful approach would be for many ecologists to investigate intensively a few appropriate model systems, using diverse approaches and extensive spatial and temporal replication.

Acknowledgments

I wish to thank J. David Allan and Charles Horton for making helpful suggestions on the manuscript. My recent research on the ecology of spiders has been supported by National Science Foundation Grants DEB 77-00484, DEB 79-11744 and DEB 79-04941 (with Charles Horton).

5.

Does Competition Structure Communities?
Field Studies on Neotropical *Heliconia* Insect Communities

RICHARD P. SEIFERT*

Department of Biological Sciences, The George Washington University,
Washington, D.C. 20052

INTRODUCTION

A major objective of community ecology is to identify the factors that determine the relative abundances of species in communities and to discern whether recurring predictable patterns of community structure exist. Identification of the structuring factors for one community may lead to high predictability about communities that have similar phyletic components and exist in similar environments. Alternatively, communities from distinct environments or with different species compositions may have differing structuring components. Studies of communities revolve around a search for underlying generalities. Ideally, a model of community structure should give valid predictions for most, if not all communities. No such model of community structure exists. If valid models of community structure can be made, they will be made for some subset of communities within an ecosystem. I believe that some very robust statements can be made about Neotropical insect communities and that these communities may be structured by factors other than those commonly associated with vertebrate and some other invertebrate communities.

Historically, ecologists have believed that competition is of prime importance in structuring communities. This belief dates from the work of Darwin (1859) for whom competition represented a driving factor in evolution and thus in determining species presence or absence in a community. More recent analyses of the importance of competition stem from MacArthur (1958), who measured the amount of utilization of different parts of forest trees by five species of warblers, discovered that those utilization functions were largely non-overlapping, and concluded that such

---

* Died December 12, 1982, age 35.

specialized use was the result of competition: all five species were maintained because of a presumed competitive superiority of each one in some aspect of its feeding behavior. Additional research, particularly with vertebrates, has continued to use field observations (Pianka, 1967, 1969, 1971, 1975; Culver, 1973, 1976; Cody, 1974a, 1975; Brown, 1975) or laboratory investigations (Park, 1948; Ayala *et al.*, 1973; Richmond *et al.*, 1975) to make general statements about the nature of competition. This work has led to extensive mathematical development (Levins, 1968, 1975; May, 1973, 1975; Roughgarden, 1974, 1976, 1979b; Schoener, 1974a) on the role of competition in natural communities. Yet observational results often rely on untested assumptions (the relationship of overlap to competition, limiting resources); laboratory results typically are based on experiments with limited resources and high population densities, which may not reflect natural systems; and mathematical models begin by assuming certain kinds of competition are occurring.

Clearly, other kinds of species interactions may be important in controlling community structure. Hairston *et al.* (1960) and Slobodkin *et al.* (1967) argued that predation reduces herbivore densities below those required for competition to occur. Paine (1980) has demonstrated how a keystone predator can cause an increase in the species diversity of marine intertidal communities by preventing potential competitors from reaching sufficiently high densities to cause resource limitation. Paine's results were not generalizable to the community of (non-arthropod) invertebrates fed upon by mosquito larvae in pitcher plant pitchers. Addicott (1974) found that mosquito predation resulted in a reduction of the number of species that otherwise would coexist.

Generalized symbiotic effects (including facilitation or commensal effects) may also be of importance in structuring natural communities (Janzen, 1966, 1969; Springett, 1968; Faegri and van der Pijl, 1971; Colwell, 1973, Seifert and Seifert, 1976a, 1979a). Unfortunately, symbiotic effects have until recently been viewed as curious and unrepresentative examples of species interactions. Theoretical analyses of symbioses have been few (Pianka, 1978; Vandermeer and Boucher, 1978; Wilson, 1980) and most researchers view symbioses as destabilizing elements in communities (May, 1973). I (Seifert and Seifert, 1976a, 1979a) have shown instances in which facilitation is of greater importance in structuring natural communities than is competition.

The insect communities that live inside the water-filled bracts of *Heliconia* inflorescences are ideal for studies on community structure. These communities are clearly defined by the physical boundaries of the inflorescence and its contained water. The insects living in these inflorescences can be easily manipulated experimentally and the plants are locally abundant, which allows for experimental replicability. Further, comparative studies of community structure can be made because different *Heliconia* species have varied floral morphology and are found in both low and mid-elevations in most of the wet Neotropics. In addition, two different kinds of insect communities live in *Heliconia* inflorescences: those dominated numerically by fly larvae and beetles and those dominated numerically by mosquito larvae. My experiments and observations on these communities have revealed the importance of species interactions, habitat age, floral morphology, floral phenology, and geographic location in determining structure and have shown that recurring patterns of community organization exist. I have been able to show that factors thought important for other natural communities, particularly competition, are of little importance for *Heliconia* inflorescence insect communities. I believe that the results of research on *Heliconia* insect communities have applicability that extends to many Neotropical insect communities, particularly those living in other kinds of plant-held bodies of water (Scott, 1914; Teesdale, 1941; Laessle, 1961; Corbet, 1964; Maguire, 1970, 1971; Frank and Curtis, 1977; Lounibos, 1978, 1981).

HELICONIA INFLORESCENCES AND THEIR ASSOCIATED INSECTS

Heliconia plants are common members of the understory of wet Neotropical low- and mid-elevation forests. Large species (up to 4 m) with erect bracts live in clumps of varying densities (Seifert, 1975) in sunlit areas including stream edges, road cuts, and light gaps in the forest left by tree falls and land slides. *Heliconia* plants may be especially common in mountainous regions, where high levels of incident radiation reach the forest floor because of the slope of the terrain. Most of the large *Heliconia* species have erect inflorescences consisting of from 4 to 20 cup-like water-filled floral bracts (Figure 5.1). In some *Heliconia* species each bract may hold up to 60 cc of water. The inflorescence produces a new bract each 3 to 10 days depending on *Heliconia* species, and an inflorescence may live for 3 to 4 months before rotting. The youngest bract on an erect inflorescence will be the top bract on the inflorescence, with successively lower bracts being successively older. Thus, each bract can be aged in relationship to other bracts and, when the rate of bract production is known, bracts can be aged by days.

An inflorescence contains an aquatic or semi-aquatic community of insects that includes species feeding on flowers, nectar, bract tissues, organic detritus, and rarely, on other inquiline insects (Seifert and Seifert, 1976b; Seifert, 1980). *Heliconia* inflorescence insect communites may be quite speciose (Seifert, 1975, 1981) but are dominated numerically either

Fig. 5.1. Diagrammatic representation of the inside of an *H. bihai* bract from Rancho Grande, Venezuela, showing the most abundant species of insects. Each insect species is represented by only a single individual per bract although a natural *H. bihai* inflorescence would have numerous individuals. Insect species and their relationship to the age of the bract are discussed in the text.

by (non-mosquito) fly larvae and beetles or by mosquito larvae. Mosquito larvae are most common in *Heliconia* species that bloom in the wet season and have large amounts of standing water in the bracts (*e.g.*, *H. aurea*, Seifert, 1980). *Heliconia* species with inflorescences that contain little water, either because of small bract size or because a large proportion of the bract is taken up by the flowers, have fly larva- and beetle-dominated communities (*e.g.*, *H. bihai*, Seifert and Seifert, 1979a). However, many fly larva- and beetle-dominated *Heliconia* insect communities have some mosquito larvae, while all mosquito-dominated *Heliconia* insect communities have some fly larvae and beetle species.

In this paper I shall consider evidence from 25 *Heliconia* collections made at 13 locations in the Neotropics. These collections indicate correlations of floral morphology and geographic location and insect species richness. In addition, I shall consider in detail the insect communities associated with *H. wagneriana* Peterson and *H. imbricata* (Kuntze) Baker from Rincón de Osa, Costa Rica, and *H. bihai* Linn. and *H. aurea* Rodriguez from Rancho Grande, Venezuela. Each site includes a dry season blooming species (*H. wagneriana* and *H. bihai*) and a wet season blooming species (*H. imbricata* and *H. aurea*). Rincón de Osa is a lowland rain-forest (tropical wet forest, Holdridge *et al.*, 1971) from 10 to 50 meters in elevation, while Rancho Grande is a mid-elevation cloud forest (tropical premontane wet forest, Ewel and Madriz, 1968) and the areas of research were from 800 to 1200 m in elevation. At Rancho Grande, *H. bihai* is most abundant at elevations ranging from 1000 to 1200 m and *H. aurea* is most common at elevations from 800 to 950 m. *Heliconia aurea* has a mosquito-dominated insect community while the other three *Heliconia* species have fly larva- and beetle-dominated insect communities. The most abundant insects in *H. aurea* inflorescences are larvae of the mosquitoes *Wyeomyia felicia*, *Culex bihaicolus*, and *Trichoprosopon digitatum*. Other members of the *H. aurea* inflorescence community include syrphid fly lar-vae (*Quichuana angustriventris* and *Copestylum roraima*) and chrysomelid (hispine) beetle larvae (*Cephaloleia neglecta*). The most abundant species of insects in *H. wagneriana* are the larvae of *Q. angustriventris*, *Copestylum ernesta*, and *Beebeomyia* sp. (a richardiid fly larva), as well as adults of *Gillisius* sp. 1 (a hydrophilid beetle). Larvae of *Q. angustriventris*, *Cepha-loleia puncticollis*, and *Merosargus* sp. (a stratiomyid fly larva), as well as adults of *Gillisius* sp. 1, are the most common insects in *H. imbricata* inflorescences. Finally, the most abundant species in *H. bihai* inflores-cences are *Q. angustriventris*, *Copestylum roraima*, *Gillisius* sp. 2, and *Ceph-aloleia neglecta* (Figure 5.1). The fly larva–beetle communities that have been studied are inhabited by only a few predators, none of which is im-portant in controlling community structure (Seifert and Seifert, 1976b). In *H. aurea*, *T. digitatum* is a facultative predator on smaller mosquito larvae. Detailed studies on the life histories of these and other *Heliconia* insects have been published elsewhere (Seifert and Seifert, 1976b, 1979a, b; Strong, 1977b; Seifert, 1980). Generally similar kinds of experiments were conducted with each of these four *Heliconia* inflorescence insect com-munities to evaluate the kinds and importance of species interactions and bract (habitat) age in determining *Heliconia* insect community structure.

INSECT SPECIES RICHNESS OF *HELICONIA* INFLORESCENCES

Between 1972 and 1980 I made 25 collections of *Heliconia* from Costa Rica (2 locations), Venezuela (5 locations), Ecuador (3 locations), Trinidad (1 location), Martinique (1 location), and Guadeloupe (1 location) (see Seifert, 1981, for details of the localities, collection dates, and *Heliconia* species). Each collection included at least 20 inflorescences. The inflores-cences were dissected and the number of insect species in each inflores-cence was recorded. Next, I computed the total species richness for each collection. Using these data, I constructed a multiple linear regression where insect species richness was the independent variable and the coded dependent variables were geographic location, floral structure, bract size, bract number, water holding capacity, season collected, and elevation of the collecting site. Of these variables only location and bract size were statistically significant (Seifert, in press).

Collections from the French Antilles had low insect species richness compared with mainland collections. Low species richness of the collections from the Antilles is at least partly due to the low vagility of some of the insect species and the inability of those insects to colonize Antillean *Heliconia*. For example, no *Heliconia*-feeding hispine beetles are found in Martinique or Guadeloupe, although in laboratory tests Ecuadorian hispines fed and grew on bracts from both species of *Heliconia* living on Guadeloupe (Seifert, 1981). Thus, the absence of these beetles from the island *Heliconia* seems to be the result of an inability to colonize the islands and not of an inability to feed on the bracts. The low species richness of *Heliconia* insects on the French Antilles is consistent with the general trend of depauperate island biota (MacArthur and Wilson, 1967; Carlquist, 1974; Simberloff, 1974).

Heliconia species with large bracts had greater insect species richness than did *Heliconia* species with small bracts. Large bract inflorescences tend to have relatively more of their total space filled with water and provide mosquito larvae with a habitat not available to them in the smaller bract species (Seifert, in press). However, the dependent variable coded for water-holding ability was not statistically significant. Alternatively, large bracts might increase some specific food source to a high enough level for an additional species to feed on a previously unused resource. If this were the case, I would have expected bract number to be correlated with species richness. No such correlation was found.

In an earlier study (Seifert, 1975) I showed how flowering phenology, *Heliconia* clump size, and species richness are correlated. I compared clumps of *H. imbricata* with those of *H. wagneriana* and I found that insect species richness increases more rapidly with clump size for *H. wagneriana* than it does for *H. imbricata*. This difference I attributed to the differences of flowering phenology between the two *Heliconia* species. *Heliconia imbricata* plants in a clump all bloom at approximately the same time so that each clump consists of inflorescences of about the same age. In contrast, *H. wagneriana* plants stagger their blooming so that a clump of inflorescences consists of plants of different ages. Organisms that are found most frequently in a particular age class of *Heliconia* inflorescences, for example young inflorescences, could remain in a *H. wagneriana* clump, where each clump would continue to have young inflorescences for several months, longer than it could in an *H. imbricata* inflorescence, where young inflorescences are found in the clump for only a few weeks.

COMMUNITY STRUCTURE OF INSECTS LIVING IN *HELICONIA* INFLORESCENCES

I have experimentally studied examples of both fly larva- and beetle-dominated (Seifert and Seifert, 1976a, 1979a) and mosquito-dominated (Seifert, 1980; Seifert and Barrera R., 1981) *Heliconia* insect communities in an effort to evaluate the importance of first-order and higher-order interactions and bract (habitat) age effect in determining *Heliconia* insect community structure. The experiments on fly larva–beetle communities involved both a dry season blooming species, *H. wagneriana*, and a wet season blooming species, *H. imbricata*, from Rincón de Osa, Costa Rica (Seifert and Seifert, 1976a), as well as a dry season blooming species, *H. bihai*, from Rancho Grande, Venezuela (Seifert and Seifert, 1979a). In these experiments, different combinations and densities of the four most common insect species from each of the *Heliconia* species were introduced into inflorescences initially devoid of insects. In all of the studies some of the experimental treatments included densities as high as or higher than the densities found in naturally occurring communities. Thus, if competition occurs in natural *Heliconia* insect communities it also should have occurred in the experimental communities. The insects were allowed to interact for from 1 to 15 days (6 to 8 days for most experiments), after which time the inflorescences were dissected and the abundances of the insect species were recorded. In the Venezuelan experiments only, the experiments were run on a per-bract basis. Next, the data were analyzed using a multiple linear regression analysis where the *per capita* change in densities was regressed against the initial densities of each insect species and, in the Venezuelan experiments only, against a higher-order term and a term indicating the bract in which the experiment was run. (*A posteriori* tests for higher-order interactions were run on the Costa Rican experiments.) These experiments measured the effects of species interactions and bract age on survivorship. Hallett and Pimm (1979) have provided mathematical justification for the use of regression techniques in estimating species interaction effects.

From a possible 36 interspecies interactions from the three communities only 6 were competitive, 6 were facilitative, and 24 were not statistically different from zero (Seifert and Seifert, 1976a, 1979a). Thus, even at densities higher than those commonly observed in the field, competition was no more common than facilitation and most potential interactions were non-significant statistically. Further, facilitation was quantitatively more important than was competition (Seifert and Seifert, 1976a). When hispine beetle larvae are found they increase the survivorship for at least some of the other community members by increasing detritus, which is used as food by fly larvae, and by forming open bract wounds, which increases feeding sites for hydrophilid beetles. I make the generalization that in *Heliconia* inflorescence insect communities where hispines exist, the hispines will facilitate some other insect species. In this view it is worth recognizing the stratiomyid larvae, which benefit from hispine presence (Seifert and Seifert, 1976a), have only been found in *Heliconia* inflorescence insect communities where hispines exist.

Higher-order interactions were rare in these studies (Seifert and Seifert, 1976a, 1979a); they occurred in only 1 of 12 possible cases. However, one would not expect higher-order interactions to be important in an insect community in which first-order interactions are uncommon.

In addition, out of 12 possible intraspecific interactions, only 7 were statistically significant and of those only 5 were competitive. Thus, the densities of these insects in the field often do not lead to intraspecific competition. Further, intraspecific competition of a species from a given *Heliconia* insect community does not imply that a similar insect species will show intraspecific competition in other communities. *Cephaloleia neglecta*, a hispine beetle, was the only insect species that exhibited intraspecific competition in the *H. bihai* experiments from Rancho Grande, Venezuela (Seifert and Seifert, 1979a). At Rancho Grande these larvae often strip the bracts of all the potential food source for hispines and as a result resource limitation occurs commonly. However, *Cephaloleia puncticollis* from Rincón de Osa, Costa Rica, feeding on *H. imbricata* bracts, did not show a statistically significant intraspecific effect and the bracts were not extensively damaged.

The age of the bract is important in determining survivorship (Seifert and Seifert, 1979a, b). Most species survive best in older bracts, but hispine beetles have greater success in younger bracts. Thus, in *H. bihai* from Rancho Grande one can see a loosely defined successional pattern of insect species as the bracts age (Figure 5.1). The top (youngest) bract might include only *Cephaloleia neglecta* larvae, the second bract might include larvae of *Cephaloleia neglecta* and *Copestylum roraima* and adults of *Gillisius* sp. 2, the third bract might include larvae of *Cephaloleia neglecta*, *Quichuana angustriventris*, and *Copestylum roraima* and adults of *Gillisius* sp. 2, while the fourth bract may have larvae of *Quichuana angustriventris* and *Copestylum roraima* and adults of *Gillisius* sp. 2. (Figure 5.1 is a diagrammatic representation of these insects in the bracts. For each bract, the first species listed in the above text is the species closest to the rachis, and successively named species are successively further from the rachis.) While such a succession is only weakly defined for *H. bihai*, a more consistent pattern of succession is shown by mosquito-dominated *Heliconia* insect communities such as that found in the bracts of *H. aurea* (Seifert, in press).

More recently I (Seifert, 1980; Seifert and Barrera R., 1981) have studied the mosquito-dominated insect community living in *H. aurea* inflorescences at Rancho Grande, Venezuela. I designed experiments to evaluate the importance of intraspecific interactions and bract age in determining both the survivorship and development rate of the three mosquito species that live as larvae in the bracts of *H. aurea*. These experiments involved only single-species studies; no experimental bracts contained individuals of two or three mosquito species. Different densities of first-instar mosquito larvae were placed in bracts free from other insects, the inflorescence was then covered to inhibit mosquito oviposition, and the larvae were allowed to interact for 14 days. After this time the bracts were collected and the number of individuals of each instar in each bract was recorded. For each mosquito species a Kruskal-Wallis test was run on the proportion of individuals remaining and on the developmental state (in instars) reached by the surviving larvae. The null hypotheses tested included no density effect, no bract effect, and no interaction effect. Only in the case of *Wyeomyia felicia* was there a statistically significant density effect: increased densities resulted in a prolongation of the larval developmental rate. For the other two species survival rate was related to bract age. *Culex bihaicolus* survived significantly better in older bracts than in younger bracts, while *Trichoprosopon digitatum* survived best in youngest bracts. These results correlate well with the oviposition preferences of the mosquito females. *Trichoprosopon digitatum* oviposition occurs almost entirely in the youngest bract, while *C. bihaicolus* oviposition occurs primarily in the oldest bracts. The results of the oviposition preferences and the bract age effects are that there is a separation of mosquito larvae into bracts of different ages. The youngest bracts contain mostly *T. digitatum*, middle age bracts contain mainly *W. felicia*, and the oldest bracts contain primarily *Culex bihaicolus* (Seifert, 1980, in press). An observational study alone might have inferred that strong species interactions resulted in the spatial partitioning of the bracts by mosquito larvae. However, the results of the experiments indicate that bract age and oviposition preferences of these specialized species cause the distributional patterns of mosquito larvae seen in *H. aurea* inflorescences. I emphasize that the results from two different kinds of *Heliconia* inflorescence insect communities show that neither intraspecific nor interspecific competition is a major structuring factor.

I conclude by pointing out that communities of highly specialized species, such as those living on *Heliconia* inflorescences, consist of members drawn from quite divergent phyletic lines: *Heliconia* insect communities have not accumulated numerous closely related competing species. Specialized utilization of resources by these insects arises not because of competitive interactions but because the insect species are sufficiently distantly related that their initial evolution has precluded overlap of resource utilization. Even in cases where we might find resource utilization overlap, the densities of insects in the bracts are seldom sufficiently high to cause either intraspecific or interspecific competition.

Acknowledgments

I am pleased to thank Douglas J. Gill, Charles Mitter, Florence Hammett Seifert, and Donald R. Strong for their constructive criticism of this manuscript. I am particularly grateful to Donald R. Strong for giving me the opportunity to publish my ideas on competition and Neotropical *Heliconia* insect communities in this volume. I thank John Prentice for contributing the art work. This research was supported financially by NSF grant DEB 79-06593.

Biogeographic Evidence
on Communities

6.

Non-Competitive Populations, Non-Convergent Communities, and Vacant Niches: The Herbivores of Bracken

JOHN H. LAWTON

Department of Biology, University of York, Heslington, York, YO1 5DD, England

INTRODUCTION

Whether or not communities of unrelated species converge in structure under similar environmental conditions is of both theoretical and empirical interest. If they do, the likelihood is that interspecific competition for limiting resources, usually food, imposes strong constraints on community structure. If they do not, then belief in strong constraints must inevitably be weakened. Obviously, different sorts of communities need not obey identical rules (MacArthur, 1972). This point has been nicely put by Cody and Mooney (1978), who, summarizing the evidence at their disposal, wrote (p. 268):

Convergent evolution (a) is best observed in patterns at the level of the community and below (species, guild), and not evidenced by patterns above the community level (*e.g.* landscape), (b) is more likely in vertebrates than invertebrates, since the latter are generally more closely tied to the specifics of their resources, such as leaf chemistry or flower phenology, while the former utilize more generalized resources (*e.g.* foliage insects) which are more likely similar across continents, and (c) is more likely to occur in small guilds of ecologically isolated species that are least likely to be affected by density compensation.

In this paper I propose to examine Cody and Mooney's generalizations with specific reference to the members of one community, the herbivorous insects feeding on bracken-fern (*Pteridium aquilinum* (L.) Kuhn). I will look for convergence in community structure in four parts of the world—Hawaii, the Southwestern United States, Great Britain, and Papua New Guinea. Then I will test for competitive interactions between species. Despite the fact that the "specifics of their resources, such as leaf chemistry" are very similar (though probably not identical) in all these places, and despite also that bracken faunas are ideal "small guilds of ecologically isolated species" showing no signs of density compensation, the results of these labors are profoundly negative. I find no community convergence, large numbers of apparently vacant niches, and on the available evidence (which is not yet experimental, and must therefore be used carefully), no significant interspecific competition. Hence, the rules that govern assemblages of bracken herbivores appear to be fundamentally different from prevailing views about the role of interspecific competition as a force structuring many animal communities (*e.g.* Cody and Diamond, 1975a; Hutchinson, 1978) but have much in common with recent voices dissenting from these views (*e.g.* Wiens, 1977a; Dunham *et al.*, 1978; Connor and Simberloff, 1979; Huey, 1979; Wiens and Rotenberry, 1979; Connell, 1980; Hairston, 1980b).

The bracken community appears to be typical of many other phytophagous insect communities (Lawton and Strong, 1981) and of "parasite communities" in the widest sense (Holmes and Price, 1980; Price, 1980). That is, like those of bracken herbivores, their populations more often do not compete than do, and resulting communities are not saturated with species, nor do they converge in structure. A clear majority of the world's biota must therefore behave this way (Price, 1980; Lawton and Strong, 1981).

This paper is organized as follows. First I describe the food plant, before establishing the number of species exploiting it in the geographical regions under study. Then I look for community convergence, comparing local sites on different continents, before moving to a detailed search for interspecific competition amongst members of the British fauna. A brief digression examines counter-evidence that might be used to support a weak case both for convergence and for interspecific competition. Finally, the bracken data are set in context alongside other studies, although I have not made any attempt to carry out an exhaustive review of the relevant literature.

THE PLANT AND THE STUDY SITES

Introduction

Bracken is probably one of the five commonest plants in the world (Harper, 1977), occurring naturally on all the continents except Antarctica, and on many islands. Not surprisingly, it is neither genetically nor phenotypically uniform throughout this vast range, although the most recent and authoritative taxonomic treatment by Tryon (1941) assigns all varieties to one of two subspecies (*aquilinum* and *caudatum*) within the single species, in the monospecific genus *Pteridium*. Page (1976) provides an up to date review, together with reasons for thinking that it may eventually prove necessary to subdivide *Pteridium* into several closely related species. However, although I am mindful of them, unlike the taxonomists whose work they bedevil I am in the main rather thankful for the small and inconsistent differences that exist between varieties. Although it is a subjective assessment, they certainly seem no bigger, and often look much smaller, than the many differences noted in the supply of resources for vertebrate consumers in ecosystems where others have expected convergence (*e.g.* Cody *et al.*, 1977; Cody and Mooney, 1978).

Bracken has probably been widespread for a long time. Fossils attributable to *Pteridium* are known from Oligocene, Miocene, and Pliocene deposits in Europe, and what also appears to be *Pteridium* is known from Tertiary beds in Australia (Page, 1976). Hence there has been more than enough time for the plant to acquire a complement of herbivores wherever it occurs (Strong, 1974; Strong *et al.*, 1977). However, it has undoubtedly become much more abundant within parts of its geographical range during historical times, as a weed of disturbed and marginal land (Page, 1976; Rymer, 1976). Thus, some of its herbivores may conceivably be recently acquired; however, the fact that only one species, *Dasineura filicina*, can be positively identified amongst Roman bracken, over 1900 years old, from northern England is probably due to the way the bracken was treated before being preserved (J. H. Lawton, unpublished data summarized in Seaward, 1976), and not to an absence of species at this time.

Bracken Characteristics in Each Study Area

(i) *Hawaii*. Bracken adjacent to Kipuka Ki on the island of Hawaii was kindly sampled for me for three months by Dr. S. L. Pimm, from shortly after the emergence of the fronds in early March until mid May 1979. The main study site was at 1300 m, and consisted of an extensive pure stand 3–400 m wide and over 1 km long.

Bracken is native on Hawaii (Zimmerman, 1948), where it is widespread between 300 and approximately 1500 m, sometimes higher (Page, 1976; S. L. Pimm, pers. comm.) The plants are referable to subspecies *aquilinum*, variety *decompositum* (Page, 1976). Frond growth is seasonal, with fronds emerging synchronously in early February (Figure 6.1).

(ii) *New Mexico and Arizona*. I sampled bracken in the southwestern U.S.A. in the spring and summer of 1979 in the mountains of New Mexico and central and southeast Arizona. In New Mexico, bracken is abundant in a clearly defined band between 8000 and 9500 feet (2400–2900 m).

Two localities were studied intensively, one a pure, dense stand 1500 m² in area, in an open meadow, the other scattered fronds under *Quercus*

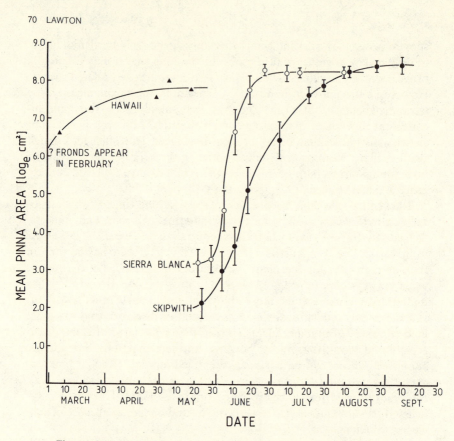

Fig. 6.1. Seasonal development of bracken fronds in three study areas: in all three the structural development, or architecture, of the frond is very similar. Mean area of the frond with 95% confidence intervals is shown for Britain (Skipwith open site in 1980) and New Mexico (Sierra Blanca open site in 1979). In Hawaii, S. L. Pimm measured the number of pairs of pinnae > 10 cm long: these data have been converted to the total area of the frond using a standard relationship based on Skipwith bracken (J.H.L., unpublished) between the sizes of individual pinnae and total area of the frond.

gambelii/Pinus ponderosa forest over an area of approximately 0.5 km². The two sites lay close together, between 8000 and 8400 feet (2440–2560 m) under the summit of Sierra Blanca in the Sacramento Mountains.

Plants in Arizona and New Mexico are referable to subspecies *aquilinum*, variety *pubescens*. To my eye they differ perceptibly but very slightly from British *Pteridium* in the angle of attachment of the pinna to the rachis and in the color of the frond-hairs and extra-floral nectaries. Frond growth is seasonal, with the fronds emerging synchronously in mid May (Figure 6.1), and dying back in early autumn.

(iii) *Britain*. I have sampled bracken throughout the British Isles since 1971 (Lawton, 1976, 1978a, b; Lawton and Eastop, 1975; Rigby and

Lawton, 1981). The plant is virtually ubiquitous below an altitude of approximately 600 m.

Intensive studies have been made at two Yorkshire localities, Skipwith Common (10 m above sea level), and a site on the North Yorkshire Moors (at approximately 200 m). At Skipwith two communities have been investigated; one since 1972 an open area of dense, pure bracken 2600 m^2 in extent, the other scattered fronds under adjacent *Betula pendula* woodland. Detailed work on this second community was only started in 1980. The Sierra Blanca (New Mexico) sites were chosen to match these two Skipwith sites as closely as possible.

The North Yorkshire Moors study area contains "islands" of bracken ranging in size from a few square meters to many hectares, set in a "sea" of heather (*Calluna vulgaris*); these two plants share no insect herbivores.

British bracken is referable to subspecies *aquilinum* var. *aquilinum*. Fronds emerge synchronously in mid to late May, and die back in the autumn (Figure 6.1).

(iv) *Papua New Guinea*. Kirk (1977, 1980, and pers. comm.) sampled bracken at 30 different localities in Papua New Guinea, finding the plant common throughout the country from sea level to 3400 m. Kirk also intensively sampled one site, Hombrom Bluff, between January 1976 and January 1977, where bracken grew in extensive stands under mixed *Casuarina/ Eucalyptus* savannah woodland at 660 m.

Two varities of bracken occur in Papua New Guinea, subspecies *caudatum* var. *yarrabense* from sea level to approximately 1600 m, and subspecies *aquilinum* var. *wrightianum* from 1300 to 3400 m. However, Kirk (pers. comm.) often found great difficulty in assigning plants to one or the other subspecies, and discerned no clear associations between particular insect species and the type of bracken.

Unlike that at the first three study areas, growth of bracken at Hombrom Bluff is aseasonal, with new fronds emerging continuously and all insect species present at the site throughout the year. However, just as in Britain, different species tend to be associated with fronds of different ages, and New Guinea bracken contains two allelochemics (cyanide and ecdysones) identical to those found in British bracken (Kirk, 1980; Lawton, 1976).

Frond Architecture

The "architecture" of mature fronds (Lawton, 1978a)—the resource space available for herbivores—is very similar in all of the study areas and follows a similar seasonal development in three of them (Figure 6.1). Since the size and growth form of plants is a major determinant of the diversity of phytophagous insects (Lawton and Schröder, 1977; Strong and Levin, 1979), just as Foliage Height Diversity determines bird diversity (MacArthur, 1972), there are no reasons to expect, arguing simply on the grounds of frond architecture, that bracken communities will not tend to converge in structure on different continents.

THE NUMBER OF SPECIES IN THE REGIONAL POOL

Throughout its world range, over 100 species of insects have been recorded feeding on bracken (Balick *et al.*, 1978; several obvious mistakes and synonyms in this list are corrected by Auerbach and Hendrix, 1980, and Hendrix, 1980). Since recent surveys have all added species to this world list, the true number of herbivores exploiting the plant is certainly well in excess of 100. As far as I know, none of these insects occurs throughout the world range of its host. Figure 6.2 shows the number of species attacking bracken in the four regions described in the previous section, together with an assessment of the geographical area occupied by the plant in each region.

Although based on only four points, the species-area relationship (see Strong, 1979, and Connor and McCoy, 1979, for reviews) is highly significant ($r = 0.987$; $F_{1,2} = 74.84$; $p = 0.013$). The lower (solid) dots, used to calculate the regression, are the number of species certainly feeding on the plant in each region; the upper (open) dots include doubtful records and casual species that appear to feed upon the plant only very occasion-

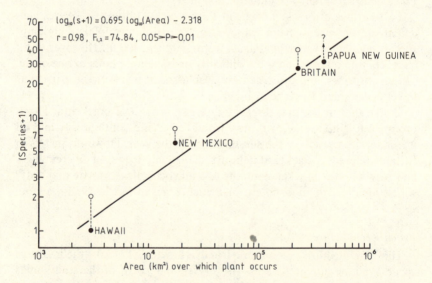

Fig. 6.2. Geographical species-area relationship for herbivorous insects on bracken. Those parts of the world with larger areas where bracken grows apparently support more species. The regression-line is fitted to the solid dots.

ally. Notice that the ordinate is "$S + 1$" because, despite extensive efforts (checked by D. S. Simberloff and E. F. Connor), Pimm failed to find any insects definitely feeding on bracken in Hawaii. A single individual of a leaf-tying Lepidopteran caterpillar represents the only possible herbivore.

Five bracken herbivores were recorded in New Mexico, with two more just over the border in Arizona; the latter are therefore included in Figure 6.1 as "possibles." The data are summarized in Appendix 1a.

The full list of herbivorous insects feeding on bracken in Britain is 39 (Lawton, 1976), plus some of very uncertain status. Since this list was published, there have been a small number of taxonomic revisions, summarized in Appendix 1b. More important in the present context is the fact that the British data draw on many more "entomologist-years" than do those from anywhere else. Hence the solid dot in Figure 6.2 is for the "category a" species in Appendix 1b, making the British sampling effort more comparable with that for other areas. The open circle includes all 39 insect species.

The Papua New Guinea species are summarized by Kirk (1980); the list is probably incomplete, particularly for Hemiptera and species from bracken at high altitude (Kirk, pers. comm.). Appendix 1c summarizes the data.

Obviously, more data are needed to establish, unequivocally, the existence of a geographical species-area relationship for bracken herbivores. But on present evidence, the number of bracken-feeding insects in a regional pool depends, rather simply, upon the amount of bracken growing in that area.

COMMUNITY STRUCTURE ON DIFFERENT CONTINENTS

Just as the herbivores attacking bracken within a geographical region are a subsample of its world fauna, so too are the species found in local communities subsamples from the appropriate regional pool.

We could test for convergence using the total species in each geographical region. But this procedure would only be justified if there were reasons for thinking that the evolutionary history of each species in the extant pool had somehow been influenced in a sustained and significant way by most, if not all, of its compatriots. Since the geographical ranges of insects may show dramatic and idiosyncratic changes back to the Pleistocene and beyond (Coope, 1978, 1979) I find this a naive assumption. Hence, we must look for convergence at the level of local communities.

Comparison of Local Communities in Large Bracken Patches

Focusing upon local communities in effect gives the convergence hypothesis a second chance. It is obvious from Figure 6.2 that regional pools do not converge in structure; they cannot, because each contains a radically different number of species.

Unfortunately, so also do local communities (Figure 6.3). These data are for the large local patches described on pages 68–71. Defining a "local patch" (community) is to some extent arbitrary, but the Skipwith open site, for example (2600 m^2), maintains self-sustaining populations of most species (see pages 87–88), and can easily be crossed by the more active insects (like adult female sawflies) in a matter of minutes. Hence I believe the choice of scale defining a local community is, biologically, a sensible one.

The feeding niches in Figure 6.3 define four major modes of exploitation: (i) chewers, which live externally and bite large pieces out of the plant, (ii) suckers, which puncture individual cells or the vascular system, (iii) miners, which live inside the tissues, and (iv) gall formers, which do likewise but induce the tissues to form galls. Insects do these four things to different parts of the plant, attacking the pinnae ("leaves"), the rachis (main stem), costae (main stalks of the pinnae, arising from the rachis), or costules (main "leaf veins" arising from the costae).

It takes only a brief inspection of Figure 6.3 to see that there are major differences between communities in the number and pattern of occupied

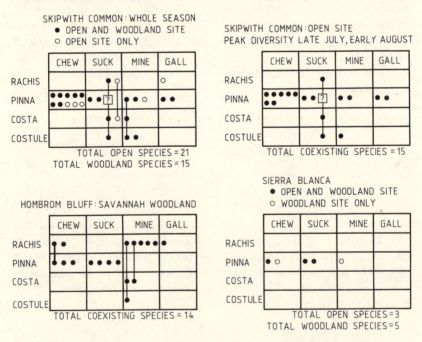

Fig. 6.3. Niche-matrices defining the feeding sites and feeding methods of herbivorous insects on bracken fronds. Feeding sites of species exploiting more than one part of the frond are joined by lines.

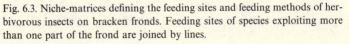

resource states. Woodland communities differ somewhat from those on open ground close by them, but in opposite directions, with more species in woodland at Sierra Blanca and fewer at Skipwith. Across continents, the differences are profound. There are no rachis miners at Skipwith, compared with five at Hombrom Bluff. The latter site is relatively depauperate in external foliage chewers, but supports two rachis chewers, a group entirely absent from Skipwith. New Mexico bracken is full of unexploited niches, and, of course, that on Hawaii (not shown) is virtually virgin territory.

A chi-squared test for homogeneity of occupied resource states between the Skipwith site and Hombrom Bluff, grouping rachis, costa and costule suckers, and all gall formers to avoid small expected values, confirms these substantial differences ($\chi^2_8 = 20.3$; $0.01 > p > 0.001$).

Figure 6.3 ignores the seasonal distribution of species. At Hombrom Bluff, all the species are present all the time, whilst at Skipwith there are marked seasonal patterns of occurrence (Lawton, 1976, 1978a). Hence another possible comparison is between the entire community at Hombrom Bluff (14 species, 18 occupied resource states) and the Skipwith open site at the height of the summer (15 species, 17 occupied resource states) (Figure 6.3). Again there is no clear evidence for homogeneity of occupied resource states ($\chi^2_8 = 16.6$; $0.05 > p > 0.01$), although several small expected values just less than one (0.97) demand cautious interpretation of this statistic.

Figure 6.3 does suggest a tendency for bracken herbivores to chew and suck pinnae wherever the plant occurs. Since most herbivorous insects, whatever species of plant they exploit, attack leaves and not stems and since most are external chewers and suckers, these broad similarities are hardly surprising, and presumably reflect the abundance and palatability of "leaves," and the more intimate adaptations demanded by mining and gall forming. Beyond this, there is no evidence in Figure 6.3 that bracken-herbivore communities are more similar than chance alone dictates. That is, there is no evidence that local communities are constrained to support the same number of species (a point returned to in Figure 6.6 and page 91), or that patterns of occupied resource states in species-rich systems converge in structure.

WHY NO CONVERGENCE?

There are three possible reasons for a conspicuous lack of convergence in community structure.

(i) Despite superficial appearances, and some evidence (Kirk, 1980) to the contrary, bracken in different geographical areas is sufficiently distinct biochemically and structurally to prohibit certain modes of feeding (Cody and Mooney's point (b), page 67).

(ii) Climatic differences between areas impose special constraints on the life histories of the resident species, overriding any tendencies for convergence.

Undoubtedly, habitat and bracken characteristics influence individual species. This influence is made plain both by the absence of some species from woodland areas, even though they are present on adjacent open sites (or vice versa, Figure 6.3), and by the tendency for certain species to occur *together* significantly more often than chance dictates (pages 80–81), suggesting that they are influenced in a similar way by unknown habitat variables. But we must be extremely careful to distinguish effects on individual species from arguments precluding, wholesale, the occupancy of certain parts of the frond, in certain ways, in particular geographical regions.

The lack of convergence between communities is due to two things. One is a marked difference in the numbers of species exploiting the plant in similar ways (*e.g.* the pinna-chewing guild). Failure to converge here cannot be due to the inability of species to exploit particular niches, because all communities have some species that do so. More interesting are the totally unoccupied niches, rachis miners at Skipwith for example. Could it be that British bracken has a peculiar stem chemistry or stem morphology, prohibiting such a mode of feeding? The answer is no: there *is* a rachis miner in Britain (*Chirosia albitarsis*), but it happens not to occur at Skipwith. In brief, I think the major reason for so marked an absence of convergence must be sought elsewhere, and is as follows.

(iii) Species do not compete for food to any significant extent, and hence experience no strong constraints on their feeding niches from other species sharing their habitat.

Evidence for a lack of significant interspecific interactions between bracken herbivores is of four sorts, outlined in the next section. None of these data are experimental, which is regrettable, but *in toto* they provide consistent circumstantial evidence for a lack of interspecific interactions between bracken herbivores.

EVIDENCE FOR A LACK OF SIGNIFICANT INTERSPECIFIC INTERACTIONS BETWEEN BRACKEN HERBIVORES

Comparison of Abundances at Sierra Blanca and Skipwith: The First Test for Density Compensation

Janzen (1973b) inferred that, because "all parts of an individual plant are connected through the medium of its resource budget," insects feeding on that plant "automatically compete with *all* other species" on the same plant. Hence each individual species at Skipwith should, on average, be much rarer than those at the faunally impoverished New Mexico site: that is, the New Mexico species should show density compensation (*e.g.* Cody *et al.*, 1977). They do not. Figure 6.4 shows average peak abundances per frond at Skipwith over five years for chewing herbivores, and Figure 6.5 over six years for sucking herbivores, compared with average peak abundances per frond for insects exploiting the plant in the same general way at Sierra Blanca in 1979. (The absence of detailed population data at Skipwith in some years is due either to lack of time or to my being away: the

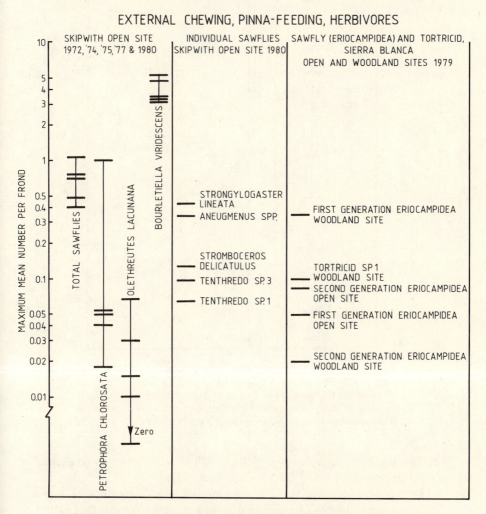

Fig. 6.4. A test for density compensation in chewing herbivores. Maximum mean population numbers per frond over five years at the Skipwith open site are shown on the left, with data for individual sawflies in 1980 in the center panel. Comparable data for the faunally impoverished Sierra Blanca site in 1979 are shown on the right. There is no sign of density compensation.

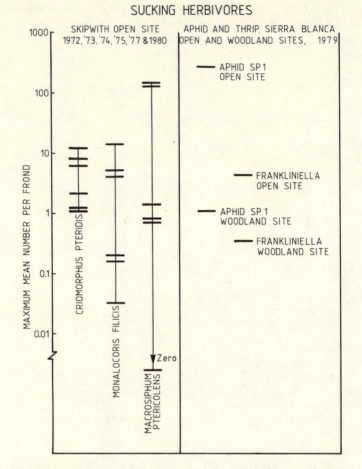

Fig. 6.5. A test for density compensation in sucking herbivores. Maximum mean population numbers per frond over six years at the Skipwith open site are shown on the left. Comparable data for the faunally impoverished Sierra Blanca site in 1979 are shown on the right. Only the aphid on the open site at Sierra Blanca achieves higher numbers than have ever been seen at Skipwith.

site has, however, been sampled well enough in most years to establish the presence or absence of species—see Table 6.5.)

Taxonomic problems with the sawfly caterpillars at Skipwith were only finally overcome in 1980; hence the data in Figure 6.4 show total sawfly caterpillars in all years, and detailed species counts for 1980.

Note that failure of the New Mexico species to be consistently more abundant cannot be attributed to the fact that New Mexico fronds are small and feeble. Over most of the growing season Sierra Blanca fronds are slightly larger than those at Skipwith (Figure 6.1).

Effects of Patch Size on the British Fauna:
The Second Test for Density Compensation

To this point, attention has been focused on large patches of bracken, quite deliberately, because an examination of well-defined bracken islands on the North Yorkshire Moors showed that large islands support more species than small islands (Figure 6.6) (Rigby and Lawton, 1981). Also plotted on Figure 6.6 are points for the open site at Skipwith and Sierra Blanca. (The other sites described on pages 69–71 are all much larger than the largest islands plotted in Figure 6.6; unfortunately I cannot measure, or do not know, the sizes accurately enough to plot them.) For its area, Skipwith—a lowland site—is slightly richer than the exposed and rather bleak upland islands on the North Yorkshire Moors. Figure 6.6 confirms that, for its size, the Sierra Blanca study area is very impoverished, but neither exposed nor bleak.

The North Yorkshire Moors species-area relationship is not quite statistically significant at the 1-in-20 level on a standard double-log plot; it is significant when species number (S) is plotted against log area ($S = 1.841 \log_{10}$ Area $+ 5.77$; $r^2 = 0.23$; $F_{1,15} = 4.52$; $p = 0.05$). Small islands

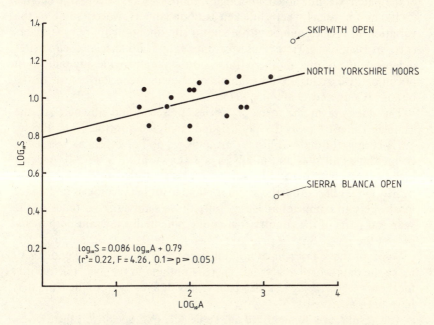

Fig. 6.6. Species-area relationship for bracken islands on the North Yorkshire Moors (Rigby and Lawton, 1981). Also plotted (open circles) are data for the Skipwith and Sierra Blanca open sites. The fitted regression is for the Moors data only. The species occurring at this site are given in Rigby and Lawton (1981), and *include* the two species of mites listed in Appendix 1b, which have otherwise been excluded from consideration in the present paper.

are therefore poor in species, which is something ecologists have come to take for granted. But none of the surviving species shows density compensation as the number of potential competitors falls. In fact exactly the opposite happens. All species get progressively rarer per frond as islands decrease in size (Lawton, 1978a; Rigby and Lawton, 1981).

A major complication with this analysis is that each frond is on average smaller on small islands (Rigby and Lawton, 1981), implying that carrying capacities decrease with island size. Hence expected levels of density compensation are difficult to predict, and the fact that we do not see simple density compensation provides only weak and equivocal evidence for a lack of interspecific competition.

Coexistence and Replacement of Species at Different Sites: A Test for Competitive Exclusion

Whenever the opportunity has arisen during the past ten years I have sampled the insects on bracken patches in a large number of places throughout Britain. Surveys usually consist of a single visit to a site, chosen haphazardly and then sampled thoroughly. In most cases the size of the patch was not noted, although I almost always select large clumps (>100 m^2) in areas where bracken is abundant. However, some of the variation in species number from site to site undoubtedly reflects differences in island size. Here, I use the data to ask: do species occupy sites independently of one another? If not, do the data provide any indication of competitive exclusion, for example by showing "checker-board" distributions (Cody and Diamond, 1975a)?

The data are in the form of species' presence and absence. All the samples cannot simply be pooled and analyzed together (e.g. Pielou, 1977) because on natural history grounds we do not expect all species to be present at all sites and in all seasons (Lawton, 1976, 1978a). Accordingly, I have employed a simple χ^2 of association on pairs of species (which, on a priori grounds, are most similar in their ecologies and hence most likely to compete) and used only those samples from times of the year and parts of the country that could, potentially, contain both members of the pair.

Some of the species have not yet been encountered often enough to make the data worth analyzing. There is nothing in the available records for these species to suggest they behave any differently from the pairs that have been analyzed.

The results are summarized in Table 6.1, and speak for themselves. Species either occupy sites totally independently of potential competitors or in fact occur *together* significantly more often than chance dictates. This latter result suggests there are important, but unknown, site or plant characteristics that influence pairs of species in similar ways.

Table 6.1. Chi-squared tests (with Yates's correction) for association between pairs of potential competitors. The data are presence and absence of species in local communities (bracken patches) throughout the British Isles (see text for details). The data are presented conventionally with the first member of each pair being designated "Species A" (see Appendix 1b for species).

Species pair	Sites with:				Expected number of shared sites if distributions are random	χ^2	p
	A only	B only	Neither	Both			
Suckers							
Philaenus & Ditropis	34	23	85	7	8.26	0.12	N.S.
Monalocoris & Macrosiphum	35	15	76	23	14.79	6.99	$0.01 > p > .001$*
Monalocoris & Ditropis	37	17	65	10	9.84	0.023	N.S.
Miners							
C. histricina & C. parvicornis	12	18	13	43	39.00	2.63	N.S.
C. parvicornis & Phytoliriomyza spp.	36	13	12	23	25.29	0.74	N.S.
C. histricina & Phytoliriomyza spp.	21	4	28	34	24.00	18.05	$p < 0.001$*
Gall formers							
D. filicina & D. pteridicola	48	2	12	37	33.49	3.16	N.S.
Lepidoptera							
Paltodora & Olethreutes	6	1	49	6	2.37	12.29	$p < 0.001$*

* Indicates that species occur together more often than chance dictates. No species occur together less often than expected.

As a program for future research on habitat selection by bracken insects, Table 6.1 is thought-provoking. As evidence for the effects of interspecific competition on community structure it is equally interesting: it says, simply, that there is none.

Species Abundances on Shared Fronds:
A Direct Test for Interspecific Competition

Data from the Skipwith open site provide a large number of samples of the abundances of species on individual fronds. Hence we can ask: are species abundances per frond influenced by those of other species on the

Table 6.2. Results of correlating the abundance of one species per frond with the abundance of a second species, at the Skipwith open site, for pairs of potential competitors. Fronds with neither species are excluded from the analysis. If species compete, a significant *negative* correlation is expected.

Species Pair	Date	n	r	p
C. histricina & C. parvicornis	29.7.80	20	+0.02	N.S.
C. histricina & C. parvicornis	12.8.80	15	+0.25	N.S.
C. histricina & C. parvicornis	27.8.80	16	−0.20	N.S.
D. filicina & D. pteridicola	21.7.80	20	+0.31	N.S.
D. filicina & D. pteridicola	29.7.80	20	+0.17	N.S.
D. filicina & D. pteridicola	27.8.80	16	−0.03	N.S.
D. filicina & D. pteridicola	12.8.80	15	+0.09	N.S.
Macrosiphum & Monalocoris	13.9.73	15	−0.13	N.S.
Macrosiphum & Monalocoris	24.8.72	25	+0.73	$p < 0.001$
Macrosiphum & Monalocoris	26.9.72	18	+0.53	$0.05 > p > 0.001$
Monalocoris & Ditropis	13.9.73	13	−0.50	N.S.
Monalocoris & Ditropis	24.8.72	20	+0.14	N.S.
Monalocoris & Ditropis	26.9.72	20	+0.18	N.S.
Ditropis & Macrosiphum	13.9.73	15	−0.21	N.S.
Ditropis & Macrosiphum	24.8.72	24	+0.17	N.S.
Ditropis & Macrosiphum	26.9.72	19	−0.03	N.S.
S. lineata & Aneugmenus spp.	1980	38	−0.47	$0.01 > p > 0.001$

Notes
C. histricina & C. parvicornis: see Fig. 6.7.
D. filicina & D. pteridicola: see Fig. 6.8.
Macrosiphum & Monalocoris: note significant *positive* correlations.
S. lineata & Aneugmenus spp.: see Fig. 6.9.
These same data are also analyzed using χ^2, *including* the non-occupied fronds, in Table 6.3. The negative correlation is then seen to be an artifact.

same frond, plotting pairs of potential competitors in phase space (Hallett and Pimm, 1979)?

This data set is very large and I have not analyzed all of it. Instead, I have picked years and seasons when particular pairs of species were both common, omitting years and seasons when they were rare, thus enhancing the likelihood of finding interspecific interactions. The results are summarized in Table 6.2 with examples plotted in Figures 6.7, 6.8 and 6.9.

Rare species should not be analyzed by this method, because most of the fronds have no individuals of either species. As a result, picking fronds containing at least one of the pair tends to give a spurious, significant, inverse relationship indicating interspecific competition (because the majority of the points are one individual of one species, and none of the other; Figure 6.9). The appropriate analysis for the rarer species is therefore a χ^2 of association, examining the probability of occurrence of one, the other, or both of a pair of species on individual fronds. For caterpillars of the two sawflies shown in Figure 6.9 and the last line of Table 6.2, there is then no sign of any significant interspecific interaction (Table 6.3). The same is true of all the other rarer species combinations (Table 6.3).

Tables 6.2 and 6.3 do not include all the species at Skipwith (see Appendix 1b), mainly because there is nothing very sensible to compare

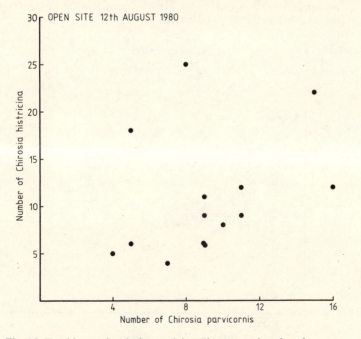

Fig. 6.7. Densities per frond of two mining *Chirosia* species when they occur together on individual fronds. There is no sign that the density of one species influences the other (Table 6.2).

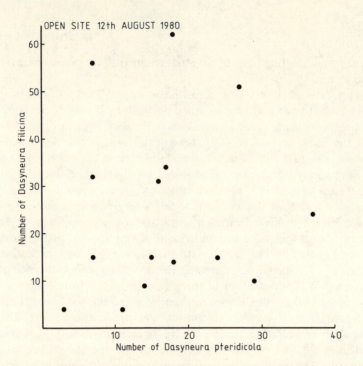

Fig. 6.8. Densities per frond of two gall-forming *Dasineura* species when they occur together on individual fronds. There is no sign that the density of one species influences the other (Table 6.2).

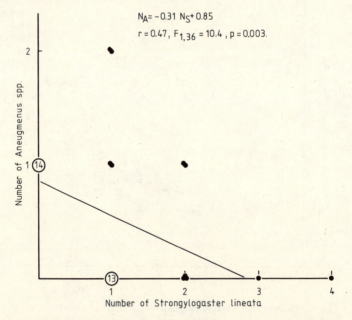

Fig. 6.9. Densities per frond of two foliage-chewing sawfly caterpillars when they occur together on individual fronds. There is no sign that the density of one species influences the other. In this case, the apparently significant inverse regression is an artifact because so few fronds contain individuals of either species (see text, and Tables 6.2 and 6.3).

Table 6.3. Chi-squared tests (with Yates's correction) for association between pairs of potential competitors (chewing herbivores). The data are for presence and absence on individual fronds at the Skipwith open site at appropriate times of the year. These species are too rare to analyze by the method used in Table 6.2. The data are presented conventionally, with the first member of each pair being designated "Species A."

Species pair	Date	Fronds with:				Expected number of shared fronds if distributions are random	χ^2	p
		A only	B only	Neither	Both			
S. lineata & Aneugmenus spp.	1980	18	14	121	6	3.01	2.74	N.S.
Tenthredo spp. & Aneugmenus spp.	1980	6	17	68	0	1.12	0.45	N.S.
Tenthredo spp. & S. lineata	1980	6	12	73	0	[0.79]	0.09	N.S.
P. chlorosata & total sawflies	1980	14	10	31	6	5.25	0.03	N.S.

Table 6.4. Results of correlating the abundance
of the aphid *Macrosiphum* per frond with
that of the thrip *Frankliniella*, at the Sierra
Blanca open site. Both achieve large numbers
in the spring. A significant negative correlation
would indicate competition.

Date	n	r	p
5.6.79	38	−0.12	N.S.
11.6.79	19	−0.22	N.S.
19.6.79	20	+0.35	N.S.

some of them with (*e.g. Bourletiella* and *Chirosia albifrons*), or data on one, the other, or both of the pair (*e.g. Olethreutes* and *Paltodora*) are too sparse to be worth analyzing by any method. Again, there is nothing in these data to suggest they are conspicuously different from the species summarized in Tables 6.2 and 6.3. None of these analyses provides any clear indication of a reciprocal influence by the population of one species upon another: indeed, the only significant results are positive associations between two species of Hemiptera.

Similar analyses can be carried out for the impoverished New Mexico fauna, with identical results (Table 6.4 provides an example). Comparable data are not available for Papua New Guinea, but Kirk (1980) notes (p. 33): "The most widely distributed species were the Diptera *D. notha* and the Coleoptera, *B. atropolita* and *P. pteridophytae*; larvae of these insects were frequently found together in the same fronds."

Conclusions

Taken together, these data provide consistent evidence for a lack of significant interspecific interactions between bracken herbivores. None of the data are experimental, and perhaps if populations were consistently pertubed away from normal levels by a sufficiently large amount (up or down), evidence for interspecific competition might be forthcoming. Rosenzweig's (1979) model is particularly worrying in this regard, because it predicts that, over a wide range of population densities in the vicinity of a two-species equilibrium, no interspecific effects will be observed, even though habitat selection is influenced by competition. Moreover, I have only measured population numbers, not population performance (*e.g.* growth or survival rates). With these important caveats in mind, I simply note that, on present evidence, all the data support the view that competitive interactions between different species of bracken herbivores are of little or no consequence in structuring the community. Hence, I would not expect communities to converge. They do not, so at least the data are consistent.

THE MAINTENANCE OF LOCAL POPULATIONS

Exactly how one chooses to generalize from these results depends to some extent upon how self-sustaining or how transitory local populations of bracken-feeding insects are. Hence it cannot be emphasized enough to those unfamiliar with insects that populations making up a bracken community appear to be every bit as permanent as breeding-bird or small-mammal populations. With the exception of *Macrosiphum ptericolens*, whose wintering habits in Britain are unknown, all the species overwinter in various stages in, and under, dead bracken, on or close to the site.

Data on the permanency of the Skipwith open-site populations are summarized in Table 6.5. Most species have been recorded every year I

Table 6.5. Apparent species-turnover at the Skipwith open site, 1972–1980. No sampling was carried out in 1979. In 1973 and 1976 sampling was not intensive, so failure to record a species is probably not indicative of its absence. Failure to record a species in these two years is indicated by "?" In all years, a positive record is indicated by "X"; failure to find a species despite intensive sampling is indicated by "0." Sawfly identification is based on samples of caterpillars collected in past years and stored in alcohol, determined in 1980. The two species of *Tenthredo* and of *Aneugmenus* cannot be distinguished confidently using alcohol-preserved specimens: adult collection requires very careful search and much effort, and for *Aneugmenus* was not done often enough to justify splitting the records. Species showing some apparent turnover are marked "∗."

Species	1972	1973	1974	1975	1976	1977	1978	1980	
Bourletiella viridescens	X	X	X	X	?	X	X	X	
Chirosia albifrons	X	?	X	X	?	X	X	X	
C. histricina	X	X	X	X	?	X	X	X	
C. parvicornis	X	X	X	X	?	X	X	X	
Dasineura filicina	X	X	X	X	?	X	X	X	
D. pteridicola	X	X	X	X	?	X	X	X	
Phytoliriomyza hilarella	0	?	0	X	?	X	0	0	∗
Ditropis pteridis	X	X	X	X	X	X	X	X	
Macrosiphum ptericolens	X	X	X	X	X	0	0	X	∗
Monalocoris filicis	X	X	X	X	X	X	X	X	
Philaenus spumarius	0	?	0	0	?	0	0	X	∗
Aneugmenus spp. (2 species)	X	?	X	X	X	X	X	X	
Stromboceros delicatulus	X	X	X	0	?	0	0	X	∗
Strongylogaster lineata	X	?	X	X	X	X	X	X	
Tenthredo spp. (2 species)	X	?	X	X	?	X	X	X	
Euplexia leucipara	0	?	0	0	?	0	0	X	∗
Oleuthreutes lacunana	0	?	X	X	X	X	0	X	∗
Paltodora cytisella	X	X	X	X	?	X	X	X	
Petrophora chlorosata	X	?	X	X	?	X	X	X	

have looked for them, although there have also, apparently, been some colonizations and extinctions. Since the bracken on the open site is within a few hundred meters of other large open patches, and adjacent to bracken growing under trees, repeated immigration and extinction is possible for the bracken-specific herbivores (*Phytoliriomyza, Stromboceros,* and *Macrosiphum*). It would be even easier for polyphagous species (*Phileanus, Euplexia,* and *Olethreutes*). However, one word of caution is in order. The open site has an area of 2600 m^2 and an average frond density of 28 m^{-2}, giving 72,800 fronds in all. An insect with an average population of only 0.01 individuals per frond would still have a total population of over 700 on the site; a population of 100 individuals (which for a larval insect must be close to imminent stochastic extinction) would have a mean density of just over 0.001 per frond, and would be very easy to overlook. Hence I do not know whether any of the apparent turnover at the Skipwith open site during the last 9 years is real.

The point of these data, however, is not to estimate turnover as such, but to show that, on a local patch like Skipwith, most species are represented by large, permanent populations, rather than by ephemeral, fly-by-night individuals.

COUNTER-EVIDENCE?

This section briefly considers evidence against my main theme—that bracken herbivore communities are non-convergent and their constituent populations non-competitive.

Niche Convergence

There are some examples of convergence. The best, involving the location and type of damage done to the plant, is provided by two insects in different orders. *Drosophila notha* (Diptera) attacks the newly emerged fronds in Papua New Guinea, the larvae tunneling in the center of the rachis. *Paltodora cytisella* (Lepidoptera) does exactly the same thing in Britain. Both expel frass through a small hole kept open in the side of the resulting gall (Figure 6.10). The main difference between the species is that *Paltodora* more or less always occurs singly, whereas Kirk (1980) sometimes found as many as thirteen *D. notha* galls on one frond.

The analogy can be extended across three continents, because in the United States *Monochroa placidella* caterpillars (Lepidoptera) also mine the rachis, causing it to swell in a similar way. However, *Monochroa* and *Paltodora* may require reclassifying as a single genus (J. A. Powell and K. Sattler, pers. comm.) so convergence here is not very surprising.

I see few other signs of convergence, except those based on close taxonomic affinity, and except for the quite staggering overall similarity be-

CONVERGENCE IN BRACKEN HERBIVORES

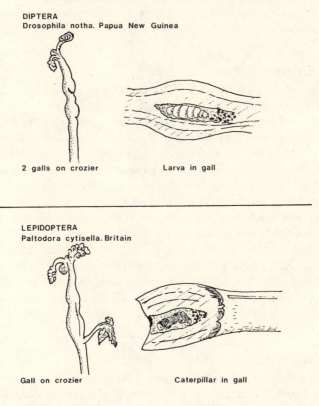

From photographs by A.A.Kirk (1980) and J.H.Lawton (unpublished)

Fig. 6.10. Convergent evolution in a fly (*Drosophila notha*) and a lepidopteran (*Paltodora cytisella*). Both induce a gall in the rachis of a young bracken plant, the one in Papua New Guinea, the other in Britain.

tween the caterpillars of typical external foliage-feeding Lepidoptera, and sawflies (Hymenoptera: Symphyta). This is certainly one of the best examples of convergence in morphology and general feeding habits in the animal kingdom, involving thousands of species on plants throughout the world. It would therefore be easy to point to any of the external foliage-feeding Lepidoptera in Papua New Guinea as good, general, ecological analogues of the British sawflies.

These examples of convergence can presumably be attributed to individual adaptations, unconstrained by interactions with other species in the same trophic level. Such is the case, of course, with many examples of convergence described by vertebrate ecologists (*e.g.* Orians and Solbrig, 1977). They certainly do not provide evidence for convergence at the level of the community.

Niche Expansion in Time?

In the absence of competitors, niche expansion has been widely reported for vertebrate consumers (*e.g.* MacArthur, 1972; Morse, 1971). *Eriocampidea arizonensis* is the only sawfly at the Sierra Blanca site, and indeed over most of New Mexico and Arizona, compared with six at Skipwith. All the Skipwith species have single, annual generations (*e.g.* Lawton, 1978a), but with sawfly larvae of some sort present throughout the spring and summer. *Eriocampidea* has two generations a year (Smith and Lawton, 1980), which might be interpreted as a classical case of niche expansion. But since there is no density compensation (pages 76–78 and Figure 6.4), this is a difficult argument to sustain: in fact, having two generations is probably coincidence. None of the other bracken-feeding species in New Mexico with discrete generations show it, and there are some non-sawfly, bracken-feeding species in Britain with two generations a year even though their close relatives all have one. I therefore see no particular significance in this case of apparent niche expansion.

The "Ghost of Competition Past"

It is possible that the reason there is no evidence for significant interactions between bracken herbivores in contemporary time is because species have evolved to avoid them. In Connell's (1980) apt and descriptive phrase, the key to understanding the structure of bracken herbivore communities may lie with the "ghost of competition past." Like Connell, I find little hard evidence for such a notion for any community, but that does not necessarily mean it does not happen, only that it is extremely hard to find (as are most ghosts) (see Gibson, 1980).

However, some tentative speculations suggest that it is not a very likely explanation. First, other things being equal, such evolution is presumably the major pathway leading to community convergence. Second, on both theoretical (*e.g.* May and MacArthur, 1972) and common-sense grounds, it seems extremely unlikely that species will evolve to avoid, totally, any interspecific effects. Hence, if competition is a potent enough evolutionary force to differentiate species' feeding niches, some evidence of competition should be manifest in contemporary time. Such evidence is conspicuously lacking. Finally, there is no evidence that species are divided across resources in a way that one might imagine would minimize niche overlap. For example, at the Skipwith open site (Figure 6.3) the distribution of realized feeding modes across resources is significantly clumped: omitting costa and costule chewers and gall formers (which are absent from all communities), there are 26 feeding positions distributed across 12 resource states in a highly clumped fashion ($\chi^2_{10} = 34.9, p < 0.001$). Similar results apply to the species at Hombrom Bluff ($\chi^2_{10} = 21.99, 0.05 > p > 0.01$), although the manner in which feeding niches are clumped is quite different at Skipwith.

A totally even spread of feeding sites across resource states is an extreme hypothesis, and therefore perhaps unfair. If so, it begs the question of how much difference those who advocate the evolutionary role of past competition would expect.

An alternative, and less extreme, test is as follows. Not all British species occur at Skipwith (Appendix 1b). Accordingly, if niches have evolved to minimize competitive interactions, perhaps the missing species are those with feeding niches too similar to invade the community. (Essentially this hypothesis has been tested on pages 80–82 for pairs of species and found wanting. Here I test it for a whole community.) The woodland community at Skipwith supports 15 species, with 12 of the species listed in Appendix 1b, category *a*, absent—a convenient ratio of residents to absentees. Suppose the number of species co-occurring in each of the elements in the niche matrix (Figure 6.3) has been constrained by the ghost of competition past. Then, the proportion of species in the national pool excluded from any one of these boxes might be expected to increase as each box fills up with residents, as shown in the insert to Figure 6.11. The data for the woodland at Skipwith (Figure 6.11) bear no resemblance to this prediction. (Remember a species can occupy more than one element of the niche matrix: the number of feeding positions or "occupied states" for the woodland is shown in Figure 6.3, and repeated as the upper figure in each box in Figure 6.11. The lower figure in each box is the potential number of takers in the national pool.)

Similar plots for the Skipwith open site and for Hombrom Bluff are equally unimpressive.

Constancy in Species Number

Well-studied local communities assembled from large regional pools (Hombrom Bluff and Skipwith) contain approximately the same number of species, particularly the wooded Hombrom Bluff site and the Skipwith woodland site (14 and 15 respectively: Appendix 1 and Figure 6.3). Similarities in species number no better than this have been taken as "strong evidence that competition acts in a similar way on both continents to regulate the species packing levels of consumers" (Cody *et al.*, 1977, pp. 166–167). Figure 6.6 suggests to me that for bracken, at least, the similarity in number of species is much more likely to be due to coincidence.

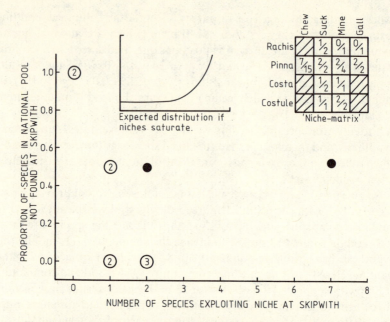

Fig. 6.11. Relationship between the number of species exploiting bracken at various points in the "niche matrix" of the woodland community at Skipwith and the proportion of species in the national pool absent from this community but potentially capable of exploiting the plant in a particular way. Raw data are the upper and lower figures in each element of the niche matrix respectively. The expected distribution of points if each element in the matrix can support only a limited number of species is shown as an insert, which assumes that species will invade the community until each element is "full." The data provide no support for this hypothesis.

Conclusions

Drawing these arguments together, I find nothing in the data currently at my disposal to suggest that the species found on bracken have so evolved as to spread themselves evenly across resource states and minimize competition, or that niche expansion occurs in response to faunal impoverishment, or that significant convergence in species number and feeding niches is imposed by competition. However, I am quite prepared to concede that, in other types of communities, evidence of the type assembled in this section may be indicative of competition and community-imposed convergence. But on its own, it is very feeble evidence.

DISCUSSION

The main results of this paper can be interpreted and summarized as follows. Over 100 species of insects feed on bracken throughout the world; the number found in one geographical region is a subsample of this total and depends upon the size of the area (and perhaps upon its isolation), with the species in each local community being a further subsample from the regional pool. Patterns of occurrence of individual species are presumably determined by differences in climate, other habitat variables and bracken itself, and by prevention of dispersal by geographical barriers, distance, chance, and accident. Moreover, species apparently occupy communities largely independently of one another. Hence, the strong inference must be that, within populations, most (though probably not all) species are kept rare, relative to their food resources, by a combination of one or more of the effects of natural enemies, disease, limited overwintering sites, and the vagaries and harshness of the environment (Caughley and Lawton, 1981; Lawton and McNeill, 1979; Lawton and Strong, 1981).

In almost all respects, the bracken system closely resembles other phytophagous insect communities—in the existence of species-area relationships at geographical, regional, and local levels (see Strong, 1979, for review), in the infrequency (but not complete absence) of interspecific competition for food (Lawton and Strong, 1981), and in the large number of apparently unfilled niches (Lawton and Price, 1979; Price, 1980). A total lack of convergence in community structure is also typical of most phytophagous insect communities, though not necessarily for the reasons envisaged by Cody and Mooney (1978), who clearly had in mind the non-convergence of insect communities, particularly ants and pollinators (*e.g.* Cody *et al.*, 1977, p. 183) on different *species* of plants. Of course, where different species of plants are involved, Cody and Mooney may be correct not to expect too much convergence. Thus, plants in the same genus (*e.g.* *Larrea*, Schultz *et al.*, 1977) or providing similar sorts of habitats (*e.g.* for grasshoppers, Otte and Joern, 1977; Orians and Solbrig, 1977) often differ sufficiently in structure, chemistry, and availability of other resources (*e.g.* "escape space") to make community convergence unlikely. However, even when the same species of plant is compared in different places, convergence is still conspicuous by its absence. Local communities of insects on birch (*Betula*) provide excellent examples (Hagvar, 1976; Koponen, 1979).

Lists of insect herbivores attacking the same species of plant in different parts of the world also provide useful evidence for an absence of convergence at the community level when obvious feeding niches are unfilled in some regions but not in others. Such is the case with several cosmopolitan species of plants, supporting rich and well-established insect faunas, primarily independently derived from the regions into which man has moved them—water hyacinth (Perkins, 1974), soybean (Turnipseed and Kogan, 1976), grape vines (Bournier, 1977), maize (Chiang, 1978), cacao (references in Strong, 1974), and sugar cane (references in Strong *et al.*, 1977). Notwithstanding that in most regions most of the major ways of exploiting each plant (chewers, suckers, miners, and gall formers) are represented by one or more species of insects, particular plant parts frequently lack particular sorts of exploiters. Hence local communities assembled from these regional pools cannot converge in structure.

In an earlier paper (Lawton, 1978a) I looked for possible convergence in bracken herbivore communities, using similar regional faunal lists and crude proportions of chewers, suckers, miners, and gall formers, irrespective of where they fed on the plant (*i.e.* the columns of Figure 6.3, ignoring the rows). With the benefit of hindsight, and the more reliable data now available, I suggest that the impression of possible convergence in broad feeding patterns seen in that paper is unlikely to be due to competitive constraints operating at the level of the community. At the moment, my guess is that the patterns are pleasing artifacts and no more.

The literature on ecological convergence in general is in a rather similar, enigmatic, state. Mooney *et al.* (1977) put it nicely when they wrote (page 10): "How similar is similar, and what level of similarity constitutes evidence for convergent evolution? The problem is particularly acute for those who learn facts second hand." Whilst I agree with these sentiments, I also think the usual way in which the convergence hypothesis is formulated makes it particularly difficult to test. The basic assumption of most studies is that competition between species is sufficiently strong and important that, "other things being equal," convergence is to be expected. When it is not found, a series of special reasons are invoked to show that, after all, other things are not equal (Peet, 1978; Roughgarden, 1979a; Mooney, 1977; Orians and Solbrig, 1977). Such arguments, though plausible, would be easier to evaluate if the initial assumption—that interspecific competition is of major importance—were tested first; obviously, if competition is feeble, non-convergence needs no special pleading to explain it away. I believe that lack of strong interspecific competition is sufficient to account for all the failures of plant-feeding insect communities to converge in structure.

The final question, of course, is how similar, or different, herbivorous-insect communities (particularly bracken insects) are from vertebrate communities (particularly birds). I would be very surprised indeed if the two sorts of communities were normally structured in identical ways, and on the face of it (*e.g.* Cody and Diamond, 1975a; Hutchinson, 1978), they do, indeed, appear to be very different, implying that we have little choice but to stop thinking about "animal communities" in general, and to define which bit of the ark we are talking about. However, before adopting such a polarized position, we ought to be sure it is necessary. Papers questioning competition's all embracing role in structuring communities have already been referred to; others provide a recurrent theme in this symposium. Hence although the difference between sets of vertebrates and invertebrates is probably real, it may actually be smaller than appears at first sight. For example, I find nothing very different and much that is identical in Orians's account (Orians, 1980) of communities of marsh-nesting birds on different continents and my own work on bracken-feeding insects—no convergence, numerous vacant niches, and a strong effect of habitat size and isolation. Abbott's review (Abbott, 1980) of island-bird communities reinforces these impressions.

Arguably more information has been assembled by those seeking convergence in Mediterranean-zone plant and animal communities than for any other type of ecosystem (Mooney, 1977; Cody and Mooney, 1978). Amongst vertebrates, the number of species in major trophic groupings on different continents differs substantially, density compensation is at best erratic, and Cody and Mooney themselves write:

> Across continents, individual species do not segregate into readily distinguishable sets of counterparts. . . . there are very few parallels at the level of individual species and especially few parallels that include a representative from all four continents. . . . [The] distribution of species over the same range of ecological opportunities is accomplished in various ways that do not produce precise counterparts at the species level.

I could have written exactly the same about bracken herbivores in particular, and about plant-feeding insects in general.

Acknowledgments

A series of technicians, A. Fisher, B. Thompson, S. Fawcett and M. Chapman, have at various times searched bracken for insects. The Royal Society of London generously supported the field work in Arizona and New Mexico with a Scientific Investigations Grant, and provided a Travel Grant towards the cost of attending this symposium. Carri Rigby, Stuart Pimm and Alan Kirk unselfishly made available their own data on bracken insects. Various entomologists provided skillful and invaluable taxonomic help, particularly Drs. L. A. Mound, V. F. Eastop, K. Sattler (British Museum), D. R. Smith (USDA), and K. A. Spencer. To all these people and organizations I am extremely grateful.

APPENDIX 1A

Bracken herbivores recorded in New Mexico and Arizona. Collections made by J. H. Lawton in period May–September 1979. (The taxonomists providing identifications are acknowledged after each species.) O: present in the open site at Sierra Blanca. W: present in the woodland site at Sierra Blanca.

DIPTERA

W   *Phytoliriomyza clara* Melander. (K. A. Spencer, pers. comm.)

HEMIPTERA

O,W   *Macrosiphum clydesmithi* (V. F. Eastop and A. G. Robinson, pers. comm.) (Referred to as "Aphid sp. 1" in Figure 6.5)

HYMENOPTERA

O,W   *Eriocampidea arizonensis* (Ashmead) (Smith and Lawton, 1980) *Aneugmenus scutellatus* Smith (Arizona only) (Smith and Lawton, 1980)

LEPIDOPTERA

       Monochroa placidella Zeller (Arizona only) (K. Sattler, pers. comm.)

W   Unidentified sp. 1 (Torticidae)

THYSANOPTERA

O,W   *Frankliniella occidentalis* (Pergande) (L. A. Mound, pers. comm.)

APPENDIX 1B

Revised list of herbivorous bracken insects occurring in Britain (modified from Lawton, 1976).

(a) Species whose status is in no doubt; feeding on the above-ground parts of the plant. W and O: present on the woodland and open sites at Skipwith, respectively.

COLLEMBOLA
O   *Bourletiella viridescens*

DIPTERA
W,O   *Chirosia albifrons* (see *C. flavipennis*, below)
       C. albitarsis
W,O   *C. histricina*
W,O   *C. parvicornis*
W,O   *Dasineura filicina*
W,O   *D. pteridocola*
O   *Phytoliriomyza hilarella*
       P. pteridii

HEMIPTERA
W,O   *Ditropis pteridis* (*Criomorphus pteridis* in Lawton, 1976)
W,O   *Macrosiphum ptericolens*
W,O   *Monalocoris filicis*
O   *Philaenus spumarius*

HYMENOPTERA
       Aneugmenus padi (= *coronatus*) (This synomymizes two species in Lawton, 1976: see Benson, 1968; Fitton *et al.*, 1978)
W,O   *A. furstenbergensis*
W,O   *A. temporalis*
W,O   *Stromboceros delicatulus*
O   *Strongylogaster lineata*
W,O   *Tenthredo ferruginea*
W,O   *Tenthredo* sp. 2. (Caterpillars not yet reared successfully and hence cannot be associated with certainty with the species recorded in Lawton, 1976)

LEPIDOPTERA
       Ceramica pisi
O   *Euplexia lucipara*
       Lacanobia oleracea (Recorded as a possible only—*Mamestra oleracea* in Lawton, 1976; now known to feed regularly in the south of England)

W,O   *Olethreutes lacunana* (Recorded as an "unidentified microlepi-dopteran" in Lawton, 1976; see Lawton, 1978b)

O   *Paltodora cytisella*

W,O   *Petrophora chlorosata*

           Phlogophora meticulosa

(b) Species listed in Lawton (1976) for which I have five or fewer records on bracken in ten years of collecting, and which on present evidence are now thought to be only casually or accidentally associated with the plant.

HEMIPTERA
Bryocoris pteridis
Aphis fabae

LEPIDOPTERA
Sphilosoma luteum

(c) Species for which adults have been found, but not the larvae: status unknown.

DIPTERA
Chirosia flavipennis (This species is misnamed in Lawton (1976, 1978a). Reference to "*flavipennis*" in those accounts should be to *C. albifrons*, category (a) above)

(d) Rhizome feeders.

LEPIDOPTERA
Hepialus fusconebulosa
H. hecta
H. sylvina

(e) Species never personally encountered as adults or larvae feeding on bracken during ten years of collecting. Recorded as feeding or probably feeding on bracken in the literature (see Lawton, 1976).

COLEOPTERA
Phyllopertha horticola

DIPTERA
Chirosia crassiseta

HYMENOPTERA
Strongylogaster macula
S. xanthocera

NOTES

(i) There is no evidence that any of the species of doubtful status listed in Lawton (1976) should be elevated to that of a regular herbivore, except *Lacanobia oleracea* (category (a) above).

(ii) The mite *Phytoptus pteridis* (Lawton, 1976) is herbivorous and is found occasionally. It does not occur at Skipwith. Mites, *Chamobates* sp., occur commonly on the fronds (Lawton, 1976), probably feeding mainly on detritus and microorganisms.

APPENDIX 1C

Herbivorous bracken insects in Papua New Guinea (after Kirk, 1980).
H: present at Hombrom Bluff.

COLEOPTERA
Tmesisternus sp. 1
Tmesisternus sp. 2
Sybra sp.
Nr. *Cornallis* sp.
Unidentified sp. 1 (Cerambycidae)
Unidentified sp. 2 (Cerambycidae)
Unidentified sp. 3 (Cerambycidae)
H   Unidentified sp. 4 (Cerambycidae)
H   Unidentified sp. 5 (Cerambycidae)
Manobia sp. 1
Manobia sp. 2
Apthona sp.
Eumolpine sp.
H   *Baris atropolita*
Unidentified sp. 6 (Curculionidae)
H   *Poecilips pteridophytae*

DIPTERA
H   *Drosophila notha*

HEMIPTERA
Shinjia pteridifoliae
H   *Pachybrachus pacificus*
H   *Balclutha* sp.
H   *Peregrinus maidis*
H   Unidentified sp. 1 (Delphacidae)

LEPIDOPTERA
Nymphula sp.
H   *Psara platycapna*
Praecedes thecophora
Unidentified sp. 1 (Tineidae)
H   *Opogona* sp.
H   *Oruza divisa*
H   ? *Adoxophyes* sp.

ORTHOPTERA
H   *Moessonia novaeguineae*

7

Experimental Tests of Island Biogeographic Theory

JORGE R. REY

University of Florida—IFAS, Florida Medical Entomology Laboratory,
200 9th Street S.E., Vero Beach, Florida 32962

INTRODUCTION

The equilibrium theory of insular biogeography has received a great deal of attention since its introduction in 1963 (MacArthur and Wilson, 1963). The model is conceptually simple, offers mechanistic explanations for a number of natural phenomena that have interested biologists for a long time, and has the potential for a wide range of practical and theoretical applications.

In its simplest form, the model predicts that the number of species on an island or similarly isolated area is a result of a dynamic equilibrium between immigrations and extinctions. Immigration rate is seen as a decreasing function of numbers of species already present on an island, whereas extinction rate is an increasing function of species number. Equilibrium occurs at the intersection of the two rate functions.

The exact shape and magnitude of the immigration and extinction curves depend upon a variety of factors. In the model, extinction rate is a decreasing function of area because larger areas theoretically support larger populations than smaller areas, and the probability of extinction of any species decreases with increasing population size. Immigration rates are inversely correlated with distance to a source of potential colonists. Secondarily, area may also affect immigration rates if islands of different sizes sample propagules at different rates (Sampling Effect), and isolation may also affect extinction rates through its effect on the arrival rates of individuals of species already established on an island. If these areas are high enough, local extinction of a species may be prevented or delayed by inputs of new individuals from outside (Rescue Effect, Brown and Kodric-Brown, 1977).

There has been some recent criticism of attempts at practical applications of the MacArthur-Wilson model on the grounds that it has not been properly tested (Abele and Connor, 1979; Gilbert, 1980; Rey and McCoy, 1979), as well as more general criticisms about some of the model's predictions and assumptions (Gilbert, 1980; Kuris *et al.*, 1980). While some of these critiques are the result of misconceptions about the model and confusion about mechanisms generating species/area relations (Rey *et al.*, 1981), there appears to be agreement that independent tests of the model under natural conditions are needed.

Below I describe field experiments with the arthropod fauna of small *Spartina alterniflora* Loisl. islands in northwest Florida, designed to examine some of the assumptions and predictions of the MacArthur-Wilson theory.

THE *SPARTINA* ARCHIPELAGO EXPERIMENTS

The experimental archipelago is located in Oyster Bay, southwest of Spring Creek in northwest Florida. It consists of a series of islands composed of pure stands of *S. alterniflora*. It was chosen for these studies because it has a large number of small islands of different sizes and various degrees of isolation, with only minimal habitat complexity differences between them. Eight islands were chosen for the experiments. They range in size from 56 m² to 1023 m², and in distance from the mainland from 38 m to 1752 m. Six of the islands were censused repeatedly during April 15–23, 1977, and were defaunated the following week (week 0) with methyl bromide gas. The other two islands and a large *Spartina* stand on the mainland were not defaunated and served as controls. After defaunation, all sites were censused weekly using non-destructive techniques that were shown experimentally to be very accurate.

Two other experiments were performed concurrently with the island colonization study. Cages measuring 8 ft on a side and covered with 20–20 mesh fiberglass screening were placed over 4 *Spartina* islands to investigate the persistence characteristics of the species trapped inside in the absence of immigration. They were also sampled weekly with the same methods used on the islands of the colonization experiment. In addition, 6 sticky traps were placed at various distances from the mainland to estimate the general movement patterns of arthropods throughout the study area independently of the immigration/extinction activity on the experimental islands. The traps consisted of clear plastic panels 0.05 cm thick and 92.3 cm square, covered with Tree Tanglefoot. The panels were suspended from 1.85-m high wooden frames and were installed on high points in the bay (*i.e.*, on top of oyster bars, sand bars, etc.). Their distances from the mainland ranged from 18 m to 2632 m. The arthropods collected on both sides of the panels were removed biweekly.

Only resident species were considered in the study. One of the greatest problems with previous attempts at estimating insular turnover rates has been the inclusion of large numbers of transients in the calculations. This practice results in highly inflated estimates of turnover rates (Lynch and Johnson, 1974; Simberloff, 1976b). The distinction of transient from resident species was based upon extensive data on the arthropod communities of these marshes obtained for three years prior to the start of this study (Rey and McCoy, in prep.), on the persistence characteristics of species inside the cages, and on previous studies of *Spartina* arthropods by other investigators. A large majority of the species collected during the study could be assigned unambiguously to one of the two categories. A total of 56 species were considered to be potential residents of *Spartina* in the area.

Immigrations and extinctions were determined as follows: An immigration was recorded when a species that was not previously present on a site appeared on the site and remained for more than one week. An extinction was counted when a species that had been on an island for two consecutive weeks was not collected for two weeks in a row. A more detailed account of the sites, methods, and conventions can be found in Rey (1981).

Colonization of the defaunated islands by resident species was rapid (Figure 7.1). The numbers of species on these islands increased steadily during the first 20 weeks, and then settled to a seasonal pattern characteristic of the area (Rey, 1981), with slight decreases in species numbers in winter and increases in late spring and early fall.

Comparison of pre- and post-defaunation species numbers among the experimental (defaunated) islands and between the experimental islands and controls are complicated by the rapid recolonization of the islands and by seasonality. Patterns can be more easily visualized if the numbers of species on the defaunated islands are expressed as proportions of the numbers on the controls. This technique standardizes the variation due to the seasonal pattern and helps identify the return of the defaunated islands to conditions normal for the area. Proportional species numbers on the defaunated islands rise to near the pre-defaunation levels in about 20 weeks and then remain close to these values during the rest of the study (Figure 7.2). The patterns are consistent for all comparisons across all study sites (Table 7.1).

Real immigrations and extinctions of resident species occurred at all sites throughout the 53-week experimental period. On the defaunated islands, turnover was significantly higher during the first 20 weeks than during the rest of the year. The rates on the control sites were low throughout, and did not differ significantly between the two time intervals. Extinction rates after week 20 ranged from 3.2 to 9.6 species $\cdot$ island^{-1} $\cdot$ year^{-1}.

Prior to defaunation, there was a highly significant ($p < 0.01$) arthropod species/area relation among the *Spartina* sites. The relationship was eliminated by defaunation, but was reestablished by approximately 20 weeks

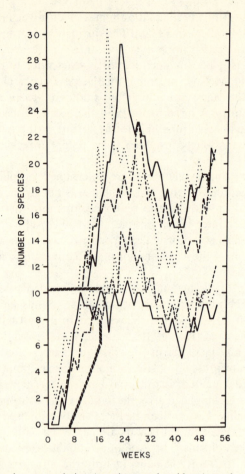

Fig. 7.1. Species accumulation (transients and residents) on the defaunated islands. Upper curves represent the three large islands, lower curves represent the three small ones. From 1 to 10 species all curves fall within the heavy dashed lines (Rey, 1981).

after defaunation, thus coinciding in time with the return of the species richness on the islands to their pre-defaunation levels, and with the stabilization of the immigration and extinction rates.

Area was also significantly correlated with extinction rates after week 20 and with immigration rates during weeks 1–20. Neither distance to the mainland nor distance to the nearest large land mass (large island or mainland, whichever was closest) was significantly correlated with numbers of species on the islands or with immigration and extinction rates.

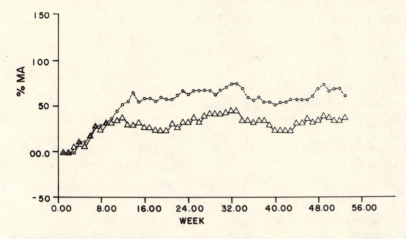

Fig. 7.2. Numbers of species on the near islands expressed as percentages of the numbers on the mainland. Triangles, near-small island; circles, near-large island.

Table 7.1. Proportion of the numbers of species on the controls attained by the defaunated islands. PD = pre-defaunation, W20 = week number 20, EQ = mean for weeks 20–25 (from Rey, 1981).

Control		*NS*	*NL*	*IS*	*IL*	*FS*	*FL*
				Experimental			
	PD	0.500	1.100	0.500	1.200	0.700	1.300
CI	EQ	0.555	1.130	0.615	1.332	0.660	0.988
	W20	0.460	1.150	0.620	1.130	0.540	1.000
	PD	0.455	1.000	0.455	1.091	0.636	1.182
CI	EQ	0.605	1.223	0.668	1.330	0.720	1.080
	W20	0.430	1.110	0.570	1.210	0.500	0.930
	PD	0.313	0.688	0.313	0.750	0.438	0.813
CI	EQ	0.308	0.625	0.352	0.680	0.372	0.547
	W20	0.230	0.580	0.310	0.650	0.270	0.500

DISCUSSION

The MacArthur-Wilson theory generates a number of predictions about patterns and processes related to the colonization of isolates. The model rests upon a set of biologically realistic assumptions about the colonization process, and about the dynamics of populations on islands, under unsaturated as well as equilibrium conditions (Table 7.2). An appropriate subset of these predictions and assumptions can be used as a framework for determining the "fit" of the model to a particular situation. The soundness

Table 7.2. Some assumptions and predictions of the MacArthur-Wilson model. The list is not all-inclusive.

Application	*Turnover*
1. Real isolates (rescue effect not important).	1. Homogeneous.
2. Comparable habitat complexity among sites.	2. Non-successional.
	3. Frequent.
3. No gross environmental change.	4. Extinction rate increasing function of species numbers.
4. Species are residents.	5. Immigration rate decreasing function of species numbers.
5. Definable species pool.	
Equilibrium	*Equilibrium Species Numbers*
1. Dynamic.	1. Influenced by area through extinction rates.
2. Approached asymptotically.	
3. Stochastic.	2. Influenced by isolation through immigration rates.
4. Obtains in ecological time.	3. Varies faster with area on distant islands.
5. Occurs at the intersection of the I/E rate functions.	4. Varies faster with isolation on small islands.

of the model's assumptions and the validity of its predictions can only be determined after numerous such evaluations are performed under natural conditions.

Equilibrium

The equilibrium question can be approached in a variety of ways. Ideally, an extinction on an island at equilibrium should always be followed closely by a successful immigration and vice versa. This deterministic approach, however, is unrealistic when one considers the inherently stochastic nature of the model (Simberloff, 1976b). A more reasonable criterion for equilibrium is the occurrence of real immigrations and extinctions in approximately equal proportions so that a given measure of diversity remains relatively constant through time (Rey, 1981). Several lines of evidence point to a reestablishment of equilibrium with turnover approximately 20 weeks after defaunation: the proportional species richness data indicated that the experimental islands attain their pre-defaunation levels in about 20 weeks and deviate little from them during the rest of the study. After this time, small seasonal variations in species numbers typical of the area influence species richness on the control and experimental sites in a similar fashion. In addition, differences in turnover rates between defaunated and control islands disappear after week 20, and significant species/area relations are reestablished at about the same time.

Area and Extinction

The MacArthur-Wilson model is only one of many that offer mechanistic explanations for the species/area relationship (Rey *et al.*, 1981). According to the theory, extinction rates should decrease as area increases because there is an inverse correlation between population size and extinction probability, and larger areas support larger populations of most species (MacArthur and Wilson, 1967). This hypothesis can be examined indirectly with the *Spartina* island data. If the hypothesis is true, then on a recently defaunated island the number of extinctions should be low at first because only a few species would be present, and therefore only a few could go extinct. As more species colonize the island the number of extinctions should rise because more species would be present and many of these would be recent arrivals with still low population sizes and therefore high extinction probabilities. As populations of the species that survive increase in size, their chances of extinction should decrease and the extinction rates on the island should drop. This exact pattern is evident in the *Spartina* archipelago. During the first 10 weeks after defaunation, the numbers of extinctions on the defaunated and control sites were low and did not differ significantly. During weeks 11–20 the numbers on the defaunated islands were significantly higher, but after week 20 they decreased again and were not significantly different from the levels on the controls. Furthermore, it can be shown that this pattern is not simply a result of changes in species numbers with time, but that it results from higher extinction probabilities *per species* on the defaunated islands during the early weeks. Comparisons of these probabilities (extinctions·species^{-1}·week^{-1}) before and after week 20 reveal that on every defaunated island this probability was higher during the first 20 weeks, when population sizes of the recently arrived colonists were still low, than during weeks 21–53. It is significant that no such difference existed on any of the control sites whose resident species populations were not disturbed by defaunation.

The extinction rate data presented above strongly support the prediction of an inverse correlation between extinction rate and population size. An alternative explanation, however, is that the pattern is a result of a succession of species in which good dispersers that colonize first persist only for a short time, and are then replaced by slower dispersers with better persistence characteristics. Five species were collected on the experimental islands only during weeks 1–20 (Group I) and 8 species during weeks 21–53 only (Group II). As expected, the mean extinction rate of the first group was significantly higher ($p < 0.05$) than that of the second group. The important question is: Is this difference a function of the time period during which the two groups happened to occur on the islands (population size effect) or is it a result of differences in species-specific extinction probabilities between the two groups (succession)? If the second alternative is true, then the differences in extinction activity between the two groups should also hold on the control sites, regardless of the time period during which the species occurred. Comparisons of extinction rates for the two groups at the control sites during weeks 1–20, 21–53, and 1–53, however, result in the opposite relation. In every case, Group I species had lower mean extinction rates than Group II species. Thus, the differences in extinction rates observed on the experimental archipelago cannot be attributed to a succession of species.

A more direct test of the effects of area on extinction rates is to compare actual extinction rates observed on islands of different sizes. A problem often encountered when making such comparisons via the usual techniques of regression and correlation is that of a statistical bias in the direction of the predicted correlation (negative) caused by the species/area relation. The comparison involves a term with numbers of species in its denominator (extinction rate = extinctions/species) and area. If, as is often the case, species and area are highly correlated, then a negative correlation between extinction rates and area is expected even in the absence of a true functional relationship between the two variables. One way to estimate the bias expected from a given species/area relation is to separate the actual extinction rates from the particular sites on which they were recorded, reassign the rates at random throughout the sites, and then calculate extinction rate/area regressions with the randomized data. The slope of this regression estimates the bias because the extinction rates thus calculated are not directly related to the particular species/area combinations with which they are coupled. This procedure was performed 50 separate times with the data from the *Spartina* archipelago and in only 6 cases was the simulated slope steeper than the actual one (all were negative). Under a binomial distribution, a result at least as extreme as this can be expected to occur by chance with a probability much smaller than 0.001, indicating a negative effect of area on extinction rates in addition to, and separate from, the purely statistical effect.

Isolation

The predicted effect of isolation on immigration rates and insular species numbers has also been difficult to test under natural conditions. The major problem has been the difficulty in defining the appropriate source area for a given set of isolates (Lawton, 1978a). For example, for islands in the *Spartina* archipelago the source of arthropod colonists may be the mainland, other large islands, or combinations of these. In addition, given non-continuous sampling methods, the proportion of immigrations detected by a given effort will be influenced by the frequency of sampling, the island extinction rates, area, and other variables unrelated to isolation (Rey, 1981). Isolation (distance to the mainland or distance to the nearest large land mass) did not significantly affect immigration rates or numbers of species on the *Spartina* islands. Preliminary analyses of the sticky trap data indicate that distance to the nearest large land mass and trap orientation significantly influenced the numbers of species collected at the traps throughout the year. The correspondence between trap data and patterns on real islands, however, is questionable.

Turnover

The equilibrium postulated by MacArthur-Wilson is a dynamic one; the number of species on an island at equilibrium remains relatively constant, but species composition constantly changes through extinction of established species and immigration of new ones (MacArthur and Wilson, 1967). Simberloff and Wilson (1969) found this to be true for arboreal arthropods of small red mangrove islands in the Florida Keys. An extremely conservative estimate of extinction rates on these mangrove islands is 1.5 species·island^{-1}·year^{-1}, and they are probably several times higher than that (Simberloff, 1976b). After week 20, turnover rates on the *Spartina* islands were surprisingly low (extinction rates ranged from 3.2 to 9.6 species per island per year). Considering the small size, simple structure, and temporal instability of these tiny islands, it is surprising to observe extinction rates within the same order of magnitude as those in the more complex and stable mangrove archipelago.

The MacArthur-Wilson Correlations

Two of the more important predictions of the MacArthur-Wilson equilibrium model are that, on islands, immigration rates will be negatively correlated, and extinction rates positively correlated, with the number of species on the island. Unfortunately, there are a number of circumstances that complicate the examination of these predictions with the *Spartina* island data. First, the sharp decline in immigrations and extinctions after week 20 makes it difficult to compare rates throughout the whole year. Second, the seasonal patterns in the area further complicate the analysis because the same number of species may occur on the same island at different times of the year, and consecutive species numbers may be observed during totally different time spans on the same island. Third, on some islands, the range in species numbers after week 20 is very small. Finally, because of the low turnover rates, the numbers of immigrations and extinctions recorded weekly after week number 20 are mostly zeros, and the rest are mostly ones and twos.

These complications, particularly the large numbers of zeros, make significance testing for the predicted correlations extremely difficult. Nevertheless, results of parametric and nonparametric correlation analyses of immigration and extinction rates with numbers of species are not inconsistent with the MacArthur-Wilson predictions. A great majority of the comparisons for the nine sites (and combinations of these by size and degree of isolation) during each of three time intervals (weeks 1–20, 21–53, and 1–53) result in correlations with at least the correct sign: positive for extinction and negative for immigration.

Homogeneity of Rates

One assumption underlying the shapes of the MacArthur-Wilson curves is that rates are relatively homogeneous among species (Figure

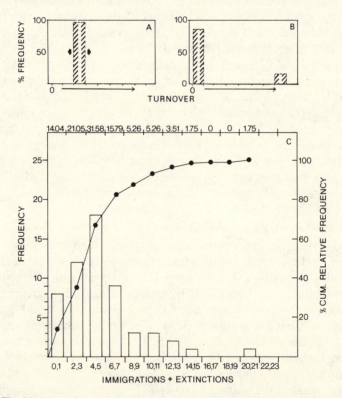

Fig. 7.3. Types of turnover. (A) Homogeneous extreme; all species contribute equally to turnover. (B) Heterogeneous extreme; only a very limited subset of the species is responsible for the turnover by continually immigrating and going extinct. (C) Immigrations and extinctions on the *Spartina* islands. Numbers on top indicate the percentage of the total accounted for by the particular category.

7.3A). The immigration/extinction curves for an island result from an almost additive combination of the immigration and extinction rates of the species on the island and in the species pool (Strong and Rey, 1982). A moderate amount of heterogeneity is easily accommodated by the model in the form of curvature of the rate functions (MacArthur and Wilson, 1967), but great rate heterogeneity (Figure 7.3B) may affect the relationship between extinction and immigration rates and numbers of species to such an extent that the MacArthur-Wilson curves no longer obtain. Turnover activity on the *Spartina* sites was intermediate between the two extremes depicted in Figures 7.3A and 7.3B (Figure 7.3C). Given the variability inherent in field experiments of this type, and the stochastic nature of the colonization process, the patterns appear much closer to the homogeneous extreme (7.3A) than to the repeating subset turnover extreme of Figure 7.3B.

CONCLUSIONS

The results of the *Spartina* island study are, in general, consistent with the predictions and assumptions of the MacArthur-Wilson theory. Equilibrium conditions were established on the experimental islands approximately 20 weeks after defaunation. Immigrations and extinctions occurred with measurable frequencies before and after equilibrium was

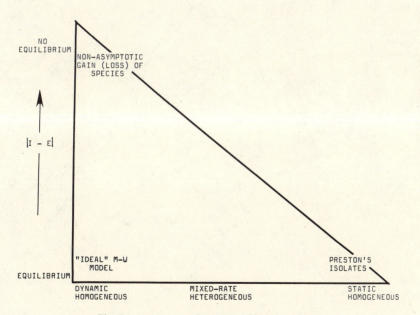

Fig. 7.4. A spectrum of ecological biogeographies.

established, but at lower rates than expected. The predicted effect of area and population sizes on extinction rates and number of species was confirmed, but the corresponding effect of isolation on immigration rates and species numbers was not. Large empty islands recruited species at faster rates than smaller islands, but the differences disappeared as colonization proceeded. Turnover was intermediate between the totally homogeneous extreme of the ideal model and the opposite, completely heterogeneous extreme. The relationships between immigration and extinction rates and numbers of species were in the correct direction, but the shapes of the rate function curves predicted by the model were not evident; low turnover rates and seasonality were partly responsible for this result.

The MacArthur-Wilson dynamic equilibrium theory is at one extreme of a spectrum of ecological biogeographies (Figure 7.4). Data from arboreal arthropods of mangroves (Simberloff and Wilson, 1969, 1970) and terrestrial arthropods of *Spartina* (this study) indicate that these communities fall relatively close to the dynamic-homogeneous equilibrium extreme. However, important deviations from theory have also been observed in these studies, many of the predictions and assumptions of the model remain untested, and there is a great paucity of experimental data on other groups of organisms.

8.

An Experimental Approach to Understanding Pattern in Natural Communities

HENRY M. WILBUR

Department of Zoology, Duke University, Durham, North Carolina 27706

JOSEPH TRAVIS

Department of Biological Science, Florida State University, Tallahassee,
Florida 32306

INTRODUCTION

Patterns of species associations, and the processes that generate patterns, are the focal points of community ecology. The study of pattern has been primarily descriptive, delimiting species associations in replicated physical environments (Wiens, 1969), along natural environmental gradients (Whittaker, 1956), or along statistically derived multifactorial gradients (James, 1971; Bloom et al., 1972; Nicholson et al., 1979). Insights into the processes that form the patterns have come through two types of experimental ecology. Microcosm studies, of varying scale, have investigated the consequences for community structure of various initial arrays of species, densities and environmental conditions (Vandermeer, 1969; Hall et al., 1970; Wilbur, 1972; Neill, 1974; Addicott, 1974; Seifert and Seifert, 1976a; Smith-Gill and Gill, 1978; Morin, 1981). Manipulations of species composition in the field have demonstrated the impact of ongoing biological interactions in systems of marine communities (Paine, 1966; Dayton, 1971), plant communities (Putwain and Harper, 1970; Allen and Forman, 1976) or associations of terrestrial vertebrates (Dunham, 1980; Hairston, 1980a; Munger and Brown, 1981). Field manipulations have also demonstrated the lack of interspecific interactions in situations in which circumstantial evidence suggested that such interactions were important (Wise, 1981b; Tinkle, 1982). Fowler's (1978) elegant study of patterns and processes in an herbaceous plant community combines all of these approaches (see N. Fowler, 1981; Fowler and Antonovics, 1981).

Observational studies, however statistically sophisticated, cannot yield conclusive insights into processes (Dayton, 1973; Connor and Simberloff, 1979; contra MacArthur, 1972; Cody, 1974a; Diamond, 1975). However, manipulative studies need not be confined to attempts to elucidate processes. Carefully designed field experiments can provide insights into the real correlates of pattern (Maguire, 1963; Simberloff and Wilson, 1969; Simberloff, 1976a).

In this report, we describe a long-term experiment designed to reveal some of the correlates of pattern in small communities of aquatic animals. Our preliminary analysis of some of our results will show that experiments designed to study pattern can lead not only to conclusions about pattern, but to conclusions about the most relevant experiments to perform to elucidate processes.

THE SYSTEM AND THE ISSUES

The Sandhills Region of North Carolina is an area on the Upper Coastal Plain characterized by rolling clay hills covered with Cretaceous sand deposits. On the broad divides, where the sand layer is deep, are pine savannahs, characterized by *Pinus palustris, Quercus laevis,* and *Aristida stricta.* On the steep slopes of the stream valleys, where the sand layer is shallow, the underlying clay makes the soil quite moist, leading to the formation of evergreen shrub bogs or pocosins, thick stands of vegetation characterized by such species as *Magnolia virginiana, Ilex glabra, Acer rubrum, Smilax laurifolia, Persea borbonia, Lyonia lucida,* and *Pinus serotina.* Excellent photographs of these habitats can be found in Wells (1932) and Martof *et al.* (1980).

This region contains a variety of aquatic habitats, from permanent impoundments and Carolina Bays (Frey, 1949) to ephemeral rain pools. Seventeen small ponds without fish have been studied during the last eight years. The ponds differ in hydrologic cycle, morphology, chemistry, temperature regime, and biota. Over 200 species of arthropods and amphibians have been collected from these study ponds.

The long-term goals of the project are to uncover the processes that underlie the observed pattern of distributions and abundances of amphibians and insects. But before the processes can be investigated, the patterns must be defined.

The determination of pattern in this context has been treated primarily as a statistical problem by other workers, whether involving only species associations in a strict sense (Jaccard, 1901; Cole, 1957; Goodall, 1969; Pielou, 1972; Hendrickson, 1979) or correlations of species occurrences or abundances with habitat variables (Bray and Curtis, 1957; Orloci, 1967; Hill, 1973; Gauch *et al.,* 1977). These techniques have two limitations in our system. First, the power to detect any pattern is limited by the observed ranges of the individual variables as well as the observed range of intercorrelations. For example, if surface area were important for determining the number of species in a pond, examining ponds with a limited range of surface areas may not provide enough power to detect the relationship. More importantly, if, for example, wooded ponds always occurred near pocosins, remaining cool because of shading, with pH largely determined by the litter from the characteristic pocosin vegetation, then we would be unable to decide whether any species association patterns were functionally correlated with all or only a few of these physical parameters, whose effects would be completely confounded. Second, any snapshot picture of pattern cannot account for historical effects that may have played a role in generating the pattern. The large number of taxa and the small number of local ponds (approximately 20) relative to the number of potentially important variables render a purely statistical approach to this system fraught with the potential for spurious interpretations.

Another tactic for the study of pattern involves focusing on a small number of variables and disregarding the others, either by eliminating them or by randomizing their occurrence over the values of the other variables in the study. This tactic requires an experimental approach that establishes replicated vacant habitats of a suitable nature close to sources of colonists. Through the experimental design, the correlates of pattern can be uncovered and attention focused on a distinct subset of variables. This approach has been used in experimental biogeography (Maguire, 1963; Simberloff and Wilson, 1969; Simberloff, 1976a) to address some general problems in community ecology:

(1) Can the structure of a community be predicted by physical characteristics of the habitat?
(2) Do communities in physically different habitats diverge as colonization and extinction proceed?
(3) Can the colonization success of a species be predicted by the structure of the established community?
(4) Can the extinction of an established species be predicted by the structure of the extant community?

These questions of pattern development and regulation are independent of questions concerning processes, but in order to be dealt with satisfactorily, they require no less an experimental approach.

EXPERIMENTAL DESIGN AND SAMPLING

Sixteen ponds were established during the summer of 1977 in Scotland County, North Carolina. Each pond is a cylindrical tank sunk flush with the ground, with a surface area of 3.58 m^2 and a depth of 0.61 m. Water was pumped from the nearest source, always less than 100 m distant, through a piston pump and coarse plankton net.

These 16 novel ponds were arranged in 4 sets. Two sets were placed near permanent streams, and two sets near temporary ponds. Within each set, two ponds were placed in a savannah, and the others were placed at the edge of a pocosin. In all cases, ponds in the savannah were further from the water source. This design allows the 16 ponds to be considered in any of 3 ways (Table 8.1): as 16 replicates, as 4 geographic sets, each of 4 replicated ponds, or as replicates of a 2(2) × 2 classification of water source (temporary or permanent), with 2 geographic areas nested within each level of water source, crossed with terrestrial plant association (savannah or pocosin).

The savannah-pocosin dichotomy may have different biological meanings for different species. For example, *Hyla andersoni* is generally considered a specialist in small ponds associated with pocosins or hardwood seeps, while *H. femoralis* is considered a frog of savannah pools (*e.g.* Martof *et al.*, 1980). For this pair, the dichotomy represents real habitat distinctions. For some insects that complete their life cycles in the water, such as most Hemipterans, this dichotomy may be more representative of distance from the source than of any distinction due to ecological habitats *per se*. Except where habitat distinctions are known throughout the range of a species, there would seem to be no *a priori* way to predict exactly what the savannah-pocosin dichotomy may mean to a given species (Colwell and Futuyma, 1971).

All experimental ponds are themselves permanent. Surface areas and depths are equal, and habitat placement and distance from water source were controlled. All ponds were established within 5 weeks of one another in late summer of 1977, equalizing historical effects and minimizing any phenological effects on variance among ponds in the initial colonization process. This experimental approach reduces the number of physical variables substantially, and permits the analysis of pattern development and community structure with far fewer confounding effects than could be obtained by analyzing only the data from natural ponds.

Table 8.1. Experimental design and pond identification.

Water Source	Local Set	Habitat			
		Pocosin		Savannah	
Pond	SB	4	1	3	2
	SQ	3	1	4	2
Stream	WC	3	2	4	1
	GP	1	3	4	2

The communities were sampled at 2–4 week intervals from August 1977 to October 1979. Sampling was done by sweeping a D-shaped insect net with a depth of 0.41 m and a maximum width of 0.33 m once around the top perimeter of the water column and by sweeping a rectangular net 0.33 m deep and 0.50 m wide of 0.4 mm mesh once around the bottom perimeter of the tank. All macroinvertebrates and vertebrates were identified and individually counted. A few individuals of each taxon were collected and preserved when first encountered or when their identities were uncertain. In addition, zooplankton sampling was begun in 1979; these data will not be discussed here.

All of our identifications could not be made at the same level of taxonomic resolution, but we attempted to separate morphological "species" within all taxa. Nearly all major taxa, those that occurred in 5 or more ponds and had a maximum abundance of 5 or more individuals, were identified to the species level. Because all growing stages were photographed, it was often possible to make unambiguous retroactive identifications of taxa, which, when first encountered, could be assigned only to the family level (*e.g.* Odonate naiads and Anuran larvae). These taxonomic decisions are not trivial ones in making inferences about community structure: several ubiquitous taxa, if in reality composed of two or more sibling species, can be the root of spurious conclusions about the nature of species association patterns in these communities.

THE DATA

Our experimental design simplified the physical variables, but also reduced the number of taxa in our sampling universe. The data from 43 sample dates between 16 August 1977 and 25 October 1979 revealed 149 taxa of macroinvertebrates and vertebrates, of which 63 taxa were considered as major ones.

Detailed natural history observations help explain why some taxa that are common in the natural ponds were absent from our experimental ponds. Small winter-breeding frogs such as *Hyla crucifer* and *Pseudacris nigrita* deposit eggs in the shallow perimeters of ponds, and our experimental ponds, because they have vertical sides, have no such microhabitat. Damselflies (Odonata: Zygoptera) and the large dragonfly *Anax junius* (Odonata: Aeshnidae) oviposit on aquatic vegetation, and our ponds have no emergent vegetation. The lack of submerged vegetation may account for the lack of creeping water bugs (Hemiptera; Naucoridae: *Pelocoris* sp.). For some of the larger species, such as bullfrogs (*Rana catesbeiana*) and giant water bugs (Hemiptera; Belastomatidae: *Lethocerus americanus*), our ponds appear too small to stimulate oviposition by females. The reasons for the exclusion of other species are unknown.

In a study of this type, in which repeated point samples are used to make inferences about the development of community patterns, the reliability of the point samples, either for detecting a taxon as present or for estimating its abundance, must be known. Because our system is a small, closed one, we can measure this reliability by taking the standard sample, and then repeatedly sampling without replacement until everything has been removed from the pond. This "exhaustive" sampling allows us to calibrate our standard sampling methodology.

Two such exhaustive samples were taken. Pond WC1 was sampled by our standard method on 8 August 1979. On 9 August the pond was again sampled by our standard method, but the standard method was then repeated for a total of 7 times, without replacement. After these 7 samples, the pond was repeatedly swept with the nets until 20 sweeps yielded no additional animals. The same procedure was repeated on pond GP1 on 8 and 14 August 1979. All animals were returned to the pond at the end of the exercise. This procedure evaluated the accuracy of the replicate standard samples (the sample taken on 8 August and the first sample of the exhaustive sampling) in view of the total census.

The proportion of taxa, known to be present in the pond, that were taken in each replicated sample, and the proportion of taxa, known to be present, that were found in at least one of those samples, were 0.45, 0.69, and 0.76 respectively for WC1 ($n = 29$ taxa), and 0.61, 0.61, and 0.91 respectively for GP1 ($n = 23$). There is an obvious bias in estimating species occurrences. We decided to pool adjacent samples in the analysis of community patterns in order to reduce this bias. The taxa that were missed in both of the two adjacent standard samples were found to have very small total populations by exhaustive census. In pond WC1, 7 taxa were discovered in the exhaustive sample that were not captured in the two standard samples. These taxa had total populations of 1 (4 taxa), 2 (1 taxon), 3 (1), and 4 (1) individuals. In pond GP, 6 taxa were missed by the two standard samples; all of these taxa had total populations of just one individual. We concluded that our decision to pool adjacent samples would reduce sampling bias such that about 86% of taxa that were present in a pond would be sampled, and that the taxa that were missed would have very low abundances.

The proportions of the total number of individuals in the pond that were obtained in the first standard sample of the censuses were about the same in the two ponds: 0.345 for WC1 ($N = 1008$ individuals) and 0.348 for GP1 ($N = 3375$). On a specific basis, individual taxa displayed wide variation in the proportions of individuals captured by the standard sample. This finding suggests that the sampling distributions of abundances varied from one taxon to the next. Until these problems are better understood, it may be very inaccurate to estimate relative abundances from these point samples. In this paper, we shall analyze only presence-absence data.

ANALYZING COMMUNITY PATTERNS

Having assayed the efficiency of the sampling method for occurrence data, we present some preliminary analyses of community patterns based on these data. Analyses of occurrence data are not powerful detectors of community structure because relative abundances of species are neglected. For example, two taxa that occur together in many communities will always be scored as present, yet if one is rare when the other is common, or one is always rare and the other always common, then there is a great deal of structure undetected by occurrence data. Thus, the patterns we analyze are "first-order" patterns. Until the sampling distributions of individual taxa are known in more detail, we must confine ourselves to this conservative analysis.

Perhaps the primary issue is the development of pattern in these communities. Note that both the total number of taxa observed and the average number of taxa per pond increased during the first year following establishment (Table 8.2). There is a seasonal component to the identity of the colonists. For example, although the dragonfly *Pantalla* was an early colonist, identifying it as a "colonizing" species would be premature, because it returned in late summer in each subsequent year.

One scenario for community development might be an initial flush of ubiquitous colonizers, followed by a gradual assortment of taxa into a pattern determined by physiological and interactional tolerances. We examined part of this scenario by testing whether the early occupants of all the ponds appeared to be a random sample of equiprobable taxa, given that different ponds may support different numbers of taxa at any one time. Cochran's Q statistic, denoted in Table 8.2 as Y^2, is the appropriate statistic to use in this test. The statistic Y^2 is compared to a chi-squared distribution with degrees of freedom equaling one less than the number of species in the total pool. This hypothesis of equiprobable occupants can be rejected with the earliest sample analyzed (Table 8.2). This conclusion could indicate different colonizing abilities among different taxa, or equivalent colonizing abilities but unequal extinction probabilities. Subsequent time periods also display too many species that are too common or too rare for acceptance of the hypothesis. Though we cannot conclude anything about process, we can conclude that pattern is established very early in the history of the community.

The degree of departure of the observed pattern from random, equiprobable association is estimated by dividing the observed value of Y^2 by the 0.01 critical level value of chi-squared for the appropriate degrees of

Table 8.2. Tests of the hypothesis that species are equiprobable among experimental ponds.

Dates of Samples		Total Taxa in All Ponds	Mean Taxa per Pond	Variance in Taxa per Pond	Y^2	N	$Y^2/\chi^2_{0.01}$
First	Second						
08 NOV 77	05 JAN 78	27	6.12	4.92	108	16	2.36
19 FEB 78	07 MAR 78	26	5.06	5.13	115	16	2.60
23 MAR 78	06 APR 78	23	4.31	2.63	130	16	3.23
18 APR 78	02 MAY 78	26	6.81	2.96	124	16	2.80
23 MAY 78	02 JUN 78	38	12.18	9.90	196	16	3.32
10 JUN 78	20 JUN 78	42	13.56	10.26	253	16	3.95
01 JUL 78	10 JUL 78	56	14.75	17.80	298	16	3.63
20 JUL 78	01 AUG 78	51	17.75	37.80	279	16	3.66
01 OCT 78	31 OCT 78	67	17.38	27.72	400	16	4.17
07 DEC 78	16 JAN 79	51	12.25	12.60	301	16	3.95
07 MAY 79	17 MAY 79	65	15.63	30.38	345	16	3.67
29 MAY 79	07 JUN 79	66	19.57	36.88	363	14	3.82
18 JUN 79	28 JUN 79	58	17.73	46.92	276	15	3.25
10 JUL 79	19 JUL 79	58	18.79	9.41	332	14	3.91
30 JUL 79	08 AUG 79	60	18.79	18.49	317	14	4.22
27 AUG 79	26 SEP 79	68	17.79	20.64	333	14	3.43

freedom (Table 8.2). The departure from randomness increased through the first year and then declines slightly. The patterns are significantly different from the null hypothesis very early in the experiment. If the decline observed in late 1979 continues it could be due to the ponds' becoming physically different by accumulation of litter or changes in water chemistry reducing the degree of replication among the initially similar ponds.

We can gain further insight into the correlates of pattern by exploiting the design of the experiment, considering the ponds as 4 geographic sets of 4 ponds each or as 4 habitat types, each replicated 4 times. We analyzed the data of 9 August 1979 in this light.

The species pool for each geographic set was restricted to only those species that were found in at least one of the ponds in that set. In the complementary analysis, the pool for each of the 4 ecological habitat types was restricted in a similar manner. These analyses assume that each geographic area or each habitat type has a partially distinct pool of species, but that within each pool, all species are equally likely to occupy the appropriate ponds. The distinct geographic species pools can reflect different historical effects or restrictions on long-range dispersal abilities, while distinct habitat pools can reflect specific differences in habitat preferences, short-range dispersal abilities, or temporal persistence in different habitat types due to physiological tolerances or biological interactions. In testing each hierarchical hypothesis, the Y^2 statistic is computed for each of the 4 separate species pools, and the overall statistic is computed as the sum of these 4 individual Y^2 values, with the total degrees of freedom equal to the sum of the 4 individual degree-of-freedom comparisons.

On this date, there was, not surprisingly, a general pattern to the taxon occurrences ($Y^2 = 288.24, df = 51, P < 0.001$). When different geographic pools were constructed, some taxa were still too common or too rare within their set ($Y^2 = 145.94, df = 111, P < 0.025$). However, when taxon pools were assembled by habitat type, the occurrence patterns could not be distinguished from equiprobable occurrences within a habitat type. Thus the associations of different species pools with different habitat types appeared to be lending structure to the data.

Note that this is an analysis of pattern, not process. There are numerous processes that could generate such a pattern of habitat associations. Different species may have evolved habitat preferences independently, competition or predation may have driven habitat selection in the past, ongoing competition or predation may be maintaining habitat segregation among species, physiological tolerances to different habitats may differ among species and cause associations, or combinations of these processes may be operating. The results do suggest that experiments on process that involve species that are found in distinct habitats may be as interesting as those involving co-occurring species.

CONCLUSIONS

By experimentally eliminating a number of naturally varying physical factors (pond size, depth range, shape, long-term history, emergent vegetation), we have been able to focus a search for pattern and its correlates. Although we are thereby forced to work with a distinct subset of the natural fauna, we have found that patterns are formed early in community development, they appear to intensify over time, and after 2 years, they are correlated with habitat associations of species.

Two sets of cautionary remarks are in order. The first set concerns future work on this system. Obviously we have presented merely an overview of the study with preliminary analysis of some occurrence data. We are currently attempting to ascertain whether clusters of similar communities arise over time, clusters that are more than simple reflections of habitat associations of species pools. Habitat associations may exist, but at a fine level communities within a habitat type may develop in quite separate and unrelated ways. This would suggest an important role for short-term historical effects. In addition, the more interesting experiments to delineate process may involve taxa that are phylogenetically and ecologically similar, but are found in distinct habitats. This experimental study of pattern has helped to bring the prospective experiments on process into sharper focus.

The second set of remarks concerns our assumptions and decisions at several critical junctures, factors that we have attempted to delineate on the previous pages. We feel that it is critically important to keep those assumptions at hand, because our conclusions rest on them. Habitat associations of taxa depend on the level of choice: there may be far more structure than we have detected if any of our taxa actually consist of several species that avoid one another or have distinct physiological tolerances in different habitats. Although the relative abundance data may reveal more pattern, we feel that such data must be approached with extreme caution. Sampling distributions of different taxa are obviously different, and rash analysis of abundance data may cause this difference to be interpreted spuriously as evidence for "community structure." The dangers are compounded if these sampling distributions themselves change with habitat type, or worse, change differentially among species in different habitat types.

We feel that the study of pattern is a central focus of community ecology, and that pattern development in many communities can be studied effectively by an experimental approach. Studies of pattern, whether observational or experimental, should be cautious in the interpretation of abundance patterns, and should be designed to suggest proper experiments to delineate process. Without such cautions and focus, studies of pattern can degenerate all too easily into mere exercises in aesthetics.

Acknowledgments

Many people devoted themselves to the tedium of sampling, but especially important were R. C. Chambers, J. T. Longino, P. J. Morin, and S. Via. Invaluable assistance in taxonomy was rendered by J. T. Longino, S. Via, and D. E. Wheeler. The experimental design, analysis, and interpretations owe much to the influence of D. S. Burdick and P. J. Morin. Discussions at the conference, particularly with A. J. Underwood, have helped clarify our presentation.

This work has been supported by NSF grants DEB76-82620 and DEB79-11539 to H. M. Wilbur.

9.

Biogeography, Colonization, and Experimental Community Structure of Coral-Associated Crustaceans

LAWRENCE G. ABELE

Department of Biological Science, Florida State University, Tallahassee, Florida 32306

Major patterns of species richness in the marine environment include a general increase in species numbers with decreasing latitude and, within the tropics, an increase in regional species number in the order Eastern Atlantic, Western Atlantic $\simeq$ Eastern Pacific, Indo-West Pacific (for a discussion on crustaceans see Abele, 1982). There is no shortage of hypotheses to explain these and other patterns of species richness (*e.g.* Ricklefs, 1979). The hypothesis considered here suggests that species in diverse environments are more specialized and have narrower niches, resulting in more coexisting species per unit habitat than in less diverse habitats. One of the few distinct marine habitats to occur over a sufficiently wide longitudinal range for a good test of this notion is the coral head habitat formed by the branching coral *Pocillopora damicornis*. This species and its crustacean inhabitants occur in the Eastern, Central and Indo-West Pacific regions. It is a habitat for crustaceans that occurs over a broad gradient of species richness.

The total number of potential crustacean species living among the branches of *P. damicornis*, including obligate symbionts, is approximately 5 to 8 times greater in the Central and Western Pacific than in the Eastern Pacific. Thus, if the increase in species richness across the Pacific results from an increase in within-habitat species richness, we should be able to detect it by comparing community structure of *P. damicornis* among these regions.

Comparisons can provide only partial answers to questions about what might structure communities, and experiments are necessary for hard evidence. Here I examine colonization of experimentally defaunated corals and the persistence of experimentally assembled communities with "too many" species placed together.

METHODS AND MATERIALS

Pocillopora damicornis is a branching coral that occurs as discrete heads on tropical, shallow-water reefs of the Eastern and Indo-West Pacific regions. Heads range in diameter from 3 to about 45 cm. At some sites heads are isolated; at other sites heads are contiguous and form a near-continuous cover. A single live coral head may be inhabited by from 1 to 25 species and 1 to 100 individuals of decapods (Abele and Patton, 1976). About one-third of the associated species are obligate symbionts of *P. damicornis*, while the remaining species will occur in a variety of habitats. Decapods account for the vast majority of species on a coral, usually about 65 to 89% of the fauna. Other organisms include fish, flatworms, molluscs, and echinoderms.

Collections were made by enclosing a coral head in a plastic bag, then breaking the head off from the substrate and bringing it to a boat at the surface. Length, width, and depth of the head were measured and the animals removed by breaking the coral apart. Care was taken to avoid collecting the dead base of the coral.

Live corals were defaunated with virtually no mortality to the head itself by placing the coral in a plastic bag and carrying it to the boat where animals were picked off with forceps. During this process, which is quick and accurate, the coral is dipped into water every 30 seconds. Large heads were broken into a number of smaller heads, with no apparent coral mortality, after teasing the animals to the dead base. Small coral heads were picked clean while whole. In one check for errors of this method, I shattered 20 coral heads after defaunation and found that only a single animal had been missed (a synalpheid shrimp about 6 mm total length that had remained in a crevice in a dead portion of the base).

Defaunated heads were used in the field for several experiments. First, they were replaced in the area from which they were collected. Then, 5 to 10 were recollected on each of the next six days, to follow recolonization.

The second set of experiments dealt with the consequences of placing "too many" crustacean species on a single head. A species-area relationship gave an estimate of the numbers of species and individuals expected as a function of coral head size. Several coral heads were broken open and the crustaceans removed and placed in small jars of water to keep them alive. These animals were measured and sexed and larger animals were marked by clipping the legs. Animals were then quickly placed on the defaunated heads. Corals with "too many" species were divided into two groups: in the first group each coral was placed in a cage (0.3 m on a side), while the second group of corals was identical in species composition and similar in size but uncaged. The mesh size of the cages (12 mm) was large enough to allow movement of the crustaceans but small enough to keep predaceous fish out. Controls were treated in the same way but were uncaged and given the "correct" number of species and individuals. The persistence of species was followed by serial collection of subsets of the coral heads.

Study Sites

Collections in the Eastern Pacific were made at the Pearl Islands and Uva Island, Panama (Abele and Patton, 1976; Abele, 1979); these areas have been described by Glynn (1976), Glynn and Stewart (1973), and Glynn *et al.* (1972).

Collections and experiments in the Central Pacific were done at Palau (~7°21'N, 134°28'W) during an *Alpha Helix* cruise of August 1979. All work was done on a shallow (1–3 m) patch reef about 300 × 450 m in size. The density of *P. damicornis* ranged from 0.08 to 1 per m² ($\bar{X}$ = 0.32/m²). Wrasses (*Thalassoma* spp.) were the most abundant diurnal fish predators.

Collections and experiments in the West Pacific were conducted at Lizard Island, Australia. Two sites were examined: one at the mouth of Blue Lagoon and the second adjacent to the inside of Palfry Island. *Pocillopora damicornis* formed a near-continuous cover at the lagoon mouth site but was patchy at the Palfry Island site (0 to 16 heads per m², $\bar{X}$ = 0.08 per m², based on 4 25-m × 1-m transects). *Thalassoma lunata*, a wrasse, was the most abundant diurnal predator at both sites.

Data from the southern Barrier Reef, Capricorn Island group, were extracted from Austin *et al.* (1980).

RESULTS

Species Richness

Comparison of within-habitat species richness among the sites is accomplished by examining the species/area relationship for individual corals at each site. This comparison will tell us if a coral head of a given size from one geographic region contains more or fewer species than one from another geographic region. A species/area relationship was determined for the samples from each region (the log-log transformation fit as well as or better than other models; Table 9.1, Figure 9.1). If the greater total species richness of the Central and West Pacific is the result of an increase in within-habitat species richness, a coral head from Palau or Lizard Island would have more crustacean species than one of the same size from Panama or Capricorn Island; the slopes of the regression lines from Palau and Lizard Island would be above those from Panama and Capricorn Island. By ANCOVA the regions do not differ in slope (F_{int} = 2.05). There are differences in intercept though none is significant and

Table 9.1. Species/area statistics for the number of decapod crustacean species on *Pocillopora damicornis*.

Region		r	p
Panama, Uva Island	$\log S = .27 \log X + .01$	.48	.001
Panama, Pearl Islands	$\log S = .37 \log X - .32$	.64	.001
Palau	$\log S = .46 \log X - .65$	.73	.001
Australia, Lizard Island	$\log S = .42 \log X - .64$	.64	.001
Australia, Capricorn Island*	$\log S = .36 \log X - .08$	—	.05

* From Austin *et al.* (1980). Not included in ANCOVA as raw data unavailable.

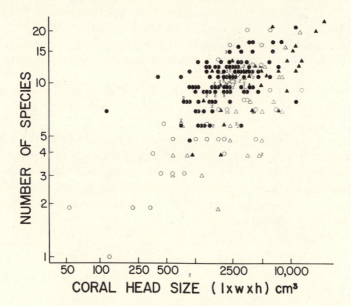

Fig. 9.1. A plot of numbers of species of decapod crustaceans against coral head size on a log-log scale. Dots = Uva Island, Panama; closed triangles = Pearl Island, Panama; circles = Palau; open triangles = Lizard Island. Non-decapod species included.

they are not in the order expected based on total species richness of the regions. Instead the intercepts decrease in this order: Uva Island, Panama; Capricorn Islands; Pearl Islands, Panama; Lizard Island and Palau. Therefore, the number of species per unit coral head size is similar in all regions; there is no evidence for increased habitat partitioning and species packing in the more species-rich regions.

The total number of species collected on all *P. damicornis* in each region is more difficult to compare as it depends on both total number of corals sampled and number of decapod individuals collected. In addition, the number of species can be inflated if any part of the dead base of the coral is collected, since species occur there that are not found on the live coral itself. For example, Austin *et al.* (1980) included *Periclimenaeus hectate*, an associate of encrusting colonial ascidians, in their total species number, although they suggested that the species may have come from the dead base.

A comparison of total species numbers based on equal numbers of corals and individuals (Table 9.1) generated by rarefaction (Hurlbert, 1971; Heck *et al.*, 1976) yields the following results. The Central (Palau) and West Pacific (Lizard, Capricorn) regions are similar in species number and are greater than the Pearl Islands site, which in turn is greater than the Uva Island site. The total number of species per similar numbers of individuals associated with *P. damicornis* is greater in the Central and West Pacific than in the Eastern Pacific only because there is variation in species composition among coral heads. That is, the faunal similarity among individual coral heads is higher in the Eastern Pacific than in either the Central or Western Pacific, with the result that any two coral heads in the Eastern Pacific will share more species than any two coral heads in the Central or Western Pacific.

Species Composition

Abundances of the most common families are given in Table 9.2 and the ten most common species and their abundances in Table 9.3. The crab

Table 9.2. Taxonomic composition of the *Pocillopora* community.

| | Panama | | Micronesia | Australia | |
	Pearl Is.	Uva Is.	Palau	Lizard Is.	Capricorn Is.*
Total number of corals sampled	35	119	40	30	40
Decapoda/Total species in fauna	89%	76%	76%	84%	65%
Total number of decapod species	57	52	47	43	65
Total number of decapod individuals	1107	4724	546	439	797
% Decapoda					
Xanthidae	24	29	49	40	38
Alpheidae	21	15	15	19	25
Majidae	10	17	6	7	3
Palaemonidae	9	12	11	12	15
Porcellanidae	21	14	0	9	9

* Data from Austin *et al.* (1980).

Table 9.3. Ten most common species of Decapoda associated with *Pocillopora* and their percent abundances.

Panama (Uva)	Panama (Pearl)	Palau
Trapezia corallina (16.5)	*Pisidia magdalenensis* (16.3)	*Trapezia* cf. *cymodoce* (15.14)
Fennera chacei (14.3)	*Harpiliopsis depressa* (9.8)	*Alpheus lottini* (10.3)
Trapezia ferruginea (12.7)	*Trapezia "ferruginea"** (8.2)	*Harpiliopsis depressa* (8.0)
Teleophyrs cristulipes (10.1)	*Fennera chacei* (8.0)	*Trapezia davaoensis* (5.0)
Harpiliopsis depressa (8.5)	*Alpheus lottini* (7.5)	*Periclimenes* sp. 1 (4.8)
Pagurus cf. *lepidus* (6.8)	*Petrolisthes haigae* (6.4)	*Periclimenes* sp. 2 (4.8)
Alpheus lottini (5.4)	*Palaemonella* cf. *asymmetrica* (5.2)	*Trapezia* cf. *cymodoce* 2 (4.2)
Petrolisthes haigae (5.1)	*Heteractaea lunata* (3.9)	*Trapezia dentata* (4.0)
Micropanope xantusii (3.4)	*Petrolisthes agassizii* (3.6)	*Thor* cf. *amboinensis* (3.7)
Harpiliopsis spinigera (3.4)	*Mithrax pygmaeus* (2.8)	*Trapezia* cf. *dentata* (3.5)

Australia (Lizard)	Australia (Capricorn)
Periclimenes amymone (24.4)	*Trapezia cymodoce* (15.3)
Trapezia cf. *cymodoce* (11.2)	*Chlorodiella* sp. A (13.4)
Alpheus lottini (10.3)	*Pachycheles* sp. A (11.3)
Trapezia areolata (6.8)	*Alpheus lottini* (7.4)
Phymodius cf. *granulosus* (6.6)	*Hapalocarcinus marsupialis* (5.5)
Harpiliopsis depressa (3.0)	*Periclimenes spiniferus* (5.1)
Galathea sp. (3.0)	*Harpiliopsis beaupresii* (4.6)
Paguridae sp. (3.0)	*Periclimenes madreporae* (4.0)
Synalpheus sp. C (2.7)	*Trapezia areolata* (3.8)
Petrolisthes sp. A (2.7)	*Cymo andreossyi* (3.5)

* Probably includes *T. ferruginea*, *T. corallina*, and *T. formosa*, which have been confused in the past.

family Xanthidae is more common in the Western Pacific than in the Eastern Pacific, while spider crabs (Majidae) and porcelain crabs (Porcellanidae) are more abundant in the Eastern Pacific. Variation in species composition is striking. The most abundant species at the Pearl Islands, *Pisidia magdalenensis*, was absent from corals at Uva Island less than 300 km away (Abele, 1979). Similarly, the most common species at Lizard Island, *Periclimenes amymone*, was not recorded at Capricorn Islands on the southern Barrier Reef (Austin *et al.*, 1980). There is also considerable variation among the other common species from site to site.

A coefficient of interspecific association, C_8 (Hurlbert, 1969), was calculated for the most frequently occurring species at Palau and Lizard Island. For the 45 comparisons (10 species) at Lizard Island there are 29 C_8's with positive signs and 16 with negative signs; a total of 26 of the values are zero. Fifty-nine of the 67 C_8 values from Palau are positive while only 8 are negative; a total of 26 of the values are zero. There is no obvious pattern of association at either locality. All species-pairs combinations for the 10 most abundant species at Palau occurred together at least once. Fourteen species of the crab genus *Trapezia* were collected at Palau, 8 species on a single coral head. All species pairs (excluding species that were collected only once) occurred together at least once.

Similar results were obtained for Lizard Island species. All species-pairs combinations for 7 of the 10 most common species occurred together at least once. There is no extraordinary morphological, ecological, or taxonomic similarity (suggesting exclusion) among those species that did not co-occur. Six species of *Trapezia* were collected and of these *T. cymodoce* co-occurred with all but a species that was represented by a single juvenile on one coral. *Trapezia aerolata* occurred with all other species except *T. guttata*, which co-occurred only with *T. cymodoce*. In contrast to Palau there is a suggestion here of possible interaction among *T. guttata, aerolata*, and *dentata*. However, presence-absence and co-occurrence data reveal little about species interactions, and experiments are necessary to test hypotheses on competition (Simberloff and Connor, 1981). The present data suggest that there are few competitive interactions at natural densities (but see *Assembly of Communities* section below).

Colonization

Tables 9.4 and 9.5 summarize the data from the colonization experiments. On day 1 all defaunated corals were set out in the field at natural densities. Then on each of 6 successive days from 5 to 10 heads were collected. This method provides data only on which species were on a coral on a particular day, and not on how much change occurred between days 1 and 6 on corals left out for 6 days.

Colonization data from Palau were gathered during August 1979 and show a general increase in the average number of species per coral head from day 1 to day 6 (Figure 9.2, Table 9.4). The data may be examined on a coral head/day basis; 48 corals were available on day 1, but day 2 had 48 available for 24 hr and 38 available for 48 hr, yielding 86 coral head days. Treating the data in this way would lower the colonization rate. Twenty-two species colonized (Table 9.4) over the 6 days, just under half of the total species collected (48). All colonists at Palau were small adults and juveniles; no settling larvae or postlarvae were found. This is consistent with the assemblages on heads collected in the field, which consisted only of adults, subadults, and juveniles. Postlarvae and very small juveniles are rarely found on corals.

Table 9.4. Comparison of colonization frequency, abundance, and frequency of occurrence* of decapods at Palau.

Species	Colonization Frequency (%)	Day First Colonized	Relative Abundance (%)	Frequency of Occurrence (%)
Trapezia cf. *cymodoce*	29.8	2	15.4	75.0
Trapezia davaoensis	17.0	2	5.0	25.0
Phymodium cf. *granulosus*	14.5	3	1.6	15.0
Paguridae	10.6	2	3.3	25.0
Trapezia sp. R	8.5	1	1.7	7.5
Majidae sp. A	6.3	1	0.9	12.5
Harpiliopsis sp.	6.3	2	8.0	47.5
Pilumnus sp. A	4.2	5	—**	—**
Xanthidae sp. A	4.2	4	1.7	15.0
Trapezia dentata	4.2	1	4.0	30.0
Galathea sp. A	4.2	2	2	2.5
Trapezia areolata	4.2	3	1.8	17.5
Synalpheus sp. NS	2.1	5	2.2	17.5
Palaemonidae sp. P	2.1	1	4.8	27.5
Portunidae sp. A	2.1	1	—**	—**
Majidae sp. B	2.1	4	0.5	5.0
Alpheus sp. MS	2.1	2	1.6	17.5
Thor cf. *maldivensis*	2.1	6	2.0	8.3
Periclimenes sp. A	2.1	6	4.8	35.0
Cymo cf. *andreossyi*	2.1	1	0.2	2.5

* Colonization frequency is the number of heads colonized out of 48; relative abundance and frequency of occurrence are based on 40 heads collected in the field before the experiment began.
** Not collected in the initial samples.

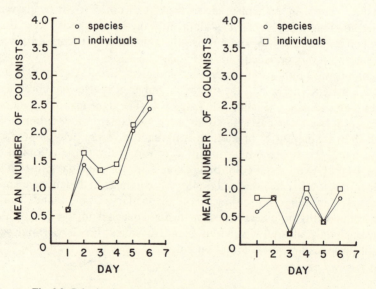

Fig. 9.2. Colonization curves for Palau (left) and Lizard Island (right).

Table 9.5. Comparison of colonization frequency, abundance, and frequency of occurrence* of decapods at Lizard Island.

Species	Colonization Frequency (%)	Day First Colonized	Relative Abundance (%)	Frequency of Occurrence (%)
Site I				
Periclimenes amymone	33.0	1	24.4	63.3
Trapezia cymodoce	10.0	3	11.2	66.7
T. areolata	10.0	2	6.8	46.7
Galathea sp.	3.3	5	3.0	26.7
Paguridae sp.	3.3	4	1.4	10.0
Megalopa (Xanthidae)	3.3	6	—**	—**
Site II				
Periclimenes amymone	20.0	2	24.4	63.3
Trapezia cymodoce	20.0	3	11.2	66.7
Chlorodiella sp.	6.6	2	6.6	46.7
Paguridae	6.6	1	1.4	10.0
Galathea sp.	3.3	1	3.0	26.7
Pilumnus sp.	3.3	2	0.9	10.0
Periclimenes madreporae	3.3	2	0.7	3.3
Leptodius sp.	3.3	6	0.9	10.0

* Colonization frequency is the number of heads colonized out of 30 at each site; relative abundance and frequency of occurrence are based on 30 heads collected in the field before the experiment began.
** Not collected in the initial samples.

Only 4 of the 10 most abundant species and only 3 of the 10 most frequently occurring species are found among the 10 most abundant and frequent colonizers (Table 9.4). A striking absence is a snapping shrimp, *Alpheus lottini*, that occurred on 70% of all corals, was abundant, yet was never observed colonizing a coral.

To determine whether colonization was nocturnal or diurnal, 5 corals were left out for 12 hr during the day and 10 were left out for 12 hr overnight. The day corals had 1 individual of 1 species, while the night corals had a total of 7 individuals and 7 species, suggesting that colonization is mainly nocturnal.

In Figure 9.3 the species/area regression line is shown for reference, and the data points represent individual corals that had been out for the number of days indicated. The number of species colonizing each coral can be read on the *y*-axis. Four of the corals have reached the expected

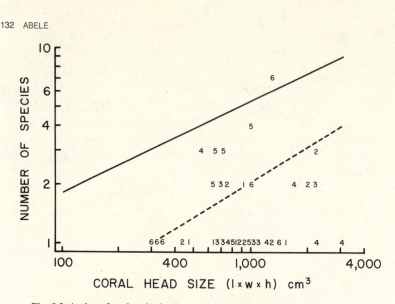

Fig. 9.3. A plot of each colonized coral head showing its size and number of species that colonized. Each number (from 1 to 6) indicates the number of days that the coral was available for colonization. The solid line is the regression line from undisturbed corals in the field. The dashed line represents the minimum number of species found on undisturbed corals from 300 to 4000 cm³ in size. For example, three corals between 300 and 350 cm³ each were out for six days and each was colonized by a single species.

species number in 4 to 6 days. The 95% confidence interval about the regression would include other points, but the variance is great (see Figure 9.1). The main point is that corals can accumulate a diverse assemblage of species quickly.

Data from Lizard Island were collected in May and June 1980 during the winter and differ strikingly from those collected during the summer at Palau. First, there is no increase in average number of species per coral (Figure 9.2) with increasing time. Second, only 6 to 8 species total colonized the corals, although 43 species were collected from corals during this study. The colonists were small adults and juveniles plus one crab megalopa out of 50 individuals.

In contrast to Palau, the most abundant and most frequently occurring species at Lizard Island were among the best colonizers. However, as at Palau, some species such as *Alpheus lottini* had a high (76%) frequency of occurrence but were not observed to colonize (Table 9.5).

Five facts emerge from the colonization data: (1) colonization is rapid, occurring within 12 hr of defaunation; (2) colonists are almost exclusively small adults and juveniles; settling larvae are extremely rare; (3) neither the abundance nor the frequency of occurrence of crustacean species at

Palau correctly predicts their coral colonization pattern, while the more common species colonize more frequently at Lizard Island; (4) colonization seems to be mainly nocturnal; and (5) many corals reach the pre-defaunation species number but not composition within 4 to 6 days.

Assembly of Communities with "Too Many" Species

The data from species addition experiments are summarized in Table 9.6. The experiments at Palau were run for 1, 3, 4, and 6 days to follow any changes in species numbers. There was a loss of species over 24 hr: for two corals beginning with 4 and 5 species the loss of species was three (-75%) and four (-80%), respectively. Four corals with 10, 8, 7, and 6 species were examined at 72 hr with a percentage loss of -70, -63, -71, and -33% respectively. The decrease in species numbers for 96 hr ($n = 2$, -90, -62.5%) and 6 days ($n = 3$, -83.3, -80, -50%) was similar. The mean species loss was -64.3% ($n = 11$) or 54 of 84 species. The decrease in species numbers is striking because it was not solely caused by disappearance of casual associates of coral; both specialist and generalist species disappeared. The controls differed substantially, losing comparatively fewer species overall and with some corals actually gaining species. For example, a coral of 1155 cm^3 began with 4 species (*Trapezia dentata*, *T. cymodoce*, *T.* cf. *cymodoce*, and *A. lottini*), lost 1 species (*A. lottini*), and gained 2 species (*Synalpheus* sp. and a hermit crab) in only 24 hr. The percentage species change (number lost or gained/original number) and time period were: 24 hr (0, $+33$, $+25$, $+133\%$), 48 hr ($+33$, -33, -100%), 72 hr (0, -50%), 96 hr ($+33$, -50%) and 6 days (0, 0, -100%). The mean species loss for control coral heads was -22.6% ($n = 14$) or 7 of 31 species. This loss is probably a result of both handling and a general movement of species among corals.

It is clear that an upper limit to the number of species on corals exists, but the proximate mechanism setting this limit is not obvious. Predaceous fish, especially wrasses, are extremely abundant on reefs, and predation

Table 9.6. Summary of "supersaturation" experiments.

	N	Total Spp. No.	Species Lost	% Change
Lizard Island				
Control	10	28	9	-32.1
Caged	10	61	31	-50.8
Uncaged	10	60	36	-60.0
Palau				
Control	14	31	7	-22.6
Uncaged	11	84	54	-64.3

seemed a possible cause for the loss of species. The above experiments were repeated at Lizard Island, using cages with a mesh size (12 mm) that excluded predators but permitted crustaceans to enter and leave. Based on the results at Palau, the experiments were run for 6 days. The results in the caged and uncaged corals with "too many" species were very similar (mean species loss $= -50.8$ versus -60.0%, $n = 20$) and differed from the uncaged controls (mean species loss $= -32.1\%$, $n = 10$; Table 9.6). Crabs and shrimps that were exposed in broken open corals and those released into open water were quickly eaten, mostly by *Thalassoma lunata*, but because caged and uncaged corals did not differ in species loss, predation alone appears insufficient as a proximate cause for the loss of species from corals with "too many" species.

Observations on corals in aquaria revealed many instances of intra-specific aggression, the most obvious among individuals of *Alpheus lottini*, but also among species of *Trapezia* and *Periclimenes*. I did not see any comparable interspecific interactions, but because the observations were made at night my ability to see them may have been limited.

DISCUSSION

The increase in species richness from the Eastern to the Western Pacific involves both obligate and eurytopic species associated with *Pocillopora damicornis*. For example, 4 species of *Trapezia* are in the Eastern Pacific and more than 20 are in the Western Pacific, and all species in the genus are obligate symbionts of pocilloporid corals. Similarly, many more pon-toniid shrimps are in the Western than in the Eastern Pacific and many of these occupy pocilloporid corals. The question of interest is whether or not more species occur in the same habitat in the Western Pacific. Comparisons of decapod communities associated with *P. damicornis* are probably as close to within-habitat comparisons as is possible. Yet no indication that more species coexist per unit habitat in the Western than in the Eastern Pacific comes from my studies. *Pocillopora damicornis* has about the same number of crustacean species on it per unit coral head size, whatever the size of the species pool. There is no evidence of increased within-habitat partitioning in this habitat.

These results are similar to those of Heck (1979), who compared inverte-brate species richness in *Thalassia* in the northern Gulf of Mexico and Panama. Although the total number of species was greater in Panama, the actual number of resident species ($=$ breeding populations) was similar. Nelson (1980) compared the number of gammarid amphipods in grass-beds along the eastern United States and found no increase in species richness with decreasing latitude. Bohnsack and Talbot (1980) reported results similar to those reported here. They set up identical artificial reef habitats in southern Florida and on the southern Great Barrier Reef. Although about twice as many fish species live in the waters around Australia, they found no greater within-habitat diversity there.

If the number of species within habitats is similar in species-poor and species-rich regions, then how is the greater total diversity attained? One way is that the region with the lowest total species number (Uva Island) had the lowest species/individual ratio and the lowest variation in species composition among coral heads. For example, at Uva Island 52 species were represented by 4724 individuals, a species/individual ratio of 0.011. The 10 most common species had a mean occurrence on corals of 63.92% ± 27.9. At Palau 48 species were represented by 546 individuals for a species/individual ratio of 0.088. The 10 most common species had a mean occurrence on corals of 38.5% ± 19.3. Similar results are obtained for the Australian samples. In species-rich regions, therefore, it appears that species are represented by fewer individuals and are more patchily distributed.

Although the species composition varies widely among regions, a few groups are held in common. *Alpheus lottini* and species of *Trapezia* are found at all sites. Local autecological conditions apparently affect species composition. In the Bay of Panama (Pearl Islands), where upwelling occurs, filter feeding porcellanids account for 21% of the species. Porcellanids are absent (or at least rare; none was collected) in Palau, where the local primary productivity appears to be low. Abele (1979) suggested that differences in species composition between Uva Island and the Pearl Islands result from local environmental differences.

The colonization experiments suggest that open space is colonized by juveniles and small adults rather than by larvae from the plankton. In fact, postlarvae are extremely rare on coral heads. At Uva Island, Panama, three species of *Trapezia* breed all year, and with the exception of a single species during June, from 71 to 100% of the females are ovigerous throughout the year. Yet settling postlarvae are often completely absent from coral head samples. Of 119 coral heads collected during four time periods (January, April, June, August) only April and June had larvae present; no larvae were collected during January or August. In addition, adjacent coral heads will differ with respect to settling larvae; one may have 10 postlarvae and the other none. It is not clear why 22 species colonized heads at Palau while only 6–8 did so at Lizard Island. Finally, there was some relationship among abundance, occurrence, and colonization at Lizard Island, where the more abundant species colonized more frequently, but not at Palau. This situation may reflect the fact that the Palau data were collected during the summer and the Lizard Island data during the winter, but it is not obvious why this should affect colonization in these tropical regions.

Observations on the abundance of predaceous reef fish and the rapidity with which they ate crabs and shrimp exposed during collecting suggest that predation could affect the number and abundance of species on coral heads. If so, the number and abundance of decapods could be limited by the availability of predator-free space within the coral head. This idea was the basis for setting up corals with "too many" species in caged and uncaged treatments. If predators were removing exposed animals, then uncaged heads should have fewer animals than heads protected by cages. However, there is not a significant difference in loss of species between the caged and uncaged treatments ($t = 0.4$, $P > 0.5$, Table 9.6), while both treatments lost proportionately more species than the controls ($t = 2.5$, $P < .05$, combining controls). Predators alone do not appear to be the proximate cause for the limit on species numbers.

Three corals with "too many" species were set up in individual aquaria and observed. The species composition and abundance were identical to those of three corals used in the field. Intraspecific interactions among some of the species were quite obvious. Two female individuals of *Alpheus lottini* fought by snapping and pinching until one was injured and left the coral. A large male of *Trapezia aerolata* picked up smaller males in the chela and rapidly shook them. Individuals of *Periclimines amymone* appeared to move off the coral without any interaction. Other species, including some xanthids such as *Trapezia guttata*, moved off the coral also without any apparent interaction. These very preliminary observations suggest that individuals maintain some inter-individual distance among themselves and that a high density results in some individuals' migrating. It is not known at present if there is a hierarchy among species or among individuals within species. I am entertaining the hypothesis that predation pressure led to the evolution of behavior that maintains a low density on coral heads. This suggestion is consistent with the results of Coen *et al.* (1981), who examined competition between two species of caridean shrimp. The competitive dominant, *Palaemon floridanus*, excluded *Palaemonetes vulgaris* from dense stands of grass and algae, thus significantly increasing the latter species' risk of predation. The competitive displacement behavior occurred in both the presence and the absence of predators, but probably evolved under the influence of predation pressure. Additional experiments with the *Pocillopora* community are necessary to test for competitive dominance among and within species.

Acknowledgements

I thank P. Castro, S. Gilchrist, P. Glynn, N. Gotelli, K. L. Heck, Jr., J. Martin, D. Simberloff, and D. Strong for comments on the manuscript. Research at Palau was supported by the National Science Foundation

Alpha Helix program and the generosity of James Cameron. Field work in Australia was supported by the Australian Museum and the Lizard Island Research Station. Special thanks are extended to Hugh Sweatman, Lois and Barry Goldman, and Allan Young for assistance at Lizard Island. Computing facilities were provided by Florida State University Computer Center.

10.

Assembly of Land Bird Communities on Northern Islands: A Quantitative Analysis of Insular Impoverishment

OLLI JÄRVINEN and YRJÖ HAILA

Department of Zoology, University of Helsinki, P. Rautatiekatu 13,
SF-00100 Helsinki 10, Finland

Assembly rules of species communities (Diamond, 1975; Cody, 1978) based on assumed interspecific interactions within guilds have been derived in order to explain details of community structure. In this paper, our approach will be somewhat different. We wish to compare island bird communities on different-sized islands with those on the mainland (or an island that is several orders of magnitude larger), and, on the basis of census information and extensive autecological data available on Northern European land birds, we attempt to answer the following questions:

(1) How much of insular impoverishment can be understood in terms of habitat differences (*cf.* Abbott, 1980)?

(2) How many of the species absent from islands would also be absent from equal-sized mainland areas, owing to sparsity of the populations?

It seems to us that these questions are too often neglected in island studies, although variation attributable to their effects should be removed before one looks for variation due to interspecific interactions; questions (1) and (2) possess logical priority in this sense. Special ecological meaning ought not to be invoked for insularity if an area of similar size and of the same habitat composition on the mainland were equally impoverished. The above questions can be posed most meaningfully if quantitative estimates of population numbers are available (see Haila and Järvinen, 1981). In view of the multiple difficulties in censusing large regions (Järvinen and Väisänen, 1981), we point out that sufficient accuracy in quantitative estimates is more often to an order of magnitude than to $\pm 10\%$.

THE ÅLAND ISLANDS

The Åland Islands (60°N, 20°E) comprise thousands of islands, their areas ranging from 970 km² (Main Åland) to only a few square meters (Figure 10.1). Inter-island distances tend to be small. A strait of about

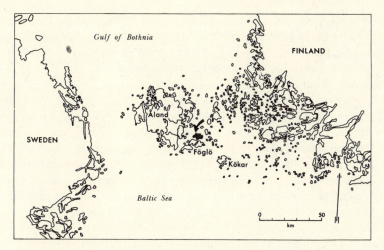

Fig. 10.1. The location of the Åland Islands (Finnish: Ahvenanmaa) between Sweden and the mainland of Finland. Notice the numerous "stepping stones" between the "mainland" of Åland and the Finnish mainland. Black arrow indicates Ulversö (blackened).

30 km separates the archipelago from Sweden, and the Finnish mainland is about 70 km away, but this distance is bridged by a dense chain of small islands. Haila *et al.* (1979) estimate that there are maximally 122 breeding bird species on Åland. More than 75% of all individuals are migrants in this part of Northern Europe (Järvinen and Väisänen, 1980).

Distribution maps indicate that more than 20 land bird species do not breed on Åland, but breed at the same latitude as Åland on the mainland (either Finland or Sweden or both). A detailed analysis is not given here owing to space restrictions, but three more or less equal groups can be distinguished, with a few species remaining as enigmatic cases.

(1) *Species avoiding the sea, i.e. those having problems with overwater dispersal.* Examples include the Green Woodpecker, *Picus viridis*, the Marsh Tit, *Parus palustris*, and the Nuthatch, *Sitta europaea*, which are common sedentary species in deciduous woodland in the adjacent Swedish mainland, but absent from Åland. Moreover, these species do not breed in Finland, where the suitable habitats occur in a region that is in a sense insular (see Järvinen and Väisänen, 1980). In general, the evidence here is straightforward. Habitats are available on Åland for many species that have never been observed on bird stations operating there; hence, overwater dispersal must be a problem.

(2) *Species absent for historical reasons.* Three species (Collared Turtle Dove, *Streptopelia decaocto*; Great Reed Warbler, *Acrocephalus arundinaceus*; and Tree Sparrow, *Passer montanus*) have recently expanded

their breeding ranges in southern Finland, but had not reached the Åland Islands by the late 1970's (non-breeding individuals of these species have been observed there, *P. montanus* being regular in winter). We expect that breeding populations will soon be established if the recent expansion continues. Four species have disappeared in recent times. Three cases (Peregrine, *Falco peregrinus*; Kestrel, *F. tinnunculus*; and Spotted Crake, *Porzana porzana*) are discussed by Haila *et al.* (1979). We include here the Partridge, *Perdix perdix*, even though it may never have been a regular breeder on Åland (Haila *et al.*, 1979). These four extirpations all refer to populations that have declined on the Finnish mainland in recent decades.

There are good reasons for supposing that changes in mainland populations are reflected in the populations breeding on Åland, *i.e.* that Åland is simply part of the mainland for most land bird species. Haila *et al.* (1979) found that (probably) 18 land populations have established permanent breeding populations on Åland during the past 50 years, and, with one exception, these species have also expanded their ranges and/or increased their numbers on the Finnish mainland. The quantitative contribution of these newcomers is, at least as yet, slight, for they contribute less than 1% of the breeding pairs (for data on Northern European newcomers in general, see Järvinen and Ulfstrand, 1980).

(3) *Species absent because of ecological impoverishment.* Åland has only tiny peatlands and very few rivers, and a number of habitat specialists seem to be absent for that reason. Examples include the Golden Plover, *Pluvialis apricaria*, and the Great Grey Shrike, *Lanius excubitor*.

In this comparison the composition of the mainland avifaunas of Sweden and Finland was regarded as given. Patterns in the mainland faunas may be produced by factors entirely different from those responsible for the impoverishment of the Åland avifauna; this kind of analysis is too weak to examine the possible role of interspecific interactions. Some examples illustrate this point. The Grey-headed Woodpecker, *Picus canus*, is more abundant on the Finnish mainland and on Åland than in Sweden, where it is a rarity. In Finland, the Grey-headed Woodpecker mostly occupies deciduous forests closely similar to those favored by the Green Woodpecker, *P. viridis*, in Sweden; the Green Woodpecker is absent from Åland and the Finnish mainland. Evidence for interspecific competition is strong in this case (see especially Svärdson, 1949; Haila and Järvinen, 1977). Another intriguing absence from Åland is that of the Ruff, *Philomachus pugnax*. At comparable latitudes in Sweden and in Finland, it breeds on shore meadows, abundantly available on Åland. It may be speculated that heavy predation on nests by a dense population of Hooded Crows, *Corvus corone*, is sufficient to prevent the establishment of the Ruff on Åland (*cf.* below).

We now make a similar comparison between Main Åland and a nearby island, Ulversö. Because extensive quantitative data are available, the analysis can be much more precise than that given above.

ECOLOGICAL CAUSES FOR INSULAR IMPOVERISHMENT: MAIN ÅLAND VS. ULVERSÖ

Ulversö has an area of about 5.8 km², while Main Åland is 970 km². Essentially everything mentioned above as regards the Åland Islands pertains to the largest island, Main Åland. We use Main Åland as a basis for our comparisons for two important reasons, one practical and the other biologically significant. First, Main Åland has been censused intensely (Palmgren, 1930; Haila *et al.*, 1979, 1980a, b; Haila and Järvinen, 1980). Second, Main Åland is certainly the most mainland-like equivalent to Ulversö, as regards climate, habitats, and other factors affecting the distribution of birds.

Breeding birds have been censused on Main Åland and Ulversö using the same methods and essentially by the same persons (see Haila and Järvinen, 1981). The censuses have been taken at similar times of the day, stages of the breeding season, and in favorable weather. Because most of our censuses from Main Åland are for 1975 and those from Ulversö mainly for 1976, control censuses were made on Main Åland in both years in order to examine the effect of annual fluctuations on the results (see Järvinen *et al.*, 1977). A more detailed presentation of these results will be presented elsewhere (Haila and Järvinen, 1983).

Rarity as a "Cause" of Absence

Small areas ought to support fewer species than large ones, for the simple reason that most rarities are expected to be absent from a small area for statistical reasons alone. However, this is not a satisfactory explanation for long-term absences; rarity as a "cause" for absence implies that, sooner or later, a species will breed on the island.

We first examined our data as a "snapshot" view of reality. We regarded a species as rare if it was observed no more than 5 times in the censuses made by us (Haila *et al.*, 1979) on Main Åland. The number of such rare species totaled 45, and 63 observations of them were made in 213.7 km of censuses. We observed many of these rarities on Ulversö, but no more than 5 pairs were included in our censuses covering 17.6 km on Ulversö (censuses made in 1976–80). This pattern conforms exactly to the statistical expectation (5.2 pairs). Ulversö thus supports as many rarities as expected. As 52 species are present on Main Åland (total 121 spp.), but absent from Ulversö (total 69 spp.), and as 38 of these 52 species are rarities, the absence of 38/52 (=73%) species, *in the short term*, can be understood as a statistical consequence of rarity; equal-sized areas of Main Åland would not be expected to support a higher number of them at a certain moment.

This conclusion is not trivial. For example, presence-absence data have been used in attempts to show that various orders, families, or genera differ in their colonization ability or extinction proneness. Because various taxa have different abundance patterns, conclusions based on taxa may be more closely related to numerical relations. Superficially examined, our data (Haila and Järvinen, 1983) suggest a significant (χ^2 test) difference in the ability of passerines vs. nonpasserines to colonize Ulversö:

	Passerines	Non-passerines
Ulversö and Main Åland	49 spp.	20 spp.
Main Åland only	22 spp.	30 spp.

But this tabulation proves nothing about differences in colonization ability. Nonpasserines tend to be less abundant than passerines; small mainland areas are therefore expected to support comparatively fewer non-passerines than passerines. Excluding the 38 rare species, the above tabulation looks entirely different (no significant difference between passerines and non-passerines):

	Passerines	Non-passerines
Ulversö and Main Åland	48 spp.	14 spp.
Main Åland only	9 spp.	5 spp.

This result warns against drawing conclusions about differential colonization abilities of different taxa of the basis of qualitative (presence-absence) data.

There is another way to check our conclusions. Data based on censusing all the major terrestrial habitats on Main Åland included 3134 pairs and 82 species (Haila *et al.*, 1980b). Our censuses indicate that Ulversö has no more than 2500 pairs of land birds and about 65 species annually. Rarefaction analysis (Simberloff, 1979) of the census data from Main Åland indicates that a random sample of 2500 pairs is expected to include about 80 species (77 to 82 species as 95% confidence limits). As Ulversö has no more than about 65 species annually, the difference (about 15 species) cannot be accounted for by "rarity."

We observed above that Ulversö lacks 52 species that are present on Main Åland, and 14 of these were not included in the class of rare species (Table 10.1). We calculated the expected numbers of pairs of these 14 species on Ulversö as the average density on Main Åland (data in Haila *et al.*, 1979) times the area of Ulversö. We then assumed that the number of pairs on Ulversö is a Poisson variate, which is a reasonable

Table 10.1. The 14 land bird species that are absent from Ulversö (1976–80) but were observed more than 5 times in the censuses made by Haila *et al.* (1979) on Main Åland. The expected number of pairs breeding on Ulversö is based on the average density on Main Åland. $P(0)$ gives, according to the Poisson distribution, the probability that the species in question is absent in one year. See text.

Species	Expected on Ulversö	P(0)
Hazel Grouse, *Bonasa bonasia*	5.2 pairs	0.005
Curlew, *Numenius arquata*	1.4	0.23
Green Sandpiper, *Tringa ochropus*	1.9	0.15
Grey-headed Woodpecker, *Picus canus*	1.9	0.14
Great Spotted Woodpecker, *Dendrocopos major*	3.5	0.03
Sand Martin, *Riparia riparia*	0.6	0.55
Meadow Pipit, *Anthus pratensis*	2.5	0.08
Jackdaw, *Corvus monedula*	2.8	0.06
Wren, *Troglodytes troglodytes*	3.3	0.03
Sedge Warbler, *Acrocephalus schoenobaenus*	5.5	0.004
Chiffchaff, *Phylloscopus collybita*	23.5	0.00
Whinchat, *Saxicola rubetra*	6.4	0.001
Mistle Thrush, *Turdus viscivorus*	2.2	0.01
Long-tailed Tit, *Aegithalos caudatus*	5.5	0.004

assumption for rare species. As our data refer to 1976–80, that is, five breeding seasons, the Poisson estimate for having 0 pairs is too conservative. On the other hand, the fifth power of the probability of having 0 pairs in a given year is not a valid estimate for having 0 pairs in all five years, because five consecutive breeding seasons are autocorrelated. Assuming that a realistic power exceeds 2 (*i.e.* that our data for five seasons are equivalent to at least two independent breeding seasons), the probability of having 0 pairs on Ulversö by chance alone is very low for 13/14 of the common absences (Table 10.1). The deviating case is the Sand Martin, *Riparia riparia*. As it is a colonial species, the assumptions of a Poisson distribution are violated anyway; therefore, *Riparia* will be included in rarities in the following analysis.

All the rare species currently absent from Ulversö may not be absent from there forever. Indeed, autecological facts lead us to expect that many rare species will breed on Ulversö in the future, while others have real biological difficulties in colonizing Ulversö. A more detailed examination (Haila and Järvinen, 1983) indicates that 17 rare species will be absent for long periods, but there seems to be nothing to prevent those species from breeding every now and then. In the following, we will examine the 13 fairly common species that are absent from Ulversö.

Historical Reasons for Absence

Since the land bird fauna of the Åland Islands has changed substantially in recent decades (Haila *et al.*, 1980b), historical reasons may account for absences; "historical" here refers to recent changes in the species range or numbers in Northern Europe. For example, the Grey-headed Woodpecker has recently colonized Åland (Haila and Järvinen, 1977). We postulate historical reasons here, for our data (Haila and Järvinen, 1983) actually document the process of colonization (probable breeding on Ulversö in 1980).

Another possible example is the Hazel Grouse, introduced to Main Åland in 1927–29. The sedentary Hazel Grouse may not yet have reached Ulversö. Among the rare species, the Pheasant, *Phasianus colchicus*, is certainly a poor colonizer. If it ever colonizes Ulversö, this will probably be a result of introduction, as elsewhere in Northern Europe.

Unavailability of Suitable Habitats

Habitat differences between islands and the mainland probably occur frequently, even though they are not often discussed. Among the several major differences between Main Åland and Ulversö, the lack of spruce forest on Ulversö is conspicuous. Also, the landscape is more mosaic-like, the undergrowth denser, and the trees are shorter on Ulversö, as compared with Main Åland. The habitat requirements (Palmgren, 1930; Haila *et al.*, 1980b) of certain species seem to explain their absence from Ulversö. For both the rare and common species, this category probably accounts for about one-half of the absences (Haila and Järvinen, 1983). Examples include the Curlew, *Numenius arquata* (fields and shore meadows too small on Ulversö), the Green Sandpiper, *Tringa ochropus*, favoring marshy woods usually dominated by tall spruce (absent from Ulversö), and the Chiffchaff, *Phylloscopus collybita*, which occurs in tall spruce in this archipelago. As regards the rare species, the habitat factors of importance include the scarcity of spruce on Ulversö, the absence of tall and thick trees, the sparse human population, insufficiently large areas of suitable habitat, etc.

Other Reasons for Absence

Among the common species absent from Ulversö, five cases are more puzzling. Competitive exclusion of a congener seems the most likely explanation for the absence of the Sedge Warbler, for the Reed Warbler, *Acrocephalus scirpaceus*, is abundant and demonstrably able to oust the Sedge Warbler from the reed-beds (Svensson, 1978). Two species breeding on shore meadows, the Whinchat and the Meadow Pipit, may be absent owing to intensive nest predation by Hooded Crows. This species, a very important nest predator in the region, is unusually abundant on Ulversö, as judged from census data. Two absences are enigmatic. The Great Spotted Woodpecker is a habitat generalist, but is usually rare in the Finnish archipelago. The Long-tailed Tit breeds in lush deciduous woods, which are abundantly available on Ulversö, but the Long-tailed Tit is absent.

Conclusions

We conclude that, of the 52 species present on Main Åland but absent from Ulversö, most are rarities that would not be expected to breed in an equal-sized area of the mainland in a given year. We tabulate below the percentage distributions for various reasons for absence. The short-term column refers to a "snapshot" view of the community on Ulversö, while the long-term view takes into account the fact that rarity as such is not a reason for permanent absence. The species in this category in the long-term column are thus expected to breed on Ulversö, sooner or later, as no special reason seems to prevent them from colonizing Ulversö. The distribution is as follows:

	Short-term	Long-term
Rarity	75%	33%
History/Dispersal	3%	9%
Habitats	12%	47%
Interspecific competition	2%	2%
Predation (of nests)?	4%	6%
Other	4%	4%

HOW FAR CAN THE REDUCTIONIST PROGRAM BE EXTENDED?

Our analysis is reductionist in the sense that we interpret insular impoverishment in terms of autecological response, without invoking any community-level explanations. It does not follow that all problems of insular distributions are amenable to a reductionist analysis, but we do not see any reasons why this kind of approach ought not to be pursued more consistently in island biogeography than is presently the case. As our data base extends to single habitats on Ulversö and Main Åland, we have been able to continue the analysis further. As shown by rarefaction, equal-sized samples from an equivalent habitat on Main Åland and Ulversö tend to differ in species richness, in favor of Main Åland. If the species absent from Ulversö are eliminated from the samples and the rarefaction analysis is repeated, only one of the six habitat pairs examined reveals a significant difference. In that case the lower height of the trees on Ulversö probably plays a role in eliminating a few extra species from this habitat (Haila and Järvinen, 1983).

Despite considerable differences in species numbers, three habitats had identical total densities on Ulversö and Main Åland, but three Ulversö habitats supported significantly higher densities than their Main Åland equivalents. This observation is not related to "density compensation" often reported from island communities, for the excess densities were entirely unrelated to the numbers or the densities of the missing species. Instead, subtle habitat differences (*e.g.* substantial edge effect on Ulversö owing to general patchiness of the habitats there) are a more plausible hypothesis.

What happens if this reductionist analysis of impoverishment is extended to smaller islands in this archipelago? We (Y. Haila, O. Järvinen, S. Kuusela, in press) have censused nearly 50 islands in the neighborhood of Ulversö, many of them in several breeding seasons. Including all data for small islands (less than 20 ha) gives a total area of almost 5 km^2 (Ulversö is 5.8 km^2). We observed 69 species on Ulversö, but only 39 on one or more of the small islands.

One-third of the missing 30 species are dependent on man either directly or indirectly. As the small islands are not inhabited and do not have fields, these absences have a simple explanation. Absence of certain forest habitats seems to account for a great number of the remaining cases. Instead of presenting a breakdown of our results, we refer to a quantitative technique developed by us (Haila and Järvinen, 1981) for analyzing minimum area requirements. We construct "prevalence functions" on the basis of average densities of a species on different-sized islands. The average densities are compared with the density on the mainland (or any suitable reference area). If prevalence equals 1, the species is equally abundant on the islands and on the mainland; deviations from unity can be tested statistically.

Prevalence functions allow us to define different colonizing strategies. There may exist species that favor small islands and their patchy habitat, and other species may indeed require large islands with continuous tracts of habitat. Such patterns are difficult to establish conclusively without quantitative data. Many rare species may look as if they were restricted to large areas, but this may be a statistical artifact only (see our above analysis on rarity as a "cause" of absence).

It is feasible to refine prevalence functions if quantitative data are available from different habitats: the functions can be so adjusted that gross differences in the percentage coverage of different habitats are eliminated. For example, the high prevalence of the Whitethroat, *Sylvia communis*, on small islands (Figure 10.2) is probably nothing but a consequence of the high percentage of its optimum habitat on these small islands. Gross habitat differences do not account for the pattern in the Blackcap, *S. atricapilla* (Figure 10.2). However, this is also a habitat

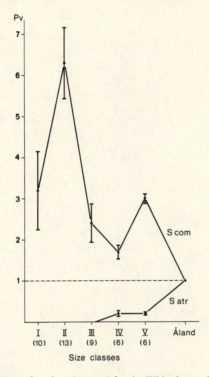

Fig. 10.2. Prevalence functions (see text) for the Whitethroat, *Sylvia communis* (S com), and the Blackcap, *S. atricapilla* (S atr), in the archipelago near Ulversö in 1976–79. Island classes were defined on the basis of the numbers of species (the group boundaries are 1–5, 6–10, 11–20, 21–30, and 31–60 spp.). Prevalence estimates, calculated for the pooled data of all islands in the same class, are shown with approximate estimates of S.D. (see Järvinen, 1976).

problem, for the tall forest favored by this species on the mainland is absent from small islands. Our results indicate that in many, probably in most, cases, habitat variation between islands or between islands and the mainland is responsible for the assembly of our northern insular communities of land birds.

Marine Community Paradigms

11

Paradigms, Explanations, and Generalizations in Models for the Structure of Intertidal Communities on Rocky Shores

A. J. UNDERWOOD and E. J. DENLEY

Department of Zoology, School of Biological Sciences, The University of Sydney, New South Wales, 2006, Australia

"The invention of alternatives to the view at the center of discussion constitutes an essential part of the empirical method." (Feyerabend, 1975, p. 41)

INTRODUCTION

Most modern investigations of intertidal ecology search for general mechanisms and processes to account for observed patterns of distribution and abundance of and the intensity of interactions among component species of intertidal communities. The early history of this endeavor consisted of detailed observation of patterns of distribution of intertidal biota (summarized by Lewis, 1964; Ricketts *et al.*, 1968; Stephenson and Stephenson, 1972). Some early workers, notably Hatton (1938), set the scene for the modern era of controlled manipulative experimentation, a methodology that has become widespread and highly productive for unraveling the complex interactions among the components of a community (*e.g.* Connell, 1961a, b, 1970; Dayton, 1971, 1975a; Paine, 1974; Menge, 1976; Underwood, 1978a) as reviewed by Connell (1974) and Paine (1977). The result of such experimental studies has been a series of generalizations and/or models about the nature and organization of assemblages of species on rocky shores (see particularly Paine, 1966, 1977; Connell, 1972, 1975; Levin and Paine, 1974; Menge and Sutherland, 1976; Lubchenco, 1978). These aim to integrate the findings of descriptive and experimental studies into a corpus of theory and, from this corpus, to propose hypotheses of wider scope or greater precision than can be achieved from individual small-scale investigations. We are in complete agreement with this intention, and believe that it is essential for the logical development of ecology as a "hard" science. Some of the proposed generalizations have already been widely accepted, and even raised to the status of paradigms of modern shallow-water benthic ecology (see particularly Peterson, 1979).

The most recent development is, however, the appearance of critical reevaluations of some of the theories, experimental designs, and hypotheses that underlie the development of these so-called paradigms of intertidal community structure. Independently derived areas of doubt are generally taking shape (see Dayton, 1979a; Denley and Underwood, 1979; Dayton and Oliver, 1980; Underwood, 1980; Underwood and Jernakoff, 1981). Dayton, in his characteristically perceptive manner, has already drawn attention to the apparent role of experimental work as a confirmation of prevailing hypotheses, rather than an attempt to falsify rigidly proposed alternatives.

Here, we examine some of the current generalizations made from small-scale experiments because we believe that the all-embracing acceptance of untested generalizations from modern experimental studies is not particularly helpful as an aid to dynamic development of theories of community structure. It may even prove as constricting to the development of our understanding of the ecology of benthic marine biota as was an earlier series of models based on physical factors as determinants of structure of intertidal communities (see discussion by Dayton, 1971; Wolcott, 1973; Underwood, 1978b). In this respect, we are influenced by the arguments of Feyerabend (1963, 1975) that alternative hypotheses, in conflict with accepted theory, are a powerful tool for advancement of knowledge and for progress in new directions. We suggest that prevailing paradigms should always be examined critically.

We are also concerned that current interpretations of the structure of intertidal communities are being developed as an area of normal science (*sensu* Kuhn, 1970), or a series of research programs (see Lakatos, 1970), constrained within rigidly accepted paradigms and subject to uncritical acceptance (*not* particularly by the proponents, but by their followers). Our opportunities to investigate intertidal communities in eastern Australia allow us to evaluate some of these generalizations on a different assemblage of biota. This investigation should provide some opportunity for independent assessment of the studies used to generate the hypotheses.

It appears that any generalization about, or model for, the structure of intertidal communities will be useful or successful (*i.e.*, accurate, predictive, justified) if it conforms to the following criteria, which will be exemplified in the body of this paper.

1. The data on which it is based must be adequate to identify the patterns and processes that are used in the development of the model. In particular, no relevant and biologically meaningful alternative hypothesis should exist that satisfactorily accounts for the existing observations and

interpretations, yet has not been considered (or tested) in the development of the model under consideration.

2. The models or generalizations should be able to withstand rigorous testing in new, as yet untested situations (*e.g.* on different continents).

3. The model should not be phrased in such general, ambiguous, or qualitative terms that it is effectively tautologous or empirically irrefutable. Under this criterion, any predictive hypothesis from the model must be potentially falsifiable in realistic (biologically natural) situations. This is a statement of our acceptance of Popper's (1963) view of falsifiability and validity of hypotheses. Dayton (1979a) has already explored some aspects of this criterion in an excellent discussion of thought processes in modern ecological investigations; see also Peters' (1976) attempts to introduce reason into the emotional appeals of neo-Darwinist metaphysics.

A further ideal for the present discussion is that it should not disintegrate into, or be taken to imply, destructive, negative criticism. Where possible, we attempt to suggest alternative hypotheses or interpretations. These will then, we hope, become subject to critical examination and experimental testing by others. The criticisms raised here are not intended to stand as any denigration of those whose work has led to a search for general models for the organization of intertidal communities. This statement is made because previous experience warns us that paranoia among researchers is more widespread than the gifts of insight and creativity in hypothesis-building. We hope the tendency will be toward improvement of the models and a more accurate understanding of nature. These can only be achieved by modification of the ideas of those who have ventured beyond the constraints of individual small-scale empirical studies.

GENERALIZATIONS ABOUT VERTICAL LIMITS OF DISTRIBUTION (ZONATION) OF A SPECIES

Upper Limits of Distribution are Primarily Determined by Physical Factors

The evidence. This statement can be found in a number of papers, particularly Connell's (1972, 1975), and is implicit in his model for the structure of natural communities. A species is prevented from extending its range to higher levels on a shore because those individuals that settle and metamorphose above the upper limit of distribution (or, in the case of motile animals, migrate there) are killed by factors such as desiccation, insolation, or high temperatures during periods of emersion. Considerable evidence reviewed by Connell supports the statement for various types of organisms, including barnacles (*e.g.* Hatton, 1938; Connell, 1961a), algae (*e.g.* Moore, 1939; Castenholz, 1963), and some gastropods (*e.g.* Frank, 1965; Sutherland, 1970; Wolcott, 1973).

This notion has become so widely accepted that it is now assumed to be correct for many intertidal species where alternatives have not been investigated. For example, Paine (1974), in his extensive study of *Mytilus californianus*, suggested that the upper limit of distribution of the mussels was probably set by desiccation, though no data were sought to test this hypothesis.

Alternative hypotheses and explanations. Other processes are confounded with the direct effects of physical factors at high levels on a shore, and quite different mechanisms have been demonstrated to limit distributions of some species. For example, animals living at high levels on a shore may be killed by starvation during periods of calm weather, because they are not submersed for sufficient periods to gain adequate supplies of food. Algal grazers may not be abundant because the algae are themselves killed by the direct effects of physical factors; the grazers may be distributed according to the supply of food (see Underwood, 1979). Some predatory animals are absent from, or inefficient at, high levels apparently because they have insufficient time whilst submersed to find and handle their prey (*e.g.* Connell, 1970; Paine, 1974). Here, the predators are limited, and the prey are safe, and thus the structure of the community is influenced by the indirect effect of physical factors on foraging; it is not that predators are unable to exist at high levels.

Apart from these considerations, there are examples of species that are not limited by mortality due to the effects of the physical environment. The Australian barnacle *Tesseropora rosea* did not settle above the upper limit of distribution of adults, nor did the spat settling at the highest levels die because of physical factors (Denley and Underwood, 1979). In that study, although we could not always count newly settled cyprids until 3 days after settlement, because of problems with identification, we could always determine, on a daily basis, whether any larvae had settled. Therefore, lack of spat in a particular area must have been due to lack of settlement, or to the complete mortality during one day of any spat that settled.

We were unable to determine whether the lack of settlement at high levels on the shore was due to an active choice of settlement sites at lower levels, or to there being a very short period of inundation at the high levels, which would reduce the probability of the larvae settling there. In general, intensities of settlement increased toward the bottom of the shore (see also Hatton, 1938; Connell 1961a), and the latter explanation is more likely. Whatever the cause, the upper limit of distribution of the newly settled barnacles was determined by processes affecting the settlement of the larvae and not mortality after settlement, unless all recently settled spat died and disappeared within one day of settlement.

This may be a more common phenomenon than is currently recognized because the process of settlement has rarely been observed. It is therefore

impossible to determine whether the ultimate distribution is due to settlement of larvae in certain places, or whether larvae settle haphazardly and
differential mortality in different areas then occurs (see the detailed discussion for gastropods in Underwood, 1979).

For some motile organisms, behavior keeps the animals in places where
they are not subject to lethal environmental stresses. Wolcott (1973) has
argued that upper limits of distribution are determined by death due to
physiological stress only for the highest species of limpets on a Califor-
nian shore. The behavior of species at lower levels confines them to "safe"
microhabitats.

Another exception to the generalization concerns low-shore algae. On
sandstone shores in New South Wales, experimental exclusion of grazing
gastropods has demonstrated that the grazers are directly responsible for
the upper limit of distribution of several species of algae (Underwood,
1980; see also Hay, 1979). These algae survived the rigors of the physical
environment when grazers were removed.

A methodological problem. In experimental studies on the upper limits
of distribution of species, evidence from experimental transplantations of
animals or plants to higher levels is often used (*e.g.* Hatton, 1938; Foster,
1971; Schonbeck and Norton, 1978). Where transplanted organisms died
as a result of physical stresses, it was concluded that physical factors
limited their distribution. Such experiments only reveal sources of mortality of organisms moved outside their normal zone and do not tell us
anything definite about reasons for the absence of organisms from such
areas (see also the discussion in Underwood, 1980). As a trivial analogy,
would the death, from excess heat, of polar bears transplanted to the
Sahara desert really reveal the causes of the southern geographical limit
of the bears? Transplant experiments must be interpreted with care and
not used to substantiate hypotheses that are not being tested. Potentially
incorrect interpretations of these transplant experiments are examples of
the role of preconceptions in experimental design and interpretation discussed by Dayton and Oliver (1980). We suggest that observations on
settling propagules (spores or larvae) and their subsequent fate at the
upper margins of distributions are essential for correct interpretations to
be made. At the moment, such information is sparse for many taxa, and
new methods are urgently needed to tackle this question. As an example,
Fretter and Manly (1977) managed to observe the settlement of a snail,
by examination of rocks between waves. Further examples would be very
welcome.

Conclusion. Thus, we are confronted by a generalization about the role
of physical factors that applies to certain species, and may, in fact, not
apply to these organisms at all times and places. Other species are limited
by different processes and, for the majority of intertidal species, no data

exist to eliminate other appropriate hypotheses. Does this imprecision matter? The answer to this question depends upon how accurately we wish to understand nature. Increased rates of mortality in harsh physical environments feature in models of structure of intertidal communities (see later discussion). The untested assumption that limits of distribution are caused by mortality due to physical factors tends to reinforce the view that it is important. Wolcott (1973) has developed a logical argument to account for differences between high- and low-shore limpets based on natural selection by physical factors operating on the behavior of the species. If we are to understand fully the processes of selection operating upon characteristics of life-histories of intertidal organisms, we must be able to identify the sources of mortality properly, and not be misled by our assumptions.

Generalizations about the causes of limits to distribution can affect interpretations of otherwise excellent experimental studies. For example, Connell (1970, p. 55) was unable to assess the relative importance of lack of larval settlement, as opposed to mortality caused by physical factors after the larvae had settled, as causal factors reducing the density of *Balanus glandula* at high levels on a shore. Yet he concluded (p. 74) that the upper boundary of distribution was probably set by desiccation. The prevailing generalization overrode the cautious interpretation made earlier in the paper.

Lower Limits of Distribution are Primarily Determined by Biological Interactions

The evidence. This is the conclusion from many studies (*e.g.* Connell, 1961a, b, 1970, 1972, 1975; Dayton, 1971; Paine, 1974; and many others). Two processes have been identified as determinants of the lower boundaries of distribution—predation and competition, as discussed by Connell (1972, 1975), who dealt fully with predation. We consider here an alternative hypothesis for some species, that the pattern of settlement of larvae determines the lower limit of vertical distribution. We then examine the nature of the competitive interactions found in experimental studies.

Alternative hypotheses and explanations. The hypothesis that the pattern of larval settlement determines the limits of distribution of some species has not been examined in many studies, and care is needed for the interpretation of data where this hypothesis has not been eliminated by appropriate data. As an example, consider Connell's (1961b) extremely important and influential work on the barnacle *Chthamalus stellatus* and its superior competitor *Balanus balanoides*. Few *C. stellatus* settled below mid-tidal levels, and Connell transplanted rocks with attached barnacles from higher levels on the shore. *Balanus* then settled on these stones, and on rocks at higher levels where *Chthamalus* was present. The *Balanus* killed many *Chthamalus* by undercutting or smothering them. Yet, when the experiment ended, after one year, many *Chthamalus* were still alive in all experimental areas except one (Connell, 1961b, Figs. 2 and 3, Tables 2 to 4). The exception only had 9 *Chthamalus* at the start. Although no data were available to confirm this conclusion, Connell concluded that all the *Chthamalus* would eventually be killed by *Balanus*, because the early crowding by *Balanus* was so deleterious to the size, shape, position, etc. of the surviving *Chthamalus* that they would be unlikely to survive subsequent periods of crowding by the superior competitor. The rates of mortality of older *Chthamalus* were considerably less than the rates of younger barnacles (Connell's Table 3), and again some *Chthamalus* survived in all areas examined. Some *Chthamalus* survived at all levels in the area studied and competition did not reduce species richness in any experimental area (except one). Thus, *Balanus* was not responsible for the lower limit of *Chthamalus* even though the former had a profound effect on the survival and growth of the latter. Because few *Chthamalus* settle below mid-tidal levels, even if *Balanus* were absent from the lower levels, the vertical distribution of the upper species would not be affected. In fact, *Chthamalus* even survived competition from *Balanus* in areas where the former did not naturally occur. Connell concluded that lack of settlement or early mortality of larvae of *Chthamalus* below mid-tidal levels, and the effects of crowding by *Balanus* were the principal causes of the lower limit of distribution of *C. stellatus*. Although Connell emphasized the role of competition in this example (*e.g.* Connell, 1961b, Fig. 5, 1971, 1972), there is no excuse for the data in this study to be extrapolated, as in some recent textbooks (*e.g.* Colinveaux, 1973, p. 306), to suggest that *Balanus* eliminated all *Chthamalus* from areas where they had settled.

Very much related to choices of settlement site and subsequent mortality of newly settled larvae is the type of competitive process that can affect limits of distribution of intertidal species. In many instances the pre-emption, or prior exploitation, of the resource (space) at lower levels by an organism prevents the settlement of other species, which are then confined to higher levels. An example is the barnacle *Tesseropora rosea*, which did not settle below the levels where adults are found, apparently because it is unable or reluctant to attach to substrata already occupied by foliose algae or sessile animals (Denley and Underwood, 1979). In this case, interference competition from other occupiers of space did not eliminate young barnacles from low-shore areas. Rather, the lower limit of vertical distribution was determined by the settlement of larvae, which only occurred above levels already occupied by other species.

This pre-emptive competition has been found in other studies. For example, the brown alga *Fucus distichus* did not settle or attach to the holdfast ("crust") of *Chondrus crispus* (Lubchenco, 1980), but grazers were also involved in the rate of settlement. Lubchenco (1980, p. 339), however, concluded that *Fucus* were outcompeted by *Chondrus*.

Sousa (1979b) drew attention to the need to distinguish pre-emption or prior exploitation from situations where one species eliminates others by interference (*e.g.* the barnacles and mussels discussed by Connell, 1975, and Paine, 1974). In Sousa's (1979a, b) experiments, pre-emption of space by *Gigartina canaliculata* was apparently the reason other algae were not able to invade the surfaces of undisturbed boulders. The importance of preemptive as opposed to interference competition is discussed below after consideration of a problem with experimental designs for investigation of lower limits of distribution.

A methodological problem. Where one species dominates the space below the lower boundary of another species, the hypothesis tested is usually that competition with the lower species prevents the downward spread of the upper one. A common procedure is to clear the lower, dominant species from patches of substratum (*e.g.* Dayton, 1971; Lubchenco and Menge, 1978; Menge, 1976; Denley and Underwood, 1979; Lubchenco, 1980). If the upper species recruits to the cleared areas (by settlement or migration) and is subsequently killed by interference from members of the dominant species that reinvade the area, it is concluded that competition eliminates the inferior, upper species. This conclusion is unarguable, but may be an artificial result. Unless such clearings occur naturally, the experimental treatment is not a valid test of the causal mechanism of exclusion of the upper species (see also Denley and Underwood, 1979). In many studies, the situation in untouched, control areas has not been described, yet the interpretation of interference competition has been given. The recruitment of an upper species to natural areas below its distribution must also be monitored, and, if cleared areas do not occur frequently in the lower areas, care is needed to interpret the experimental data. Even if clearings do occur naturally, they would have to be the dominant type of habitat before the processes occurring in them could be claimed as the cause of the lower limit of distribution of some organism.

Conclusion. We argue that larval settlement may be an important contributor to the causes of lower limits of distribution of some species, and this hypothesis needs to be tested by appropriate observations and experiments for any particular species. Direct observation of the settlement of larvae is necessary, as for the upper limits discussed earlier.

Further, we draw attention to the role of pre-emptive competition, or prior exploitation of resources, as a very different mechanism from interference competition, as a process eliminating species from some areas. This is not a trivial, semantic difference of mechanisms. Where interference by dominant species removes competitively inferior species, the latter are killed (by smothering, undercutting, etc.; see Connell, 1961b; Dayton, 1971; Paine, 1974). Such competition may occur even if there is bare space available, provided only that the superior competitor is adjacent to the inferior one. Where pre-emption occurs, however, space is not available, because it is already occupied, and the "inferior" species cannot recruit. It is usually impossible to determine whether recognition of occupied space actually occurs in the field such that larvae or spores do not attach to the substratum. If this recognition does occur it is reasonable to conclude that some, if not all, propagules encountering an occupied area of substratum will recruit to free space elsewhere if it is available. Pre-emption may therefore be of less significance as a source of natural selection for dispersive species than is interference. It may even be a relationship having no deleterious effect on the numbers of a competitively inferior species, if its propagules can all find an alternative, suitable place to settle.

Pre-emption of space, leading to no recruitment of putatively inferior species, may not be a competitive process at all. Although a necessary resource, free space, is not available to subsequent arrivals, predation is sometimes the mechanism preventing them from invading the occupied substratum. Dayton (1971) suggested that the anemone *Anthopleura elegantissima* could cover an area, and might be an efficient competitor by preventing recruitment of potentially superior species. Given that anemones often eat small planktonic animals, one method of preventing recruitment would be the ingestion of the recruits. Similarly, some barnacles consume large numbers of barnacle larvae (D. T. Anderson, pers. comm.) and mussels are extremely efficient consumers of planktonic larvae (*e.g.* Thorson, 1950). Thus, dominance of substrata by filter-feeders or anemones may be followed by predation, not competition, as a means of keeping subsequent invaders out.

The "grab-and-hold" strategies (Sousa, 1979b) of predatory users of space and of non-predators, such as algae, may be very different, and may have quite different consequences for the structure of communities. Mussels, for example, may require certain species to be on the substratum for successful settlement (*e.g.* Seed, 1969; Dayton, 1971; Paine 1974; Menge, 1978b). Invasion of algal beds may be easier than recruitment to mussel beds.

The order in which species colonize, as well as the mechanism for preventing subsequent invasions, is likely to be an important component of the dynamics of intertidal communities. Where pre-emptive competition is an important process, the order of arrival of colonists is of major importance—the first arrival can prevent other species from invading, unless the later arrivals are able to outcompete an earlier colonist by interference.

GENERALIZATIONS ABOUT ABUNDANCES, MORTALITY, AND AGE-STRUCTURE OF POPULATIONS

Several interrelated generalizations are considered here. These form the basis for models for the structure of intertidal communities developed by Connell (1975), Menge and Sutherland (1976), and Menge (1976). In these models, the structure of communities is the end result of mortality of potentially dominant species (*i.e.* the most abundant, or those occupying a major portion of the available space). Mortality due to physical factors, predation and competition, and the interactions between these processes are integrated into the formation of the models, leading to the following generalizations:

A. In habitats with relatively benign physical environments, predation is the dominant biological interaction structuring intertidal communities.

B. With increasing environmental harshness, the effectiveness of predation is decreased and competition becomes a major process structuring communities.

C. When environmental harshness is much greater, competitive interactions are of decreased importance, and physical stresses are of major importance in determining community structure.

D. Local escapes from controls on abundance by predators (in benign environments) or by physical stress (in harsh environments) are a major source of patchiness, and can determine the dominance of different patches, or the presence of dominant age-classes in a population.

Two major gradients of environmental harshness have been considered by Connell (1972, 1975) and Menge (1978a, b). The first is the increasing harshness, defined in terms of physiological stress at increasing heights on the shore. Thus, the effects of desiccation, temperature, and other associated factors generally increase with increasing height above low water (*e.g.* Hatton, 1938; Connell, 1961a; Wolcott, 1973; Branch, 1975; and many others). Low levels on the shore are generally considered to be benign, relative to higher levels. The second gradient of harshness is that of increasing wave energy, from sheltered or protected areas (usually considered to be benign) to wave-exposed (harsh) areas.

Here, we consider exceptions to some of the generalizations about predation and competition along environmental gradients, and discuss the particular circumstances where these generalizations are likely to be valid.

Predation in Benign Environments

The evidence. The abundance and/or efficiency of intertidal predators tends to be greater at lower than at higher levels on a shore (Connell,

1961a; Paine, 1969, 1971, 1974; Dayton, 1971; Feare, 1971; Menge, 1976). Predators are apparently prevented from consuming prey at higher levels because of constraints on the period of foraging due to the pattern of tidal rise and fall. Thus, prey species often have a refuge above the levels where predators are efficient. Unless the abundances of the prey at these higher levels are reduced by the direct effects of physical or other factors, competition may occur among them for the limiting resource of space (see above references).

The gradient of physical harshness associated with increasing wave-shock has been studied intensively and experimentally by Dayton (1971), Lubchenco and Menge (1978), and Menge (1976, 1978a, b). The role of predatory whelks, which eat barnacles and mussels, has been examined in detail by Menge (1978a, b).

Alternative situations and factors modifying the impact of predation in benign environments. There are examples where consumers are able to forage more efficiently and with most impact in the harsher parts of the prey's environment, but have little effect where conditions are benign, in contrast to the generalization (C) above. The spores and small sporelings of algae are consumed by gastropods at mid-tidal levels where conditions are harsh for the algae (Underwood, 1980). At lower levels on the shore, conditions are considerably more benign for algae, which are able to grow more quickly and rarely suffer from desiccation or other factors associated with conditions during low tide. Here, limpets are relatively ineffectual at consuming the plants, and are apparently excluded by the rapid colonization and growth of the algae (Underwood and Jernakoff, 1981). A similar phenomenon has been reported by Dayton (1975a) and Sousa (1979b), although predation reduced the density of grazers at low levels on the shore in these latter studies.

As a second consideration, predation can be more effective in benign environments only where predators have narrower ranges of physiological tolerance to conditions during low tide than do their prey. Otherwise, the prey would not have a refuge from predation in the harsher areas. In most studies cited earlier the predators were apparently unable to remain with their prey in harsh areas (*e.g.* the starfish *Pisaster ochraceus* retreats to lower levels on the shore when the tide ebbs; Paine, 1974), or cannot continue to feed during harsh periods (*e.g.* the whelks studied by Connell, 1970). Connell (1971) has suggested that this is a general phenomenon, and that predators may be more deleteriously affected by fluctuations in weather than are their prey. Where predators can coexist with, and feed on, their prey in harsh parts of the prey's distribution, predation will not necessarily be more important in benign areas.

Also, for predation to be of greater importance in benign environments, the distribution of the predators must overlap with that of the prey into the benign end of the environmental gradient. Considering a gradient of exposure to wave-action, Rice (1935) and Connell (1970) have reported that adult barnacles, *Balanus glandula*, are only found in a narrow band at high levels where wave-action is moderate to great on shores of the north-west Pacific coast of the United States. In sheltered bays, *i.e.* the benign end of the gradient, the adults are found at all levels. Connell's (1970) experiments demonstrate that predation, mostly by whelks, is the primary explanation for the lack of adult barnacles at low levels outside sheltered areas. Connell (1972) suggested that predators of these barnacles were scarce in protected bays. Clearly, predation was not a major factor influencing barnacles as components of intertidal communities in the areas most protected from waves (*i.e.* the most benign environments in the gradient under consideration).

Finally, Menge's (1978a, b) observations and experiments demonstrate that the efficiency of predation along a gradient of wave exposure is confounded with the effects of desiccation on the predators during low tide. Menge discussed the increased risk for whelks due to desiccation at protected sites, and the increased probability of their being swept away by waves in exposed areas (see also Kitching *et al.*, 1966). In both types of habitat, whelks were often found sheltering in crevices and/or under a canopy of fucoid algae. These topographic features provide protection from waves and desiccation, and greatly influence the effects of predatory whelks on surrounding populations of prey. Increasing wave-shock caused no changes in the rates of consumption of prey by whelks in experimental cages (Menge, 1978a). In all areas, the efficiency of whelks was considerably reduced by desiccation stress.

Thus, the influence of predation on community structure is not related in any simple manner to a single gradient of environmental harshness (see also Menge, 1978b, for discussion of other complications). Whatever the force of waves, predatory whelks were most efficient near crevices or in areas with a canopy of large algae, which moderated the effects of waves and/or desiccation. Such topographic features of the habitat are a major influence on the activities of intertidal predators (see also Paine, 1974) and variations in the availability of such shelters will have marked effects on the intensity of predation. As an example, Moran (1980) filled the crevices on a shore with cement, and caused a marked reduction in the density of the predatory whelk *Morula marginalba* compared with control areas. This decrease was followed by a reduction in rates of mortality of prey species (several barnacles) and altered patterns of community structure in the experimental area. Unless the availability of topographic shelters is itself correlated with environmental stress, there will be no particular gradient of importance of predation from benign to harsh environments.

Methodological problems. If stress from desiccation can affect the activities of a predator, there is a problem of interpretation of experimental data where predatory whelks are confined in, or excluded from, patches of prey by the use of cages. The cages cast shade, and may drastically reduce the intensity of desiccation, and thus act like crevices. Rates of predation will probably be overestimated inside cages. Where a roof is used to control for the effects of shading, whelks may aggregate under the roof, treating it as a shelter from desiccation and, perhaps, waves. Grazing gastropods concentrated in unnaturally large densities under experimental roofs (Underwood, 1980), and Menge's (1976, Fig. 5) photograph shows enhanced numbers of *Thais* under roofs. As discussed elsewhere (Underwood, 1980), if untouched control areas are immediately adjacent to roofed areas, and whelks aggregate in great numbers under roofs, such experimental treatments run a serious risk of overestimating the impact of predators. A better design would have completely independent treatments, with cages, roofs, and controls sufficiently separated to prevent artifacts due to the treatments themselves. In particular, if the benign environments are those characterized by protection from wave-shock, yet are harsh in terms of the effects of desiccation on the predators, roofs will have dramatic effects on the rate and intensity of predation. Such effects will themselves be correlated with the environmental gradient.

The use of fences, which enclose areas without shading them from above, rather than cages, may go some way toward eliminating the artificial reduction of stress due to desiccation. Such fences do not require controls for the effects of shading, but may introduce other artifacts, such as interference with water currents or increased siltation (see Dayton and Oliver, 1980).

Further, as Menge (1976) has pointed out, the recruitment of prey (barnacles and mussels) may be directly affected by the experimental manipulations. He found greater intensity of recruitment within cages than outside, in some areas. In any experiment, this artifact could again cause overestimation of the role of predation, simply because the comparison of density of prey in areas with and without predators is already biased. Fewer prey are present where predators are active outside the cages, even if predation is trivial. This artifact was carefully examined by Menge (1976) and did not alter the conclusions of his studies.

We have found the opposite effect, that fewer barnacles (*Tesseropora rosea*) settled in areas surrounded by experimental fences than in unfenced control areas (Denley and Underwood, 1979). Clearly, the artifacts introduced by an experimental manipulation will vary according to the species investigated, and in general, more appropriate controls need to be considered (see also Dayton and Oliver, 1980).

Conclusion. The relative importance of predation in benign, as opposed to harsh, intertidal environments is not subject to simple generalizations. For predation to increase with decreasing physical harshness of the environment, the following conditions must be met:

(i) Prey must have ranges of tolerance to the effects of physical factors (such as desiccation or wave-shock) wider than the ranges of conditions over which the predators can forage efficiently. Otherwise, predation will be important everywhere, and the prey will not have refuges in the physically harsh parts of their distribution.

(ii) A decrease of harshness of the environment must not be correlated with sufficient increases in the rates of growth of the prey, such that they can rapidly become too large for, and thus safe from, their predators (*e.g.* the algae described above). Otherwise, predation will be of increased importance in harsher parts of the prey's distribution.

Where the physical environment is not a gradient of a single factor, interpretations will be difficult. A decreasing gradient of wave-shock is often correlated with increasing stress from desiccation for the predator (*e.g.* Menge, 1978a, b). Where these two gradients are negatively correlated, predation can only be more important in areas protected from waves if:

(i) local topography provides refuges from the effects of desiccation in areas sheltered from waves, and

(ii) the predators have the same range of distribution toward the sheltered end of the gradient of wave-action as do the prey. Otherwise, predators will be absent, and therefore unimportant, in the environments that are benign with respect to wave-action.

These four modifying statements weaken the value of any generalization. Exceptions to the generalization exist, and further studies in other intertidal communities are needed to evaluate generalizations about variations of predation along gradients of the physical environment.

Increased Competition When Intensity of Predation is Reduced

The evidence. Connell (1975) has argued that competition is a common phenomenon when predation is reduced. His model for the domination of patches in intertidal communities depends upon competitive interactions, which are prevented in some areas and at some times by predation or by mortality due to harsh physical conditions. Connell cited a number of examples where the abundances of intertidal organisms were increased when predators or grazers were removed experimentally, or otherwise reduced in number. This increase in itself does not support the hypothesis that reductions in numbers of consumers will lead to competition amongst the prey. Resources are not necessarily limiting simply because predators are absent, and competition among the prey may not occur.

There are, however, some well-documented examples, some of which were discussed by Connell (1975), where the experimental removal of predators led to increased abundances of organisms, which then competed for space (*e.g.* Dayton, 1971; Lubchenco, 1978; Lubchenco and Menge 1978; Menge, 1976; Paine, 1971, 1974). These studies provide support for the notion that competition can occur when natural predators are absent, reduced in abundance, or less effective. For example, Lubchenco (1978) has discussed the different ways in which grazing snails can influence the intensity and outcome of competitive interactions in algal communities. If grazers prefer a competitively dominant alga, moderate intensity of grazing will decrease the intensity of competitive interactions, and competitively inferior species of algae can persist.

Alternative Situations and Factors
Modifying the Impact of Competition

There are examples where a lack, or reduction in abundance of, predators does not lead to increased intensity of competition, or where competition occurs but has little effect on the structure of the intertidal community.

Keough and Butler (1979) removed predatory starfish from pier pilings in South Australia, with no particular consequences for the ensuing structure of the invertebrate communities present. Sousa (1979b) found no increase in competitive interactions amongst low-shore algal species when grazing molluscs were removed from boulders.

In New South Wales, on sheltered shores, a number of species of grazing gastropods were abundant, and apparently subject to little mortality from natural predators (*e.g.* Underwood, 1975; Moran, 1980; Underwood and Jernakoff, 1981). When in sufficiently great densities, these animals compete, probably for food, and the snail *Nerita atramentosa* can seriously reduce the density of the limpet *Cellana tramoserica* (Underwood, 1978a). If the diversity of the community is defined in terms of species richness (*e.g.* Menge and Sutherland, 1976), this competition does not lead to local exclusion of the limpet, and therefore there is no reduction in species richness, nor apparent change in the structure of the community. This inconsistency apparently results because the inferior competitor, *Cellana*, can recruit into any area from other shores where limpets may be free from, or not excluded by, the effects of *Nerita* (Underwood, 1978a).

Significant effects of interspecific competition have been found in experiments on the limpets *Cellana tramoserica* and *Siphonaria denticulata* (Creese, 1978; Creese and Underwood, 1982). The former can cause increased mortality of the latter when in sufficiently great densities, by depriving *Siphonaria* of food. *Cellana*, however, also suffers from intraspecific competition, which reduces its density to low levels at which it cannot eliminate *Siphonaria* from experimental areas.

In these examples, the absence, reduction, or experimental decrease in intensity of predation has not resulted in any increased role of competitive interactions as determinants of community structure. Even where competition occurs, it may not lead to change in structure of a community.

Of much greater importance in the present context is the fact that temporal and spatial vagaries of settlement of the planktonic propagules of potential competitors have a major influence on the prevalence of competition. When large numbers of propagules arrive, the resources of space, food, light, access to the water column, etc. can quickly become limiting, unless natural enemies or the physical environment thin out populations (*e.g.* the review by Connell, 1975). Thus, when Paine (1974) removed *Pisaster* from an area, large numbers of mussels colonized the rock, and because they were free from a major source of predation, eliminated other species and formed a virtual monoculture. When Dayton (1971) repeated this experiment, he did not obtain similar results, largely because mussels did not settle in great numbers (see also Paine, 1977).

In some areas of shores in New South Wales, the whelk *Morula marginalba* can drastically reduce the density of the barnacle *Tesseropora rosea*, which overlaps in distribution with the smaller barnacle *Chamaesipho columna*. When whelks are experimentally excluded from areas of the shore, there is no significant interference competition between the barnacles. Where they coexist, both barnacles settle too sparsely for space to be a limiting resource and they have different preferenda for microtopographical settlement sites. In habitats where either species settles in sufficient densities to occupy most or all of the space, the other species does not settle, either because of pre-emption of the space by other sessile species or because of the requirements of the settling larvae (Denley, unpublished experimental data).

In these examples, variations in the intensity of recruitment of potential competitors will determine whether competition occurs. The presence, absence, or efficiency of predators may be an unimportant or secondary factor, and may be an unreliable basis for making predictions about the structure of communities. This problem will be discussed later after consideration of other generalizations that may be affected by patterns of recruitment.

Direct Effects of Harshness of the Environment

The evidence. Harshness of the physical environment, due to waveshock or desiccation, may influence the structure of intertidal communities by completely eliminating species that are too delicate to withstand the rigors and stresses of a harsh habitat. Alternatively, mortality due to the physical environment may reduce the abundance of those species more tolerant to the stresses in harsh areas. The effects of physical stresses on upper limits of vertical distribution were discussed earlier.

There are examples of animals that are washed away from areas subject to intense wave-action (*e.g.* Harger, 1970; Menge, 1978a, b; Moran, 1980; reviews by Connell, 1972, 1975). Dayton (1971) transplanted anemones (*Anthopleura elegantissima*) to protected shores, where they succumbed to desiccation and all of them died. The comments made previously about interpretations from experimentally transplanted organisms also apply to animals transplanted horizontally along a coast. Although Dayton's experiments tell us why anemones are not likely to be found in protected sites, they do not actually identify the cause of their absence from such areas under natural conditions.

An alternative case. Despite general acceptance of the importance of direct effects of physical factors in the structure of intertidal communities, there is an alternative hypothesis for the absence or reduced density of some species in harsh areas. As discussed earlier for upper limits of distribution, the organisms may not recruit or may only recruit in small numbers to such areas. This sparse recruitment could occur because of behavior during settlement of the propagules or because patterns of water flow, waves, etc. do not bring the recruits to some areas.

As an example, in New South Wales, the shores most protected from wave-shock tend to be dominated, at mid-tidal levels, by grazing gastropods (Underwood, 1975). Where there is some wave-action, the barnacle *Chamaesipho columna* covers extensive areas of rock. Where wave-action is greater, *Chamaesipho* is confined to high levels, above the level of the larger barnacle *Tesseropora rosea* (Dakin, 1969; Dakin *et al.*, 1948; see Figure 11.1). In such areas, *Chamaesipho* are rarely as abundant as in more protected sites. Neither *Chamaesipho* nor *Tesseropora* settles in the very sheltered areas dominated by gastropods, even if the grazers are removed. *Chamaesipho* does not settle, even in cleared areas, at levels dominated by *Tesseropora* on more exposed shores. *Tesseropora* rarely settle on cleared patches (*i.e.* where *Chamaesipho* have been removed) in areas normally dominated by the smaller barnacles.

In all these cases (Denley, in prep.), experimental manipulations and extensive monitoring indicate that lack of recruitment is primarily some function of the behavior or availability of larvae. The lack of recruitment of each species described above is not attributable to mortality of newly settled larvae due to physical factors, nor is it due to predation, competition, "bull-dozing" by limpets (*e.g.* Dayton, 1971), or any other agent of mortality acting after the larvae arrive. Thus, the distributions of these

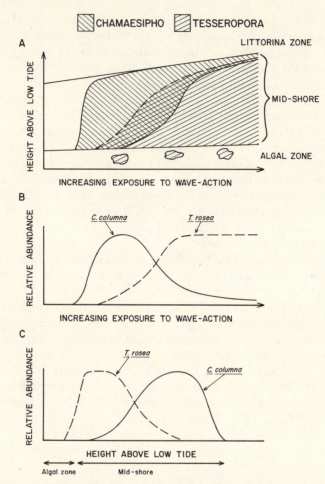

Fig. 11.1. Diagrams illustrating the general distribution of the barnacles *Tesseropora rosea* and *Chamaesipho columna* on rocky shores in New South Wales. The diagrams are simplified and there is much variability and patchiness. Data and further details may be found in Creese (1978), Denley and Underwood (1979), Moran (1980), and Underwood (1981).

A. Vertical and horizontal distribution of each species. Shaded areas indicate the presence of the two species. The unshaded area, protected from waves (to the left of the figure) is dominated by grazing gastropods; barnacles are absent. Occasional patches of *T. rosea* at low levels among algae are indicated.

B. General trends in relative abundance of the two barnacles at mid-shore levels along a gradient of wave-exposure. *T. rosea* is more abundant and *C. columna* less abundant where wave action is greater.

C. Vertical trends in relative abundance of the two barnacles in areas where they occur together. *C. columna* is more abundant and *T. rosea* less abundant at higher levels on the shore. The positions, areas of overlap, and height of the modes of these curves vary according to the wave exposure.

barnacles along a gradient of physical harshness (due to wave-action) are primarily determined by patterns of larval settlement involving habitat selection and/or any process influencing the availability of larvae. This example may not be unique, but few species have been examined to determine the patterns of settlement in the field.

Patchiness in Time and Space and Escapes from Predation and Weather

The evidence. Connell (1972, 1975) and Menge and Sutherland (1976) have considered the roles of temporal and spatial heterogeneity (patchiness) in stabilizing predator-prey interactions, in determining dominance of areas of a shore, and as agents influencing the age-structure of populations. Both approaches to a model for the structure of intertidal communities stress the importance of occasional escapes by prey from their consumers or from the stresses of the physical environment. Because many organisms are safe from their predators once they reach a certain size, and because some marine organisms are less vulnerable to physiological stress at larger sizes, such escapes can influence the appearance and organization of a community for long periods (see examples in Connell, 1972, 1975).

In particular, Connell (1975) has suggested that occasionally a mild season will allow survival of organisms that would usually be killed in harsh physical environments. Similarly, in benign environments, where predation is generally supposed to be the major regulator of abundances of organisms (as discussed earlier), there are occasional periods when natural enemies are absent or reduced in number and the prey remain abundant and can grow to invulnerable sizes before the predators return.

An alternative hypothesis. Recruitment of many intertidal organisms that have planktonic dispersive stages in their life histories is very variable, in both time and space. This variability has not been documented in many studies, but there are some excellent examples (Burkenroad, 1946; Coe, 1956; Loosanoff, 1964, 1966; Bowman and Lewis, 1977). The causes of differences in the numbers of recruits from site to site and year to year can only be guessed, but are likely to include variations in the numbers of larvae in the plankton and in the numbers of larvae settling, as well as changes in processes causing mortality after the larvae have settled. It has not been possible to distinguish between these in the studies cited above. Fluctuations in wind, currents, turbulence, silt, nutrients, and predators will all contribute to very variable survivorship during the planktonic stage of life. The large number of factors involved suggests that the numbers settling in any area or at any time will have great variance. The influence of processes such as habitat selection on the numbers of settling larvae are very poorly known, because most experimental analyses of behavior during settlement have been done in the laboratory (*e.g.*

G. B. Williams, 1964; Meadows and Campbell, 1972; Crisp, 1974), and may not apply well to field situations.

This potential variability in larval settlement suggests that many species may "escape" from their predators, or from the effects of weather, by virtue of being present in great numbers in some areas at some times. Such "escapes" may occur sporadically in harsh environments without concomitant mild seasons, provided that the effects of physical stress are not so severe as to kill all the recruits that arrive. Then, dense recruitment will be followed by large numbers of survivors, which will subsequently occupy resources for long periods and have a profound effect on the community. In areas where predators normally control the abundance of some species, a season of heavy settlement will be followed by the survival of numerous prey, provided that predators are not so numerous or efficient that they can cope with all the recruits, whatever the number that arrives (which is unlikely given the usual functional response of intertidal predators, *e.g.* Murdoch, 1969).

Two examples will illustrate this mechanism. Dayton (1971) drew attention to the patchiness of settlement of the barnacle *Balanus cariosus* and suggested that the barnacle occasionally overwhelms the ability of predatory whelks to kill it. Some barnacles then grow too large to be eaten by the whelks, and can persist for long periods in the presence of these predators. During 1978, the settlement of larvae of *Tesseropora rosea* was very heavy over many areas of New South Wales, in contrast to less dense settlement in 1977, 1979, and 1980. Sampling and monitoring of many areas until 1980 reveals that the 1978 year-class constitutes most of the barnacles in many areas (Denley, unpublished data, and see Table 11.1). These barnacles are the dominant part of the adult population, and occupy extensive areas of the shore, despite the effects of limpets, whelks, and harsh periods of weather, which were not decreased during their earlier life. They have "escaped" simply because of their huge density at the time of settlement. We predict that they will continue to dominate local shores for some time to come.

Conclusions. Throughout the preceding discussion, we have argued that the importance of competitive interactions and the effects of predators on them, are likely to be very variable because of fluctuations in settlement of the propagules of most components of intertidal communities. Generalizations about predation and competition are weak, and have very limited predictive value, if the numbers of prey or potential competitors are themselves unpredictable. Variations in the settlement of intertidal species are not well documented (with the notable exceptions of the studies cited earlier). Where settlement has been examined quantitatively, great variances from place to place and year to year have been found. Such stochasticity has often been ignored in interpretations and generalizations

Table 11.1. Settlement and survival of barnacles in areas subject to predation and natural disturbances. Note the variations from place to place and year to year, and that the dense settlement during 1978 dominates the population two years later.

Area	No. quadrats (20 × 20 cm) Sampled	Mean density at:	Year-class of barnacles 1978*	1978	1979
1	13	Settlement	—	309.5	7.3
		June 1980	4.2	14.5	0.0
2	5	Settlement	—	615.0	193.4
		June 1980	2.2	39.8	1.6
3	5	Settlement	—	624.8	288.4
		June 1980	0.0	179.2	11.0
4	16	Settlement	—	172.8	16.3
		June 1980	0.0	13.6	0.0

* Settlement of barnacles before 1978 was not recorded in these areas.

made from experimental studies. Sometimes, variations in settlement are considered to be a "nuisance" or "inconvenience" in the interpretation of experimental data, particularly when some species fails to arrive in a study area. As Paine (1974) has pointed out, the irregularity of appearance of a potentially dominant competitor can contribute greatly to temporal (and spatial) heterogeneity in the structure of a community, and is an important attribute of the community. Yet, in contrast to this statement, he considered irregularity of settlement to introduce an artifact into his experiments; *i.e.* Paine (1974) and Dayton (1971) obtained different results from essentially the same experimental manipulation. It is unclear what is "artificial" about this result. In an earlier study, Paine (1971) considered that an experimental removal of a predatory whelk "failed," because its prey, a mussel, did not settle in the area. We believe that the experiment did not fail, but provided data to suggest that mussels do not always become dominant users of space when their predators are removed. Menge (1976, p. 384) omitted from discussion some experimental areas because of a lack of settlement of barnacles.

We do not agree that lack of settlement is an artifact, a failure, or something to be ignored. Occasional periods of zero, or very intense, recruitment of important species may have long-term effects on a community. We agree with Dayton (1979a) that ignoring such variability in our interpretations of nature may be very restrictive. If some numerically important, or competitively dominant, or actively predacious species occurs in irregular densities in an intertidal community, we should expect a variety of possible outcomes from any experimental study. If the experiment were repeated at some later date, or were done in different areas at the same time, variations in recruitment of organisms would probably lead to variations in intensity of biological interactions and to different results. Pooling results from experimental plots in different areas and years will produce average, but not necessarily realistic pictures of nature. The structure of intertidal communities is not a simple function of characteristic interactions of predation and competition. The functions are likely to be complex because of enormous temporal and spatial heterogeneity in the abundances of interacting species, a direct result of the dispersive phase of their life histories.

Regularity, predictability, and generality of interpretations of the structure of communities will probably only be perceived where the starting numbers of component species are somehow regular or predictable. Some studies preclude spatial variations in recruitment from influencing the results, because relatively homogeneous areas are chosen for study (*e.g.* Connell, 1961a, b) or because experiments are done in "typical areas" (*e.g.* Menge, 1976). This practice is often necessary because of logistic constraints. Connell (1974) has concluded that the effects of major experimental treatments can often be revealed with few replicate manipulated plots. Where there are few replicates (in time or space), or replicates are in close proximity, manipulative experiments are unlikely to provide much information about spatial and/or temporal variations in the intensity and outcome of ecological processes. The interpretations of such studies should be confined to the homogeneous, typical, or whatever, circumstances investigated.

We believed that various patterns and processes were important in the ecology of the barnacle *Tesseropora rosea*. As foreshadowed in an earlier paper (Denley and Underwood, 1979), subsequent observations and experiments during a year of very dense settlement of cyprids have forced us to accept some alternative views (Denley, in prep.). Ours is probably not a unique experience (see also Dayton and Oliver, 1980), but we have some difficulty in reconciling it with the desire to generalize that is evident from most short-term or small-scale experimental studies.

GENERALIZATIONS ABOUT DIVERSITY OF SPECIES IN INTERTIDAL COMMUNITIES

Apart from patterns and processes affecting the distributions and abundances of, and interactions among, intertidal species, a number of general statements have been made about processes regulating diversity of species. Notably, following earlier work by Paine (1966), there have been descriptive and more mathematical models by Menge and Sutherland (1976) and

Levin and Paine (1974). These are based primarily on experimental studies in intertidal communities, but the authors indicated that their conclusions may have wider applicability.

We believe that some of the predictions or generalizations in these papers are premature, may be oversimplified, and in some cases have not been borne out by subsequent experimentation. In particular, we draw attention to the following generalizations:

The relative importance of competitive interactions is dependent on the trophic complexity of the community (Paine, 1966; Menge and Sutherland, 1976) and there is an inverse relationship between the relative importances of competition and predation as determinants of community structure, dependent upon the trophic level examined (Menge and Sutherland, 1976).

Relationships between Trophic Complexity, Predation and Competition

The model. The relationships between predation, trophic complexity, and species diversity of intertidal communities were discussed by Paine (1966), who concluded that the number of species in a community is dependent on the number of predators present, and on their ability to prevent competitively dominant species from eliminating other species. This theme was pursued for algal communities by Lubchenco (1978), who stressed the importance of preferences among different species of prey by consumers. Where they prefer competitively dominant prey, the consumers (*i.e.* predators or grazers) tend to increase species diversity. Where they prefer competitively inferior species, consumers tend to decrease diversity.

Menge and Sutherland (1976) reviewed evidence for a complex set of hypotheses (or model) that holds that high diversity tends to be maintained by competition among species at higher trophic levels in a community, while at lower trophic levels maintenance of high diversity is by predation. Also, predation was considered to be the most important organizing interaction in trophically complex communities, but competition was more important in trophically simple communities. These relationships are summarized here in Figure 11.2.

Problems with the model. One problem with Paine's (1966) arguments about relationships between species diversity and the trophic complexity of a community is that alternative hypotheses are given little attention and some of the data used to support the predation hypothesis are very weak. Diverse systems were considered to have disproportionately more predatory species than those found in communities with fewer total species. The data presented do not support this claim. The temperate area, Mukkaw Bay, had 2 predators out of 13 species in the food web discussed by Paine (1966). In contrast, the subtropical area in the Gulf of California had 11 predatory species out of a total of 45 species. The null hypothesis that the proportion of predatory species is equal in the two

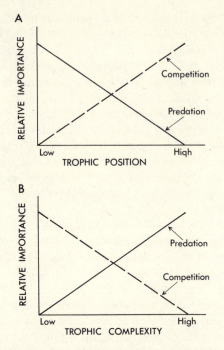

Fig. 11.2. Qualitative models for relative importance (A) within a community and (B) among communities of predation and competition in the organization and maintenance of diversity of communities. Trophic position refers to status as primary producer, grazer, predator, etc. Trophic complexity is an unspecified function of various factors, including the number of trophic levels, number of species in each level, and abundance of each species, etc. This figure is taken directly from Menge and Sutherland (1976) and presented again here for convenience.

communities cannot be rejected by Fisher's exact test ($P = 0.39$), and thus the data do not confirm the prediction of Paine's hypothesis, as he claimed they did (1966, p. 73).

Second, comparisons between the different areas examined by Paine (1966) and Menge and Sutherland (1976) reveal that a major difference in both the trophic complexity and the importance of predation is the presence of higher-order carnivores in some areas (starfishes in the cases examined). The presence of such carnivores automatically increases the impact of predation (see Paine, 1966, p. 71) and considerably increases the complexity of a food web (*e.g.* Dayton, 1971; Paine, 1966; Menge and Sutherland, 1976). Paine proposed three hypotheses to account for the absence of high-order carnivores from some areas. The first was that such predators had not evolved in some areas because of historical accident.

This hypothesis was considered to be "unapproachable," and therefore to have no "reality," and was ignored, leading Paine to the conclusion that the evolution of high-order carnivores was related to the stability of annual productivity in different areas. Areas with high productivity thus have high-order carnivores, which in turn help maintain high diversity. We do not agree that historical accidents should be ignored simply because hypotheses about them are not amenable to current experimental methods. The evolutionary history of an area may be an important determinant of community structure (*e.g.* the high diversity of patellid limpets in South Africa, Branch, 1976), and may be unique to each area, and not a function of simple "rules" allowing comparisons between areas to reveal causal mechanisms. This unpalatable conclusion cannot, in our view, be dismissed by the available evidence. It does, however, complicate the comparisons of areas by Menge and Sutherland (1976) in the development of their model.

Relevant to all of this is the fact that a correlation between increased diversity of species in a community and increased numbers of predators, or importance of predation, or increased complexity of food webs does not demonstrate causality. The greater diversity of intertidal communities on the west coast of the U.S.A. than on the east coast (Paine, 1966; Dayton, 1971; Menge, 1976; Menge and Sutherland, 1976) is due to more species of barnacles, whelks, and grazing gastropods, in addition to the presence of high-order predatory starfish. It is possible to argue that increased diversity of prey species, for whatever reason, may allow increases in the number of predators in the system. Increased complexity of food webs and intensity of predation may be the result, not the cause, of increased species diversity (see also Dodson, 1970; Menge and Sutherland, 1976). We shall not pursue this discussion further here.

A third problem with the Menge and Sutherland hypothesis lies in its testability. For comparisons between communities, the axis of trophic complexity (Figure 11.2) was not quantified. Thus, any case considered may fall anywhere along the axis, and any revealed pattern of predation and competition will, perforce, fit on the graph somewhere. This problem was discussed by Keough and Butler (1979, see below) with possible reasons for the failure of their experimental data to conform to Menge and Sutherland's hypothesis.

Worse is the nature of the hypothetical relationships between competition and predation. Menge and Sutherland predict that competition becomes relatively more important, and predation relatively less important, towards higher trophic levels within a community. In part this prediction is a tautology, because of the definition of trophic level; organisms at the higher trophic level in a community have no predators, and predation on such organisms is of zero importance (absolutely and relatively). Assuming that predators have some effect on lower trophic levels (otherwise they do not contribute to trophic complexity at all), the absolute importance of predation is a decreasing function of trophic level (Figure 11.3A). Following Menge and Sutherland, the precise shape of the function can be considered unimportant.

When only predation and competition are considered, if competition is equally important at all trophic levels, contradicting the intention of the hypothesis, the relative importance of competition will nevertheless appear as an increasing function of trophic level (Figure 11.3B). The absolute importance of competition as an organizer of communities would be difficult, if not impossible, to measure. Where only the relative impor-

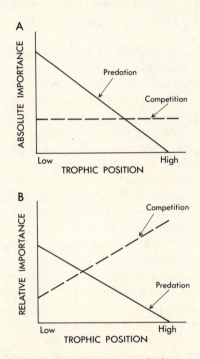

Fig. 11.3. The qualitative relationship between the importances of predation and competition in the organization and maintenance of diversity within a community.

A. The absolute importances of competition and predation as organizers at difference trophic levels, under the null hypothesis that competition is equally important at all levels.

B. The relative importances of competition and predation, calculated from (A), when only these two processes are considered. Compare (B) with the corresponding graph in Fig. 11.2A. The absolute importance of competition is of hypothetical value only; see text for further details.

tance of predation and competition are considered, some situations contradicting the hypothesis will apparently confirm it, making it difficult to test in any adequate manner.

Analyses of communities relevant to the model. Since the Menge and Sutherland (1976) hypothesis was proposed, two relevant studies appear to contradict it, despite the difficulties inherent in testing it discussed above. Keough and Butler (1979) removed predatory starfish from pier pilings in South Australia, and concluded that predation was of little importance in determining the organization of the community. This conclusion contradicts the prediction of Menge and Sutherland that predation would have a great influence on the trophically complex community investigated (see Figure 11.2, here). Keough and Butler suggested several explanations for the difference between this system and those considered by Menge and Sutherland (1976).

The second study was on the dietary overlap of two sympatric species of starfish in low-shore areas of New England coasts (Menge, 1979). These starfish are among the most important predators in the system, and would appear to be the top predators in the food web, yet they do not compete with each other, despite considerable overlap in diets and body size. Although Menge did not discuss the data in this context, it is clear that competition is not of increased importance at the top trophic level in this community. Competition does, however, occur among algae at lower trophic levels (Lubchenco and Menge, 1978). This example violates the prediction of Menge and Sutherland (1976) for comparisons within a community (see Figure 11.2, here).

Thus, two different studies provide contradictions to hypotheses about the relative importance of predation and competition in intertidal communities. As far as intertidal systems are concerned, the hypotheses were proposed after consideration of particular studies (references in Menge and Sutherland, 1976). Their generality is not confirmed by subsequent investigations in other communities. Circumstances in which the relationships between trophic structure, predation, and competition conform to the model will become apparent only after a greater number of intertidal communities has been investigated.

CONCLUSION

Throughout this discussion we have concentrated on alternative hypotheses and some counter-examples to generalizations about intertidal communities. These do not detract from the worth of attempts to generalize, but may provide a note of caution. Experimental intertidal ecology offers a self-congratulatory gesture to workers in other fields, based on the ease and success of many simple manipulations on the shore (see Connell, 1974; Paine, 1977). The theme appears to be that such manipulative experiments provide insights into mechanisms operating in other systems where experimentation is impossible, difficult, or has not been attempted. Thus, if and when others do such experiments, they will obtain results generally similar to our own. We have shown here, by counter-examples, that *we* do not always obtain results generally similar to our own. This result is our justification for restricting our attention in the present paper to intertidal research—it produces sufficient variation in outcome of experiments that we need not venture beyond this narrow field to see the inherent dangers of premature generalization from particular studies.

Much of our discussion has been concerned with hypotheses about patterns of larval settlement, because we believe that this source of variation has been overlooked in the design or interpretation of many experimental analyses of intertidal communities. We have followed a theme proposed by Dayton (1979a), Dayton and Oliver (1980), and Denley and Underwood (1979). The prevailing paradigm, as proposed by Dayton, is that the key to understanding the structure of intertidal communities is the identification of major biological interactions occurring after the organisms have arrived on the scene from the plankton. Thus, experiments mainly consist of the removal of single species (predators or competitors) or of clearing away the entire community and observing the recolonization of denuded rock. The tendency in such studies is to be much less concerned with the invasion of natural areas (true controls), but there are exceptions (*e.g.* Lubchenco, 1980; Underwood, 1980).

We are convinced that considerable variations in intensity and outcome of processes will occur because of the vagaries of larval settlement of the species in a system. No amount of elegance in experimental design, nor of creative and inspired interpretation, will reveal these components of the system if the hypotheses being considered ignore such processes. The larger paradigm will continue to prevail, and will appear to be strengthened by disallowal of the temporal and spatial heterogeneity caused by fluctuations in numbers of the interacting species.

This is not a claim that the results of all the papers discussed are somehow incorrect. For example, Menge (1976) discussed differences in recruitment of barnacles in some of his experimental plots, and concluded that these were of little importance in the long run. This has not been our experience (Underwood *et al.*, 1983). Where does this contradiction lead? To us, the interesting feature of counter-examples to some generalizations is not simply that they disprove generality. They are much more interesting, because they provide a basis for determining the conditions under which some process is important, or what "families" of conditions yield different mechanisms. They demonstrate that similar or identical patterns of structure on seashores are not always caused by the same mechanisms, and reveal alternative evolutionary histories leading to apparently similar life histories of the organisms. For example, Paine (1971, 1974) found a broad similarity of intertidal zonation and the effects of an asteroid predator on shores in New Zealand and Washington. No such large, generalist predator occurs on shores in New South Wales, yet the general patterns of zones are quite similar (Underwood, 1980, 1981). We stress that differences are important, but they will not be revealed if we are constrained by the results of previous investigations and do not consider other hypotheses.

A further note of caution seems necessary. Before we are caught up in too many idealized generalizations or models, we should note that most of the experimental studies used as a basis for the models (particularly Levin and Paine, 1974; Paine, 1974; Connell, 1975; Menge and Sutherland, 1976) were done on shores with very obvious competitive dominants and very voracious predators. Perhaps these shores are also characterized by relatively regular patterns of recruitment of the animals (*e.g.* Connell, 1970) (although for mussels this is clearly not the case). In other parts of the world, similar experiments have not been done, and attention has focused on different but related questions (*e.g.* Branch, 1976; Underwood, 1978a), partially, at least, because the communities do not obviously contain such clear-cut competitive dominants for the resource of space, and such universally important predators (Branch, pers. comm.; Keough and Butler, 1979; and our own observations). Before we attempt to explain patterns of distribution in other ecosystems (*e.g.* Lubchenco, 1980; Paine, 1971), should we not reach agreement about different intertidal systems?

Finally, what sort of models should we be attempting? First, as far as the dynamics of the system go, we must be concerned with considerable environmental stochasticity in numbers, and not be trapped into models based on simple relationships between numbers in one generation and those in the next. The vagaries of the planktonic "mystery stage" (Spight, 1975) are too great for that, and numbers of recruits of many species are very variable (see the earlier references). Second, but related, we are in need of models that have different end-points, dependent upon the starting conditions. The detailed experimentation of Dayton (1971) indicates a clear competitive hierarchy involving several species. The potential winner fails to dominate because its numbers are reduced by predators or non-biological disturbances. The process of competition along such chains will also be influenced by the numbers of the different competitors that arrive in any area. Where recruitment is stochastic, the starting numbers of any member of the hierarchy can be, unpredictably, from zero to saturating. Where pre-emptive competition is important, the temporal sequence of arrivals will also have profound effects. These will both influence the outcome of competitive interactions, regardless of predation or disturbances, and also perhaps have an effect on the process of predation. We therefore need models that predict what will happen on a shore if very different numbers of the interacting species arrive from the plankton. To develop such models, we shall need data from experiments concerned with different starting densities. The simple "all-or-none" approach that has been so successful to date may not be adequate. Instead of the manipulative removal of all barnacles, or mussels, or snails, or algae, etc., leaving the population intact in control areas, greater insight might be gained from experiments involving a range of densities of each of these species. For example, the experiments and observations of Lubchenco (1978) indicate that the effects of intermediate densities of grazers are in no way predictable from areas with large or small densities.

We believe that this slight shift of focus in experimental design will cover the variations in numbers caused by variable recruitment. It will also require the proper quantification of variations in recruitment, and their consequences, so that realistic manipulations will be possible. When such data become available, comparison of different geographical areas and the validation of subsets of general principles of intertidal ecology should be a realistic endeavor.

Acknowledgments

The preparation of this paper was supported by funds from the Australian Research Grants Committee and a University of Sydney Research Grant (to A.J.U.) and an Australian Commonwealth Scholarship (to E.J.D.). We have benefited from discussion and arguments with students, colleagues, and friends including: D. T. Anderson, L. C. Birch, A. J. Butler, M. G. Chapman, A. C. Hodson, P. Jernakoff, M. Keough, and G. Russ. Particular help was gained from R. G. Creese, M. J. Moran, P. A. Underwood, and P. F. Sale (who disagreed with everything we said) and J. H. Choat (who did not). The encouragement of the latter and correspondence with J. H. Connell and P. K. Dayton were invaluable. These two and L. G. Abele substantially improved the manuscript.

12.

Processes Structuring Some Marine Communities: Are They General?

PAUL K. DAYTON

Scripps Institution of Oceanography, University of California, San Diego, La Jolla, California 92093

INTRODUCTION

This paper is a brief review of mechanisms structuring some marine communities. I address three related questions. First, are the distributions and demographic patterns of the populations regulated by general and/or predictable biological interactions? Specifically, are the habitats, resources, and risks modified in a predictable way by other populations in the community? Second, is the organization of the very different communities structured by similar mechanisms, or are different processes emphasized? Finally, what are the most efficient means of studying these processes?

When appropriate, I will discuss these questions with regard to hard- and soft-bottom and plankton communities. Most of the benthic communities tend to be mosaics of seral stages or distinct patches that have considerable internal resilience. It is important to understand the roles of short- and long-term episodic events causing the mosaic, and it is especially enlightening to develop a mechanistic understanding of the biological processes by which these patches resist invasions or by which they change through time. Plankton communities in boundary currents seem structured by many physical processes and tend to be unpredictable in space and time (McGowan, 1977); in contrast, the plankton assemblage in the North Pacific gyre is predictable (McGowan and Walker, 1979).

This review focuses exclusively on the population dynamics approaches to community organization. I argue that a functional approach to such community study results in a much better mechanistic understanding than other alternatives and I will emphasize the ubiquity in benthic systems of particular species that have functional roles more important than their abundance and/or biomass would predict.

TEMPERATE ROCKY INTERTIDAL COMMUNITIES

Mechanistic processes of competition and predation in rocky intertidal communities were outlined almost 50 years ago (Hatton, 1938) and have been refined in temperate habitats in England (Connell, 1961a), the north-eastern (Menge, 1976; Lubchenco and Menge, 1978) and northwestern (Paine, 1966; Dayton, 1971) United States, Japan (Hoshiai, 1964), Chile (Paine, pers. comm.), and New Zealand (Luckens, 1975a, b; Paine, 1971). The complex suite of processes summarized by Price, Irvine, and Farnham (1980) and Paine (1980) revolves around the fact that primary space can be dominated by a few species. Competitive dominance usually is expressed in a rather patchy manner that is integrated with selectivity, recruitment variability, predation, several levels of biological disturbance, and the competitive hierarchies. Superimposed on these levels of biological inter-actions are varying types of physical stresses and disturbances resulting from desiccation, wave surge, freezing, and damage from drift material and cobbles (Dayton, 1971; Osman, 1977; Sousa, 1979a). Underwood and Denley (this volume) challenge the absolute nature of the generalizations of this legacy and they argue for consideration of alternate hypotheses.

Many of the important processes in the upper intertidal appear also to occur in the algal-dominated lower level component of rocky intertidal habitats (Dayton, 1975a). Once again, there appear to be reasonably con-sistent guilds of algae that are dominated by clear competitive hierarchies, usually resulting in a particular species or guild capable of monopolizing the habitats via competition for light or space or via whiplash (Dayton, 1975a). In addition, I found a positive relationship in which an overstory canopy actually protects understory plants from desiccation. Again, there are relatively clearly defined guilds of herbivores that have varying roles influencing the algal guilds and interrupting the competitive monopoli-zation of the habitat (Paine, 1980). While the molluscan effects are often weak (Paine, 1980), they too can be important (Underwood, 1980). Fi-nally, the effects of sea urchin grazing usually are very strong (Lawrence, 1975). Wave exposure is always an important component of the environ-ment (Dayton, 1975a, b); indeed *Hedophyllum sessile*, the competitive dominant in most exposure regimes, loses its dominant role in the wave-exposed physiologically optimal area, as this is the only habitat physio-logically available to *Lessoniopsis littoralis*, a much stronger dominant. Geographic evidence suggests that the same pattern holds for the massive *Durvillea antarctica* in Chile, which appears to be displaced upward into more marginal areas by *Lessonia migrescens* (Santelices *et al.*, 1980; pers. obser.).

While much intertidal research is done in attractive wave-exposed habi-tats, a considerable amount of the world's temperate rocky intertidal occurs in less glamorous but more protected areas in which the fauna and flora are very strongly influenced by desiccation and/or salinity stress (Lewis, 1977; Underwood and Denley, this volume). In many such habitats grazers have complicated competitive (Branch and Branch, 1980) and mutualistic relationships (Paine, 1980); usually spore or larval nurseries are important but little studied, as are asexual reproduction and growth.

Another common but ignored algal community is that represented by the extensive flats of turf composed of small ephemeral algae and, often, articulated coralline algae. Joan Stewart (1982) identified 67 species of algae, most of which grow epiphytically and depend upon "anchor species," which in San Diego, California, are 6 species of articulated coralline algae that survive the desiccation and seasonal sand inundation and furnish the habitat for all of the other species in the community; turf habitats are frequently exposed to desiccation and seasonal sand inundation. While some kelps are adapted to moderate sand inundation (Markham, 1973; Daly and Mathieson, 1977), those systems exposed to more severe inundation tend to be marked by seasonal blooms (Joan Stewart, pers. comm.).

One very important feature of temperate rocky intertidal communities that is virtually ignored in the literature, but that is of some interest to this symposium, is that a majority of the species in almost all intertidal communities have a very limited functional role. The densities of particular species could be doubled or the species could be completely eliminated and most of the other species would not, in an evolutionary sense, "perceive" the difference. However, often there are species that have functionally important roles disproportionate to their abundance or biomass. In such situations these "foundation" species (Dayton, 1972) strongly influence most of the other species.

This brief summary of temperate rocky intertidal communities shows that the existence of disproportionately important species depends upon several general factors: (1) the presence of a population capable of monopolizing the spatial resource and sufficient recruitment to insure that the monopolization can be realized; (2) a predator or disturbance that limits the expression of the monopolization; and (3) various refugia and larval nurseries important to the recruitment of various species, especially the important ones (Dayton, 1971, 1975a; Paine, 1980).

SUBTIDAL HARD-BOTTOM COMMUNITIES

These communities tend to be dominated by sessile animals or, in temperate areas, by kelps. The encrusting communities represent a gradient of fouling-encrusting-boring types, each of which exhibits different processes and rates.

The fouling communities recently reviewed by Sutherland (1980), Osman (1977), and Schoener and Schoener (1981) are composed of ephemeral species such as barnacles, hydroids, bryozoans, and thin compound ascideans. These founder communities (Yodzis, 1978) often seem characterized by unpredictable settlement and usually low survivorship; the reviews also emphasize that even fouling communities exhibit considerable effects of biological interaction as their development includes aspects of facilitation and inhibition (Sutherland, 1978; Dean and Hurd, 1980). However, relative to most other communities, fouling communities are unstable and unpredictable in time and space.

Encrusting communities of long-lived bryozoans, compound ascideans, colonial cnidaria, and especially sponges (Rützler, 1970; Reiswig, 1973; Jackson and Buss, 1975; Jackson, 1977; Ayling, 1978; Bergquist, 1978; Karlson, 1980) must be considered in a much longer time frame (decades rather than months). These are dominance communities marked by lottery recruitment and a strong reshuffle (Yodzis, 1978). The populations, especially of sponges, tend to have low dispersal, and the patches are extremely resistant to competition, invasion, and predation. In many cases the mechanisms of resistance involve powerful chemical defenses (Ayling, 1978; Bergquist, 1978; Bakus, 1981; J. Thompson, pers. comm.).

Finally, many hard-bottom communities, especially those of softer rocks and limestone, involve situations in which the substrata are extensively burrowed by sponges (Rützler, 1975) and scraped or bored by bivalves. This activity creates biogenous structures which are habitats for many species of clams, crustacea, sipunculids, polychaetes, echinoderms, etc. (Warme and Marshall, 1969; Warme et al., 1971; Palmer, 1976). The development rates of these communities are unknown, but probably range from many decades to centuries. It is important to note that these common communities depending upon biogenous habitats are much more complex than the two previous types.

Coral communities are reviewed in this volume by Abele and Sale. See other reviews in Endean and Jones (1973), Glynn (1976), and Jackson (1979). These communities are among the most diverse in the world; their organization emphasizes biological habitats, lottery recruitment, competitive cross interactions, and physical and biological disturbances of intermediate intensity as well as many remarkable mutualistic relationships.

ANTARCTIC SPONGE COMMUNITY

I discuss briefly an Antarctic sponge community existing on a sponge spicule matrix for which we (Dayton et al., 1974; Dayton, 1979b) have many data relative to the three community approaches discussed by Paine (1980): food webs, productivity, and functional community roles. Figure 12.1A shows a food web derived from Dayton et al. (1974) plus unpublished recent observations; each of the lines could be quantified with an estimate of Kcal/100 m²/yr, but it is hard to see what important understanding could be gained. Figure 12.1B summarizes the energy flows documented in Dayton *et al.* (1974). The three hexactinellid sponges plus *Tetilla leptoderma* are circled as being important because they contribute almost 96% of the sponge biomass. Most of the sponge consumption is by the asteroids *Acodontaster conspicuus* and *Odontaster validus*, which represent strong and weak links (*sensu* Paine, 1980). *Odontaster validus* nonetheless gets most of its energy from primary producers. The biomass and percent cover of the other sponges are relatively minor, as is the consumption of the other sponge predators. The dotted line indicates a very rare occurrence of *Odontaster validus* consuming an *A. conspicuus*; although rare, this predation is adequate to control the *A. conspicuus* population. Finally, Figure 12.1C shows the strong functional relationship of a relatively rare sponge (*Mycale acerata*), which is the competitive dominant in the sponge community; its rare status despite its fast rate of recruitment and growth (Dayton, 1979b) is maintained by another relatively rare asteroid, *Perknaster fuscus antarcticus*, which shows extremely efficient functional and numerical responses to any increase in *Mycale* density (Dayton *et al.*, 1974). That is, juvenile *Perknaster* have a general diet allowing them to persist in the absence of *Mycale*, but upon encountering a *Mycale* they become absolute specialists that are extremely efficient foragers (functional response). In addition, they almost triple their body size and become fecund (numerical response). Thus, both predators and prey are rare, but both have very important functional roles. In addition, the four sponge species dominating the biomass enjoy a release from *Acodontaster conspicuus* because of an extremely rare predation by *Odontaster validus*, the population of which is in turn regulated by primary productivity (Dayton *et al.*, 1974).

This brief summary reinforces Paine's (1980) contention that analysis of connectedness and energy flow approaches contribute relatively little important biological information about evolutionary relationships within biological communities. Because the functionally important species often are rare, the latter approaches do not necessarily give any hints of important population relationships. I argue that this conclusion is generally true in nature.

KELP COMMUNITIES

Most hard-substratum habitats in shallow, subtidal areas and temperate waters are conspicuously dominated by marine algae. The structure and organization of a few of these communities are reasonably well understood. They often have guilds of kelps forming distinct canopies, including

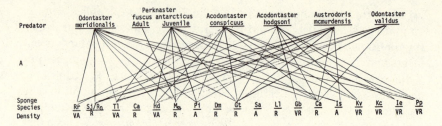

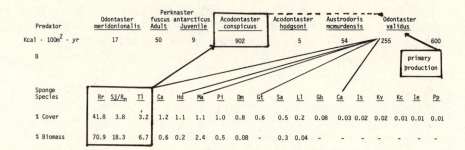

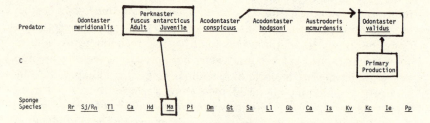

Fig. 12.1. Food webs, productivity, and functional community roles of an Antarctic sponge community. (A) Sponge-eating predators and their sponge prey, after Dayton *et al.* (1974) plus unpublished subsequent observations. Densities of sponges are estimated from unpublished observations as very abundant (VA), abundant (A), rare (R), and very rare (VR). (B) Strong and weak (*sensu* Paine, 1980) productivity linkages showing the total estimated calories per 100 m²/year for each predator population. (C) Functionally important species; the relatively rare but dominant sponge *Mycale acerata*; its rare status is maintained by the unusually efficient predation by *Perknaster*. See text for details. The names of the sponges are, left to right: *Rossella racovitzae, Scolymastra joubini* and *Rosella nuda, Tetilla leptoderma, Cinachyra antarctica, Haliclona dancoi, Mycale acerata, Polymastia invaginata, Dendrilla membranosa, Gellius tenella, Sphaerotylus antarcticus, Leucetta leptorhapsis, Gellius benedeni, Calyx arcuarius, Isodictya setifera, Kirkpatrickia variolosa, Kirkpatrickia coulmani, Isodictya erinacea, Pachychalina pedunculata.*

floating canopies, stipitate algae in which the canopy is held off the substratum, prostrate canopies in which the fronds move back and forth
across the substratum, and various turf-forming associations. The patterns seem constant in many habitats (see Kain, 1979; Chapman, 1979, for
reviews), and they parallel patterns I saw in intertidal tide pools (Dayton,
1975a). In most kelp associations, there are clear dominance hierarchies in
which the higher canopies intercept the light and dominate the plants
below. However, in areas exposed to stronger wave stress, the stipitate
canopies were the competitive dominants, and the floating canopies,
which are more susceptible to damage from the waves, were behaving as
fugitive species (Dayton, 1975b). This conclusion is supported by unpublished work in California in which we found that the lower-level canopies
can dominate wave-stressed habitats. In almost all well-studied kelp habitats, it appears that competition between the canopy guilds can be a
pronounced and critically important facet of the community organization
(North, 1971; Dayton 1975b).

In addition to the important cross linkages (*sensu* Paine, 1980) involving competition, almost every temperate kelp community has been found
to be particularly sensitive to over-exploitation by grazing sea urchins,
which potentially eliminate most macroalgae and then persist indefinitely,
producing an alternate steady state (Lawrence, 1975). This susceptibility
to sea urchin grazing seems especially pronounced in higher-latitude kelp
communities such as those of Alaska (Estes and Palmisano, 1974; Dayton,
1975b), Nova Scotia (Mann, 1977), and, to a certain extent, Great Britain
(Kain, 1979). Certainly, such sea urchin barrens dominated many of the
potential kelp habitat areas we visited in southern Chile (Dayton, unpublished) and appear to be potentially important in Tasmania (T. Dix, pers.
comm.).

Additional strong vertical linkages (*sensu* Paine, 1980) are thought to
exist in Nova Scotia and in the northeastern Pacific where the potential
over-exploitation by sea urchins tends to be mitigated by predation on
the sea urchins by lobsters and crabs (Mann, 1977; Bernstein *et al.*, in
press), and sea otters, *Enhydra lutris* (Estes and Palmisano, 1974). The kelp
habitat south of Pt. Conception, California, includes many carnivores
that probably functionally replaced the sea otters after they were over-
exploited in the early 1800's. These carnivores include asteroids and especially fishes and spiny lobsters (Tegner and Dayton, 1977, 1981). The same
predator control of echinoids has been proposed to have existed in Nova
Scotia, where lobsters, crabs, and fishes reputedly controlled sea urchin
populations until each was over-exploited by human fishing (Mann, 1977;
Bernstein *et al.*, in press). At this time, much of the exposed coastline from
areas of Newfoundland south to Maine and New Hampshire in the northwestern Atlantic has seen the destruction of the kelp beds by echinoids,

which have established a new and apparently resilient biological community with very low diversity and almost no algae (K. Mann, R. Welsford, L. Harris, pers. comm.). Recently the sea urchins have experienced a disease that has killed small patches of the urchins along the central east coast of Nova Scotia; the patches, however, recover, and the urchins continue to dominate (B. Bernstein, pers. comm.)

Hard-bottom communities can be summarized as follows: Fouling communities are characterized by frequent disturbance, unpredictable recruitment, and low survivorship (usually months rather than years); they are constantly renewed, often with different suites of species and are relatively unpredictable in time and space. Others such as upper and lower intertidal and kelp communities are characterized by somewhat unpredictable recruitment but relatively predictable patterns of succession, usually determined by competitive hierarchies that are relatively predictable in given habitats. The composition of these communities often is modulated by patterns of disturbance. Many communities of very long-lived organisms such as sponges persist for decades, as they are very resistant to disturbance or invasion. While the recruitment and development of these communities is little understood because of their slow rates, it is likely that the recruitment is relatively unpredictable, and that competition plays a large part in development. Finally, some communities are structured around the activities of boring animals, which often create considerable spatial heterogeneity and provide habitats for many species.

SOFT-BOTTOM COMMUNITIES

Only recently have mechanistic questions of communities in soft substrata been approached from a functional perspective. These communities are, at least superficially, very different from those on hard substrata because most soft-substratum communities are composed primarily of small infaunal species. While hard-bottom communities also have small organisms, which are usually ignored, the soft-bottom infaunal organisms are similar in size to their substratum and have a much more intimate relationship with it: they eat it, they lick it, they move through it, and they modify it in many physical/chemical ways. These populations are difficult to study experimentally because they are composed primarily of small, hard-to-manipulate animals that live beneath the surface. Excellent reviews by Gray (1974), Rhoads (1974), Pearson and Rosenberg (1976), Rhoads et al. (1978), Peterson (1980), Virnstein (1980), and review volumes edited by Livingston (1979) and Tenore and Coull (1980) describe recent efforts to derive a functional understanding of the many processes structuring these soft-substratum communities. A summary of these reviews reveals that well-documented correlations between infaunal species and sediment types date at least from the turn of the century (Petersen, 1913); the idea that predation has important structuring roles has also been long appreciated (see Blegvad, 1928, and Thorson, 1966, for reviews). More recently, however, the existence of various functional groups has received considerable attention. This is true especially with regard to suspension and deposit-feeding and burrowing and tube-building associations (Sanders, 1958; Rhoads and Young, 1970; Levinton, 1972; Myers, 1977; Jumars and Fauchald, 1977; Woodin, 1978; Brenchley, 1979). In addition, it is becoming well established that particular species have important roles in providing nutrient regeneration and structure to an otherwise relatively homogeneous substratum (Rhoads and Young, 1971; Keegan and Konnecker, 1973; Gray, 1974; Woodin, 1976, 1977, 1978).

The search for evidence of interspecific competition has been frustrating for many reasons (Peterson, 1979, 1980). There are hints of its importance (Woodin, 1974; Levinton, 1977; Fenchel, 1975, 1977) but these hints are usually indirect, relying on cage studies that may include sediment artifacts and/or on indirect inferences such as character displacement. In most infaunal systems, there are no data supporting strong competition for a potentially limiting resource and, indeed, recent reviews (Peterson, 1979, 1980) argue that most infaunal communities can have very large increases in densities without any evidence of competitive exclusion, but that diffuse competition may be more important than we realize. In most cases, the best evidence for competition rests with much larger organisms such as clams and crabs (Peterson, 1977; Dayton and Oliver, 1980; Peterson and Andre, 1980).

In hard-bottom communities, predation usually interrupts the competitive monopolization of space; in communities in soft substrata, however, there is little evidence for exploitative competition for space, and experimental documentation of the role of predators has proven very difficult. This documentation is difficult partly because of the difficulty of determining diets and partly because most experimental work has utilized cages, which can result in serious interpretive errors (Virnstein, 1977a, b; Peterson, 1979; Dayton and Oliver, 1980). Nonetheless, this old emphasis on the importance of predation (Blegvad, 1928) has recovered popularity (for example, Arntz, 1980; Holland et al., 1980) because there are many naturalists with an intuitive and theoretical appreciation of the interplay between disturbance and succession (see Woodin, 1978). There are many types and scales of disturbance that often intergrade but should be distinguished. These include physical disturbances such as wave action (Oliver et al., 1980), ice scour (Dayton et al., 1969; Oliver, 1980), hypoxia (Santos and Simon, 1980), and temperature stress (Gerdes, 1977).

They also include natural (Davis and Spies, 1980) and man-caused stress (Pearson and Rosenberg, 1976, 1978).

Perhaps more interesting, however, are the local disturbances resulting in patches of successional phenomena. These disturbing agents include fishes, especially rays, which dig pits in the sediment to procure prey; as they dig they displace small infauna, which are eaten by smaller "picker" fishes. The overall effect is that the bottom becomes a mosaic of successional patches, which are common components of many benthic communities (VanBlaricom, 1978; Dayton and Oliver, 1980).

The actual pattern of succession depends upon the temporal and spatial scale of the disturbance, because recruitment is commonly seasonal (Thorson, 1950) and infaunal species have great variation in larval dispersal. For example, VanBlaricom (1978) found that the colonization of small ray pits was very different from that of a large slump, as the former were colonized by adult amphipods while the latter was colonized by polychaete larvae. Coastal habitats are exposed to a wide array of disturbances, including storms (Kuhn and Shepard, 1981) and hurricanes (Chabreck and Palmisano, 1973); large-scale slumps are probably common around the heads of submarine canyons and the mouths of large rivers. Indeed, strong turbidity currents may be common along the outer continental shelf and beyond (Kerr, 1980).

While the deep sea is often thought to be a homogeneous and stable habitat, most of the above phenomena, especially the predictability and scale of disturbances, are almost certainly important to the structure of the deep-sea benthos. The major difference is that the disturbance rates and the rates of succession are very much slower in the deep sea (Grassle, 1977). Various scales of disturbance are probably important in the deep sea, as it is becoming appreciated that there are large-scale turbidity currents that undoubtedly result in large disturbance areas. In addition, the monster camera photographs (Dayton and Hessler, 1972) show that one important consequence of the deposition of large food items such as dead squid, fish, or mammals is a local feeding frenzy of scavenging fish and invertebrates. These more local disturbances scour and turn up the bottom and almost certainly also result in disturbed patches that undergo succession; similarly, the deposition of a large piece of wood results in a community of boring clams (R. Turner, 1973 and pers. comm.), the feces and byproducts of which result in an enriched patch dominated by early successional infaunal species (F. Grassle, pers. comm.). While these disturbances are rare and certainly unpredictable in time and space, their evolutionary importance is magnified by the quick responses of scavenging species (Stockton and DeLaca, 1982) and of the successional species. In contrast, rates of recruitment of other species are so slow that we have as yet no realistic measure of rates of recovery by this process if, in fact, much succession is involved (Grassle, 1977). Thus, the model of sustained disequilibrium (Dayton and Hessler, 1972) remains plausible in the deep sea.

Temporal patterns of disturbance and succession in soft bottoms are particularly difficult to understand, as they include seasonal variation, year-to-year differences, and almost always very infrequent but probably important episodic events. That is, there are predictable pulses of larval availability that are very tightly tied to season (Oliver, 1980), but in most cases it is extremely difficult to understand large-scale year-to-year differences in recruitment patterns. These temporal patterns are often superimposed on much longer scale patterns of episodic recruitment, probably resulting indirectly from climatological shifts and changing nearshore currents (Coe, 1956; G. Kuhn, pers. comm.; Soutar and Isaacs, 1974). There are a few documented examples of year-to-year variation for functionally important species such as *Mytilus californianus* and *Balanus cariosus* in intertidal communities (Dayton, 1971; Paine, 1974) and *Dendraster* and *Renilla* in the subtidal (Merrill and Hobson, 1970; Kastendiek, 1975; Timko, 1975). There are also a few documentations of very rare but heavy settlements of worms, *Owenia* (Fager, 1964), and especially of bivalves (Coe, 1956). These seasonal, year-to-year, and long-period episodic events are reviewed by Coe (1956) and are summarized by Dayton and Oliver (1980).

In summary, it would appear that there are differences in emphasis between mechanisms structuring communities on hard substrata and those in soft substrata. For example, the diversifying processes in communities on hard substrata seem often to revolve around the role of disturbance, which interrupts the competitive monopolization of primary and/or secondary space; refugia from stress and predictability of recruitment may also be significant. The important heterogeneity often results from biological as well as physical stresses. The lack of recruitment predictability results from uncertainties in the availability and size of the planktonic larval pool and from larval habitat choice. In contrast, except for dense monospecific assemblages of organisms such as phoronids, amphipods, etc., the soft-substratum communities exhibit little evidence of competitive monopolization of resources, and the role of predation and disturbance does not involve disruption of clear competitive dominance. The diversifying phenomena in soft-substratum communities seem more often related to biogenous phenomena, especially the effects of biologically induced heterogeneity provided by tubes or large infauna species, which produce niches for small species (see Woodin, 1978; Peterson, 1979; Dayton and Oliver, 1980, for recent reviews).

PLANKTON COMMUNITIES

I discuss oceanic plankton communities last because the nature of their structuring mechanisms may be fundamentally different. These communities do not have a common, potentially limiting resource such as primary or secondary space; if there is a common limiting component of the community, it must involve nutrients. Inorganic nutrients such as phosphorus and nitrogen are limiting only to the autotrophs and are not directly relevant to the planktonic heterotrophs (although it has long been understood that excretion by the many heterotrophs is an extremely important component of nutrient regeneration). One of the most fundamental differences between oceanic planktonic communities and benthic and terrestrial communities is that the food webs of pelagic planktonic communities seem influenced only by the relative sizes of the food particles, and hence are unstructured (Isaacs, 1973) with respect to other, better-known systems. Indeed, the relatively small reproductive products may involve an important part of the food web (Isaacs, 1976). Thus, the copepods are important predators on the larvae of fishes that, should they survive, become important predators on copepods. Indeed, it is reasonable to imagine that small plankters such as copepods, euphausids, chaetognaths, etc. are important predators on the larvae of even very large carnivores such as tuna. A terrestrial analogy would involve the specter of, say, tigers or wolves releasing thousands of tiny tigerlets or wolflets, which were largely consumed by spiders, lizards, birds, shrews, etc. For these and many other reasons, population and food-web approaches to community function in the pelagic realm have proven extraordinarily difficult, and most biological oceanographic research has focused on questions of primary and secondary productivity, mass flux, and nutrient regeneration. These processes are closely tied to the population dynamics of all the species, but a synthesis of population and productivity approaches has proven to be extremely elusive. Thus, I will confine myself to a summary of a few papers describing the structure of pelagic systems. One further caveat that impugns much of the pelagic literature is that it begins to appear as though the dynamics of populations within currents along the boundaries of the major oceanographic basins are heavily influenced by advection, upwelling, and mixing processes strongly mediated by large-scale climatogolical patterns. While sweeping biological generalizations on boundary currents are very problematic, these currents nevertheless influence most of the world's fisheries, and because of their economic importance and the relative logistic ease of working in the nearshore, they are very well sampled. Summarizing books include Parsons and Takahashi (1973), Steele (1974), and Cushing and Walsh (1976). The fishery literature, as well as these summaries, does offer evidence that the

continued exploitation of particular species or size classes of animals susceptible to fishing regimes might have important community-wide effects
(Steele, 1974). The specific types of interactions are little understood and
are almost certainly influenced by the wide-spread effects of climatological
shifts resulting in important changes in currents, upwelling, eddies, and
formation and fate of large "rings" (Wiebe *et al.*, 1976) that spin off
from boundary currents such as the Gulf Stream. In brief, although there
is much biology going on, the state of the system is seemingly driven
by physical processes.

Rather than attempt to review and summarize these very confusing
coastal boundary currents and associated gyres, I wish briefly to summarize some recent research in one of the most stable gyre communities
of the major ocean basins, the North Pacific central gyre. The North
Pacific gyre is a very large, relatively homogeneous and geologically old
(150 million years) system. While the gyre communities integrate at the
edges with the boundary currents, they are essentially isolated with regard
to species composition and the mechanisms, timing, and magnitude of
nutrient input (McGowan, 1977); they thus represent a self-contained
climax ecosystem. The community is extremely speciose but, for the most
part, its structure and organizational processes must be inferred from
samples. Thus, the study of the small plankters in the central gyre represents perhaps the best test of a neutral null hypothesis of a real community, because the sampling regimes are essentially unbiased. That is,
replicated, vertically stratified net tows are taken at regular intervals
through a 24-hour day without regard to particular preconceptions of
community structure.

Perhaps the most exhaustive published study of the structure of such a
community is that of McGowan and Walker (1979), which focused on
the structure within the copepod component of the North Pacific central
gyre. Copepods are the most numerous components of the macrozooic
plankton fraction of the community, and in this study were represented
by at least 125 species that were regularly present. Samples in this study
were taken through 6 depth ranges from 550 meters to the surface at noon,
sunset, midnight, and dawn for 10 days over 180 km. This regime was
repeated in three summers and one winter. There is a considerable vertical
pattern in many physical variables in the central gyre; however, copepods
and other zooplankters are very mobile and many homogenize the physical pattern with vertical migration. That is, there may be some patchiness
or heterogeneity of resources (actually very little in a gyre) as well as
vertical physical gradients, but the extensive movement of these small
animals assures that they are exposed to a relatively predictable environment. The intensive analysis of McGowan and Walker found a considerable structure in this group. Long-term sampling programs have

found that these species groups are remarkably constant from place to place within the gyre and over many years. In addition to the fidelity of the groups, there is remarkably constant overall pattern in the order of abundances of these species.

How general is this structure? McGowan (pers. comm.) has very long-term (15 years) data describing the numerical dominance structure for the gyres in the North and South Pacific, which have long been isolated from each other by the large equatorial water mass. The South Pacific gyre has essentially the same diversity and includes many of the same species and genera, so that one can argue that the functional components are basically the same. He has found (pers. comm.) that the order of numerical dominance is extremely constant in both the North and South Pacific gyres, but that there is a very interesting difference: the South Pacific gyre species have an order of dominance different from that of their North Pacific analogs.

Given that these communities have considerable structure and constancy, how resilient are they to perturbation? There are two types of perturbation that might be expected to have relevance to such a community. It is a well established premise that nutrient regeneration and primary productivity are critically important to the population dynamics of most zooplankton species. Thus, one might expect that a very enlightening experiment would be to alter the nutrient influx into the community and evaluate the reassortment of the many species of zooplankton. Such an experiment occurred naturally in 1969 (Hayward and McGowan, 1979) when the vertical turbulence in the thermocline apparently increased, allowing a considerable influx of nutrients from below. This influx resulted in an approximate doubling of the primary productivity, and Hayward and McGowan found that the zooplankton responded to the increased productivity with a proportionate increase in biomass. That the trophic levels are food-limited can be inferred from the observation that the heterotroph standing crop was higher following the event, but the autotroph standing crop was not. Most importantly, the order of dominance of the zooplankton remained essentially constant despite the approximate doubling of the biomass with a marked increase in the population of each species. A similar pattern was observed by Weiler (1980), who sampled dinoflagellates over time in the gyre and found a stable order of numerical dominance. She found that cell division times and doubling rates did not follow the dominance order, as some of the rare species turned over as fast as the abundant ones. The response to increasing nutrients in the laboratory was similar to that seen by McGowan and Walker. Thus, both copepods and dinoflagellates maintain their rank order of abundance when responding to a potential increase in productivity, despite the fact that the species of the two groups have relatively similar reproductive rates but may differ in abundance by orders of magnitude. This resilience is undoubtedly related to an extremely efficient response to nutrients. Yet these nutrient perturbations tend to negate a hypothesis of scramble or exploitative competition, which predicts population increases proportional to their reproductive rates following an enrichment.

A second type of perturbation is related to the roles of predators and/ or disturbance. In this case, too, an event has occurred in the equatorial current system that can be considered a natural experiment. An intensive, relatively recent tuna fishery has very much decreased the abundance of the top-order predator. Fortunately, there are before and after plankton samples, and McGowan (pers. comm.) reports that preliminary analysis shows no obvious before-and-after differences. He speculates that all of the patterns will remain unchanged despite the fact that abundance of the top predator has been much reduced.

The role of smaller predators would appear to be another logical place to look for structuring mechanisms. McGowan and Walker (1979) and Hayward and McGowan (1979) summarize intensive efforts to correlate predator effects with many taxonomic groups of predators, including mesopelagic fish, squid, shrimp, euphausids, amphipods, ostracods, and chaetognaths, as well as carnivorous copepods. In most cases, the predators appear to be food generalists, and there is no evidence that the predators are important functional factors in this community. There is certainly no evidence for a keystone type of predator, but the sum effect of all of these predators is probably a critically important part of the community maintenance.

In summary, marine plankton communities are composed entirely of motile entities, and considering their diversity and stability, they differ markedly from most other biological communities on earth. As Yodzis (1978) puts it, they may play by entirely different rules. Certainly, they are among the world's most interesting but bewildering and difficult communities to study.

DISCUSSION

In benthic communities, or probably in any potentially space-limited community, the majority of the species appear to have little effect, if any, on each other's populations. Thus, a neutral null hypothesis tested with a properly devised sampling program would rarely be falsified. Yet it is clear that the distributions and demographic patterns and, ultimately, the evolution of the component populations are strongly influenced by biological interactions with functionally important species. However, the different communities are organized by very different types of interactions. For example, many space-limited systems, such as the rocky intertidal and kelp communities, have very important competitive relationships that are modified by stress and disturbance. Fouling communities and some intertidal communities (Underwood, 1980; Denley and Underwood, 1979) are characterized by a lottery recruitment and subsequent interactions, resulting in variable community compositions. This unpredictability and the relatively fast turnover rates result in community composition that is unpredictable in time and space. Encrusting communities seem very stable, possible tautologically so, because they are composed of long-lived species (Frank, 1968). They appear to resort to allelochemical interference competition, which also effectively renders them relatively immune to invasion and predation. In contrast, boring communities seem to exhibit biogenous habitat modification and many positive or commensal relationships. Soft-bottom communities also are structured by biological habitat modifications, but in contrast to the situation in the above communities, exploitative competition among soft-bottom infaunal species is extremely rare and seemingly inconsequential. Processes influencing larval recruitment and several levels of disturbance seem to be among the more important soft-bottom community mechanisms.

The repeated conclusions from almost all the diverse types of benthic communities is that they exhibit species with community roles that are disproportionately important inasmuch as they affect the growth and regulation of most of the component populations in the community. These species include those that usually have the potential of developing clear-cut competitive monopolies of primary space and those guilds of species that disrupt the hierarchical competitive relationships resulting in the monopolies. They also include those species such as boring clams, kelps, or "anchor species," or other species that furnish important biogenous habitat or important settling substrata. The individual character of these various communities is defined by a multitude of lower-level but nonetheless important relationships including, especially, mutualistic patterns of habitat requirements, larval settling patterns that often emphasize particular nursery microhabitats, and the presence or absence of higher-order carnivores (Underwood and Denley, this volume).

In contrast to the benthic communities, marine plankton communities include very loosely organized assemblages in coastal boundary currents, structured by unpredictable advective and upwelling processes, or apparently well-organized, self-regulated assemblages in cyclonic or anticyclonic gyres, some of which are very predictable. The stable North Pacific central gyre shows no evidence of resource allocation or keystone community structuring types of predation. Certainly, there is no evidence for competitive dominants or keystone predators. The intriguing possibility exists that marine plankton communities are organized around processes basically different from those that structure benthic communities.

Acknowledgments

I appreciate the extensive patient advice of Lisa Levin, Dave Thistle, Janice Thompson, Tony Underwood, and especially John McGowan. I also very much appreciate the cheerful secretarial assistance of Cynthia Brockett. My research was supported by NSF grants from Biological Oceanography and Division of Polar Programs.

Morphology,
Species Combinations,
and Coexistence

13.

Interspecific Competition Inferred from Patterns of Guild Structure

PETER GRANT and DOLPH SCHLUTER*

Division of Biological Sciences, University of Michigan, Ann Arbor, Michigan 48109

INTRODUCTION

Different views prevail on the extent to which communities can be considered structured, and the role of biotic interactions, most notably competition and predation, in determining community structure. Plant ecologists have been preoccupied with these questions for a long time (Jackson, 1981), but the current concerns of animal ecologists (Connell, 1980) appear to stem from the work of David Lack with Darwin's Finches (1940, 1945, 1947). Lack made an argument for considering interspecific competition to be an important process in the adaptive radiation of the finches. In this paper we will briefly review Lack's argument, the work that it stimulated in others, the challenges it has received, and our own work designed to test the competition argument.

Competition and the Adaptive Radiation of Darwin's Finches

Lack's aim was to offer a comprehensive explanation for the evolutionary diversification of Darwin's Finches from a presumed single ancestral species. He devised an allopatric model of repeated speciation and adaptive radiation of the finches. Differentiation, he argued, occurred initially in allopatry and was completed when secondary contact (sympatry) was established. Since he could not see how differentiation could proceed very far in allopatry, because he believed all the islands to be very similar in their habitats, he was forced to invoke interspecific competition in sympatry as an important driving force of the differentiation (Grant, 1981b).

As evidence for this view, he pointed to regularities in morphological and distributional data (*e.g.* Figure 13.1). Bill sizes of congeneric species are approximately regularly spaced along a size axis with virtually no overlap. If diet differences correspond to bill differences, the species do

---

* Order of authorship determined by competitive ability.

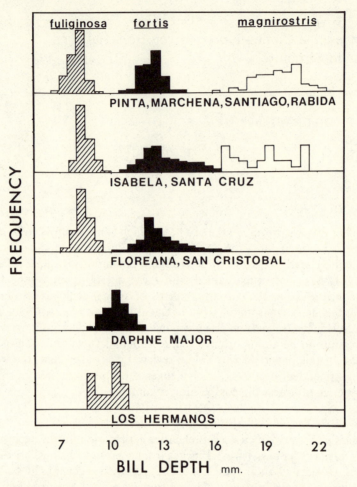

Fig. 13.1. Beak depth frequency distributions of *Geospiza fuliginosa*, *G. fortis*, and *G. magnirostris* on selected Galápagos islands. Based on Lack (1947); see Harris (1973) for correspondence of English and Spanish names of the islands. The top three rows of histograms represent the combined samples from the two to four islands specified beneath the rows, and therefore no significance should be attached to the shape of the frequency distributions. Unaware of the pooling of data, at least one investigator has been misled into believing that departures from normality reflect current responses to natural selection (Wilson, 1965).

not now compete for food. Lack interpreted this regularity as resulting from either competitive exclusion of ecologically similar species or evolutionary shifts in bill sizes and presumed diets. For example, some species with similar bill morphology, such as *Geospiza conirostris* and *G. scandens*, do not co-occur. As a second example, in the absence of *G. fuliginosa* on Daphne Major Island, *G. fortis* has a bill size intermediate between those of *G. fuliginosa* and typical *G. fortis* elsewhere (Figure 13.1); *i.e.* it has sequestered part of the *G. fuliginosa* niche in the absence of interspecific competition. To use modern language, the first is evidence of competitive exclusion, the second is evidence of competitive release.

The Stimulus of Lack's Work

Influenced by Julian Huxley (Lack, 1973a), Lack explained and synthesized, but did not test his ideas (Grant, 1977). He saw patterns of regularity in the occurrences of congeneric species in Europe, and extended his views on interspecific competition to account for them (Lack, 1944). But he then turned his attention to matters of population regulation and breeding characteristics of birds, and only returned to questions of competition late in his life (Lack, 1969a, b, 1971, 1973b, 1976) when the seeds he had sowed had yielded a harvest in North America.

Hutchinson (1957, 1959, 1965) and MacArthur (1958) further developed the idea that interspecific competition could structure communities by determining the limits to similarity of coexisting, ecologically similar, species, and hence the number of such species in a community. A large amount of effort has been devoted since then to the theoretical investigation of conditions permitting coexistence, the species packing problem, and the stability properties of communities structured by competition.

MacArthur (1972) once wrote, "To do science is to search for repeated patterns, not simply to accumulate facts." This is exactly what Lack did, and more recently what Diamond (1975) has done, using the idea of interspecific competition both as guidance in the search for patterns and as explanation for them.

Science involves more than just searching for repeated patterns, of course, and testing the explanations or hypotheses against alternatives is a logical next step in a scientific investigation. Perhaps because it is more difficult than pattern searching, until recently this step has been applied infrequently to the hypothesis that interspecific competition has structured communities. An early study of birds on islands used this approach, and rejected an obvious alternative to the competition hypothesis (Grant, 1965a). Some consequences of the hypothesis were then tested and upheld (Grant 1966b, 1968; but see Simberloff, 1970; Grant and Abbott, 1980), but in other cases explanations for apparently obvious manifestations of competition in birds have not been supported by the results of tests (Grant, 1972b; Abbott, 1973, 1980).

All this says that one can assume competition and explore its possible consequences, or test it against alternatives to see if it is the best explanation of observations. Some combination of the two is desirable. Certainly some check on the theory is needed. The patterns themselves, like beauty, may exist more in the eyes of the beholder than in reality (Grant, 1977), and the explanations for patterns may be simply ingenious contrivances that are completely wrong. And so we return to Darwin's Finches and to tests of the hypothesis that competition has caused some of their properties.

The Challenge to the Competition Hypothesis

Bowman (1961) disputed Lack's claim for interspecific competition on the grounds that all of the differentiation, not just the initial stages, could have occurred in allopatry. Bowman argued that islands were very different in their floras; that when birds dispersed from one island to another they would encounter different feeding conditions; that adaptation to the new food supply would occur; and that, when members of this population established secondary contact with the original population through subsequent dispersal, competition between them would not occur because they would be substantially different in their feeding habits.

Abbott *et al.* (1977) established a method for comparing islands in the physical properties of food supply, principally seeds and fruits, of the ground finches (genus *Geospiza*). They were able to show that islands do indeed differ in food supply characterized in this functionally meaningful manner. This result supports Bowman's view that adaptation to local food supply occurred in allopatry. So is it necessary to invoke interspecific competition to account for species differences?

Our approach to this question is to use quantitative data to see if (a) combinations of species on islands are non-random, (b) morphological differences between species are non-random in a direction predicted by an interspecific competition hypothesis, and (c) ecological characteristics of selected pairs of species conform to the expectations of a competition hypothesis or to alternative hypotheses. Our use of ecological data to test the competition hypotheses with Darwin's Finches does not make our arguments circular, because Lack did not have quantitative data when he formulated his views on competition a few years after his visit to the Galápagos (Lack, 1973a). Ironically, while he was on the Galápagos the methods by which he might have obtained such data were just beginning to be developed at the place in England that was to become his professional home (Colquhoun, 1941).

The difficulties of constructing realistic null hypotheses will be considered as we proceed. We restrict attention to the six species of the genus *Geospiza*. All are ground-feeding, seed-eating species, and thereby constitute a feeding guild (Root, 1967). The tree-finches and warbler finch are clearly different in their feeding habits (Lack, 1947; Bowman, 1961), but with the exception of the warbler finch (Grant and Grant, 1980b), we do not yet have quantitative data to use a statistical criterion for recognizing guild membership. The only other members of the guild are doves (*Zenaida galapagoensis*) and mockingbirds (*Nesomimus* spp.). These are ignored here because they are morphologically very different, and feed in very different ways (Grant and Grant, 1980b). They both occur on all major islands and many small ones so they constitute an almost constant element in the environment of the ground finches.

DISTRIBUTIONS OF SPECIES COMBINATIONS

We consider a community to be structured if there are non-random patterns in the components, in the sense that components deviate significantly from some particular mathematical model of randomness. While the observation of non-randomness is not in itself proof of the importance of any particular structuring force, competition often provides a compelling explanation. Abbott *et al.* (1977) applied a statistical test to finch distributions on islands to decide whether the ground finch species composition of the various island communities could be considered random or not. They concluded that chance alone could not explain the prevalence of certain combinations of species (see also Power, 1975).

Here we reanalyze distributions of finch species. First we correct a few errors in the computations of the Abbott *et al.* (1977) analysis. Then we subject a more complete set of data to analysis.

Are ground finch species distributed randomly among Galápagos islands? An example will illustrate how we answer this question. While there are six *Geospiza* species in the archipelago, at most five occur together on an island (Table 13.1). Thus there are

$$\binom{6}{5} = \frac{6!}{5!1!} = 6$$

possible 5-species combinations. Only one of these is observed in the archipelago, on each of three islands. The question of random distributions may then be rephrased: What is the probability that, given six equiprobable combinations, the same combination (and no other) occurs three times? The calculated probability is small,

$$P = \binom{6}{5}\frac{1^3}{6} = 0.028,$$

Table 13.1. Probability (*P*) that the observed number of *Geospiza* combinations is a random sample of the possible combinations. Revised species list 1 uses only those populations known to be breeding currently or those known to have gone extinct recently. List 2 adds those populations probably present once and now extinct (see Appendix, p. 232).

		Abbott *et al.* (1977) data		
No. Geospiza species/island	*No. possible combinations*	*No. islands*	*No. observed combinations*	*P*
5	6	3	1	.028
4	15	3	1	.0044
3	20	8	2	<.0001
2	15	7	4	.072
1	6	2	1	.167

		1		
No. Geospiza species/island	*No. possible combinations*	*No. islands*	*No. observed combinations*	*P*
5	6	3	1	.028
4	15	4	2	<.0001
3	20	8	2	.017
2	15	7	4	.072
1	6	10	1	<.0001

		2		
No. Geospiza species/island	*No. possible combinations*	*No. islands*	*No. observed combinations*	*P*
5	6	6	1	.003
4	15	5	2	.004
3	20	4	2	.017
2	15	7	4	.072
1	6	10	1	<.0001

and we therefore conclude that the observed combination is not simply a random sample of the possible combinations.

We generalize this procedure with the following formula (Feller, 1950). Given n possible choices and r islands, the probability that $n - m$ combinations *or fewer* are found is

$$P = \sum_{x=m}^{n} \binom{n}{x} \sum_{v=0}^{n-x} (-1)^v \binom{n-x}{v} \left(1 - \frac{x+v}{n}\right)^r$$

(x and v are dummy variables of summation). Employing this equation, we determined the probabilities for the remaining species combinations.

Results are shown in Table 13.1 for three data sets. In the first we present corrected probability values for numbers of islands and combinations considered by Abbott et al. (1977). The second analysis uses a slightly different species list derived from more recent surveys (see Appendix). We include populations of species that are known to breed now, or that once bred but are now extinct. The third data set includes also populations that are suspected of having bred before and that have now gone extinct.

Results of 3-, 4-, and 5-species combinations are similar (Table 13.1). The probability that observed combinations are a random sample of possible ones is extremely low in each case, whether original or revised data sets are used. Of 15 possible 2-species combinations of *Geospiza*, only 4 are represented; the probability of this pattern's occurring by chance is small, but it is greater than 0.05. All 10 single-species islands used in the second and third analyses are occupied by *G. fuliginosa*. This pattern is unlikely to be due to chance alone ($P = 1 \times 10^{-7}$; Table 13.1). Thus we conclude that distributions of *Geospiza* in the Galápagos are decidedly non-random—indeed more so than previously reported (see Abbott et al., 1977, and Simberloff, this volume).

MORPHOLOGICAL DIFFERENCES BETWEEN SPECIES

Certain combinations of species are highly under-represented, especially on islands with 3 to 5 species present. Habitat characteristics of islands may be an important element here (Abbott et al., 1977). However, if interspecific competition for food has been partly responsible for this non-randomness, we would expect species occurring together to be more dissimilar in morphology than those not occurring together. This expectation rests on the assumption, supported by data (Abbott et al., 1977; Grant, 1981a, b; Grant and Grant, 1980b; Schluter, 1982b; Smith et al., 1978), that morphological differences between species reflect dietary differences. The prediction was tested by an analysis of measurements of museum specimens of the finches made by I. Abbott (see also Grant, 1983).

Means of nine morphological dimensions (Abbott et al., 1977) were computed for males of each population represented: three body dimensions (length of wing, tarsus, and hallux) and six beak dimensions (length, width, and depth of upper and lower mandibles). Body measurements were included, since for some species (*e.g. G. difficilis*) these variables appear to be important determinants of foraging position, and hence diet. We later repeated some of the analyses with beak dimensions only, and obtained the same results. With these measurements we obtained the first two principal components from the correlation matrix (Cooley and Lohnes, 1971) based upon population means. Together these account for

97.6% of the variance among populations. We also computed the mean vector for each species in the resulting two-dimensional morphological space (Figure 13.2) and the Mahalanobis distance, D^2 (Cooley and Lohnes, 1971), between pairs of vectors.

Employing these values we determined D^2 values for nearest neighbors in morphological space for all species. These were averaged to yield a mean nearest-neighbor D^2 for each possible combination. Overall average D^2 values of combinations that occur and that do not occur are compared in Table 13.2. If interspecific competition has influenced the formation of species combinations then we expect the mean D^2 for observed combinations to be greater than the mean D^2 of unobserved combinations. This is in fact the case in four of four sets (Table 13.2), though the result is not statistically significant in any one of them (Mann-Whitney U tests, $P > 0.1$).

The preceding analysis does not identify the species that are responsible for the differences observed. To rectify this shortcoming we list all pairs of species and the morphological distance between them in Table 13.3, and then we perform an analysis of co-occurrence of species taken two at a

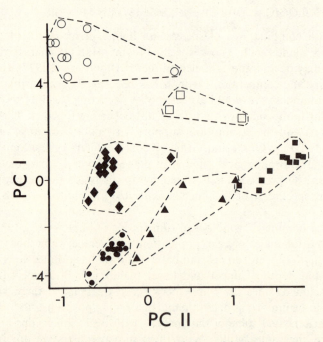

Fig. 13.2. Principal component plot of morphology of the six *Geospiza* species. See text for details. The symbols represent *G. magnirostris* (○), *G. conirostris* (□), *G. scandens* (■), *G. fortis* (◆), *G. difficilis* (▲), and *G. fuliginosa* (●).

Table 13.2. Comparison of mean nearest-neighbor distances (D^2) between observed and unobserved species combinations, based on species list 1 of Table 13.1.

No. Geospiza species/island	No. possible combinations	No. observed combinations	Mean Nearest-Neighbor D^2		
			Observed combinations		Unobserved combinations
5	6	1	27.0	>	25.1
4	15	2	36.9	>	29.9
3	20	2	41.2	>	39.8
2	15	4	80.3	>	64.1
1	6	1	—		—

Table 13.3. Morphological distance (D^2) and number of islands shared for *Geospiza* species pairs (list 1 of Table 13.1). Each nearest neighbor is indicated by direction of the bold arrow (*e.g. G. conirostris* is the neighbor of *G. magnirostris*). The total numbers of island populations sampled for the morphological analyses are 10 (*G. magnirostris*), 17 (*G. fortis*), 27 (*G. fuliginosa*), 7 (*G. difficilis*), 14 (*G. scandens*), and 3 (*G. conirostris*).

Share more than 5.7 islands		
Species	No. islands shared	D^2
G. magnirostris — G. fortis	11	72.4
— G. fuliginosa	11	184.1
— G. difficilis	7	163.4
— G. scandens	10	181.4
G. fortis — G. fuliginosa	15	26.0
— G. scandens	15	54.6
G. fuliginosa — G. scandens	13	58.2

Share fewer than 5.7 islands		
Species	No. islands shared	D^2
G. magnirostris → G. conirostris	1	53.3
G. fortis — G. difficilis	4	22.1
↔ G. conirostris	0	21.4
G. fuliginosa ↔ G. difficilis	4	10.8
— G. conirostris	2	77.2
G. difficilis ← G. scandens	3	19.0
— G. conirostris	1	43.8
G. conirostris — G. scandens	0	38.4

time. Pairs are classified on the basis of number of islands shared: if all combinations of species are equally probable then, for example, the probability that any two particular species are present on a 5-species island is $4/6 = 0.67$. Since there are 3 islands with 5 species present (Table 13.1, list 1) a given species can expect to share $3(0.67) = 2$ of these with a second given species. By repeating this procedure for islands with 4, 3, and 2 species present and adding together the expectations, we obtained the value 5.7, which is the expected total number of islands shared by any two particular species.

The data in this form allow another test of the prediction of the competition hypothesis that morphologically similar species should co-occur less frequently than more different species. It can be seen (Table 13.3) that D^2 values of the two groups are different: pairs of species occurring together on fewer than 5.7 islands are morphologically more similar generally than those co-occurring on more than 5.7 islands. The trend is difficult to verify statistically since pairwise distances are not independent. Nevertheless, we compared the two groups using a Mann-Whitney U-test, for lack of a better solution. The null hypothesis was rejected ($P = 0.014$, one-tailed test). Following a suggestion from R. E. Ricklefs (pers. comm.) we calculated Euclidean distances between species as an alternative that eliminates the possible complication of unequal variances within groups. We repeated the analysis and obtained the same result (Mann-Whitney U-test, $P = 0.037$, one-tailed).

Table 13.3 also shows that all six of the species occur together with their own nearest neighbors on fewer islands than expected. Again, this pattern is not likely to be due to chance, and the competition hypothesis is supported.

In our tests we have assumed that all combinations of species are equally probable. We have not relaxed this assumption for two main reasons. First, it is a reasonable one because all *Geospiza* species are widespread in the archipelago (Lack, 1947; Harris, 1973). Second, it is difficult to imagine just what assumption would be more realistic in this case. Other authors (*e.g.* Simberloff, 1978a) have used the number of islands occupied by a given species as an estimate of "colonizing ability" and have incorporated this estimate into calculations of expected co-occurrences. Not only is this procedure likely to be self-defeating in any attempt to detect non-randomness (Grant and Abbott, 1980), but the weighting procedure probably rests on false premises for *Geospiza*. The two rarest species, *G. difficilis* and *G. conirostris*, are thought to be older than the others on morphological grounds (Lack, 1947), and they each occur on islands that are distant from one another and, in the case of *G. conirostris*, distant from the center of the archipelago (see Abbott *et al.*, 1977). Hence the distributional data suggest that despite their rarity these species have superior colonizing ability (see also Alatalo, 1982): they may have colonized all the islands, then gone extinct on most.

Connor and Simberloff (1978) and Strong *et al.* (1979) similarly attempted to eliminate the complication of differential occurrences by holding fixed the total number of islands occupied by a species. By this procedure they were unable to reject the null hypothesis of random colonization (but see Hendrickson, 1981). However, while this procedure may tell us whether distributions are random or not *given* the differential occurrence of species, the analysis is restrictive, as it evades the question of why certain species are uncommon to begin with. They may be uncommon because they have been competitively excluded from several islands by other species. Our data support this interpretation; for example, we regard it as more than just coincidence that one or the other of these two rare species is the nearest morphological neighbor to each of the remaining four, common *Geospiza* species (Table 13.3). Thus an analysis that utilizes the above weighting procedure may not be able to detect the phenomenon it is designed to detect because the phenomenon has been at least partly factored out. We conclude from our analyses that *Geospiza* species assemblages and morphology are non-random in the direction predicted by interspecific competition.

BEAK MORPHOLOGY IN SYMPATRY

Implicit in the above arguments is the assumption that the mean vector of a species in morphological space is sufficient to characterize all of its populations. However, while this is suitable as a first approximation, considerable morphological variation exists among populations of each *Geospiza* species. Might the morphology of individual populations be adjusted to the characteristics of sympatric congeners? Lack (1947) believed so, and presented evidence to support his claim that dissimilarities between coexisting species were in part due to evolutionary shifts impelled by competitive interactions. In this section we test his claim.

Grant (1981b, 1983) drew attention to an interesting feature of beak dimensions in sympatric populations of *Geospiza* (Figure 13.3). Measurements of museum specimens show that all coexisting populations of the six *Geospiza* species differ by at least 15 percent in mean beak depth and/ or length. The phenomenon occurs despite inter-island variation in beak size and despite a high frequency of species pairs that differ by less than 15 percent in one or the other dimension. This "empirical rule" (Grant, 1968, 1972a) suggests some manner of spacing along a beak-size continuum that is consistent with a competition hypothesis and consistent with the concept of a limiting similarity. To proceed further we ought to test the observed data against a null hypothesis, but we have not done

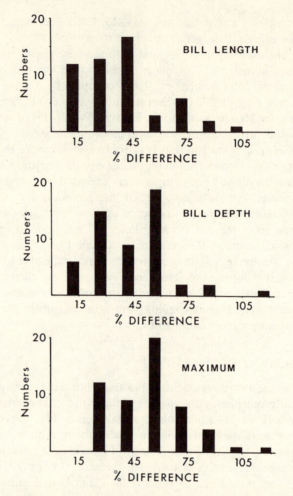

Fig. 13.3. Differences in mean beak dimensions between all pairs of sympatric populations of *Geospiza* species. The difference between any pair is expressed as a percentage of the smaller mean (grouped into 15 percent classes). The lower histogram represents the dimension giving the largest difference between pairs of populations, in either bill length or bill depth.

so because we have been unable to devise an expected frequency distribution of beak size differences based upon a realistic model of no interaction between species.

Grant (1983) also showed that 11 of 13 pairs of *Geospiza* species were more different, on average, in sympatry than in allopatry. In order to investigate this aspect of coexistence further, we designed a procedure that compares beak-size differences of sympatric populations of species pairs with those of randomly combined pairs of populations of the same species. We restricted the test to species occurring together on a relatively large number of islands (Table 13.3) to ensure reasonable sample sizes. However, it should be noted that Lack (1947) suspected there was also a very strong relation between morphology and the presence in pairs of congeners that coexist infrequently (*e.g. G. fuliginosa* and *G. difficilis*; see also previous section). We used three species pairs of similar beak dimensions: *G. fortis–G. scandens*, *G. fortis–G. fuliginosa*, and *G. fortis–G. magnirostris*. Details of the procedure are summarized here for one of the three comparisons.

First, we observed that the two beak dimensions under consideration, beak length and depth, are related among populations in an allometric fashion. Figure 13.4 illustrates this relation for populations of *G. scandens*. The distribution of populations of *G. fortis* is likewise allometric, but with a different coefficient of static allometry. Second, we assumed that in both species the mean and variance of the two beak dimensions and the covariance between the dimensions characterize a bivariate normal

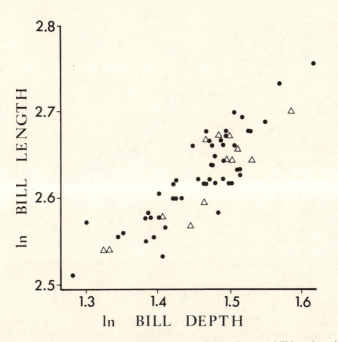

Fig. 13.4. The allometric relationship between adult male mean bill length and mean bill depth among 13 populations of *G. scandens* (△). The solid circles represent 50 points drawn randomly from a bivariate normal distribution as described in the text.

distribution of possible population morphologies. A random sample of
50 points was then selected from each distribution using a Gaussian simulation algorithm in MIDAS (Michigan Interactive Data Analysis System).
These are shown in Figure 13.4 for *G. scandens*. By randomly combining
pairs of these points, one from the *G. scandens* random sample with one
from the *G. fortis* random sample, we constructed 50 artificial "sympatric"
pairs of these two species.

If species pairs are hyper-distributed along a beak-size axis in sympatry, we would expect measurements of pairs from observed populations
to be more different than those of randomly constructed pairs. Figure
13.5 compares the differences in beak length and depth (log-transformed)
between these data sets. Interestingly, an inverse relationship results; this
is a consequence of the very different allometries in *G. fortis* and *G.
scandens* between the variables beak length and beak depth. To compare the difference in beak morphology between real and random pairs
we used the variable Z, which is the straight-line distance between points

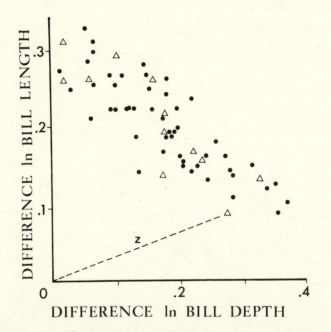

Fig. 13.5. The differences along two bill dimensions between 12 sympatric
populations of *G. fortis* and *G. scandens* (△) and between 50 randomly generated pairs of these species populations (●) (see text for details). Differences
along two dimensions can be collapsed onto a single dimension by calculating
the distance (z) between each point and the origin, as illustrated here with a
single example.

in Figure 13.5 and the origin. Cumulative proportions of individual points are plotted against Z-values in Figure 13.6 (top) for both data sets. The observed distribution is found to be statistically indistinguishable from the random one (Kolmogorov-Smirnov test, $P > 0.05$). We thus conclude that, in this case, there is no evidence of adjustment of beak morphology in sympatry.

However, there is a hidden problem. Measurements of 13 populations of each species produced the bivariate normal distribution from which random samples were derived, but 12 of the populations of *G. scandens* are sympatric with 12 of the *G. fortis* populations. Hence the characteristics of random communities in this case are highly dependent upon those of the actual pairs they are to be compared with, which is a severe limitation of the technique (*cf.* Grant and Abbott, 1980). We therefore cannot say from Figure 13.6 (top) that one species has no effect on the beak characteristics of the second. The results merely show that a re-shuffling of already sympatric populations produces no discernible difference in beak length and depth differences.

The situation is different in the second comparison, *G. fortis–G. fuliginosa* (Figure 13.6 middle). We repeated the test procedure using 19 populations of *G. fuliginosa* to characterize the random samples for this species. Seven of these points are derived from islands where *G. fortis* is not present. Thus the distribution of the Z-values for the random pairs is based upon measurements of the two species both in allopatry and in sympatry. The Z values for real species pairs are, of course, only from sympatric populations. The difference between these curves is significant at the 0.05 level (Kolmogorov-Smirnov test). Randomly constructed pairs of *G. fortis* and *G. fuliginosa* are more similar than actual pairs, and therefore the hypothesis of regular size-spacing is supported in this case.

The test was repeated once more with *G. fortis* and *G. magnirostris*. Cumulative proportions of Z-values for actual and random pairs are shown in Figure 13.6 (bottom). In this case nine populations of *G. magnirostris* and 13 populations of *G. fortis* generated the random pairs, and these are compared with dimensions of only six sympatric pairs. The sample size is small, but there is no detectable difference between them ($P > 0.05$).

Hence in one of three comparisons, observed beak differences between pairs of species in sympatry are significantly greater than those of randomly constructed pairs. Notably, this one pair is morphologically the most similar of the three, and is the most similar of all species pairs that occur together on a greater number of islands than expected (Table 13.3).

We emphasize that we cannot be certain a comparable effect does not also occur in the other species pairs investigated. The problem of non-independence of actual and random populations is present in all three

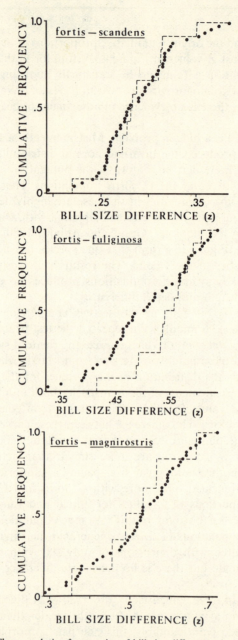

Fig. 13.6. The cumulative frequencies of bill size differences (z, see Fig. 13.5) of observed sympatric pairs of populations (broken line) and of randomly generated pairs (solid circles) for (a) *G. fortis* and *G. scandens*, (b) *G. fortis* and *G. fuliginosa*, and (c) *G. fortis* and *G. magnirostris*. The difference between observed and randomly expected frequency distributions is significant only in the case of *G. fortis* and *G. fuliginosa* populations (Kolmogorov-Smirnov test, $P < 0.05$).

analyses. There could not be a very great difference between the actual and random pairs in the mean distance Z in such cases as *G. fortis–G. scandens*, where most populations of one species are sympatric with the other, regardless of any possible structure inherent in the real community. The test is therefore nearly powerless. The characteristics of randomly combined *G. fortis–G. fuliginosa* pairs are less restricted by the original data, and in this case the randomly combined pairs were quite different from actual pairs. However, even here the characteristics of random pairs are not free of dependence.

An alternative procedure is to test the variances of the Z-values for equality, with the expectation that if competition has constrained the bill differences between species the variance will be lower for the observed than for the randomly generated values. The expectation is realized in all three interspecific comparisons, but statistical significance is approached only in the case of *G. fortis* and *G. fuliginosa* ($F_{49,11} = 2.23$, $P = 0.07$, for a one-tailed test). A second problem of dependence occurs here, however, since *G. fortis* enters all three tests. We see no solution to this dilemma with *Geospiza* species; the problem is reduced to an acceptable minimum only when sympatric populations are a small subset of total populations (Schoener, this volume).

In conclusion, the hypothesis that the beak morphology of a species is influenced by the characteristics of sympatric congeners has some support. Results corroborate those of Grant (1983). All coexisting populations of the six *Geospiza* species conform to the 15-percent rule, and in one of three species pairs tested, beak differences in sympatry are significantly greater than in randomly combined populations.

THREE EXAMPLES SUBJECTED TO DETAILED ANALYSIS

Character Release in Geospiza fortis *on I. Daphne*

Probably the most frequently published set of data from David Lack's (1947) book on Darwin's Finches is the histogram of beak depths, reproduced again in Figure 13.1. The small ground finch (*G. fuliginosa*) and the medium ground finch (*G. fortis*) are discretely different in bill dimensions on the several islands where they occur together, but each species has an intermediate beak size on one island where it occurs alone. Lack (1947) interpreted the morphological intermediacy in allopatry as a response to the release from competitive constraints, a phenomenon since named character release or competitive release (Brown and Wilson, 1956; Grant 1972b). Bowman (1961) was critical of much of Lack's evidence for competitive shifts, but left this example alone.

In planning our research in 1972, Ian Abbott and one of us (P.R.G.) could not see why *G. fortis* should become smaller on I. Daphne in the absence of *G. fuliginosa* when, to us, it seemed just as reasonable to expect it to become larger in the absence of *G. magnirostris*. A closer reading of Lack revealed that some specimens of *G. fuliginosa* had been collected on this island on more than one occasion; furthermore, a pair of *G. magnirostris* had been observed by Beebe (1924) in 1923. We know now that both these species immigrate regularly, and some die on the island; most probably emigrate, and neither species has a sustained breeding population there (Grant *et al.*, 1975). The problem remains; why has *G. fortis* become smaller on Daphne?

Lack's competitive release hypothesis assumes a similar food supply on the different islands and presupposes that competitive interaction between *G. fortis* and *G. fuliginosa* on islands other than Daphne prevents either species from occupying a position of morphological intermediacy at which it can exploit elements of both niches as one of them does on Daphne. Bowman's (1961) views on the importance of food supply *per se* suggest an alternative; the nature of the food supply alone dictates that smaller than typical *G. fortis* have an advantage over typical *G. fortis* on Daphne. This hypothesis has the merit of simultaneously explaining why *G. magnirostris* is not present; large seeds are lacking on the island. However, it does not explain the absence of *G. fuliginosa* without the modification that the small seeds eaten elsewhere by this species are lacking on the island.

Part of the Ph.D. dissertation research of Peter Boag has been directed toward testing the competition and food-supply hypotheses (Boag, 1981). This research has involved an estimation of food supply and a characterization of feeding behavior of finches on I. Daphne and on the nearby large island of Santa Cruz, in both the early and the late dry season (Boag and Grant, 1983).

The results do not support the food-supply hypothesis (Figure 13.7). Daphne is not a "small-seed island," compared with the north shore of I. Santa Cruz, so *G. fortis* has not become smaller on Daphne simply because the seeds are smaller (Boag and Grant, 1983). There is some evidence that *G. fuliginosa* is absent because small seeds, although present, are not always common. The absence of *G. magnirostris* might be attributed to the relative scarcity of large seeds, but this possibility is not suggested by the data. Unfortunately, *G. magnirostris* was not present at the Santa Cruz study site or in its neighborhood so we cannot make the necessary quantitative comparison between islands.

Lack's assumption of a similar food supply in the two islands is only approximately correct. His hypothesis of competition is supported to the extent that the small *G. fortis* on Daphne exploit the periodically large supply of small-medium seeds there. Is this a simple response to the absence of *G. fuliginosa* or do *G. fortis* competitively exclude *G. fuliginosa*?

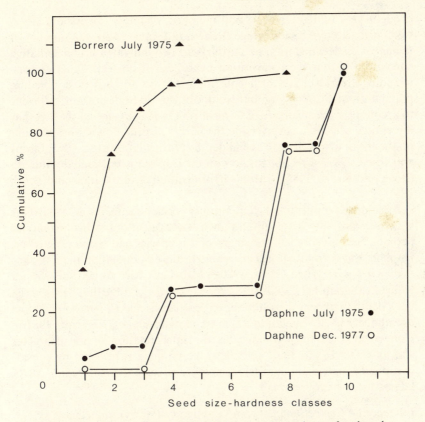

Fig. 13.7. Cumulative frequency distributions, based on volume, of seeds and fruits on Isla Daphne Major and on the neighboring large island of Isla Santa Cruz at Bahía Borrero (from Boag and Grant, 1983). Seeds and fruits are classified by their size (D, depth) and hardness (H) (see Abbott *et al.*, 1977). There is less separation among frequency distributions based on seed and fruit numbers, but their relative positions are the same (Boag, 1981).

Geospiza fuliginosa certainly immigrate, but most leave before breeding. Although breeding attempts of *G. fuliginosa* were recorded on Daphne in 1974 (Grant *et al.*, 1975) and 1981 (T. D. Price, pers. comm.), five of the six individuals breeding in the period 1976 to 1980 were females paired with *G. fortis*. Thus it is the failure of immigrant male *G. fuliginosa* to stay to breed that has resulted in the failure of the species to become established recently as a breeding population. Observations suggest that this failure may be caused by interference from male *G. fortis*, which prevents male *G. fuliginosa* from setting up territories (Ratcliffe, 1981). Exploitative competition with *G. fortis* for food may also occur, because during the drought year of 1977, when small members of the *G. fortis* population were at a selective disadvantage (Boag and Grant, 1981), the number of *G. fuliginosa* declined from about thirty to two individuals. Our tentative conclusion is that the competition hypothesis is correct. Other factors, including hybridization, also contribute to the absence of a breeding population of *G. fuliginosa*, but we doubt that hybridization is sufficient to account for the small size of *G. fortis* on Daphne in view of the fact that directional selection can override the effects of gene flow (hybridization) on the mean of a trait such as bill depth (Ehrlich and Raven, 1965; Grant and Price, 1981): directional selection on Daphne is occasionally intense (Boag and Grant, 1981), whereas hybridization is rare (Grant and Price, 1981).

An added complication is that another species occurs on Daphne! This is *G. scandens*, the cactus ground finch. Lack left it out of his histogram, perhaps because very few specimens of this species had been collected on Daphne. But currently it has a sizeable breeding population there (Grant *et al.*, 1975; Grant and Grant, 1980a). Had Lack tried to add this species to his histogram he would have experienced great difficulty, because the beak depth frequency distributions of *G. scandens* and *G. fortis* on Daphne are almost exactly superimposable (Abbott *et al.*, 1977; Grant, 1981b)! Convergence certainly complicates the interpretation of the character shift of *G. fortis*, although we believe the paradox is resolved by consideration of a second bill dimension, length, and its functional feeding correlate as we discussed earlier.

Interspecific Competition between G. fuliginosa and G. difficilis

Lack (1947) observed that the two smallest ground finch species, *G. fuliginosa* and *G. difficilis*, are essentially allopatric. Low-elevation, dry islands are occupied by either one species or the other, never both. High-elevation islands have both species, but here they are apparently segregated; *G. fuliginosa* occupies the lowlands, and *G. difficilis* occupies the highlands. Noting the morphologial similarity between the species, Lack supposed that *G. difficilis* was excluded from low islands or restricted to the highlands through competition with *G. fuliginosa*. Here we briefly describe a test of this hypothesis based on inter-island comparisons of the behavior and density of the species. A detailed version is presented elsewhere (Schluter and Grant, 1982).

An investigation of the distribution, abundance, and diets of the two species was conducted on one island, Pinta, over a full calendar year (Schluter, 1982a). Surveys showed that on this island *G. difficilis* and *G. fuliginosa* are not allopatric. Instead their ranges overlap, although they are not identical (Figure 13.8). Detailed study revealed that, in sympatry,

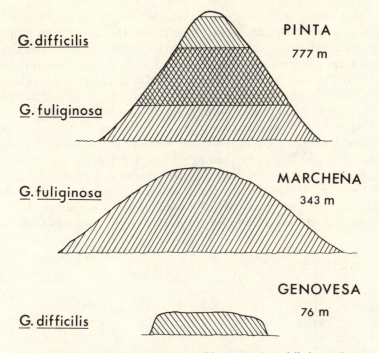

Fig. 13.8. The distributions of *Geospiza fuliginosa* and *G. difficilis* on three northern islands, drawn only approximately to scale and with elevation in meters indicated. On Isla Pinta the two species overlap altitudinally to a large but not complete extent (based on data in Schluter, 1982a).

interspecific competition might currently be weak or absent. Interspecific territoriality and other forms of aggressive interference were not observed, and diets and feeding behavior of the two species were very different. Moreover, the distributions and abundance of *G. fuliginosa* and *G. difficilis* on Pinta could be explained simply by the availability of their different foods. Competition between them was not detected even when food was relatively scarce.

Present patterns in use of resources and altitudes by the two species might nonetheless by the result of past competition. Lack's hypothesis may therefore still account for the observed distributions in the archipelago. We may use the information gained from Pinta on the diet, density and food supply of each species of predict the attributes of the same species in allopatry. This procedure forms the basis for the inter-island test. In this study we used populations of *G. difficilis* on Genovesa (where *G. fuliginosa* is absent) and of *G. fuliginosa* on Marchena (where *G. difficilis* is absent; Figure 13.8).

Three alternative hypotheses to account for the distributions of *G. fuliginosa* and *G. difficilis* are here considered. The first on (H1) proposes that the unique and unvarying food requirements of species suffice to explain the pattern of distribution of the species among islands. This hypothesis therefore predicts that the diet of a given species in allopatry (Genovesa or Marchena) should be the same as its diet in sympatry (Pinta). Densities in relation to food supply should be the same, too.

A second hypothesis (H2; actually a modification of the first) suggests that food requirements of individual species may be adapted to local conditions, and hence vary among islands, but that the distribution of species is nevertheless explained by the distribution of their acceptable food supply. H2 thus predicts that if a species does not occur where the other is present it is because of some food deficiency in the habitat. These alternatives (H1 and H2) are two restatements of Bowman's (1961) sug-

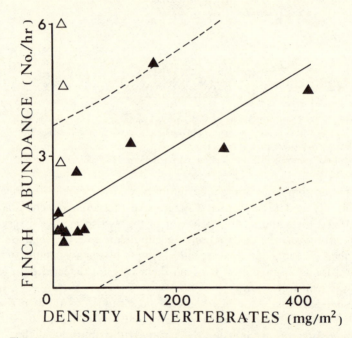

Fig. 13.9. Regression relationships (solid lines) of finch abundance on preferred food density on Isla Pinta in the dry season. Dashed lines are 95% confidence limits for Y-estimates. (a) Calculated relationship for *G. difficilis* on I. Pinta (▲), with three points for *G. difficilis* on I. Genovesa (△) superimposed. (b) Calculated relationship for *G. fuliginosa* on I. Pinta (●), with points for *G. fuliginosa* on I. Marchena (○) and *G. difficilis* on I. Genovesa (△) superimposed. The food axes differ in (a) and (b) because on I. Pinta *G. difficilis* feeds predominantly on invertebrates and *G. fuliginosa* feeds predominantly on seeds (based on data in Schluter, 1982a).

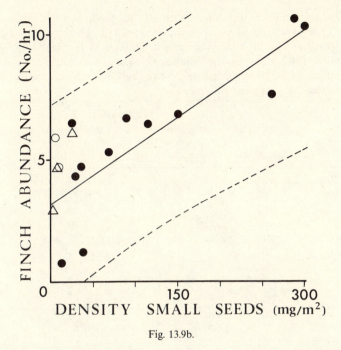

Fig. 13.9b.

gestion that food supply is the major determinant of finch distributions (see also Abbott *et al.*, 1977).

The third hypothesis (H3), that of competitive exclusion, argues that the distribution of species is not explained by food supply alone. Instead it invokes competition for food and predicts convergence in allopatry: on low dry islands (*e.g.* Genovesa and Marchena) where the food supply should be similar, the feeding ecology of the two species would be similar, too.

All three hypotheses are in agreement regarding *G. fuliginosa* on Marchena: its behavior and density should correspond to those in lowland sites on Pinta. Therefore only the data from Genovesa can be used to discriminate among the three hypotheses. We present Marchena information as a check on the testing procedure.

H1: This hypothesis explains the distributions of finches by the distributions of species-specific food requirements. To test it we first compared the density of *G. difficilis* on Genovesa with that expected from its density in relation to food supply on Pinta (Figure 13.9a). The results show that at two of three sites on Genovesa the density of *G. difficilis* is considerably greater than predicted, which contradicts the hypothesis. As a check on the procedure we note that *G. fuliginosa* density on Marchena is quite accurately predicted by a similar Pinta regression (Figure 13.9b).

Table 13.4. Diet similarities among populations of G. *fuliginosa* and G. *difficilis* in the late dry season of 1979. Diet items were classified into categories of arthropods, seeds, and pollen/nectar plus berries/arils. Similarities were determined by the Renkonen-Whittaker formula (Hurlbert, 1978). Comparisons involving G. *difficilis* from Genovesa (G) involve an unweighted average of the diet proportions in three grids on that island, all at about 20 m altitude. P = Pinta, M = Marchena.

		SIMILARITIES			
		difficilis (G) v	*fuliginosa* (P) v	*difficilis* (G) v	*fuliginosa* (M) v
Sites	Altitude (m)	*difficilis* (P)	*difficilis* (P)	*fuliginosa* (P or M)	*fuliginosa* (P)
P1	110	.21	.18	.94	.92
P2	180	.69	.63	.91	.91
P3	225	.27	.25	.93	.90
P4	285	.29	.28	.80	.76
P5	400	.10	.05	.72	.46
P6	510	.09	—	—	—
M1	10	—	—	.82	—
M2	170	—	—	.84	—

To perform a second test we compared the diets of G. *difficilis* on Genovesa and on Pinta (Table 13.4). The two island populations of G. *difficilis* are very different in their diet—as different as G. *fuliginosa* is from G. *difficilis* on Pinta (Table 13.4). These diet differences are not accounted for by variation in the relative availability of foods. Figure 13.10 shows that G. *difficilis* on Pinta feeds on invertebrates and gastropods more commonly than on seeds at most sites. However, on Genovesa the same species infrequently consumed invertebrates despite a greater relative availability of these in the dry season. These data also contradict the hypothesis.

We repeated the diet analysis and found that seasonal changes in the absolute abundance of food also fail to account for diet differences between Genovesa and Pinta populations. We also noted that on Genovesa G. *difficilis* frequently consumes nectar, whereas on Pinta the same species never does so despite the abundance of nectar there. The food categories used in the diet comparison (Table 13.4, Figure 13.10) were broad; more detailed categorization of diet incorporating prey type and feeding position only enhances the difference between Pinta and Genovesa G. *difficilis*. In short, species-specific food requirements do not seem to exist in this species. Consequently, we reject the first hypothesis.

H2: The second hypothesis predicts that Genovesa is different in food supply from other low islands in a manner that accounts for the absence

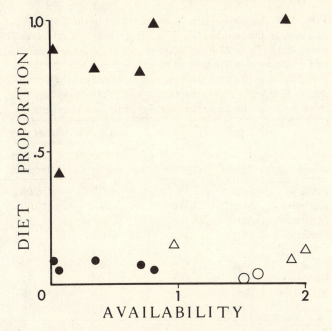

Fig. 13.10. The proportion of invertebrates in the dry season diets of *G. fuliginosa* and *G. difficilis* in comparison with their availability. Diet represents amount of time spent feeding on invertebrates as a proportion of total time spent feeding on invertebrates and seeds and fruits. Availability is the ratio of invertebrate to small seed dry weights converted to percentage of the maximum value recorded and then $\log_{10}$-transformed. Symbols represent *G. difficilis* on I. Pinta (▲) and I. Genovesa (△), and *G. fuliginosa* on I. Pinta (●) and I. Marchena (○).

of *G. fuliginosa* and the presence of *G. difficilis* there. Table 13.5 shows the abundance of the plant food sources at low-altitude sites on the three islands studied. Seed density on Genovesa corresponds closely to that at Marchena sites, and flower supplies are similar to those on Pinta at comparable dates. Clearly, therefore, sites on Genovesa are not deficient in these important foods for *G. fuliginosa*.

Species composition is also similar among islands. *Croton scouleri* and *Trichoneura lindleyana* are among the most abundant small seeds on Genovesa; the first of these predominates in the diet of *G. fuliginosa* on Marchena, and the second is highly preferred by *G. fuliginosa* at low-altitude sites on Pinta. The most abundant nectar-producing flower at low altitudes on Genovesa is *Waltheria ovata*; it is intensively exploited by *G. fuliginosa* on Pinta and Marchena.

Thus there is evidence that conditions are sufficient on Genovesa to allow *G. fuliginosa* to survive there. Breeding requirements also seem to

Table 13.5. Density of plant foods important in the diet of *G. fuliginosa* in lowland sites in 1979. Pinta values are averages of sites 1 and 2 (Table 13.4). Small seeds refer to those seeds preferred by *G. fuliginosa* on Pinta or Marchena (see Schluter, 1982b, for definition of preference) plus *Eragrostis cilianensis*, a preferred seed of *G. difficilis* on Genovesa. Flowers refer to those exploited by *G. fuliginosa* for nectar or pollen on Pinta or Marchena.

| | | | Mean density of food types | |
| | | | Small seeds (mg/m^2) | Flowers $(no./m^2)$ |
Island	Month	Season		
Pinta	Feb.	Wet	340.7	0.00
Genovesa	March	Wet	109.4	0.00
Marchena	April	Wet	100.6	.88
Pinta	May	Wet	600.9	.48
Pinta	Aug.	Dry	46.8	.24
Genovesa	Sept.	Dry	9.4	.27
Marchena	Oct.	Dry	5.6	1.36
Pinta	Nov.	Dry	29.5	1.58

be met: arthropods, chiefly caterpillars and spiders, are abundant in the wet season (Grant and Grant, 1980b). In addition, we can identify nothing unique about Genovesa that would explain the presence of *G. difficilis* there but its absence from Marchena. These data contradict the second hypothesis and therefore we reject it.

H3: Lack's competition hypothesis predicts that, on Genovesa, *G. difficilis* should exhibit niche convergence with *G. fuliginosa* in the absence of the latter species. Thus we expect the diet of *G. difficilis* there and, when food is limiting, its density to be similar to the diet and density of *G. fuliginosa* elsewhere. To test these expectations we first compared the dry season density of *G. difficilis* on Genovesa with that expected for *G. fuliginosa* on the basis of the food supply (Figure 13.9b). At all three sites on Genovesa we found that *G. difficilis* occurs in approximately the density predicted, supporting Lack's hypothesis. We next compared the diet of *G. difficilis* on Genovesa with that of *G. fuliginosa* on Pinta and Marchena (Table 13.4). In each case the diets are very similar—more so than the diets of different populations of *G. difficilis* and as similar as different populations of *G. fuliginosa* (Table 13.4).

Figure 13.10 reveals a partial basis for such similarity between the species: on Genovesa *G. difficilis* (like *G. fuliginosa*) consumes small seeds much more frequently than arthropods and gastropods. The species also takes nectar, a very important food for *G. fuliginosa* on Pinta and Marchena. Moreover, the feeding behavior of *G. difficilis* on Genovesa is similar to that of *G. fuliginosa* on other islands: seeds are frequently removed from rock surfaces or vegetation, and arthropods are taken from the foliage. These data clearly indicate convergence of diets in allopatry, in parallel with a convergence in bill form (Lack, 1947), and the competition hypothesis is further supported.

We conclude that the two non-competition hypotheses, based upon differences in food requirements between species or in food supply among islands, are not sufficiently supported to account for the distributions of *G. difficilis* and *G. fuliginosa* in the Galápagos. We observe that the best predictors of *G. difficilis* behavior and density on Genovesa are those of *G. fuliginosa* on other islands. Since similarity in allopatry in similar habitats is expected under the hypothesis of competitive exclusion, we reject H1 and H2 against this superior alternative.

Competition and Niche Expansion by Geospiza conirostris

Geospiza conirostris has three breeding populations: one on Isla Española, another on its satellite, I. Gardner, and a third on I. Genovesa. In beak size and shape *G. conirostris* resembles species not present on those islands, and this finding led Lack (1945, 1947, 1969b) to suggest that *G. conirostris* had competitively excluded the missing species and appropriated their niches. Thus the populations of *G. conirostris* on Islas Española and Gardner were supposed by him to have appropriated the resources exploited elsewhere by *G. magnirostris*, *G. fortis*, and *G. scandens*; on I. Genovesa *G. conirostris* was supposed to have taken over the niches of absent *G. scandens* and *G. fortis*. The fact that *G. conirostris* populations were not the same on all three islands was attributed to the presence of *G. magnirostris* on I. Genovesa (only), which had resulted in a competitive displacement of the *G. conirostris* there toward a *fortis/scandens*-like species.

The hypothesis makes four predictions, which we have recently tested with ecological data (Grant and Grant, 1982):

1. Española *G. conirostris* should have a broad feeding niche, which is more similar to a combined *magnirostris-fortis-scandens* niche than to the niches of these species considered alone. This prediction was found to be correct. The niche of Española *G. conirostris* is unusually broad.
2. Genovesa *G. conirostris* should have a feeding niche similar to a combined *fortis-scandens* niche. This prediction was not correct. However, the *G. conirostris* niche is more similar to a *G. scandens* niche than to the niche of any other species or to the niche of Española *G. conirostris*.

3. The niches of *G. conirostris* and *G. magnirostris* on Genovesa should be substantially different, and the difference should reside in that *G. magnirostris* takes food items exploited by *G. conirostris* on Española but not on Genovesa. This prediction is correct.

4. *Geospiza conirostris* on Española should be more efficient than *G. magnirostris* at dealing with foods on that island. This prediction was tested by comparing the performance of the two species on a single but probably critical food type, *Cordia lutea* stones, on different islands (*G. conirostris* on Española and *G. magnirostris* on Genovesa). This prediction was not correct; *G. magnirostris* was the more efficient species. However, *G. conirostris* on both islands were found to be more efficient at dealing with the foods normally taken by *G. fortis* and *G. scandens* than were those two species.

Thus, there is substantial support for the competition hypothesis, but it does not explain why *G. fortis* is absent from Genovesa or why the apparently more efficient *G. magnirostris* is absent from Española and Gardner. The two major alternative explanations for the absence of these species, and others, are (1) the missing species have never reached those islands and (2) their food supply is absent from those islands. Neither of these alternatives is supported by data.

Movements of the missing species between islands seem to occur often, and three of the five have been recorded as vagrants on the islands in question. Our quantitative data on food supply show that the food items normally consumed by the missing species elsewhere are present on Española and Genovesa (Grant and Grant, 1982). Since the alternative hypotheses are unsatisfactory, we accept an imperfect competition hypothesis and seek *ad hoc* explanations for the discrepancies. They are easy to find! First, we suggest that *G. fortis* is absent from Genovesa because it has been competitively excluded by the combined influence of *G. conirostris*, *G. magnirostris*, and a third congener, *G. difficilis*. The food available to *G. fortis* on Genovesa is currently exploited by these three species. Second, we suggest that a breeding population of *G. magnirostris* is lacking on Española because whenever *G. magnirostris* arrive they do so in small numbers, and then they die, depart, or hybridize with *G. conirostris*. The only condition under which competitive exclusion is likely is the immigration of immature *G. magnirostris*, because immature birds are likely to be less efficient at dealing with hard seeds than adult resident *G. conirostris*. Although this is an *ad hoc* explanation, it is reasonable, because most immigrant *G. magnirostris* on another island, I. Daphne Major, are immatures (Grant *et al.*, 1975).

Note that the hybridization-exclusion hypothesis differs from a similar one applied to the absence of *G. fuliginosa* from Daphne (above).

Resident *G. fortis* males are individually dominant to the males of the smaller immigrant species, *G. fuliginosa*, on Daphne; in contrast, resident *G. conirostris* males are unlikely to be individually dominant to the males of the larger immigrant species, *G. magnirostris*, on Española. On Genovesa *G. magnirostris* are individually dominant to the smaller *G. conirostris* (Grant and Grant, 1980b). Therefore we are unable to identify a single reason for the absence of *G. magnirostris* from Española, and suggest, as mentioned above, more than one factor, including exploitative competition.

CONCLUSIONS

The last major symposium on communities produced a publication edited by Cody and Diamond (1975a). Both of the editors contributed papers on their studies of bird communities, and both argued for the importance of interspecific competition as a structuring force. Partly in reaction to this viewpoint and the methods by which it was supported, several authors have since posed this role of competition as a testable hypothesis and have often successfully rejected it (*e.g.* Connor and Simberloff, 1978, 1979; Simberloff 1978a; Strong *et al.*, 1979; Ricklefs and Travis, 1980; Rotenberry, 1980b; Rotenberry and Wiens, 1980a; Wiens and Rotenberry, 1980a). In contrast, we have failed to reject it, except in a few cases. Why is there a difference between our results and theirs?

One reason lies in the tests and analyses. Strong *et al.* (1979) are explicit in calling for appropriate null hypotheses, which, in their view, have been neglected. We have argued here and elsewhere (Grant and Abbott, 1980) that Connor and Simberloff (1978, 1979), Simberloff (1978a), and Strong *et al.* (1979) formulated inappropriate hypotheses and then tested them by inappropriate means. Our results differ from theirs largely because we constructed our tests differently. We then applied them to almost the same biogeographic data and found that the assemblages of finch species were structured.

Another reason is that ecological studies have been conducted in different ways. Rotenberry (1980b), Rotenberry and Wiens (1980a), and Wiens and Rotenberry (1980a), in impressively thorough studies of grassland communities of birds, have found little reason to believe that interspecific competition has influenced species composition, habitat use, and diets. We might have reached the same conclusion if our studies had been restricted to the breeding season, as theirs largely were. But it is clear from our seasonal comparisons of food supply and diets that the non-breeding dry season is the time when resources are limiting and effects of competition are most likely to be manifested (Grant and Grant, 1980a, b; Smith *et al.*, 1978). Temperate-zone bird communities disintegrate in the non-breeding season through departure of the migrant component, and this phenomenon makes it difficult to obtain data comparable to ours.

Perhaps Darwin's Finches are unique, by virtue of a monophyletic origin and restricted mobility, and they alone have experienced interspecific competition. We see no reason why this should be so. Competition has been demonstrated in the other major groups of vertebrates, notably mammals (Brown, this volume; Grant, 1972c, 1978), lizards (Dunham, 1980), frogs (Inger and Greenberg, 1966), salamanders (Hairston, 1980a; Jaeger, 1971), and fish (Werner, this volume), and between different taxa such as rodents and ants (Brown and Davidson, 1977).

While there are limitations to each of our biogeographic, morphological, and ecological analyses, they yield a consistent result, so our overall conclusion is that competition has been partly responsible for present-day properties of Darwin's Finch guilds within communities. We further suggest it has been important in other guilds, including those of birds, elsewhere.

Since this conclusion and suggestion will, correctly, invite critical attention, we conclude by making explicit what we do and what we do not claim. The tests we have applied yield support for the competition hypothesis and no support for the alternatives we considered; therefore we accept it as the best available explanation of the data. We do not claim that a better hypothesis is not possible. For example the non-random combinations of species may be caused entirely by non-random distributions of food supply across the Galápagos archipelago, and not by competitive exclusion. This possibility is rendered unlikely by the results of our morphological and ecological studies. But the differences in bill size between sympatric species may not function solely to effect an ecological separation between the species; they may also effect reproductive isolation. Lack (1945) originally interpreted beak differences solely in these terms. Therefore it can be argued that ecological differences between coexisting species are an incidental byproduct of selection for divergence in species-specific reproductive signals, *i.e.* bill dimensions. Some implications of the reproductive isolation hypothesis have been tested experimentally, and the hypothesis has been supported by the results (Ratcliffe, 1981).

While plausible, this reproductive isolation hypothesis does not account for much of our data. For example, it does not account for the seasonal divergence of feeding niches of coexisting species as food abundance declines (Grant and Grant, 1980b; Schluter, 1982b; Smith *et al.*, 1978) or for the particular beak characteristics of species in allopatry. The chief merit of the competition hypothesis is its broad applicability.

So far it has been consistently the better explanation of patterns in Darwin's Finches.

Acknowledgments

We appreciate comments received from I. Abbott, A. E. Dunham, B. R. Grant, T. D. Price, B. Rathcke, and T. W. Schoener. Our recent research has been supported by grants from N.S.F. (DEB 77-23377 and DEB 79-21119).

Appendix. Revised island list for *Geospiza* species. Breeding populations are indicated by B. E indicates populations that are now extinct, and (E) indicates populations suspected of having been present in the last century and having since gone extinct. Sources for this compilation are Harris (1973), Abbott *et al.* (1977), Grant *et al.* (1980), Sulloway (1982), D. Day (pers. comm.), and our unpublished observations. Either we or our associates have visited all islands except Pinzón, Eden, Beagle, and Caldwell. Obvious clusters of islets, such as Rocas Bainbridge, have been treated as single islands. We have deleted *G. fortis* from the Los Hermanos list given by Abbott *et al.* (1977) because on two visits we have not found it. *Geospiza magnirostris* was observed breeding on Santa Fe in the 1960's (Tj. de Vries, pers. comm.), but may have become extinct. *Geospiza fuliginosa* has not been included in the list for Daphne Major because recent breeding attempts have not been sustained for more than a year (Grant and Price, 1981).

	SPECIES					
	G. magnirostris	*G. fortis*	*G. fuliginosa*	*G. difficilis*	*G. scandens*	*G. conirostris*
ISLANDS						
Seymour		B	B		B	
Baltra		B	B		B	
Isabela	B	B	B	(E)	B	
Fernandina	B	B	B	B		
Santiago	B	B	B	B	B	
Rábida	B	B	B		B	
Pinzón	B	B	B		E	
Santa Cruz	B	B	B	E	B	
Santa Fe	B	B	B		B	
San Cristóbal	E	B	B	(E)	B	

Española	E		B		B	B
Floreana	B	B	B	(E)	B	B
Genovesa	B	B	B	B	B	
Marchena	B	B	B	B	B	
Pinta	B	B	B	B	B	
Darwin	B		B	B	B	
Wolf	B		B			
Plazas		B	B			B
Gardner (by Española)		B	B			
Bartolome			B			
Daphne Major		B	B		B	
Daphne Minor		B	B			
Tortuga			B			
Hermanos			B			
Eden			B			
Bainbridge			B			
Beagle			B			
Cowley			B			
Gardner (by Floreana)			B			
Enderby			B			
Champion		B	B		B	
Caldwell						

14.

Properties of Coexisting Bird Species in Two Archipelagoes

DANIEL SIMBERLOFF

Department of Biological Science, Florida State University, Tallahassee, Florida 32306

INTRODUCTION

Strong *et al.* (1979) concluded that size ratios of coexisting bird species on islands provide "little, if any, evidence of the phenomenon [community-wide character displacement]" and therefore do not support the hypothesis that interspecific competition was a key determinant of sizes. Grant and Abbott (1980) and Hendrickson (1981) criticize this paper on various grounds, and the former authors find fault with several of my efforts in a similar vein: Simberloff (1970, 1978a) and Connor and Simberloff (1978). Here I reexamine avifaunas of the Tres Marias Islands and Galápagos and attempt to deal with the criticisms and to incorporate the authors' suggestions in expanded analyses.

The gist of the criticized papers was to ask whether the sets of bird species on the islands differed in some characteristic of interest, such as taxonomic relationships or sizes, from sets drawn randomly from the species pool. My colleagues and I have suggested that where differences are not significant, one need not invoke a cause, namely interspecific competition, for the observed sets. We have attempted to be scrupulous in *not* claiming that our results indicate either random colonization or absence of competition, only that certain patterns adduced in support of competition support a non-competitive alternative just as strongly. Grant and Abbott (1980) view this entire enterprise as likely to be bedeviled by high type II error, and I concede this. It may well be that particular statistical tests associated with our procedures lack power. I would have viewed the efforts as successful if they had simply fostered the practice of formulating null and alternative hypotheses and stating what observation would *not* be taken as supporting a particular view.

A related general criticism, that many of our null hypotheses are *a priori* flawed since their underlying models are unrealistic, I do not find compelling. A model is an abstraction, and ecological models are *always* unrealistic because some aspects of the components and/or their inter-actions are not incorporated. We cannot reject a hypothesis because it is insufficiently realistic unless we subject the alternative to the same scrutiny. And it must be clear in rejecting either hypothesis on grounds of realism that the unrealistic features affect the model's predictions and bear on issues of interest. Finally, since the model that best fits a set of points is the set of points itself, it is not clear how curve-fitting with an undefined but large set of parameters can tell us much about why certain sets of species are found together and others are not, or generally how to distinguish which is the more veracious of two models that both fit a set of data well. An n-degree polynomial can fit a set of n points perfectly, but usually lacks biological explanatory power. We are more impressed if an equation with fewer parameters fits n points reasonably well, espe-cially if the parameters have a ready biological interpretation. But we can usually produce several models with approximately the same degree of complexity, all of which are more or less biologically reasonable and all of which fit the data quite closely. A choice among them, at least with the given data set, is quite arbitrary. At the risk of flogging a dead horse I would venture that experimental manipulation would be more likely to enlighten us about how species affect one another. Abbott (1980) and Grant and Abbott (1980) also feel that this is imperative for island bird studies, and that "natural experiments" are insufficiently controlled to be convincing.

For both the Tres Marias and Galápagos, two sorts of analyses are at issue: taxonomic distributions and distributions of sizes of coexisting birds. For the Galápagos, we must also discuss the number of com-binations of species, irrespective of any particular trait like taxonomy or morphology.

THE TRES MARIAS ISLANDS

Taxonomic Distribution

In 1966 Grant (1966b) suggested that islands have fewer potentially competitive species than the mainland, and that this paucity should be reflected in the presence of fewer island congeners. Comparing the Tres Marias Islands to an equivalent area of adjacent Mexican mainland, he found this conjecture confirmed. But the number of species in the main-land comparison area (121*) far exceeds that in the islands (34), and in 1970 I showed by simulation that both percent congeneric species (Grant's statistic) or mean-number-of-species-per-genus (the statistic used in much

*Grant (pers. comm.) has kindly provided a slightly revised list of the birds in the main-land comparison area; this list is available from me.

Table 14.1. Observed numbers of genera in the Tres Marias Islands and some Mexican mainland regions, plus number expected for random subsets from various pools. Integers are observations. First row is San Juanito, second Maria Cleofas, and third Maria Magdalena and Maria Madre. Maximum difference is 1.12 standard deviations.

| | | Number of species in pool | | |
		121	185	445
Number of species in subset	14	14.00	14.00	14.00
		13.66	13.68	13.70
	31	29.00	29.00	29.00
		29.30	29.41	29.55
	34	32.00	32.00	32.00
		31.96	32.09	32.26
	121		99.00	99.00
			100.07	102.30
	185			141.00
				145.89

earlier literature) are highly correlated with number of species for *random* subsets of species drawn from any mainland pool, and that for the Tres Marias Islands both statistics do not differ significantly from their expectations if the avifaunas had been random subsets from all Mexican species found within 300 miles of the islands.

Grant and Abbott (1980) level two criticisms. First, they suggest that my species pool was so large that non-random colonization might not have been detected had it occurred, because there might be some systematic geographic variation in the distribution of species into genera over a large area. In addition to Grant's pool of 121 species, Hendrickson (1981) provided a definition of an intermediate-sized source pool—all species resident in Nayarit, Sinaloa, and Jalisco below 3,000 feet—from which, using Grant (1965b), Ridgway and Friedmann (1901–1950), Friedmann *et al.* (1950), and Miller *et al.* (1957), I was able to construct a set of 185 species. My original pool was 445 species. Using expected number of genera for a given number of species (Simberloff, 1978b) as test statistic, I found (Table 14.1) that island generic compositions do not differ from random expectations whether expectations are derived using Grant's species pool, Hendrickson's, or mine. Further, numbers of genera in the smaller pools do not differ from those expected if the small pools were random subsets of the large ones. None of this says that island species are but random subsets. Nor does this say competition is not occurring. All I have shown is that the distribution of species into genera does not reflect competition or non-random (with respect to taxonomic position) colonization for this particular archipelago, and that using larger species pools in this particular system for this trait (generic affiliation) does not obscure a pattern that would otherwise have been manifest.

Grant and Abbott's second suggestion is that the model underlying my null hypothesis envisions all species as equally likely to colonize, which is unrealistic. Further, they state that "when equal weight is given to species with low or zero probability of dispersal and to species with high probability of dispersal, the analysis is tipped in favor of accepting a null, random, hypothesis" (p. 334). I cannot see how a categorical statement can be made in this matter. In my study of number of species shared between islands (Simberloff, 1978a), I found that a hypothesis based on an equiprobable model is *more* likely to be rejected than is one based on different likelihoods of colonization. For species-to-genus ratios or related statistics, bias could be increased for equiprobable models if species with different probabilities of colonization tended to belong to different-sized genera (Simberloff, 1970; Grant and Abbott, 1980). Hairston (1964) first raised this possibility, but the only systematic examination I know was my analysis of monotypic versus other genera (1970), which showed no difference. In any event, my original null hypothesis was that island species do not differ in distribution into genera from equal-sized random subsets of mainland species, which was also Grant's original hypothesis ("It is important to ask if there is a paucity of homogeneric species even in relation to the paucity of total species," 1966b, p. 452), so I fail to see why some subsets should be weighted more than others. I am only testing the original assertion.

Sizes of Coexisting Birds

As has been pointed out by Grant (1969), Strong *et al.* (1979), and Grant and Abbott (1980), species on islands may differ extraordinarily from one another for either or both of two reasons. First, species that survive on islands may be a non-random subset of the species pool. Second, species once on an island could have evolved to differ more or less than their mainland conspecifics do. The claim (Grant, 1966b, 1968; Grant and Abbott, 1980; Hendrickson, 1981) that Tres Marias congeners differ remarkably can be examined in either vein.

As for whether surviving congeners are a non-random subset of the mainland pool, we may first examine the two genera represented (on three islands) by two species: *Myiarchus* and *Vireo*. Each genus is represented in the intermediate-sized species pool by four species, so for each genus there are six species pairs, of which only one is found on the island. For each genus, I ranked species from largest to smallest in both bill length and wing length (data from same sources as used to construct the pool),

Table 14.2. Differences in size between mainland populations of sympatric island congeneric pairs, and tail probabilities that randomly drawn pairs from same genera would differ by as much. B = bill length, W = wing length.

Genus	Trait	Percent Difference	Probability of Difference as Large as Observed
Myiarchus	B	37.9	.333
	W	32.1	.167
Vireo	B	20.3	.500
	W	27.0	.333

then calculated percent differences for all six pairs, and asked whether the observed difference was extraordinarily large. The answer (Table 14.2) was "No." Fisher's combined probability test for these data gives tail probabilities of .465 for bill length and .216 for wing length.

Grant and Abbott (1980) are correct that one would be less likely to detect effects of competition within families than within genera because congeners are more likely to be ecologically similar than are confamilials. Since Hendrickson (1981) has followed Strong *et al.* (1979) in examining sizes of confamilials, it is perhaps worth repeating the above test for whether surviving confamilials are non-random (with respect to size) subsets. There are nine island families with confamilials, and the analysis is complicated since two have more than two species on some islands: four columbids and five tyrannids. For other families I located the observed size ratio in the distribution of all ratios. For these two families I compared the sum of all pairwise ratios (6 ratios for columbids, 10 ratios for tyrannids) to the distribution of sums of ratios for all 15 quartets of columbids and all 3,003 quintets of tyrannids. The latter distribution I derived by 100 simulated random draws. Table 14.3 gives the results: there is no evidence for extraordinarily different confamilial survivors. Fisher's combined probability test (with San Juanito data omitted where other islands have more species) gives tail probabilities of .400 for bill length and .560 for wing length.

These results may be compared to those of Hendrickson (1981), who followed Strong *et al.*'s procedure (1979) of ranking island confamilials from largest to smallest, calculating size ratios between contiguous species in these rankings, and comparing these ratios to expected ratios in random subsets of the same sizes from the mainland pool. Hendrickson corrected our error of comparing island ratios to means instead of medians (for the Tres Marias data two of 38 comparisons are affected (Strong and Simberloff, 1981)), and used a different battery of tests. None of his three tests shows significant bill-length differences between observed and ex-

Table 14.3. Null probabilities of size ratios as large as observed for mainland populations of sympatric confamilial island birds, relative to all mainland populations. J = San Juanito, C = Maria Cleofas, G = Maria Magdalena, D = Maria Madre. B = bill length, W = wing length.

Family	Species in Pool	Species on Island	Trait	Probability of Ratio or Sum of Ratios as Large as Observed
Columbidae	6	2-J	B	.067
			W	.200
		4-D,G,C	B	.133
			W	.400
Psittacidae	4	2-D,G,C	B	.167
			W	.167
Trochilidae	5	2	B	.900
			W	.600
Tyrannidae	15	2-J	B	.229
			W	.248
		5-D,G,C	B	.420
			W	.250
Mimidae	3	2-D,G	B	1.000
			W	.667
Turdidae	5	2-D,G,C	B	.300
			W	.700
Vireonidae	4	2-D,G,C	B	.500
			W	.333
Parulidae	9	2-D,G,C	B	.639
			W	.528
Fringillidae	19	2-D,G,C	B	.099
			W	.322

pected values, but two tests show that wing-length differences are significantly larger on islands than expected. Strong and Simberloff (1981) suggest that wing lengths are much less likely to reflect competition for food than are bill features, and I presume from their more frequent references to bill features that Grant and Abbott agree, but in any event perhaps our results differ because our original method conflates the questions of whether colonizing species are a non-random set and whether subsequent evolution serves to differentiate them further.

Another contention of Grant (1966b, p. 458), repeated by Grant and Abbott (1980, p. 339), concerning initial colonization rather than subsequent evolution is that certain mainland congeners of island species are absent from the islands because they are too similar to the island species, and so would have sustained intense competition. Eleven species that occur with one or more congeners on the mainland are alone on the islands (Grant, 1966b, p. 458). These mainland species comprise fourteen congeneric pairs, and of these fourteen pairs that are absent on the islands, Grant (1966b) and Grant and Abbott (1980) observe that twelve differ in bill length by less than the 20.3% that separates the most similar of the two island pairs, the vireos. Data on the same species from Grant (1965b) and Ridgway and Friedmann (1901–1950) differ somewhat from those that Grant used to construct his Table 7. Using these data I find that five absent pairs, not two, differ in bill length by a greater percentage than the vireos do, and an explicit calculation of the distribution of Mann-Whitney's U-statistic for fourteen mainland pairs and two island pairs shows the observed ranks of the two island pairs to be in the .125 tail. However, Ridgway and Friedmann (1901–1950), whose data both Grant (1966b) and I examined, measure different parts of the bill for different species and it is likely that comparable percent differences in bill length cannot be gotten from them.

So far, then, I have dealt with whether colonizing species are a random subset. The matter of subsequent evolution is not as readily addressed. However bad family-level data are in a search for competitively induced non-randomness, for the Tres Marias they at least permit for the two families with more than two island species (Columbidae and Tyrannidae) straightforward tests of whether island sizes are more displaced than mainland sizes. Simberloff and Boecklen (1981) describe these tests. The Barton-David test determines whether three or more points on a line are more evenly spaced than if they were drawn from a uniform random distribution, by comparing the ratio G_{ij} of the ith smallest segment to the jth smallest, where $i < j$, to the distribution under a uniform random hypothesis. The Irwin test determines whether the minimum resulting segment (g_1) is larger than expected if the points had been drawn from a uniform distribution. Since numerous papers in the evolutionary ecology literature contend that size *ratios* tend to be either constant or above some minimum because of interspecific competition (references in Simberloff and Boecklen, 1981), one may ask of *logarithms* of species' sizes whether they differ significantly from a set drawn randomly from a uniform distribution. Overly constant ratios would be detected by the Barton-David test, while the Irwin test would find extraordinarily large minimum ratios. Since one would not wish to reject or accept a hypothesis because of one aberrant ratio, I have used three G_{ij} statistics in the Barton-David test: G_{1n}, $G_{1,n-1}$, and G_{2n}. These are, respectively, the ratios of smallest to largest, smallest to second largest, and second smallest to largest ratios.

Of course the uniform distribution has the highest expected minimum and greatest expected constancy (Mosimann *et al.*, 1981), so these tests are conservative with respect to patterns that competition is predicted to

Table 14.4. Probabilities that confamilial birds would produce ratios as constant or with as large minima as those observed. Statistics explained in text. B = bill length, W = wing length, M = mainland, I = islands.

Family	Site	Number of Species	Trait	Observed Statistic	Probability of Result as Extreme
Columbidae	M	6	B	$G_{15} = .164$	.205
				$G_{14} = .195$	.341
				$G_{25} = .169$	.560
				$g_1 = .030$	.271
			W	$G_{15} = .016$	.848
				$G_{14} = .038$	.814
				$G_{25} = .105$	.753
				$g_1 = .007$	.832
	I	4	B	$G_{13} = 0$	1.000
				$G_{12} = 0$	1.000
				$G_{23} = .582$	.406
				$g_1 = 0$	1.000
			W	$G_{13} = .170$	.474
				$G_{12} = .462$	.437
				$G_{23} = .369$	.657
				$g_1 = .096$	.446
Tyrannidae	M	15	B	$G_{1,14} = .034$	.244
				$G_{1,13} = .128$	.026
				$G_{2,14} = .040$	.508
				$g_1 = .019$	.096
			W	$G_{1,14} = 0$	1.000
				$G_{1,13} = 0$	1.000
				$G_{2,14} = .012$	.902
				$g_1 = 0$	1.000
	I	5	B	$G_{14} = .007$	.950
				$G_{13} = .017$	.945
				$G_{24} = .370$	.358
				$g_1 = .005$	.951
			W	$G_{14} = .207$	.238
				$G_{13} = .258$	.403
				$G_{24} = .669$	.075
				$g_1 = .055$	.330

produce. That is, a value of g_1 or G_{ij} that would be improbably large for points drawn from another distribution might be insignificant for points drawn from a uniform distribution. So there may be high type II error.

Results, with the intermediate source pool, are given in Table 14.4. Since the probability of a G_{ij} or g_1 as extreme as a particular observation changes with number of species, the proper comparison is between probabilities, for mainland vs. island, of results as extreme as observed. If the notion of intensified competition's leading to morphological divergence is correct, island statistics should be more extreme than mainland ones. For both families the bill-length statistics are exactly opposite to this prediction: mainland statistics are more extreme. Wing-length statistics, however, conform to the prediction.

Since no genus is represented on the islands by more than two species, the above tests cannot be run, and I cannot frame a falsifiable null hypothesis for evolution after colonization. Grant (1966b, 1967, 1968) has found that species in the two genera—*Myiarchus* and *Vireo*–represented by pairs of species on the islands have not diverged in either bill length or wing length.

THE GALÁPAGOS

Taxonomic Distribution

Connor and Simberloff (1978) observed that, for the fifteen islands discussed by Harris (1973), one might ask whether land birds on each island could be construed as a taxonomically random subset of the 23 species (in 14 genera) in the entire archipelago. To test this we calculated by rarefaction (Simberloff, 1978b) how many genera one would have expected. We found that thirteen of the fifteen islands had more genera than expected (two significant). Noting that this is opposite to the pattern usually observed for island biotas (*cf.* Simberloff, 1970), we speculated that the result might simply reflect the peculiar genus-size distribution of the Galápagos avifauna. Grant and Abbott (1980) observe that the result is also compatible with a hypothesis of interspecific competition, since it means there are fewer coexisting congeneric species than chance would have predicted. They tested this hypothesis with a sign test: if as many islands should have more genera than expected as have fewer genera than expected, the result of thirteen more vs. two fewer is in the .011 tail.

Since the number of genera in random subsets is not symmetrically distributed about its expectation, I replaced the sign test with a simulation. One hundred times, I randomly drew from the species pool fifteen subsets of species, sizes of the sets corresponding to the respective numbers of species on the fifteen islands. For each random draw, I tallied how many islands had more genera than expected and how many had fewer. The distribution is skewed, but a result as extreme as that observed occurred only once, so Grant and Abbott's conclusion is sustained: the observation is in the .01 tail.

Sizes of Coexisting Birds

Strong *et al.* (1979) followed a procedure similar to that used for the Tres Marias, only the Galápagos do not have a relevant mainland pool. Consequently, there is no clear way to raise the first question asked of Tres Marias birds: are island subsets non-random with respect to bill size? Our method of testing Abbott *et al.*'s assertion that bill sizes on the islands manifest competitively induced displacement was therefore to view all Galápagos ground finches in one instance and tree finches in another as equiprobable members of a species pool. For each island we randomly drew from this pool the number of species on the island, then for each drawn species randomly (equiprobably) drew a race. We then size-ranked the species and calculated ratios of birds contiguous in the rankings, for comparison to observed values. As noted above for Tres Marias birds, this approach confounds the matter of non-random colonization and subsequent evolution.

Grant and Abbott (1980) feel that this source pool introduces circularity into the procedure, since the randomly generated subsets and the real subsets are both drawn from the same set. This is the single criticism that I do not understand, since this procedure is a textbook example of the classical statistical method of randomization (*e.g.* Bradley, 1968, p. 84); Hendrickson (1981) agrees that the procedure is valid. I can find no reference in any statistics text, nor is one given by Grant and Abbott (1980, p. 334), for their assertion that "the statistical bias that this kind of nonindependence causes is not random; it consistently favors the acceptance of the null hypothesis, and hence maximizes the risk of making a type II error. . . ." All that we and Hendrickson ask is whether the particular subsets of birds on islands differ characteristically from those we would see if available species and races were randomly and independently assigned to islands. Surely if competition leads to non-randomness in terms of either which species colonize or how colonists subsequently evolve, coexisting birds should differ from a random set. We are not disputing that sizes of races that comprise the species pool might have been strongly influenced by competition. We are asking only if the species of *coexisting* races offer additional evidence that competition has been important.

We looked at culmen length and beak depth separately for both ground finches and tree finches, and by the binomial test (but using means when we should have used medians) found significant differences for none of the four cases between observed and expected ratios (Strong *et al.*, 1979). Hendrickson (1981), using medians, still finds no significant differences by the binomial test, but for ground finch culmen length, by two other tests he finds that island size ratios are extraordinarily large.

A second area of interest has been the relationship of minimum beak depths on islands with different numbers of *Geospiza* species. Abbott *et al.* (1977) observed that the minimum beak-depth ratio on islands with two, three, and four species tended to decrease as species number increased. This trend they associated with interspecific competition, reasoning that smaller islands, albeit with fewer species, would constitute more stringent competitive regimes and thus generate greater displacement. Strong *et al.* (1979) pointed out that even if species and races on islands were random subsets, one would *expect* larger minimum beak-depth ratios the fewer the species present. Hendrickson (1981) does not dispute this, but observed that twelve of fifteen minima are above medians for random sets; the probability he finds by the binomial test to be .02. A correction for continuity (Snedecor and Cochran, 1967) raises this to .04. However, there is disagreement between Lack (1947), whose data Hendrickson used, and Abbott *et al.* (1977) on what *are* the minimum beak-depth ratios on the different islands. In particular, Abbot *et al.* (1977) show two minima (for Santa Cruz and James) below the medians tabulated by Hendrickson, while for the same two islands Lack's data give values above the median. Of course the medians themselves would also be changed, in a direction I cannot predict, but if they were not, Abbott *et al.*'s data would show ten of fifteen minima above the median, which by the binomial test has an associated probability > .30. Grant (pers. comm.) feels that for this particular comparison Lack's data are superior since they are from adult birds only. In any event, neither Lack's nor Abbott *et al.*'s data show differences between observed minima and medians for random draws to be correlated with number of species on islands, so whether or not the differences are consistent with a competition hypothesis, they are not consistent with a hypothesis that competition is more intense on smaller islands.

Grant and Abbott (1980) present a related analysis, based on data from Grant (1983). For all pairs of size neighbors on an island they calculate morphological Mahalanobis distances based on beak depth, length, and width of all species. They then construct, by randomly drawing species sets from the pool, expected mean distances for such pairs on islands with two to five species. This curve is a decreasing function of species number. For islands with two, three, four, and five species, respectively, they then calculate the mean over all islands of mean observed distance. For all four classes of islands, the observed mean exceeds expectation; for two classes (two- and four-species islands) it is outside 95% confidence limits. However, there is no relationship between number of species on the islands and difference between observed and expected. So although this analysis may be construed as supporting the notion that competition has contributed to size differences between coexisting species, it does not support the contention that such competition is more intense on smaller islands.

Grant (1983) has performed another relevant analysis with the Mahalanobis distance data. He has found that average distance between sympatric populations of *Geospiza* species pairs is greater for eleven of thirteen observed species pairs than expected (one possible pair, *conirostris* with *scandens*, is never found, and another, *conirostris* with *fortis* on Española, is likely not germane since *fortis* is transient there: Grant, pers. comm). As Grant observes, this is compatible with the hypothesis of competitive character displacement. He adds, however, that the two species pairs that buck the trend—*difficilis* with *fuliginosa* and *difficilis* with *fortis*—are precisely the two morphologically most similar pairs. That these seem to be more similar where they co-occur he suggests need not invalidate a competition hypothesis, and he proposes that either latitudinal variation of some environmental factor or habitat differences between species may account for these anomalies. Strong and I (unpublished data) reached a similar result for *magnirostris* and *fortis*: observed sympatric races are more similar for both culmen length and beak depth than randomly assembled races would have been.

To see whether effects of competition were evident in post-colonization evolution, I performed Barton-David and Irwin tests on *Geospiza* and *Camarhynchus* separately to see first if coexisting species' bill sizes manifested unusually large minima or regular spacing (as contended by Grant, 1983) and second if any such effects were more pronounced on small islands. Results, for culmen length and beak depth, are portrayed in Figures 14.1 and 14.2. For neither species is there a trend vs. number of species. Degree of size ratio regularity and size of minimum ratio are unrelated to number of species present. A summary, with probabilities combined by Fisher's method (Table 14.5), shows the only significant result to be for minimum *Geospiza* beak-depth ratios, which are improbably large; this result is consistent with the competition hypothesis.

Bowman (1961) suggested that interspecific competition need not be invoked to explain Galápagos finch morphological variation, and that different food regimes on the different islands, a reflection of the islands' different physical environments, selected for certain bill characteristics for all species irrespective of which other species were present. To examine this idea, Strong *et al.* (1979) ranked, within each species, each race from smallest to largest for both culmen length and bill shape (bill depth/culmen length). Then, for each island, one race from each species (both tree and ground finches together) actually present was randomly drawn, and ranks of drawn races were summed. This procedure was repeated 250 times, and the observed rank sum for races actually present was compared to the expected rank sum.

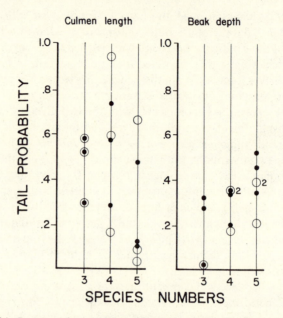

Fig. 14.1. *Geospiza* culmen-length and beak-depth ratios for islands with different numbers of species, plotted as tail probabilities for hypothesis of independent, uniform random distribution. Dots represent constancy of ratios, circles represent minimum ratios.

Our result (Strong *et al.*, 1979, Tables 5 and 6) was consistent with Bowman's hypothesis, since we found a weak tendency toward character *convergence* for both morphological traits. There appeared to be too many islands in both tails of both distributions. That is, some islands tend to have species with long bills while other islands tend to have species with short bills; a similar pattern obtains for bill shape. Hendrickson (1981, Tables 7 and 8 and Figs. 1 and 2) substituted an explicit calculation of expected rank sum and variance, corrected several errors in our tables, and computed new values of the standard normal variate Z to look for whether there are indeed too many islands in the tails of the rank sum distributions. He says there are not, but I disagree; Strong and Simberloff (1981, Fig. 1) plot Hendrickson's Z-values and show that for both bill length and bill shape there are too many islands in the tails.

Combinations of Species

Abbott *et al.* (1977) and Grant (1983) contend that there are too few combinations of *Geospiza* to be consistent with a hypothesis of independent colonization, and that interspecific competition causes at least some

Camarhynchus

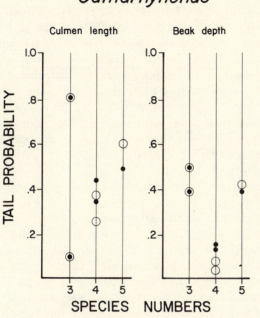

Fig. 14.2. As in Figure 14.1, but for *Camarhynchus*.

Table 14.5. Results of Fisher's combined probability test for whether finches on all islands produce unusually large or constant size ratios. For ratio constancy, individual island probabilities are means of the three G_{ij} statistics discussed in text. L = culmen length, D = beak depth, C = ratio constancy, M = unusually large minimum ratios.

Genus	Trait	Pattern	χ^2	df	Probability
Geospiza	L	C	20.002	18	.333
		M	22.497	18	.211
	D	C	27.462	18	.071
		M	29.479	18	.043
Camarhynchus	L	C	10.570	10	.392
		M	11.060	10	.353
	D	C	13.222	10	.212
		M	16.895	10	.077

absences. I have dealt with this matter elsewhere (Simberloff and Connor, 1981) and provide a summary here. Their contention was based on a Maxwell-Boltzmann analysis, which views combinations as equal-sized buckets into which islands (balls) are randomly thrown. Maxwell-Boltzmann probabilities were incorrectly computed, but even were they done correctly this does not seem to us an appropriate test of whether species colonize independently. The species have different frequencies, so that *ceteris paribus* one would not expect all combinations to be equally likely to occur. Consequently the Maxwell-Boltzmann model, where the combinations are viewed as equally likely to colonize any particular island, should be replaced by one in which combinations with infrequent species (such as *conirostris*) are less likely to colonize than are combinations composed of frequent species. Of course this is not to say that differences in species' frequencies might not be due to competition, only that the resulting combinations' frequencies cannot automatically be viewed as additional evidence for competition.

We therefore constructed an alternative, "weighted Maxwell-Boltzmann" model, in which sizes of buckets (combinations) into which islands are thrown are made proportional to the products of their component species' frequencies. Consequently combinations with rare species are *a priori* less likely to colonize any particular island. This done, there is the matter of whether the species pool for any class of islands (say, three-species islands) should comprise only species found on islands in that class (the convention adopted by Abbott, 1977, in a similar study) or simply all six *Geospiza* (as in Abbott *et al.*, 1977). The convention is of more than academic interest, since Abbott *et al.* show only one four-species combination and one five-species combination, so that Abbott's (1977) convention would predict in the independent colonization model exactly the result observed for these two classes of islands, with probability 1.000. However, since Abbott *et al.* (1977) do not feel this convention is reasonable for the finches, it is omitted in further analyses. Finally, Grant (pers. comm.) suggests that at least two changes are necessary in Abbott *et al.*'s assignment of combinations to islands before reanalysis.

Results of 100 simulated colonizations are presented in Table 14.6, and indicate that there are indeed fewer combinations than independent col-

Table 14.6. Tail probabilities for weighted Maxwell-Boltzmann model (see text) that as few combinations occur as are observed on islands with given numbers of species. Data from Abbott *et al.* (1977) and Grant (pers. comm.).

Number of Geospiza	Number of Combinations	Number of Islands	Observed Combinations	Tail Probabilities
1	6	2	1	.19
2	15	7	5	.42
3	20	8	2	<.01
4	15	3	1	<.01
5	6	3	1	.05

onization ought to have produced. For three-, four-, and five-species islands, observed numbers of combinations are in the .05 tail; there appears to be a tendency for fewer species combinations than expected. This is certainly consistent with a hypothesis of interspecific competition, though as Pielou and Pielou (1968) pointed out, it is equally consistent with a hypothesis that islands are physically different in ways that make them differentially suitable for different species, but the gradient of suitability is different for different species. Finally, two species (*conirostris* and *difficilis*) have restricted geographic ranges, so that not all islands really have species pools of six species. Though adjusted pools would raise tail probabilities of observed numbers of combinations somewhat, there would likely still be a clear tendency for the number of combinations to be too small to sustain an independent colonization hypothesis.

Two of the fifteen *Geospiza* species pairs, *conirostris-fortis* and *conirostris-scandens*, are never found together, either alone or embedded in larger species sets (Abbott *et al.*, 1977; Grant, 1983; Grant, pers. comm.). Two other pairs that co-occur on three islands but were thought to be elevationally segregated there (Abbott *et al.*, 1977) are now known to overlap (Grant and Schluter, this volume; Schluter, 1982a; Grant, pers. comm.). Even if species are randomly and independently placed on islands, there will almost always be missing combinations, including pairs (Connor and Simberloff, 1979). So the existence of two missing pairs does not automatically constitute evidence for competitive exclusion. Especially since *conirostris* occurs on the fewest islands it is unnecessary to posit competitive exclusion as the reason for the absence of two combinations that include it. Its limited frequency may be caused by competition, but the absence of two pairs that include it does not provide added evidence of exclusion.

Grant (1983, p. 9) states that the missing pairs, *conirostris-scandens* and *conirostris-fortis*, resemble one another in bill length and bill depth, respectively, and that this makes competitive exclusion a reasonable inference. To test this assertion I examined whether observed allopatric pairs are remarkably similar in bill length or depth. Bill dimensions are taken as means for all races listed by Lack (1947, Table XXIII), and my test is identical to that for whether Tres Marias congeneric colonists are extraordinarily different relative to other pairs from the same genera (Table 14.2), but here I am seeking extraordinary similarity. For both bill length and bill depth I calculated all fifteen ratios, then ranked them. One allopatric pair, *conirostris-scandens*, is second most similar in bill length and tenth most similar in bill depth. Four of the fifteen pairs would have produced as extreme a result as this one in at least one dimension ($Pr = .267$). The other allopatric pair, *conirostris-fortis*, is seventh most similar in bill length and fourth most similar in bill depth. Eight pairs would have produced as extreme a result in at least one dimension ($Pr = .533$). In sum, morphological differences of both allopatric pairs are not extraordinarily small, so the fact that these pairs are allopatric does not support a competition hypothesis on morphological grounds.

SUMMARY AND CONCLUSIONS

The Tres Marias

The distribution of species into genera does not differ from that expected for a random draw from any species pool. Mainland populations of insular congeneric species do not differ particularly from one another, in either bill or wing length, relative to others in the genus. Nor do mainland populations of island confamilials differ unusually from one another relative to others in the family. Confamilial island species do not manifest different bill-length ratios than if species were random draws from mainland pools. For the two families with island confamilials, neither minimum bill- and wing-length ratios nor the constancy of bill- or wing-length ratios are extraordinary on either island or mainland, relative to a uniform random, independent distribution of logarithms of sizes. For bill length, if anything, the ratios are more extraordinarily large and constant on the mainland than on the islands, while for wing length, the results are in the opposite direction. Probably neither trend is significant. Finally, both pairs of island congeneric species are not more or less different in either bill or wing length than the same pairs are on the mainland. These results argue against the notion that competition has affected sizes or coexistence of related species. They do *not* argue against competition *per se*, only against the contention that its effects are manifested in these particular traits.

One result counters the above. Confamilial wing-length ratios on islands are significantly greater than those for random subsets, by two statistical tests but not by the binomial test. The significance of wing length to competitive interactions seems to me more obscure than that of bill size.

The Galápagos

For coexisting sets of ground finches, beak-depth ratios do not differ significantly from those of random subsets. Neither do culmen-length and beak-depth ratios for tree finches. Minimum beak-depth ratios for ground finches are not more extraordinarily large on small islands than they are on large ones, nor is the dispersion of bill sizes by a Mahalanobis distance based on culmen length, beak depth, and beak width. The two most morphologically similar pairs of *Geospiza* are more similar by Mahalanobis distance where they occur sympatrically than are allopatric races of the same species. For coexisting tree finches, neither beak-depth minimum ratios nor constancy of ratios is extraordinary relative to a uniform, random model, and the same is true for culmen-length ratios. For coexisting ground finches, culmen-length and beak-depth ratios are not extraordinarily constant, nor are minimum culmen-length ratios remarkably large. For neither ground nor tree finches and neither culmen length nor beak depth is there a tendency for minimum ratios to be especially large or for ratios to be expecially constant on small islands. There is a weak tendency for all finches on an island to have bills that vary in size and or shape in the same direction from medians—bills that are too short, or too fat, etc., relative to those of random subsets. Finally, the fact that two pairs of *Geospiza* species are nowhere found, on any size island, might not have been unexpected even had species been randomly arranged on the islands, and the specific pairs that are allopatric are not particularly similar morphologically, relative to all possible pairs. These results run counter to specific hypotheses that competition has likely generated some pattern. They do *not* imply that competition has not occurred, only that these particular morphological features or aspects of geographic distribution and coexistence do not necessitate a competitive explanation.

Against these results are the following: There are too many islands with too many genera, relative to expectations if species were independently distributed on islands. For coexisting *Geospiza*, culmen-length ratios are significantly larger than in random species sets. Minimum *Geospiza* beak-depth ratios may tend to be larger than would have been expected for random sets of species and races, though different measurements for the same species yield different results. Bill-feature Mahalanobis distances between pairs of sympatric *Geospiza* on all size islands exceed those expected for randomly assembled species sets, though not always significantly. Eleven of thirteen *Geospiza* pairs (but not the two most similar morphologically) tend to be more distant in a Mahalanobis plot for sympatric races than for allopatric ones. For coexisting *Geospiza*, minimum beak-depth ratios exceed expectations if logarithms of beak depth were uniform randomly and independently distributed. Finally, islands with four, five, and probably three species contain fewer four-, five-, and three-species *Geospiza* combinations, respectively, than a random, independent colonization hypothesis would predict.

All these results are compatible with competitive hypotheses, and boil down to three clear-cut aspects of community structure. First, coexisting species are distributed into too many genera. Whatever this result means, Grant and Abbott (1980) and I (Simberloff, 1970) agree that so many forces likely act on species/genus ratio and related statistics, and they summarize data on so many species, that their biological interpretation is difficult. Second, there are too few observed combinations of *Geospiza* to sustain a hypothesis that species colonize randomly and independently.

This is exactly what one would have expected if competition precluded certain groups of species from coexisting. An alternative hypothesis that predicts precisely the same result (Pielou and Pielou, 1968) is that islands are physically different and different species rank the islands' suitabilities differently. The combination data alone do not permit a choice between the two hypotheses, and the two forces could both be operative. I think the second must certainly be true, but this does not preclude the first. Finally, a variety of different approaches show bill-feature differences for coexisting *Geospiza* to be unusually large. The different methods find non-randomness for different features, and about as many approaches show differences not to be more pronounced than expected, but the positive results are predicted by various forms of competition hypothesis. Further, I do not see an attractive alternative hypothesis.

On balance, there *are* aspects of Galápagos avifaunal taxonomy, morphology, and coexistence that are inconsistent with simple models of randomness and independence, and are consistent with competitive hypotheses, though for the latter two traits one must note that there are at least as many aspects that *are* consistent with random, independent models. I have not addressed other evidence (*e.g.* in Abbott *et al.*, 1977) that bears on competition hypotheses, and I take none of the negative results I have found here to indicate either random colonization or absence of competition, but only to demonstrate that certain classes of evidence that have been used to indicate competition are in fact neutral. To me the clearest support for competition in the welter of data I have examined here comes from the large *Geospiza* bill-size differences, and I observe that this is the salient result of Grant and Schluter's contribution (this volume) to this symposium. So we appear to agree that Galápagos ground finches are not random subsets, that interspecific competition is likely at least partly responsible for this, and that the strongest evidence resides in bill features of coexisting races and species.

A final point that I would make, however, is that statistical evaluation of various hypotheses is becoming increasingly difficult, for reasons elaborated by Selvin and Stuart (1966). First, to examine a number of related hypotheses with correlated significance tests is not invalid, but is difficult because of problems in calculating the null probabilities of different numbers of positive and negative results. Second, to examine data in order to choose which of several candidate variables to include in an explanatory model does not automatically preclude significance tests of the model, but does condition the probabilities, and not in a simple way. All one can say here is that probabilities of an observed extreme result are very likely to be raised when the winnowing of variables is considered. It is not legitimate to focus on only those variables that support one position. Finally, whereas the above two procedures preselect tests and variables, respectively, before significance tests are made, some approaches (such as much curve-fitting) do neither, and for these, statistical tests are impossible in principle (Selvin and Stuart, 1966). One example would be scanning all data and examining many relationships to find some worth testing. Another would be to examine the same hypothesis with several sets of data, but to report and proceed only with confirmatory sets.

I believe I can cite examples of the first two procedures in literature I have examined here, and this renders statistical analysis doubly difficult. The third procedure may be implicit in much competition literature, though in the subset I have reviewed it is not obviously present. In general, all three procedures are relevant primarily to surveys of non-experimental data, and though experimental results are not immune to the attendant problems, they are far less likely to be affected. Which returns me to my earlier suggestion, with Abbott (1980) and Grant and Abbott (1980), that an experimental approach if at all possible is worth a major effort.

Acknowledgments

I thank Donald Strong, Peter Grant, and Dolph Schluter for illuminating discussion on many issues raised here and for comments on an earlier draft of the manuscript, and Peter Grant and John Hendrickson for preprints. Finally, I note that Peter Grant and Ian Abbott have scrupulously published most data on which they based their analysis, and have provided much relevant unpublished data.

15.

Size Differences Among Sympatric, Bird-Eating Hawks: A Worldwide Survey

THOMAS W. SCHOENER

Department of Zoology, University of California, Davis, California 95616.

Closely related sympatric species, although similar in most ways, can differ strikingly in some aspect of size. Enough such cases were noted by Huxley (1942) and Lack (1944) for them to suggest that the size differences resulted from selection to avoid interspecific competition. Hutchinson (1959), in giving "homage to Santa Rosalia," began the tradition of computing size ratios between adjacent members of size-ranked associations of sympatric species. He listed a small number of such ratios, most of which fell between 1.2 and 1.4, and wrote that 1.3 "may tentatively be used as an indication of the kind of difference necessary to permit two species to co-occur in different niches but at the same level of a food web." This guarded statement apparently inspired a wealth of similar calculations. Some of these analyses according to their investigators confirmed Hutchinson's generalization, whereas others found ratios very different from 1.3 but still substantially above 1.0. The latter studies, according to their investigators, frequently indicated a minimal difference between species, though its magnitude sometimes varied from study to study. The whole body of literature has recently been critically reviewed by Simberloff and Boecklen (1981) in their "reconsideration" of Santa Rosalia. These authors reassess the published data with several statistical tests having in common the null hypothesis that species sizes are drawn randomly from a uniform distribution. Many of the data sets, which typically consist of a small number of species, fail the test; a few, including one very large data set, pass.

Related to this statistical examination of small sets of taxonomically close species are recent attempts to assess community-wide character differences. Here the issue is whether the species that actually co-occur are more different in size or other morphological aspects than would be expected were they shuffled together from all localities and drawn randomly from the resulting pool. Sometimes (Strong *et al.*, 1979; Ricklefs and Travis, 1980), but not always (Gatz, 1979; Gilpin and Diamond, this volume), the "random" or "neutral" hypothesis cannot be rejected.

While the attempt to be statistically rigorous in assessing size differences certainly has its merits, I would like to express several reservations about the approaches so far used. First, statistical significance is hard to achieve no matter how narrowly defined taxonomically (and by implication ecologically) a group is if the number of its species is small. Second, it seems likely that a group composed of species that are sufficiently dissimilar ecologically will appear chaotic no matter how many there are—too few species will actually interact with one another. Very possibly our intuition about morphological differences between species is not based on any single, small group of species in one place, and certainly not on differences between all species in a broadly defined group. Rather, we might expect some ecologically tightly defined group of species to have representatives in a variety of different places that more often than not show major size differences. This is what some have claimed goes on in certain archipelagos, but the problem is that there just aren't likely to be enough ecologically similar species in small archipelagos to falsify a null hypothesis (see also below).

The approach in this paper differs from previous ones in that it considers an ecologically very restricted group everywhere, that is, all over the planet Earth. The group chosen comprises bird-eating hawks with accipiter-like habitats. The three species of *Accipiter* sympatric over much of North America have become a textbook case of size differences (MacArthur, 1971; May, 1973), and the group, because of its elevated trophic position, should show competitively caused niche overdispersion (Schoener, 1974b). However, virtually nothing has made its way into the general ecological literature concerning size differences between most other sympatric accipiters. In this paper, I ask whether the species in *all* sympatric associations of these hawks, taken together, differ more in size than would be expected from certain null models. I also discuss some statistical and methodological problems associated with the null-model approach in general.

STATISTICS AND METHODS

Null Models and Statistical Comparisons

The basic procedure used to test the proposition that sizes of sympatric species are more different than random is the following. A species pool is formed, consisting of all the species in the world belonging to the ecological group considered. Each species is assigned a mean size. Then all possible pairs, trios, quartets, and quintets are computed for this pool. Sizes are ordered for each of these units, and ratios (larger to smaller) are computed between adjacent species. In this way, a number of null distributions are generated. For pairs, there is a single such distribution. For trios, two null distributions are generated, one for the ratio of the largest to the intermediate-sized species, the other for the ratio of the intermediate-sized species to the smallest species. Two distributions are distinguished because it could not be assumed *a priori* that the null distributions would be invariant of species position in the size-ranked association. Similarly, for quartets three null distributions, and for quintets four distributions, were generated. Because of the huge number of possible associations and the small number of real ones, associations with more than five species were not considered.

Each null (expected) distribution was then compared to the appropriate empirical (observed) distribution using the Kolmogorov-Smirnov One-Sample Test. This test compares the cumulative-frequency distributions of the expected and observed ratios and is based on the maximum difference, D, between them.

Three variations on the above test (Test 1) were performed.

Test 2 weights each real association by the area of its geographical range. This weighting can be justified in two ways. First, associations appearing to occupy very small ranges may not be real but may simply result from inadequate information, *i.e.* poor resolution of the species' geographic distributions. Second, one might be somehow more impressed by striking size differences were these to occur among species distributed over a wide portion of the earth rather than over a limited one. A counterargument to this is that size differences among species on islands, where competition is sometimes hypothesized as potentially being especially intense and effective, are weighted differentially less. However, for the species considered here, islands if anything have smaller ratios. That is, if one compares island to mainland ratios by a median test (doing separate comparisons for each ratio possible in each of pairs, trios, and quartets), islands always have smaller ratios. Only in one case, however, is the difference significant (one of the two possible types of ratios for trios, $P = 0.026$).

In Test 2, the cumulative frequency distributions of the unweighted observed and weighted observed data are compared. No statistical conclusion is drawn, but the direction of the difference is noted—does the weighted distribution have more or fewer large ratios than the unweighted one?

Test 3 results from the following criticism of Test 1. Real associations have certain members in common and in that sense are not totally independent. Thus, for example, a certain species very different in size from the rest could be wide-ranging and occur in a number of associations. The reason it is wide-ranging may have nothing to do with competition (though it of course *may*, also), yet its presence could generate an unusually high number of large ratios. Cursory observation did not reveal any obvious such species, but the following procedure was carried out nonetheless. An algorithm was developed for selecting a subset of real associations, each association of which has no species in common with any other. Of the many possible such algorithms, I used the following. Order the associations according to the area of geographic range. Select the association with the largest area. Now select the association with the second largest area. If it has no species in common with the first, group it with the first as a "selected" association; otherwise, discard it. Proceed in this way through the ordered associations until all are either added to the selected associations or discarded. The selected associations can then have their cumulative frequency distribution compared (1) statistically to the expected distribution (as in Test 1) and (2) qualitatively to the distribution of all real ratios to see which has more large ratios.

Test 3 totally eliminates dependence of the data but drastically reduces sample size. A less extreme test in both respects, Test 4, can be performed. For associations with more than two species, a given category of ratios, say largest to second-largest species, might include the same species pair more than once. This situation would be particularly likely for two widespread species that overlapped broadly in geographic range; within their area of overlap they might occur with several different sets of less widespread species. Hence they might occur in several associations of a given species number, and the ratio between them might be counted more than once, even though the reason for their distribution might be related to something other than avoiding size similarity. Of course, the reason for their geographic distribution could, in fact, have much to do with avoiding size similarity, which is justification for Test 1. In any event, this test deletes certain ratios for trios, quartets, and quintets, such that each real ratio for a given comparison (quartets, largest to second-largest, for example) is between a unique pair of species. Comparisons to expected ratios were done as in Test 1.

The Associations

All of the above tests were performed using two groups of species. The first group (Group I) comprises all members of the genus *Accipiter*; 47 species are recognized by Brown and Amadon (1968). Wattel (1973), in a somewhat later study limited to *Accipiter*, recognized four fewer species, but to be consistent with Brown and Amadon's taxonomy for other hawks, I have used their scheme throughout. The genus *Accipiter* forms a rather well-defined ecological unit so far as genera go; most species feed primarily on birds, and the species frequent wooded habitats, typically hunting about edges, clearings, and similar units of space. However, five species of *Accipiter* are reported by Brown and Amadon to feed primarily on prey other than birds. These are *A. trinotatus* (mainly lizards and snakes), *A. poliocephalus* (mainly insects, lizards, and snakes), *A. soloensis* (mainly frogs), *A. butleri* (mainly lizards) and *A. francesii* (mainly reptiles). Furthermore, certain species not in the genus *Accipiter* feed primarily on birds and occur in habitats very similar to those of *Accipiter*. Certain of these species are very close taxonomically to *Accipiter*, *e.g. Erythrotriorchis radiatus*, *Urotriorchis macrourus*, *Melierax gabar*, and *Megatriorchis doriae*. In addition, two major genera of raptors have some or all members ecologically similar to *Accipiter*: *Hieraetus* (certain species) and *Micrastur* (all species). The second group considered, Group II, differs from Group I in excluding the five non-bird-eating accipiters and including thirteen non-congeneric, ecologically similar species. Appendix I gives all species in Groups I and II.

To determine geographic region of overlap, breeding ranges of all species were compared using the maps in Brown and Amadon (1968). We used the largest such map for each major geographic region and traced or drew species ranges from all maps of that region onto a single sheet. The ranges form a mosaic, each piece of which was noted as to species contained. Each piece containing at least two species and larger than 0.006 in^2 (in map units) was measured with a planimeter to determine its area. Although this cutoff is arbitrary, inclusion of pieces much smaller than 0.007 in^2 would in my opinion greatly increase the danger of designating associations that in fact do not exist but are artifacts of mapping imprecision. Of course, having a single cutoff for each map means that the actual areal cutoff varies from region to region. In fact, such variation is not very great, and in certain regions (Australasian, New Guinean, Nearctic, Asian), our convention includes all visible areas of overlap. In the Palearctic, one clear region of overlap was missed by this procedure, a tiny piece of North Africa uniquely holding two species. Because the scale of the Palearctic was so coarse, and because there is no doubt about the overlap here (the region is *ca.* 9000 mi^2), this two-species combination was included. All other areas of overlap for the Palearctic visible to the eye measured greater than the cutoff and so were included to begin with. The areas of possible overlap excluded by our procedure are from Africa and South America. Both of these regions contain an extremely large number of species, and here, the possibility of designating associations that do not exist, particularly in view of the uncertainty of some of the range maps, is greatest.

The list of associations so generated was then culled to remove doubtful cases of overlap according to Wattel (1973), whose study is more recent than Brown and Amadon's.

Finally, I attempted to classify the species by broad habitat category, thereby considering only those species in similar habitats as being sympatric. I used two spatial axes: vegetation type and altitude. The former particularly applies to Africa, for which I classified each species as occurring in open, closed or both types of vegetation. Vegetation type separated a few non-African species as well. Altitude was especially important in separating species from New Guinea and South America. Where precise information was available, only species whose altitudinal ranges overlapped more than 500′ were considered sympatric. In some cases, especially South America, I simply used the broad categories (lowland, montane) given by Brown and Amadon.

Overall, I tried not to draw fine habitat distinctions between species, but to use coarse categories. If the species are overdispersed on several ecological dimensions, this procedure is conservative: it will result in smaller apparent size differences between sympatric species than actually exist.

Appendix II lists the associations resulting from the above procedure.

Measure of Size

The measure of size used in this study is wing length. Wing length has been suggested (Lack, 1946; Schoener, 1965) as a more reliable measure of feeding differences for raptors than bill length, the characteristic most commonly used for birds.

For the sake of standardization, all measurements were obtained from Brown and Amadon (1968). Because I wished to use a single size measurement per species, an averaging procedure had to be devised.

Males and females are frequently very different in size in *Accipiter*, whose North American species are noted for their high degree of dimorphism (Storer, 1966; Snyder and Wiley, 1976). I weighted the sexes equally in computing an average.

Certain species have sizes given separately for different geographic regions, *i.e.* for different subspecies. Rather than average all subspecies together, thereby disproportionately weighting subspecies with smaller geographic ranges, I averaged the subspecies with the largest ranges only. I began with the subspecies with the largest range, then added others until the combined ranges comprised half the total range of the species. Occasionally, two subspecies had about equal ranges, and either one if added to the total would cause it to exceed the 50% cutoff; then both subspecies were included. Each subspecies in the resulting set was weighted equally in computing the average.

Sizes resulting from the above procedure are listed in Appendix I.

RESULTS

The Species Pool

The distribution of sizes of the world's species of *Accipiter* (47 species, Group I) is roughly unimodal and asymmetrical, having a long right tail (Figure 15.1). The distribution of sizes of bird-eating hawks and eagles of wooded habitats (55 species, Group II) is similar, though especially because of certain eagles the right tail is somewhat longer. Thus the distributions of "available" sizes are roughly lognormal in shape.

The Real Associations

Table 15.1 gives the number of real and simulated associations of 2, 3, 4, and 5 species for Groups I and II. In all cases, the number of possible associations is much larger than the number of existing ones. For Group I there are 1081 possible pairs and 31 existing ones; thus one of every 35 possible pairs actually exists. For quintets, 3 of the possible 1,533,939 associations, or one of every 511, 313, actually exists. For Group II, 45 of the possible 1485 pairs exist, or one of every 33. This is the largest such ratio. For quintets, 21 of the possible 3,478,761 associations exist, or one of every 165,655. Incidentally, the small number of realized associations, combined with the decline in the ratio of actual to possible

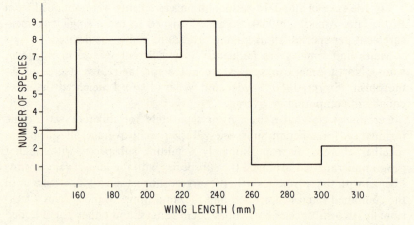

Fig. 15.1. Size distribution for the world's 47 species of *Accipiter* (Group I). Note the lognormal-like shape.

Table 15.1. Number possible and real associations.

	Number Possible Associations	Number Real Associations
Group I		
pairs	1,081	31
trios	8,596	19
quartets	178,365	9
quintets	1,533,939	3
Group II		
pairs	1,485	45
trios	26,235	39
quartets	341,055	29
quintets	3,478,761	21

associations from pairs to quintets, might itself suggest some sort of competitive limitation on the number of species per area. I have no null model to test this possibility against, and dispersal ability would probably be relevant here.

The Null Distributions

Figure 15.2 shows null distributions for 2, 3, and 4 species, Group I. The distributions are strikingly asymmetrical. In all cases, the mode occurs within the first category (ratios of 1.00 to 1.05), and means are substantially larger than medians (Table 15.2). The smallest ratios for associations with three species are between the middle and smallest species. For associations with more than three species, the smallest ratios are the intermediate ones, and the largest ratios are between the largest and second largest species. The locations of the smallest ratios are expected, given the roughly lognormal shape of the species-pool size distribution. Finally, of course, ratios are smaller the more species in an association, a result expected from a variety of null models (Strong *et al.*, 1979). Null distributions for Group II resemble those for Group I in the above qualitative properties; quantitatively, variances are greater and ratios are larger (Table 15.2) than for Group I.

Test 1

This test compares the distributions of all real ratios to those generated from all possible combinations of species.

The real pairs of Group I have significantly larger ratios than expected (Figure 15.3). The maximum difference in cumulative frequency, D, is

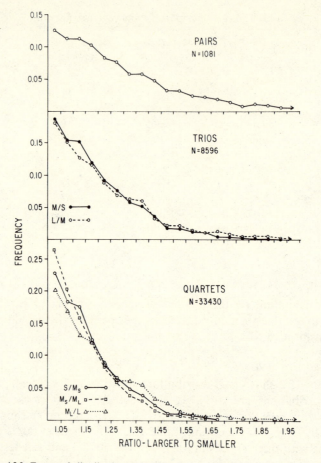

Fig. 15.2. Expected distributions of size ratios for pairs, trios, and quartets, comprising all possible combinations of the 47 species of Fig. 15.1. Abbreviations as in Table 15.2.

0.253, *i.e.* a 25% difference (Table 15.3). Moreover, as Figure 15.3 shows, ratios smaller than 1.15 are notably lacking, even though 35% of the possible ratios lie here. This absence suggests a kind of limiting similarity for pairs.

Real trios of Group I have larger ratios than expected, but only the difference for ratios between the intermediate-sized and smallest species is significant ($D = 0.336$, intermediate to smallest; $D = 0.165$, largest to intermediate). For the first type of ratio, limiting similarity is again suggested; no ratio is smaller than 1.08, as compared to an expected 29%; however no striking minimal difference is evident for the other type of

Table 15.2. Expected means and medians, and number of real ratios above and below the median.[1]

		Expected Mean	Expected Median	GROUP I Number Above Median	Number Below Median	Number at Median
pairs		1.29	1.23	23	7	1
trios:	L/M	1.22	1.18	11	8	0
	M/S	1.20	1.16	14	5	0
quartets:	L/M_L	1.19	1.15	6	3	0
	M_L/M_S	1.14	1.11	7	2	0
	M_S/S	1.16	1.13	3	6	0
quintets:	L/M_M	1.17	1.14	3	0	0
	M_L/M_M	1.12	1.09	1	2	0
	M_M/M_S	1.11	1.09	2	1	0
	M_S/S	1.14	1.11	1	2	0

		Expected Mean	Expected Median	GROUP II Number Above Median	Number Below Median	Number at Median
pairs		1.38	1.29	29	16	0
trios:	L/M	1.31	1.22	28	11	0
	M/S	1.23	1.19	25	14	0
quartets:	L/M_L	1.27	1.20	22	7	0
	M_L/M_S	1.18	1.14	17	12	0
	M_S/S	1.18	1.15	23	6	0
quintets:	L/M_M	1.25	1.19	13	8	0
	M_L/M_M	1.15	1.13	9	12	0
	M_M/M_S	1.13	1.10	13	8	0
	M_S/S	1.15	1.13	16	5	0

[1] Abbreviations: L = largest species; M = middle species; S = smallest species; M_L = larger(-est) middle species; M_M = middle middle species; M_S = smaller(-est) middle species.

ratio. Because the expected distributions of the two types of ratios for trios are not very different (Figure 15.2), a statistical test on combined ratios was also performed. This test gives $D = 0.236$ ($P < 0.05$).

Very few real associations of 4 or especially 5 species exist for Group I (Table 15.1), and statistics were computed only for quartets. D's were large and positive for two of the three types of ratios ($D = 0.385$, $P > 0.10$; $D = 0.297$, $P > 0.10$), but D was actually negative (expected

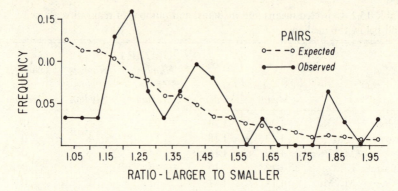

Fig. 15.3. Expected and observed distributions of size ratios for pairs, Group I, Test 1.

ratios larger) for the ratio of the second-smallest to smallest species ($D = -0.294$, $P > 0.10$).

D values for Group II are comparable in magnitude to those for Group I, being larger in two and smaller in four cases. Statistical significance, however, was generally higher, and in no case was D negative (Table 15.3). Unlike Group I, D's were also computed for quintets; these were large and positive, ranging from 0.249 to 0.519, and three were significant or nearly so at the 5% level (Table 15.3). While more significant, the data for Group II did not as strongly suggest a limiting similarity as the data for Group I.

Table 15.2 also gives the mean and median ratios for the expected distributions as well as numbers of real ratios above and below the median. The latter in general parallel patterns in cumulative frequencies.

Test 2

This test weights each ratio by the relative geographic range occupied. These weighted real ratios are compared in Table 15.3 to the unweighted real ratios.

For pairs, Group I, weighted ratios are substantially larger than un-weighted ones ($D = 0.347$); Figure 15.4 shows how the weighted distribution compares to the expected one. Note that smaller ratios are even more de-emphasized than in Test 1. The other weighted distributions for Group I (trios, quartets) all have larger ratios than the weighted distributions; D's range from 0.158 to 0.407 (Table 15.3).

The situation is more ambiguous for Group II. Pairs behave much the same as for Group I. However, the maximum difference in cumulative frequency is negative (unweighted ratios larger) for trios (both types of

Table 15.3. Comparison of real and all possible communities.[1]

	Test 1[2]		Test 2	Test 3		Test 4		
	D: Unweighted Obs. vs. Expected	Significance	D: Unweighted Obs. vs. Weighted Obs.	D: Selected Obs. vs. Expected	N	D: Selected Obs. vs. Expected	Significance	N
Group I								
pairs	.253	<.05	.347	.350	12	—	—	—
trios: L/M	.165	ns	.169	.203	6	.143	ns	15
M/S	.336	<.05	.407	.285	6	.317	≃.06	17
combined	.236	<.05	—	—	—	.233	<.10	30
quartets: L/M_L	.385	≃.10	.180	−.458	4	.289	ns	7
M_L/M_S	.297	ns	.158	−.350	4	.255	ns	8
M_S/S	−.294	ns	.369	−.282	4	−.252	ns	8
Group II								
pairs	.247	<.01	.233	.268	13	—	—	—
trios: L/M	.246	<.05	−.165	−.132	6	.221	ns	28
M/S	.252	<.05	−.181	−.248	6	.250	<.05	33
combined	.190	<.01	—	—	—	.242	<.01	55
quartets: L/M_L	.351	<.01	.085	.385	5	.327	<.05	17
M_L/M_S	.235	≃.10	−.131	−.293	5	.162	ns	22
M_S/S	.315	≃.01	−.088	.423	5	.275	≃.05	23
quintets: L/M_L	.247	ns	−.172	−.311	3	.236	ns	9
M_L/M_M	.288	≃.06	−.241	−.467	3	.289	ns	15
M_M/M_S	.330	<.05	.239	.841*	3	.315	ns	13
M_S/S	.519	<.01	.215	.459	3	.319	ns	12

[1] Abbreviations as in Table 15.2: N = sample size; D = maximum difference in cumulative frequency (minus indicates second distribution has larger values than first).

[2] Sample size as in Table 15.1.

* $P < .05$

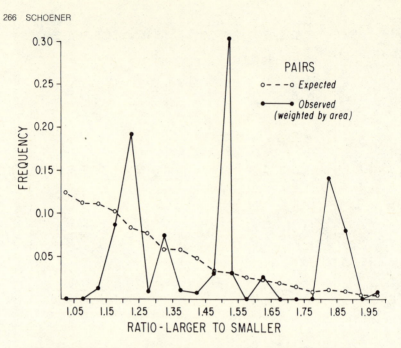

Fig. 15.4. Expected and observed distributions of size ratios for pairs, Group I, Test 2. This test weights each observed ratio by the area of the geographic range of its pair before computing an expected frequency distribution.

ratios), for two of the three types of ratios for quartets, and for half the types for quintets. In general, however, absolute D's are small compared to those of Group I: positive D's range from 0.081 to 0.239, and negative ones range from -0.088 to -0.241. Thus in all cases for Group I, but only 40% for Group II, do associations with relatively large geographic ranges have relatively large size ratios.

Test 3

This test selects, according to the algorithm described above, a subset of real associations none of which has a species in common with any other.

As Table 15.3 shows, sample size is drastically reduced using this procedure. Given such samples sizes, values of D would have to be enormous to be statistically significant, and in only one case is there a value this large ($D = 0.841$). We can also, however, compare the magnitudes of D produced by Tests 1 and 3.

For Group I, pairs and trios have D's larger for selected than complete sets; D's range from 0.203 to 0.350 (Figure 15.5 compares selected to

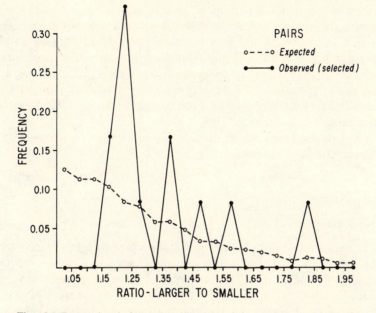

Fig. 15.5. Expected and observed distributions of size ratios for pairs, Group I, Test 3. This test selects, according to the algorithm given in the text, pairs none of which has any species in common with any other.

expected for pairs). Quartets, however, are for all three types of ratios the opposite: D's are negative (expected ratios larger than observed) and range from -0.282 to -0.458. For Group II, half the possible types of ratios for selected associations are smaller, and half are larger, than ratios of inclusive associations (Table 15.3). Moreover, in nine of ten cases, a positive D for Test 2 is associated with a positive D for Test 3 and vice versa. This relation, if any, is to be expected from the nature of the two tests; both emphasize, though in different ways, associations with relatively large geographic ranges.

Test 4

This test eliminates ratios between the same species before comparison to expected ratios; for pairs, it is equivalent to Test 1. Sample sizes of the selected sets were typically 50–80% those of full sets (Table 15.3).

In most cases, D's computed using selected ratios did not differ much from those computed using all real ratios (Table 15.3). However, the trend of change was definitely toward smaller absolute values, whether positive or negative D's are considered: only two of sixteen increased in absolute value. The net result, especially as sample sizes were reduced, was to diminish the number of comparisons giving statistical significance. For Group I, comparisons that were marginally significant become marginally insignificant, and for Group II, of the eight comparisons originally significant at the 5% level or better, only four were significant using Test 4. Quintets, whose sample sizes were especially drastically reduced, were especially hurt by this procedure, even though D's for the selected associations ranged from 0.236 to 0.319.

In no case did Test 4 change the sign of the difference between observed and expected distributions.

DISCUSSION

General Significance

The major result of this study is that, for cases with moderately large sample sizes ($N = 20$–50), ratios are significantly larger than expected from a null model. Typically, maximum differences in cumulative frequency between real and expected distributions of ratios are 0.20 to 0.35. Differences are, however, for the most part statistical rather than absolute. That is, except for pairs and possibly trios of *Accipiter* (Group I), a limiting similarity or minimum permissible difference is not apparent in the data. Conceivably, a limiting similarity would appear everywhere were habitat and other ecological differences between the species better known and taken into account, but for now, the results must be taken at face value. Moreover, conspicuous exceptions to species having large ratios exist and seem unexplainable from present knowledge. For example, four small species of *Accipiter* overlap in the Celebes; one eats mainly lizards, but this still leaves three species with ratios of 1.01 and 1.11. Wattel (1973), who states that Gause's Principle is "normally . . . fulfilled" in *Accipiter*, is unable to suggest how the Celebes species might differ. A second group that currently appears anomalous is *Micrastur*, the forest falcons of tropical America. Though not as similar in size to one another as are the Celebes accipiters, they still do not separate very nicely on the basis of size, even when altitudinal differences are taken into account. Of course, these species are all highly unstudied; possibly more differences will appear upon thorough examination (see below).

The fact that both these exceptional cases are tropical recalls Klopfer and MacArthur's (1961) hypothesis that tropical species may have smaller ratios than temperate ones, implying that more ecological overlap is allowable there. Moreover, the Nearctic trio of *Accipiter*, as mentioned above, is a textbook example of substantial size differences between species. However, the genus is primarily tropical, reaching a much greater diversity there. Part of the reason tropical ratios may seem smaller is simply because tropical associations have on the average more species, and so the trend is expected statistically (Figure 15.2). Nonetheless, for the *same* species number and the same category of ratio, temperate associations always have larger ratios. Relatively few temperature associations exist, however, and median tests are never significant; the smallest P value is 0.106.

While real associations have larger ratios than expected, little support exists for Hutchinson's (1959) suggestion that ratios might be constant. The situation depicted in Figure 15.3 for pairs is typical; moderate-sized ratios range from 1.02 to 2.10. On the other hand, more ratios occur in the interval 1.2 to 1.4 than in any other similarly sized interval. Because the number of ratios is expected to decline with ratio size, this result boils down to the observation that ratios are commoner in the 1.2–1.4 interval than in the 1.0–1.2 interval, *i.e.* to the result on limiting similarity. As the above discussion implies, real data cluster about somewhat lower ratios than 1.2–1.4 for associations of more than two species. A tendency for ratios not to be particularly constant was also noted for birds in general (Schoener, 1965) and *Anolis* lizards (Schoener, 1970), though a variety of investigators have claimed otherwise for their data (review in Simberloff and Boecklen, 1981).

Types of Errors

While the above analysis concludes that accipiter-like hawks are 20–35% more different in size than expected by chance, a number of possible types of errors may have biased this judgment one way or the other.

The most important such bias is a general flaw in the procedure of using the species being tested for size differences to construct the pool from which to generate null combinations (Case, 1983, Colwell and Winkler, this volume). As dramatically illustrated by Colwell and Winkler's simulations, this procedure is biased against rejection of the null hypothesis of no competition: species too similar to other species may have been exterminated everywhere, making them "unavailable" for construction of the pool. As a result, some combinations comprising very similar species could not be included in the null distribution. Where the arena of competition is the entire earth, as assumed for the hawk study, extinction everywhere is perhaps less likely than were the area an archipelago; *i.e.* Colwell and Winkler's bias is stronger, the smaller the number of geographic units considered. Using a somewhat different scheme (size variation within species was considered) and hence for somewhat different but related reasons, Case (1983) also illustrated the pitfalls of forming a null expectation by sampling from the survivors of a competitive process. Nonetheless, the bias, inasmuch as it exists for the hawk analysis, goes in the direction opposite to the results actually found; despite it, species are significantly dissimilar in size.

Most decisions that had to be made concerning size measurements and delineations of associations if anything probably bias the study away from rejecting the null hypothesis as well.

An average size was used for each species. Lumping sizes from several geographical regions could reduce apparent size differences if character displacement were important. Averaging male and female sizes may lead to a variety of biases. This procedure always decreases the size differences between species from those obtained when the sexes are distinguished. Specifically, suppose we have two species and we average the sexes' sizes and then compute a size ratio. This ratio will always be smaller than the mean ratio computed by keeping the sexes separate and averaging the four possible interspecific ratios. The argument generalizes to any number of species (I am indebted to M. Turelli for this latter point). An alternative procedure is to consider each sex as a separate morph. Then both observed and expected associations would have smaller ratios between adjacent morphs. In certain real associations, *e.g.* the three North American accipiters, adjacent interspecific morphs belong to different sexes, but this need not be the case. A future analysis keeping the sexes separate may be sensible, but one would have to decide whether to look at adjacent morphs regardless of species, or only adjacent morphs of different species, or something more complicated, such as each morph's closest interspecific morph. The outcome of such an analysis is hard to guess, but were sexual-size dimorphism in part determined by interspecific competition as is often suggested, lumping the sexes might obscure an overdispersion pattern.

Habitat differences recognized in this study are very coarse; species might differ in how they use space within a vegetation zone or altitudinal band. Indeed, since my simulations were performed, E. Temeles has informed me for *Micrastur* and J. Diamond for Southwest-Pacific *Accipiter* that their unpublished data show that certain combinations of similarly sized species judged sympatric from published information in fact do not occur in the same habitats.

Except for eliminating five species in the formation of Group II, no attempt was made to order the species according to how different their diets actually are—though data here are often scanty. Moreover, the size measurement used, wing length, may not be the best one for indicating differences in prey size or type; foot characteristics (E. Temeles, pers. comm.; Wattel, 1973); or body weight (Johnson, 1978) may be better. Again, good data are scanty.

One convention used to construct the species pool may unambiguously bias results against the null hypothesis, however. The individual species, rather than the phylogenetically determined species group, is our basic unit of study. Hence, depending to some extent upon taxonomic opinion, a single phylogenetic stock may be split into a number of allopatric species. Each such species is considered a member of the species pool, and the typically small ratios between them are all included in the expected distribution. If the species were not biologically valid, they would not be really "available" for forming expected associations. In fact, some doubt exists about certain species; as mentioned above, the list we used, that of Brown and Amadon (1968), has a few more species of *Accipiter* than does Wattel's (1973) list. Even if all species were real, however, they might be recent; it is perhaps unfair to consider incipient species as "available" in the same sense as older species. A more conservative procedure might be to use broader phylogenetic stocks such as species groups; these, however, are sometimes difficult to identify. Moreover, that procedure would overlook cases of closely related species not overlapping in range because of great similarity, and such cases can be important support for the hypothesis of competition.

As discussed above, the basic test used, Test 1, can be biased because of dependencies among the species or ratios. Such dependencies can magnify a small tendency in the data (supportive or not); if data are weakly inclined against the null hypothesis, dependence can produce a stronger level of significance than is valid.

In this study, Test 4 reduces dependence for associations of more than two species, and Test 3 eliminates it entirely for all associations. Sample size is substantially reduced in Test 4, but slightly fewer than half the cases are still significant. Test 3 reduces sample size drastically, and almost nothing is significant. Moreover, the degree of difference is sometimes less, sometimes more, than for Test 1. For pairs, it is more and especially so for Group I. For associations with more than two species, it is mostly less, and sometimes D is negative. Thus for trios, quartets, and quintets, Test 3 eliminates all trace of pattern. Possibly, of course, other algorithms for generating sets of associations with no species in common would give different results. However, because of the greatly reduced sample size, the likelihood of statistical significance is low, no matter what the algorithm.

Finally, both tests that select associations de-emphasize ratios of widespread species pairs that overlap several different combinations of other species. Test 4 excludes all but one entry for such pairs. A fifth type of test would more often include such species pairs: select pairs within each association of more than two species on the basis of the total range occupied *by the pair* (rather than by the association, as in Test 3).

Relation to Other Studies

A variety of other studies have addressed the question of whether morphological differences between species are more than would be expected from a null model. Strong *et al.* (1979) asked whether bill dimensions of Darwin's finches sympatric on the various Galápagos Islands are more different than expected from randomly drawing from a species pool; they concluded they are not. Grant and Abbott (1980; see also Grant, 1983) varied the methodology and concluded otherwise. Using eight morphological characters, Ricklefs and Travis (1980) computed the geometric distance between species of passerines from Cody's (1974a) grassland bird communities. They concluded that average minimal distances for real communities were no greater than expected from random selection from a species pool containing all species in all communities. No evidence for overdispersion of characters was found, but species with relatively extreme morphologies seemed to be added in going from simple to more diverse communities. Simberloff and Boecklen (1981) analyzed the published data on size differences between various kinds of species. They concluded that, in the majority of but not in all associations, differences did not differ from a null model. Karr and James (1975) concluded for various kinds of birds that congeners differed more from one another in a set of morphological variables than would be expected by chance. Finally, Gatz (1979) concluded that real communities of fishes had more large and small differences between species than intermediate ones; he simulated this result assuming that species differed regularly and substantially from one another along certain ecological dimensions and randomly along others.

Results presented here appear to be at variance with some of those just cited. This discrepancy might suggest that the respective species differ in the degree to which competition structures communities. Certainly accipiters and other bird-eating hawks are especially likely to have their communities structured by competition, because they are at the top of the food web and have few predators (Hairston *et al.*, 1960; Menge and Sutherland, 1976; Schoener, 1974b). However, the importance of competition has been claimed frequently for Darwin's finches, and Cody also makes this claim for his grassland birds. And in fact, the present paper's approach is different from other ones in ways that might at least partly account for the discrepancies. I now discuss the major differences.

A. *Number of species, number of associations, and ecological similarity.* Compared to the present study, most of those cited above suffer from one or more of the following features. First, too few real associations are available, making it difficult to distinguish their properties from those of randomly generated associations. This problem was cited by Ricklefs and

Travis, who had available only eleven associations, and was also considered a problem by Grant and Abbott. The present study, which attempts to identify all associations in the world, has available substantially larger sample sizes. Second, too few species are available to form the species pool, so that the null distribution of associations necessarily contains a major fraction of the real ones. As Grant and Abbott point out, this situation makes it very difficult to distinguish the two statistically; in the Galápagos study, a relatively small number of species was available. In the present study, the problem does not exist; the number of possible associations is larger, and usually vastly larger, than the number of real ones. Third, species with feeding ecologies that are too diverse are sometimes lumped. Perhaps this in part results from too strong a desire to achieve a respectable sample size. Ricklefs and Travis include all passerines in their analysis; while habitats are similar in Cody's grassland communities, feeding methods may not be. Of course, the hope may be that the variety of morphological characteristics used will reflect the diversity of ecological types, but the reflection may be weak. Also, Grant and Abbott criticize Strong *et al.* for lumping different genera of Galápagos birds in one of their tests. In the present study, care was taken to include species very similar in food type and foraging habitat. Composition of Group I followed the most frequent approach, that of considering species in the same genus only. Group II species are perhaps more similar ecologically: five accipiters were deleted because they do not eat primarily birds, and non-congeneric bird-eaters of wooded habitats are added. Because all species in the world were used, this ecological restriction still allowed a moderately large sample size. To examine the problems for this type of analysis that ecological dissimilarity might produce, W. Schaffer has suggested to me that nested species pools be analyzed, each having more species, but species of less ecological similarity, than the last. This procedure would be extremely interesting to carry out, but the difficulty of determining all the associations from maps of intersecting ranges would quickly become extreme, and some sort of computer technique would probably be needed.

B. *Null models.* In studies of size or other differences between species, the most appropriate null model might be a matter of severe contention. Simberloff and Boecklen's null distribution of sizes, from which they were unable to distinguish much of the data, was a uniform (log scale) distribution bounded by the minimum and maximum observed sizes. In strong contrast, the null distribution generated above from existing species is lognormal-like in shape. The same is true for size distributions of many other kinds of species (Hemmingsen, 1934; Schoener and Janzen, 1968). Of course, these distributions differ from lognormal in that they are necessarily truncated. Any truncated distribution that tends toward uni-modality will generate a lot of small ratios relative to a uniform distribution. Thus, by choosing a uniform distribution, Simberloff and Boecklen have chosen a null model that is relatively hard to reject.

To investigate the importance of the null model further, I performed the following experiment on the hawk data. Simberloff and Boecklen give two kinds of tests whereby species sizes are distinguishable or not from values drawn randomly from a uniform distribution. The first tests for uniform spacing between species (constant ratios); the second tests for a minimal difference between species (no ratio less than a certain number). Each association is run through each test separately. Because the claim in the present paper is that species have larger ratios than expected by chance, I performed the second test only, which applies to associations of more than two species.

Of the 19 trios for Group I, 2 give significance at the 5% level, 6 at the 10% level and 11 at the 30% level (Simberloff and Boecklen use the 30% level as their most generous assessment). For quartets and quintets, Group I, nothing is significant at the 5 or 10% level, and 3 of 9 and 1 of 3 are significant at the 30% level. For trios, Group II, 1 of 39 is signifi-cant at the 5% level, 6 at the 10% level, and 14 at the 30% level. For the 29 quartets, Group II, the figures are 4, 7, and 13; for the 21 quintets, they are 0, 0, and 2. Overall, this is a very poor performance.

Why the discrepancy between the above conclusions and those of Simberloff and Boecklen? First, the assumed null distributions are differ-ent. Second, Simberloff and Boecklen's minimal-difference test is really very similar to their first test: it asks what the probability is of obtaining the smallest observed ratio if sizes were scattered randomly on a log-scaled line. For small numbers of species, this probability will be large unless sizes are very regularly spaced. Moreover, each association is treated separately, so that the particular ratio that would contradict the model varies with each association. For example, the ratios 1.01 and 1.01 for trios will pass the minimal-difference test, whereas the ratios 1.75 and 2.50 will fail! While it is true, as Simberloff and Boecklen point out, that the claimed minimal ratio can vary with the investigator, it surely doesn't vary this much, nor is it claimed to vary from one association to another for the same kind of animal. In short, considering each associa-tion separately, Simberloff and Boecklen's test in a somewhat misleading manner picks them off one by one. In the present paper, minimal differ-ences or larger-than-expected ratios are investigated in the aggregate, and in that aggregate, the hawks usually pass the test.

Choice of a null model can be absolutely crucial, yet it is not at all obvious how one goes about it. Where a species pool is available, as in the present study, the model constructs itself, though it has the problems

Table 15.4. Frequency of associations in which ratios increase from smaller to larger sizes.

	Obs.	Exp.	P^1	Obs.	Exp.	P
trios	.368	.530	.881	.487	.565	.776
quartets	.444	.187	.070	.103	.213	.114
quintets	.333	.049	.143	.00	.058	.273

[1] Binomial probability of a difference as or more extreme.

of failing to include species that have undergone competitive extinction and of non-independence from real data discussed above. Perhaps a better approach would be somehow to deduce a null expectation from knowledge of evolutionary and ecological processes. On this basis, in my opinion, a uniform distribution is one of the least sensible models that could be selected. In the absence of effects of all other species, for example, a model of optimal feeding (Schoener, 1969a) predicts a single or at most two optimal body sizes for a given set of ecological conditions. Were ecological conditions identical, the null distribution might comprise ratios of 1.00 only, or (for two optima) two ratios! Additional variance in the null distribution could then be related to differences in ecological conditions between species and between localities and to less deterministic character change. To predict such variance, much information on those conditions and much faith in the model are required. Alternatively, one could proceed empirically, comparing sizes of species occurring by themselves (solitary species) with sizes of similar sympatric species. I have done this for Caribbean *Anolis* lizards (Schoener, 1969b); the distribution of solitary species differs significantly from that for species on the most diverse islands. Barring ecological differences between the islands (and these exist, but are mostly not extreme except for island size), this test can be considered to have falsified the hypothesis that species do not affect one another's sizes.

For certain birds and lizards, I found that larger species in an association have larger size ratios than do smaller ones (Schoener, 1965, 1970). An argument for the lognormal nature of the species-pool size distribution may follow similar lines, as suggested for such associations, *i.e.* because their preferred food is relatively scarce, and for reasons of handling efficiency, larger species take a large range of food sizes and space themselves more widely. However, in this study fewer associations showed increasing ratios with increasing size than expected. Table 15.4 gives the fraction of real associations with increasing ratios and the fraction expected from the lognormal-like species pool. Expected fractions do not differ much from the case in which all size orderings are equally likely.

For Group I, the observed fraction is smaller than expected for trios, but larger for quartets and quintets. For Group II, the observed fraction is always less than expected. No difference is significant at the 5% level (binomial test with variable P), though that for quartets, Group I, is nearly so. Thus the most significant comparison does support the trend previously found, but in general the data do not.

C. *Other statistical problems*. A number of minor statistical problems exist with some of the tests comparing null models to data.

First, when comparing number of values (*e.g.* ratios) above and below some mean expectation, a binomial test with $P = 0.50$ is only valid when the expected distribution is symmetrically distributed about its mean. In the present study it clearly is not: distributions of ratios are highly skewed, and mean ratios are all well above median ones (Table 15.2). Rather than means, medians should be used, or one should perform some entirely different test, such as the non-parametric Kolmogorov-Smirnov test used above. Unfortunately, in their analysis of Galápagos finch ratios, Strong *et al.* misused the binomial method just mentioned. They do not give the distribution of ratios from their random samples, but ratios in principle are likely to be positively skewed, because they are bounded by 1.00 on the left. Because median ratios are less than mean ratios, their test favors rejection of the non-random hypothesis. The test performed by Grant and Abbott (1980), in which the variable is the Mahalanobis D^2 (calculated from three morphological variables), could suffer from a similar problem; one needs to know the distribution of expected D^2 to be sure. Moreover, when comparing mean geometric distances, one is making the same sort of assumption about symmetry (Ricklefs and Travis, 1980) or normality (Strong *et al.*, 1979).

Second, which data to lump with which is always a problem. Ideally, one wants the data to be independent and homogeneous. Grant and Abbott lump all D^2 values for all types of ratios within associations of a given species number. In the present study, as in the study of Strong *et al.*, expected ratios between different positions in the size rankings are kept separate for at least some tests; as Figure 15.2 shows, not to do so might make a difference. However, Strong *et al.* lump all data from all associations, regardless of species number, when doing their binomial test; they do separate tests for ground finches and tree finches. Yet one might lump across taxa rather than across associations with different species number, or one might lump everything. In this case it makes no difference, but it might have.

Third, if one is not computing all possible combinations of species, one must make certain the number of random communities generated is sufficiently high. In the present study, as in the study of Grant and Abbott, all possible combinations are generated. However, in the studies of Strong *et al.* and Ricklefs and Travis, they are not, sampling being employed instead. A way of deciding what sample size is sufficient is given for a certain class of null models by Pimm (1983).

A HISTORICAL COMMENTARY

The recent wave of critical papers flowing from Florida State has as its common theme that investigators have improperly drawn conclusions about their data, largely because they failed to state the proper null hypothesis. The discussion in Strong *et al.* (1979) is typical: "We propose another possibility ... that other hypotheses must first be tested against, but that is rarely considered at all by ecologists. This is the null hypothesis that community characteristics are ... random. ..." The novitiate reading the recent literature might get the impression that ecology has just discovered this null hypothesis. To go on, Strong *et al.* claim that Schoener's (1965) null hypothesis in comparing ratios of island and mainland birds was character displacement. This is an intriguing guess, and in a peculiar way perhaps even a complimentary one, but it is unfortunately incorrect. In fact, Schoener (1965, p. 202) stated: "Although it could be argued that larger ratios of character difference on islands might be produced mainly by random sampling from all the mainland species, it is not likely that random sampling alone of an area where the great majority of species are similarly sized small birds could be responsible for the much greater frequency of large ratios ... on islands." The null hypothesis is stated here, but it is not tested. The reason is simply historical: I wrote the paper as an undergraduate and was never told that I had to test such things statistically. Mathematicians in evolutionary biology at that time were often regarded as good or evil shamans, exalted or reviled with equal passion, but in neither case particularly well understood. In short, I plead guilty of being sophomoric but to nothing more profound.

In 1966 Grant (1968) did test exactly the above null hypothesis and rejected it. The major contribution of Strong *et al.* to this particular issue was to redo the test and fail to reject the null hypothesis. The reasons for the discrepancy are not clear, but Grant and Abbott (1980) reexamined the problem and came to the same conclusions that Grant (1968) had earlier.

Other statements of Strong *et al.*'s null hypothesis were made well before 1979. In 1974, I wrote, "What evidence demonstrates that the pattern of resource utilization among species results from competition? Mere presence of differences is not enough, for even if niches were arranged randomly with respect to one another, differences would exist," and, later in the same paper, "Where niches are regularly and widely spaced over one or more dimensions, the alternative or 'null' hypothesis of randomly generated differences must be rejected." A second, more specific example is to be found in Diamond (1975), where the probability "by chance" that particular combinations of species occur on no island is calculated (though he did not calculate the expected number of such combinations over all species; Connor and Simberloff, 1979).

Concerning habitat differences, Strong et al. write, "It will be more difficult [than for morphological differences] to frame null hypotheses for particular instances of habitat displacement because habitat data are fewer and less detailed. . . ." This statement is made after they cite Schoener (1975), in which many tests show that habitat-shift patterns are non-random.

I am sure there are other examples; I have flagrantly emphasized my own work, which I know best, and I am not going to pursue the point any further. The fact is, null hypotheses have been stated, some in a fashion nearly identical to recent versions such as that of Strong et al., and they have sometimes, though not often enough, been tested. But it seems to me that one area of real difference lies now not in stating or testing null hypotheses, but rather in the nature of the appropriate procedure. What set of statistical and biological assumptions is most acceptable? What data should be included in a single statistical test? Most importantly, what alternative hypotheses and what null hypotheses are most appropriate? This paper has attempted to show that how these questions are answered can make a difference.

Acknowledgments

I thank J. M. Diamond, P. R. Grant, D. Schluter, C. H. Stinson, E. Temeles, and C. A. Toft for valuable comments on an earlier draft. Participants in the conference giving rise to this volume also made some excellent suggestions.

APPENDIX I.

Species and sizes used in this study.

Species	Wing length (mm)	Species	Wing length (mm)
1. *Accipiter gentilis*	334.7	31. *A. castanilius*	171.6
2. *A. melanoleucus*	314.3	32. *A. ovampensis*	233.3
3. *A. trivirgatus*	218.4	33. *A. luteoschistaceus*	202.8
4. *A. grisieceps*	184.5	34. *A. imitator*	191.5
5. *A. meyerianus*	321.0	35. *A. poliocephalus*	202.8
6. *A. fasciatus*	258.0	36. *A. rhodogaster*	186.0
7. *A. henstii*	298.5	37. *A. henicogrammus*	230.8
8. *A. poliogaster*	257.0	38. *A. collaris*	161.8
9. *A. novaehollandiae*	254.3	39. *A. minullus*	150.0
10. *A. griseogularis*	245.8	40. *A. melanochlamys*	239.0
11. *A. trinotatus*	160.5	41. *A. virgatus*	179.6
12. *A. cooperii*	245.5	42. *A. nanus*	165.8
13. *A. bicolor*	228.0	43. *A. badius*	189.5
14. *A. gundlachi*	258.3	44. *A. soloensis*	198.8
15. *A. rufiventris*	225.0	45. *A. buergersi*	304.7
16. *A. francesii*	165.2	46. *A. princeps*	270.0
17. *A. striatus*	182.8	47. *A. erythropus*	158.6
18. *A. nisus*	222.5	48. *Erythrotriorchis radiatus*	379.0
19. *A. tachiro*	218.2	49. *Urotriorchis macrourus*	312.8
20. *A. cirrhocephalus*	220.1	50. *Megatriorchis doriae*	308.0
21. *A. brachyurus*	185.8	51. *Melierax gabar*	197.0
22. *A. haplochrous*	216.5	52. *Parabuteo unicinctus*	349.8
23. *A. rufitorques*	215.8	53. *Hieraaetus kienerii*	374.0
24. *A. albogularis*	231.3	54. *H. fasciatus*	472.3
25. *A. erythrauchen*	193.8	55. *Micrastur ruficollis*	176.8
26. *A. gularis*	178.0	56. *M. buckleyi*	219.5
27. *A. brevipes*	226.5	57. *Hieraaetus dubius*	362.8
28. *A. butleri*	177.3	58. *Micrastur plumbeus*	172.0
29. *A. madagascariensis*	202.8	59. *M. mirandollei*	224.3
30. *A. superciliosus*	140.0	60. *M. semitorquatus*	268.5

Group I: Species 1–47
Group II: Species 1–60,
except species 11, 16, 28, 35, 44.

APPENDIX II

Real associations used in this study.

Group I—pairs:
  1–17; 1–18; 1–26; 1–27; 2–15; 2–19; 2–39; 2–47; 3–41; 5–24; 6–9;
  6–22; 8–13; 8–30; 9–20; 9–24; 9–35; 10–25; 12–17; 13–17;
  13–30; 13–38; 14–17; 15–39; 16–29; 18–41; 18–43; 19–39; 19–43;
  32–43 and 39–43.

Group I—trios:
  1–12–17; 2–15–39; 2–19–31; 2–19–39; 2–19–47; 5–9–24;
  5–9–25; 6–9–20; 6–9–25; 7–16–29; 8–13–30; 9–20–35; 9–24–34;
  9–35–45; 10–25–37; 13–17–38; 19–32–43; 19–39–43 and
  32–39–43.

Group I—quartets:
  2–15–19–39; 2–19–31–47; 2–19–39–47; 4–11–36–42;
  5–9–20–35; 5–10–25–37; 6–9–20–35 and 19–32–39–43.

Group I—quintets:
  2–15–19–31–39; 2–19–31–39–47 and 5–9–21–33–46.

Group II—pairs:
  1–17; 1–18; 1–26; 1–27; 1–54; 2–15; 2–47; 2–54; 3–41; 3–53;
  3–54; 5–24; 6–9; 6–22; 6–54; 7–29; 9–20; 9–24; 10–25; 12–17;
  13–17; 13–30; 13–55; 13–60; 14–17; 15–54; 17–52; 17–55; 18–41;
  18–43; 18–54; 19–54; 26–54; 27–54; 32–51; 38–55; 39–51;
  39–54; 40–50; 41–53; 41–54; 43–51; 43–53; 43–54 and 55–60.

Group II–trios:
  1–12–17; 1–18–54; 1–27–54; 2–15–54; 2–19–54; 2–39–54;
  2–54–57; 3–41–53; 3–41–54; 3–53–54; 5–9–20; 5–9–24; 5–9–25;
  6–9–20; 6–9–25; 6–9–54; 9–20–50; 9–24–34; 9–45–50; 10–25–37;
  13–17–55; 13–30–59; 13–38–55; 13–55–60; 15–39–54; 18–41–54;
  18–43–53; 18–43–54; 19–39–51; 19–39–54; 19–43–54; 19–51–54;
  19–54–57; 30–55–60; 32–43–51; 32–43–54; 43–51–54; 43–53–54
  and 55–59–60.

Group II–quartets:
  2–15–39–54; 2–15–54–57; 2–19–31–57; 2–19–39–54;
  2–19–39–57; 2–19–47–54; 2–39–54–57; 2–47–49–54;
  2–47–54–57; 3–41–53–54; 4–36–42–53; 5–9–21–33;
  5–10–25–37; 6–9–20–48; 6–9–20–50; 8–13–55–60;
  13–17–38–55; 13–30–55–60; 13–38–55–58;
  13–55–59–60; 19–32–43–54; 19–39–51–54;

19–39–54–57; 19–43–54–57;
30–55–59–60; 32–43–54–57; 32–43–51–54; 39–43–51–54
and 43–51–54–57.

Group II–quintets:

2–15–19–39–54; 2–15–19–39–57; 2–19–31–54–57;
2–19–39–47–54; 2–19–39–54–57; 2–19–47–49–54;
2–19–47–49–57; 2–19–47–54–57; 5–9–21–33–46;
8–13–30–55–60; 8–13–55–59–60; 8–30–55–59–60;
13–17–38–55–58; 13–30–55–58–60; 13–30–55–59–60;
19–32–43–54–57; 19–39–43–51–54; 19–39–43–54–57;
32–39–43–51–54; 32–43–51–54–57 and 39–43–51–54–57.

16.

Patterns and Processes in Three Guilds of Terrestrial Vertebrates

JAMES H. BROWN and MICHAEL A. BOWERS

Department of Ecology and Evolutionary Biology, University of Arizona, Tucson, Arizona 85721

INTRODUCTION

The early development of community ecology was stimulated in large part by the suggestion that morphological and geographical patterns among closely related species of terrestrial vertebrates were related to ecological mechanisms of coexistence or competitive exclusion (*e.g.* Lack, 1947; MacArthur, 1958; and much subsequent work including that of Pianka and Schoener on lizards; Cody, Grant, and Diamond on birds; and Rosenzweig and Brown on mammals). The majority of these studies supported the thesis that interspecific competition plays a major, but not necessarily exclusive, role in determining the organization of guilds of closely related, ecologically similar species (*e.g.* MacArthur, 1972). In recent years this conclusion and the evidence purporting to support it have received increasing criticism (*e.g.* Connell, 1975; Wiens, 1977a; Strong *et al.*, 1979; Connor and Simberloff, 1979; Simberloff and Boecklen, 1981).

We have neither the time nor the ability to review this extensive and controversial area in detail in this symposium. Instead, in the present paper we reconsider the relationship between patterns of morphology and geographic distributions and processes of interspecific interaction for three guilds of terrestrial vertebrates: chipmunks, hummingbirds, and desert heteromyid rodents. These three guilds share several features in common. All comprise taxonomically related homeotherms of relatively small body size (< 150 g). These animals spend much time and energy foraging for energy-rich food items that are sparsely distributed in discrete units. The foraging of these vertebrates has a major impact on their food supply, substantially reducing the standing crop; although the evidence for this influence is somewhat inferential for chipmunks, it is clear and direct for hummingbirds (*e.g.* Kodric-Brown and Brown, 1978; Brown *et al.*, 1981) and heteromyid rodents (Brown et al., 1979). Finally, we chose these guilds because the senior author has worked intensively on them over the last 15 years, because they have been studied independently by many other workers, and because they provide excellent systems for investigating ecological relationships among closely related species (*e.g.* see Hall, 1946; Heller, 1971; Sheppard, 1971; States, 1976; Chappell, 1978; Patterson, 1981, for chipmunks; Pitelka, 1942; Grant and Grant, 1968; Snow and Snow, 1972; Lack, 1973b, 1976; Feinsinger, 1976; Carpenter, 1978, for hummingbirds; and Rosenzweig and Winakur, 1969; Rosenzweig and Sterner, 1970; Rosensweig, 1973; Schroder and Rosenzweig, 1975; Hoover *et al.*, 1977; M'Closkey, 1978; Price, 1978, for heteromyid rodents). While there is some legitimate disagreement among investigators about the details of patterns and processes in these guilds, there is widespread agreement about the major features of their organization.

CHIPMUNKS

North American representatives of the rodent genus *Eutamias* comprise 21 currently recognized species (Hall, 1981). These small, dorsally striped members of the squirrel family, Sciuridae, are found in shrub and forest habitats throughout most of western North America. Chipmunks are active, diurnal rodents that spend much time both on the ground and climbing in woody vegetation. They feed largely on seeds, which they collect in large quantitites in late summer and fall and store in their burrows to sustain them during hibernation. Chipmunks vary about threefold in body weight (30–100 g). Inter- and intraspecific variation in body size is in general positively correlated with habitat productivity, so that the smallest forms are found in desert and alpine areas whereas the largest chipmunks inhabit mesic coastal forests.

Patterns

Although chipmunk species often have broadly overlapping geographic ranges (see Hall, 1981), they typically occur in virtually mutually exclusive habitats (Hall, 1946; States, 1976; Chappell, 1978). The failure of these species to coexist locally has long been attributed to competitive exclusion (Hall, 1946). The most convincing evidence comes from "natural experiments" in which a species is absent from insular patches of habitat where it would normally be expected to occur, apparently because it never had the opportunity to colonize these areas or because it colonized but subsequently went extinct. There are at least 18 isolated mountain ranges in the southwestern United States where either the species characteristic of low-elevation, xeric, open woodland (usually *E. dorsalis*) or the species that normally inhabits higher elevation mesic forest (usually a member of the *E. quadrivittatus* species group) is absent. In every case the other species has expanded its altitudinal and habitat distribution to occupy virtually the entire range of habitats normally inhabited by both species (Hall, 1946; Brown, 1971; Patterson, 1981). Interestingly, when these chipmunks expand into habitats normally occupied by congeners, they seldom attain such high population densities as the species that usually occurs there (Brown, unpublished observations).

Apparently because the modest variation in body size within the genus *Eutamias* is related to productivity, and because adjacent habitats do not differ strikingly in productivity, those species that come into contact often differ by less than 10% in mean body weight or in linear measurements (Brown, 1971; Patterson, 1981). Despite these slight differences in body size and other morphological traits, Patterson (1981) has shown by means of multivariate analyses that species populations that expand their habitat and altitudinal distributions in the absence of missing congeners change morphologically as well. Populations that have expanded into habitats normally occupied by other species are convergent in morphology toward their missing congeners. These morphological shifts document character

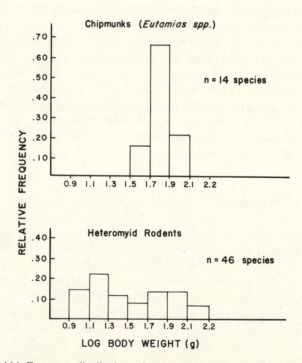

Fig. 16.1. Frequency distributions of body weights for species of chipmunks and desert heteromyid rodents. Note the small variation in body size in chipmunks and the much greater variation in heteromyid rodents.

displacement among chipmunk species that occur together on the same mountain ranges. Although these patterns provide only indirect evidence for interspecific competition, they are completely consistent with the interpretation that the non-overlapping local distributions of chipmunks are due to competitive exclusion.

Processes

A number of independent studies (Brown, 1971; Heller, 1971; Sheppard, 1971; Chappell, 1978) have concluded that direct interference competition plays a major role in maintaining the non-overlapping local distributions of *Eutamias* species. Because chipmunks are diurnal and often both abundant and relatively tame, their behavior can be observed directly in the field, and results of interspecific encounters can be quantified. Chipmunks are not strictly territorial, but they vigorously defend their burrows and patchy food resources within their home ranges against both conspecifics and congeners. These aggressive interactions result in generally non-overlapping local distributions of species, because one species is able to exclude others from habitats where two or more could otherwise live.

For example, Brown (1971) studied interactions between *E. dorsalis* and *E. umbrinus* in the narrow band of habitats where they come into contact on isolated mountains in eastern Nevada. The more terrestrial, lowland species, *E. dorsalis*, was much more aggressive and won 80% of all interspecific encounters. However, the aggression of *E. dorsalis* became ineffective and self-defeating in densely forested habitats where the abundant *E. umbrinus* readily escaped through the trees. Chappell (1978; see also Heller, 1971) studied a more complex situation where several species came into contact in the Sierra Nevada of California, but the results were generally similar. Within each habitat one species tended to dominate and exclude all others, but the outcome of aggression was mediated by habitat. One species, *E. speciosus*, was consistently subordinate to *E. amoenus* in laboratory encounters, but these aggressive relationships were reversed in the field, where *E. speciosus* excluded all other species from mesic coniferous forest habitats. *Eutamias minimus* was subordinate to all other species, but it possessed physiological adaptations for living in hot, dry desert shrub habitats where none of the others could occur.

HUMMINGBIRDS

The avian family Trochilidae contains more than 100 genera and 300 species (Peters, 1945). Hummingbirds are widely distributed throughout the New World, but the greatest diversity is found in the tropics. The entire family is specialized to feed largely on floral nectar, and all members share a common set of characteristics that includes long, slender bills, long, grooved tongues, tiny feet, and wings adapted for hovering flight. Although they vary tenfold in body weight (2–20 g), even the largest hummingbirds are among the smallest homeothermic vertebrates.

Patterns in Temperate North America

Eight species in four genera have geographic ranges that extend well into the temperate regions north of the Mexican border. Although some of these species have broadly overlapping geographic distributions, they usually occupy different habitats or the same habitats at different seasons, so they rarely coexist locally in both time and space.

The most striking characteristic of temperate North American hummingbirds is their great similarity in morphology. Compared to the family as a whole, or to representatives from many local tropical areas, the temperate hummingbirds are remarkably similar in body weight, bill length, and wing length (Figure 16.2). They are so similar in fact that, if we draw measurements of eight species at random from a pool representing 203 species of Trochilidae for which we have been able to find morphological data, none of 1000 random draws have variances in these three traits as small as those actually observed among the eight temperate species (Brown and Bowers, in prep.). This procedure should not be taken as a rigorous statistical test, because the temperate hummingbirds probably are more closely related to each other than they are to some other members of the family. Nevertheless, the fact remains that the temperate hummingbirds have either converged toward or failed to diverge from a remarkably constant morphology, and hummingbirds of divergent morphology have not successfully colonized temperate North America.

Processes in Temperate North America

The apparent explanation for this morphological similarity is that these hummingbirds have coevolved to forage from and to pollinate specialized flowers that themselves have converged to similar morphologies and rates of nectar secretion (Grant and Grant, 1968; Brown and Kodric-Brown, 1979). Similarities in these floral traits require that the hummingbirds use similar cues for locating flowers, have similar behavior and foraging apparatus for extracting nectar, and have similar energetic requirements so they can profitably harvest the nectar. The interrelated similarities among temperate hummingbirds on the one hand, and temperate hummingbird flowers on the other, represent an interesting pattern of community organization, but they are of interest here primarily because they constrain interspecific interactions among the hummingbirds. The birds have similar traits that reflect highly overlapping requirements for virtually identical resources.

HUMMINGBIRDS

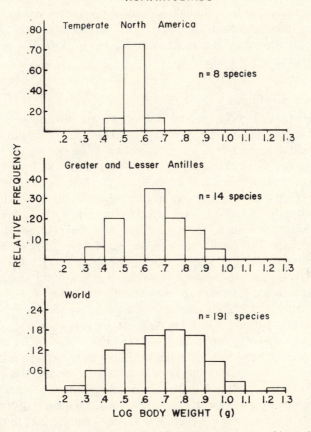

Fig. 16.2. Frequency distributions of body weights for species of humming-birds from temperate North America, the Greater and Lesser Antilles, and the world. Note the small variation among temperate species and the bimodal pattern for Antillean species compared to the world species pool.

Temperate hummingbirds are highly aggressive and territorial, both intra- and interspecifically (*e.g.* Pitelka, 1942; Carpenter, 1978; Kodric-Brown and Brown, 1978). This behavior results in exclusion by direct interference competition, so that species normally have non-overlapping local distributions in space or time. The outcome of interspecific aggression appears to depend on climate, elevation, habitat, and flower density (*e.g.* Kodric-Brown and Brown, 1978). Significant differences among species in relative wing length (expressed as wind disc loading) and slight differences in body size and perhaps other characteristics seem to be sufficient to account for the ability of particular species to exclude others from certain habitats, including the seasonal changes in distribution of these migratory birds. Although additional research is needed to clarify details of these mechanisms, interspecific territoriality and aggression clearly play a primary role in determining the organization of the temperate hummingbird guild.

Patterns in the Antilles

Fourteen species belonging to nine genera are distributed among the Greater and Lesser Antilles, tropical islands in the Caribbean Sea. Two interrelated patterns, one in geographic distribution and the other in body size, were first noted by Lack (1973b, 1976), although he did not test them against appropriate null hypotheses. The first pattern concerns the number of species per island: two or more species inhabit 36 islands, and there are no islands with only one species, but at least four islands (Mona, Grand Cayman, Little Cayman, and Cayman Brac) have no hummingbirds. This highly non-random distribution of numbers of species (observed distribution differs from a Poisson, $\chi^2 = 21.44$, $p < 0.01$) is particularly interesting because two of the islands without hummingbirds (Mona and Grand Cayman) are substantially larger and not significantly more isolated than islands supporting two or more species.

The second pattern is in body size: species that coexist on the same island tend to fall into two distinct size categories, which differ by a factor of approximately two in body weight. Although this pattern is complicated by the fact that the absolute sizes of birds differ among islands (*e.g.* the larger of the two species on Cuba is barely larger than the smallest of the five species on Puerto Rico), it can be demonstrated quantitatively. If measurements for all species are expressed as ratios with respect to the smallest species, frequency distributions for body weights and culmens of the hummingbird biotas are bimodal (Figure 16.3). Furthermore, this pattern is significantly different from those that would be expected if species were distributed randomly among islands in the frequencies in which they occur. The observed distribution of body sizes has fewer similar-sized hummingbirds and more pronounced bimodality than the distribution expected on the basis of random placement of species on islands (Figure 16.2; Brown and Bowers, in prep.).

Processes in the Antilles

The sizes and distributions of Antillean hummingbirds apparently reflect the influences of two processes: close coevolution between hummingbirds and hummingbird-pollinated flowers, and intense interspecific competition among birds of similar size (Kodric-Brown *et al.*, in press). Based on intensive work on Puerto Rico and short trips to Jamaica,

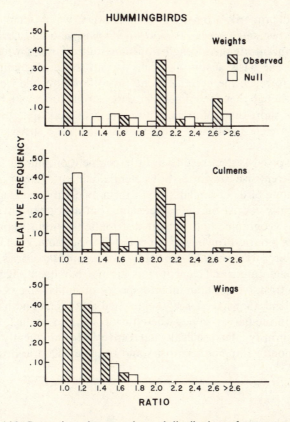

Fig. 16.3. Comparisons between observed distributions of measurements for Greater and Lesser Antillean hummingbird species and those measurements predicted from a null model that assumes species are distributed independently among islands in frequencies in which they occur on all the islands. Measurements are expressed as ratios with respect to the smallest species. Patterns for observed communities differ significantly from those expected from the null model for both weights and culmens (Kolmogorov-Smirnov test, $p < 0.025$). From Brown and Bowers (in prep.).

Culebra, Vieques, Mona, and Grand Cayman, we find evidence for extensive coevolution between hummingbirds and flowers. Red, tubular flowers specialized for hummingbird pollination are abundant on islands with hummingbirds, but absent from islands where no hummingbirds occur. On Puerto Rico there are two size categories of red, tubular flowers, each of which is highly specific in attracting a particular size category of hummingbirds. The three large hummingbirds (*Anthracothorax dominicus*, *A. viridis*, and *Sericotes holosericeus*), which are remarkably similar in size, feed primarily from long-tubed flowers, which secrete copious nectar. The two small species (*Chlorostilbon maugaeus* and *Orthorhyncus cristatus*) concentrate their foraging on short-tubed flowers, which secrete less nectar. The basis of this specificity is that the small hummingbirds are unable to reach the nectar of long-tubed flowers while feeding legitimately (although they sometimes pierce the corolla tubes with their bills), whereas the short-tubed flowers, whose nectar is accessible to all hummingbirds, secrete insufficient quantities for the large species to forage profitably.

This coevolved system results in little competition among hummingbirds in different size categories, but potentially intense interspecific competition among birds of similar size. Birds in different size categories not only frequently coexist on the same island, but also often have broadly overlapping geographic and habitat distributions within islands (Lack, 1973b, 1976; Kodric-Brown et al., in press). Furthermore, even on a local scale these birds frequently have broadly overlapping territories and traplines. Although interspecific encounters are not infrequent, aggression is rare. In contrast, when hummingbird species of similar size co-occur on the same island, they have largely segregated altitudinal and habitat distributions. In the narrow zones where they come into contact, these species are vigorously interspecifically aggressive and consequently they tend to occupy nonoverlapping territories and traplines (Kodric-Brown et al., in press).

HETEROMYID RODENTS

The desert and semiarid regions of southwestern North America are the center of abundance and species diversity of the rodent family Heteromyidae. The desert-dwelling representatives of this family belong to three genera: *Perognathus* (pocket mice), which contains 25 species characterized by generalized, mouselike morphology and scansorial, quadrupedal locomotion; *Microdipodops* (kangaroo mice), which is represented by two species of bipedal, hopping rodents that are confined to the cold Great Basin Desert; and *Dipodomys* (kangaroo rats), which contains 19 species specialized for bipedal, ricochetal locomotion.

Perognathus and *Microdipodops* species hibernate, but *Dipodomys* are active through the year. Although they differ significantly in some traits, including about twentyfold variation in body weight (7–150 g), these rodents share a suite of characteristics including nocturnal activity, external, fur-lined cheek pouches, and kidneys capable of secreting a concentrated urine, which reflect common adaptations for collecting and eating the dry seeds of desert plants.

Patterns

There is much overlap in the distribution of desert rodents on both geographic and local scales. Although local species diversity varies with habitat productivity and structure, as many as four species of heteromyids may be abundant in small areas of uniform habitat, where they may also coexist with several species of cricetid rodents that are less exclusively granivorous (Rosenzweig and Winakur, 1969; Brown, 1973, 1975).

Several investigators have noted that coexisting heteromyid species tend to differ substantially in body size (Rosenzweig and Winakur, 1969; Rosenzweig and Sterner, 1970; Brown, 1973, 1975; Blaustein and Risser, 1976; M'Closky, 1978; Hutto, 1978; Price, 1978). A recent quantitative analysis by Bowers and Brown (1982) compared observed ratios of body size among coexisting heteromyid species with those expected if species associated at random with respect to body size. Species of similar size (with ratios < 1.5) coexist much less frequently in local habitats and overlap less in their geographic ranges than expected on the basis of the null hypothesis of random associations (Figure 16.4). These results support Simberloff and Boecklen's (1981) analysis showing that body size ratios of desert rodent species that coexist in diverse communities are often significantly more uniform than expected on the basis of chance.

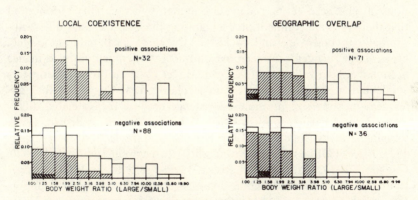

Fig. 16.4. Frequency distributions of pairwise associations among desert heteromyid rodent species plotted against body weight ratios on a logarithmic scale. Positive and negative associations represent pairs of species that occur together more or less frequently, respectively, than expected if they were distributed independently of each other. Rodent species of similar size (body weight ratios < 1.5) occur together in local habitats and overlap in their geographic ranges less frequently than expected on the basis of chance. Shading indicates phylogenetic affinities: unshaded, species in different genera; lightly shaded, congeneric species in different species groups; darkly shaded, members of the same species group. From Bowers and Brown (1982).

Colwell and Winkler (this symposium) have argued cogently that any null model of phyletic radiation that assumes no interspecific interactions predicts a unimodal distribution of character states. In fact, frequency distributions of body sizes among species within large taxonomic groups tend to be lognormal (Hutchinson and MacArthur, 1959; May, 1978). Consequently it is interesting that desert heteromyids exhibit a distribution of body sizes that differs from a lognormal distribution ($\chi^2 = 11.22$, $p < 0.01$) but is not statistically different from a uniform distribution ($\chi^2 = 4.52$, $p > 0.5$; Figure 16.1).

These patterns can be summarized by the statement that desert rodent species frequently overlap in both their local and geographic distributions, but that some deterministic process has influenced the radiation and dispersal of heteromyids so that species that occur together, especially those that coexist within local communities, differ substantially in body size.

Processes

These patterns are consistent with the explanation that interspecific competition is most intense among rodents of similar size because they use similar resources in similar ways, and that such competition has played a major role, not only in determining the organization of local communities, but also in influencing the evolution and biogeography of the entire guild. Abundant evidence (reviewed in Brown *et al.*, 1979) suggests that populations of these seed-eating rodents are limited by availability of food and that there is much competition even among those species that differ substantially in body size and coexist in the same local habitat. Largely because desert rodents are secretive and nocturnal, much of this evidence is circumstantial rather than based on direct observation or experiments conducted in the field. Different investigators have suggested that differences in body size promote coexistence because of associated differences in food particle size (Brown and Lieberman, 1973; Brown, 1975), foraging strategy (Reichman, 1979b), microhabitat selection (Rosenzweig, 1973; Price, 1978), and aggressive ability (Rosenzweig and Sterner, 1970; Congdon, 1974; Blaustein and Risser, 1976; Price, 1978). While there is legitimate disagreement among investigators about the relative importance of these mechanisms of interaction, there is almost universal agreement that interspecific competition has played a major role in determining the organization of the heteromyid rodent guild.

Perhaps the clearest and most direct demonstration of competition is provided by the "semipermeable membrane" experiment recently reported by Munger and Brown (1981). They used fences with different-sized gates to exclude large seed-eating rodents (three species of *Dipodomys*) differentially from fenced experimental exclosures, while allowing them free access to similarly fenced control plots. Munger and Brown observed densities of small seed-eating rodents (primarily two species of *Perognathus*) 3.5 times higher on the experimental plots than on the controls where *Dipodomys* were present. These results indicate that even those species that differ greatly in size and coexist in the same habitat still compete strongly.

DISCUSSION AND CONCLUSIONS

An idealized research program for investigating ecological interactions among a group of closely related terrestrial vertebrates might proceed as follows. It could begin with the discovery of an apparently non-random pattern of morphology or spatial distribution among species. The discoverer, who could be a systematist or biogeographer rather than an ecologist, might suggest that the pattern is caused by some process such as interspecific competition.

At this point, both the existence of the pattern itself and the correctness of the proposed causal mechanism should be regarded as hypotheses. The apparent pattern should be evaluated to determine whether it differs significantly from that expected on the basis of chance alone. This evaluation can be accomplished most rigorously by statistical comparison with results predicted by an appropriate null model that assumes no influence of the proposed causal mechanism (*e.g.* Strong *et al.*, 1979; Connor and Simberloff, 1979; Simberloff and Boecklen, 1981). At least two major problems complicate the process of constructing such null models and using them to evaluate empirical data. First, there are many possible null models that differ in their statistical and biological assumptions. Often it is not clear which model is statistically most appropriate and biologically most realistic for evaluating a particular data set. Second, it is possible that the apparent pattern is the result of inadvertent preselection by the discoverer of a small ordered subset from among a much larger set of random data. Tests against null hypotheses are unlikely to detect such an error, so it is important to regard the pattern with skepticism even if it differs significantly from the predictions of a null model.

The existence of a pattern implies the operation of a general causal process and therefore calls for a mechanistic explanation. This explanation should be viewed as a hypothesis that should yield new predictions that can be tested by new observations and experiments that go beyond the original data. For example, hypotheses about the role of interspecific competition in limiting distribution and abundance can be tested most definitively by experimental manipulation of densities in areas where the species come into contact (*e.g.* Hairston, 1980a).

The main criticism leveled against traditional vertebrate community ecology appears to be that processes are inferred from patterns and then treated as demonstrated facts rather than as still untested hypotheses.

Certainly it is true that in many cases patterns were claimed, and causal explanations proposed, before the data were tested against null hypotheses and before the empirical investigations required to test the mechanistic hypotheses independently had been conducted. Often the organisms that exhibited the pattern were not suitable for detailed observation or experimental manipulation. It is hardly surprising that our understanding of vertebrate guilds is based largely on a few systems, which can be viewed as empirical models because they not only show particularly clear patterns of organization but also have proven suitable for the kinds of observations and experiments necessary to elucidate the causal processes.

We regard the guilds of chipmunks, hummingbirds, and heteromyid rodents as such empirical models. Each of these guilds has been subjected to the entire program of investigation outlined above. Sometimes the work has not been performed in the recommended sequence, perhaps because it has been done by several independent investigators over a period of a decade or more. Nevertheless, there is now much evidence to support a direct, causal relationship between patterns of guild organization, as expressed in the morphology and distribution of species, and processes of interspecific interaction, especially interspecific competition.

Perhaps even more importantly, comparisons of these guilds reveal similarities in organization that appear to reflect common causal mechanisms. Within each guild, species of similar body size rarely coexist in the same local habitat, apparently because they compete intensely, primarily by aggressive interference. The outcome of such competition is mediated largely by the availability of resources and the physical structure of habitats, so that closely related species of similar size sometimes occupy different habitat patches within a local area and sometimes inhabit non-overlapping altitudinal and geographic ranges. These are the predominant patterns and processes of guild organization in chipmunks and in temperate North American hummingbirds.

Competitive exclusion of species of similar size also occurs in Antillean hummingbirds and in desert heteromyid rodents, but guild organization is more complex in these animals because there is more diversity in body size, and species of different size frequently coexist in local habitats. In both of these guilds there appears to be a limiting similarity (MacArthur and Levins, 1967) in body size, expressed as body weight ratios in the range 1.5–2.0, that must be exceeded in order for two species to coexist locally. We can speculate that this ratio reflects the difference in body size at which interspecific aggression becomes uneconomical and ineffective, because the costs of interference for the larger, presumably dominant species outweigh the benefits obtained from aggressing against smaller competitors that use fewer resources. Although species of such different body size coexist, observations and experiments show that they still overlap somewhat in diet and they still compete sufficiently to have substantial effects on local population density.

For guilds of chipmunks, hummingbirds, and heteromyid rodents, we conclude that non-random patterns of species morphology and distribution can be demonstrated by rigorous statistical analysis and that these patterns can be attributed largely to the effects of interspecific competition.

In closing, it seems appropriate to consider the limitations, both of the approach outlined above and of the general conclusions drawn from the case studies of the three guilds. Patterns discerned among closely related species are useful because they suggest hypotheses about the causal processes and suggest possible systems for testing the hypotheses, not because they show that other processes are unimportant and can therefore be ignored. Consequently, although interspecific competition can be shown to play a major role in the organization of these guilds, this result cannot be taken as evidence that other processes such as predation, mutualism, and interactions with the physical environment are unimportant. It is difficult to study predation on these kinds of vertebrates, but that is no reason to assume it does not influence community patterns and processes. Mutualistic interactions between hummingbirds and flowers clearly affect the organization of hummingbird guilds, but an investigator who was unaware of these relationships would still be able to detect patterns of body size and spatial distribution and to observe the interspecific aggression that is in part responsible for them.

Conversely, failure to observe in other communities patterns that are similar to those in these vertebrate guilds cannot be taken as evidence that interspecific competition is not important. We suspect that patterns of morphology and distribution are particularly apparent in these vertebrate species because they are phylogenetically constrained to use similar resources in similar ways and because they are major consumers of their food resources and can have large effects on each other. There is increasing evidence of substantial competition between such distantly related organisms as hummingbirds and nectar-feeding insects (Brown *et al.*, 1981) and heteromyid rodents and seed-eating ants (Brown and Davidson, 1977). But there is little or no pattern in the morphologies and distributions of these distantly related taxa to indicate how they interact in natural communities. As a result, it is possible that interspecific competition plays a major role in the organization of communities, but in many cases the patterns that suggest its importance in guilds of closely related vertebrates will be absent. Such competition will be detected only by ecologists who are sufficiently good naturalists to notice overlapping resource requirements among diverse kinds of organisms; and the importance of such competition will have to be assessed with direct observations and experiments.

Acknowledgments

Over the years many people have collected data and discussed ideas that have contributed to the analyses and interpretations presented here. Of these A. Kodric-Brown deserves special thanks. R. L. Zusi and R. K. Colwell kindly provided unpublished data on hummingbird measurements. T. S. Whittam gave valuable statistical advice. The National Science Foundation has generously supported J. H. Brown's research, most recently with grant DEB-8021535.

17.

Are Species Co-occurrences on Islands Non-random, and Are Null Hypotheses Useful in Community Ecology?

MICHAEL E. GILPIN

Department of Biology (C-016), University of California, San Diego, La Jolla, California 92093

JARED M. DIAMOND

Department of Physiology, University of California Medical Center, Los Angeles, California 90024

INTRODUCTION

Questions of structure lie at the heart of any science. That ecological systems have structure due to their abiotic environments and also due to limitations involving their own internal energy flows has long been recognized. Beyond this, there is a tradition, extending back to Darwin, that interspecific interactions such as competition, predation, parasitism, and mutualism regulate membership and abundance of species in local communities.

During the 1960's and early 1970's, the expected character of interspecific competition was given a strong theoretical formulation by the late Robert MacArthur and his associates. Since then, many field and experimental ecologists have described various phenomena that they interpreted as manifestations of competition. Much of the evidence has come from taxa high on the trophic pyramid, such as lizards and birds. This school of ecologists does not take interspecific competition as universally dominant, for they recognize that predation, physical disturbance, disequilibrium, or other factors may also be dominant. Furthermore, they recognize that past evolutionary adjustments to competition may minimize present ecological manifestations.

One form of opposition to these studies has come from scientists who study taxa for which predation and/or disturbance are of overriding importance (*e.g.* Connell, 1975). This opposition has been healthy and has generated new empirical data for diverse taxa. Recently a second form of opposition has appeared in a series of papers by Daniel Simberloff and colleagues. These authors claim that analyses of data in community ecology must begin with construction of a so-called "null hypothesis," which compares the actual data to simulated data generated by a model of randomness and lacking the factor whose explanatory value is being tested. If the "null hypothesis" adequately explains the data, this school argues with seeming plausibility that the data fail to demonstrate the importance of the posited factor. These papers of Simberloff and colleagues consist of reanalyses of other authors' data, conclude that the "null" test fails to support the explanatory factor postulated by the original authors, and avoid producing evidence for some different explanatory factor. In most of these papers, the explanatory factor attacked is competition (Simberloff, 1970, 1976c, 1978a; Connor and Simberloff, 1978, 1979; Strong *et al.*, 1979), but other factors have been attacked in a few recent cases (Connor and McCoy, 1979; Simberloff *et al.*, 1981). As philosophical justification for this approach, Simberloff and colleagues cite a particular philosophy of science developed by Popper, stressing hypothesis falsification, and formerly more in favor among philosophers of science than it is today (*cf*. Suppe, 1977).

In the present chapter we examine one such critique, that in which Connor and Simberloff (1979) challenge Diamond's (1975) conclusions that species are non-randomly distributed with respect to each other on islands and that competition is part of the explanation for these community assembly rules. Connor and Simberloff (1979, p. 1132) claimed to show that Diamond's data largely conformed to their "null hypothesis" and that "every assembly rule is either tautological, trivial, or a pattern expected were species distributed at random." We first point out that Connor's and Simberloff's own analysis refutes their conclusion: out of eleven data sets, seven compellingly reject their "null hypothesis," usually at extreme levels of $p < 10^{-8}$, while four sets were calculated wrongly and yield meaningless tests even when corrected. We then demonstrate seven serious weaknesses in the Connor-Simberloff approach. We develop a different approach, which demonstrates for two data sets that species co-distributions are radically non-random, as one would expect, and which lets one identify seven biological factors involved. Examining other "null hypotheses" by Simberloff and colleagues, we find them to exhibit the same types of errors inherent in the Connor-Simberloff approach. In particular, the term "null hypothesis" is a misnomer that we place in quotes throughout this chapter, for these models are not at all null with respect to the posited factor but contain it, lack the logical primacy claimed for them by Simberloff and colleagues, and are merely one more competing hypothesis. Finally, we explain why we feel that in community ecology the strategy of using "null hypotheses" to test for the effect of a single factor will generally be unworkable, so that its seeming plausibility is

illusory. The present chapter summarizes a more detailed analysis published in two recent papers (Diamond and Gilpin, 1982; Gilpin and
Diamond, 1982), which may be consulted for further information.

COMMUNITY ASSEMBLY AND INCIDENCE

The question considered by Diamond (1975), Simberloff (1978a), and
Connor and Simberloff (1978, 1979) is as follows: are species randomly
distributed with respect to each other over the islands of an archipelago?

A little reflection and biological intuition make it obvious that the
answer to this question is "no," for two types of reasons. First, species
individually are non-randomly distributed on islands in many respects:
e.g. because species colonize from particular directions (some South
American recent colonists of Caribbean islands are still confined to those
islands near South America), or because species may require particular
habitats present only on certain islands (*e.g.* ducks are confined to islands
with water habitats), or because certain species may require large islands
(*e.g.* species with large territories). Species similar or dissimilar in such a
respect could tend to have coincident or exclusive distributions even if
the species had no direct interactions. Second, direct interactions may
tend to cause additional non-randomness in species co-occurrences: *e.g.*
competitors might tend to have exclusive distributions, and mutualists
might tend to have coincident distributions.

This preliminary reflection points to a severe and probably insoluble
dilemma in constructing an appropriate "null model" to test patterns of
species co-occurrence for evidence of direct interspecific interaction. What
is an appropriate "null model"?—*i.e.* what is a model that excludes the
posited factor but includes all other important factors structuring the
data? The model must somehow incorporate factors that cause species
individually to be non-randomly distributed, else one would mistakenly
take a mutually exclusive distribution of a duck (on wet islands) and a
desert hummingbird (on dry islands) as evidence of competition. If one
attempts to deal with this problem by incorporating some empirical
features of actual distributions into one's "null model," such as island
species numbers or species occurrence frequencies, how can one avoid
thereby also incorporating features that do result from competition,
making one's "null model" no longer null? Is there any empirical feature
of distributions that is determined by species' individual properties but
not by species interactions?

With this as background, let us now summarize the conclusions about
community assembly reached by Diamond (1975). His data base then
consisted of lists of resident, breeding, non-marine land-bird species on
50 ornithologically surveyed islands (52 islands in the later analysis of

Diamond and Gilpin, 1982) of the Bismarck Archipelago in the tropical southwest Pacific Ocean: *i.e.* an *m*-species by *n*-island presence/absence matrix to be examined for evidence of ecologically significant patterns. Diamond began by noting that species individually were non-randomly distributed with respect to island area A or species number S. If islands are divided into different size classes by their S values and the frequency of occurrence of a given species in a given size class is defined as its incidence (abbreviated J), graphs of J against S would be linear if species were randomly distributed with respect to S, but are in fact nonlinear for almost all species. For most species J increases steeply and sigmoidally from 0 over a certain range of low S values towards 1 over a range of high S values, but the S range over which J rises varies among species (Figure 17.1). For certain species of high dispersal ability and low competitive ability, termed supertramps, J decreases again towards 0 at high S. Diamond was able to interpret these patterns, which describe for each species its area requirements or else ability to fit into communities of a given species richness, in terms of a species' biological attributes, such as its habitat preference, territory size, dispersal ability, and competitive ability. In particular, a simple model based on the independently known

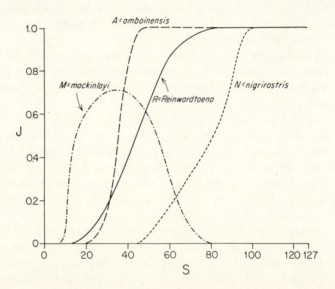

Fig. 17.1. Examples of how species' incidences vary with island species number S. For four species of cuckoo-doves on islands of the Bismarck Archipelago, the fraction of islands occupied by the species in a given island size class of S values (ordinate) is plotted against mean S of the class (abscissa). Note that species M is a supertramp confined to species-poor islands, while species N is confined to species-rich islands.

area dependence of a species' immigration and extinction probabilities accounts for the form of J–vs.–A graphs for most species (Diamond and Marshall, 1977) and is related to the newly developing "molecular" theory of island biogeography (Gilpin and Diamond, 1981).

To analyze non-randomness in species co-occurrences, Diamond (1975) confined attention to four guilds of species defined by shared resource utilization (shared habitat, diet, and foraging technique). Only within guilds can one expect to find strong interspecific competition, if it exists. With few exceptions the species of the guilds studied were geographically distributed throughout the Bismarcks rather than confined to one end adjacent to a recent colonization source. The species were ones of forest, which is the dominant habitat on all Bismarck islands. Thus, non-random co-occurrence patterns arising from restricted geographical distribution or from habitat requirements met on few islands were unlikely. It turned out that, of all the possible combinations of species in each of the four guilds analyzed, only certain of these combinations were actually observed as sets of species co-occurring on an island. Statistical tests showed that the surveyed islands were sufficiently numerous, and the species studied sufficiently widely distributed, that the failure to observe certain other combinations in nature was unlikely to be for reasons of chance. Diamond summarized these patterns of coexistence in a set of rules that he termed assembly rules. These patterns were interpretable in terms of ecological attributes of the species: on the one hand, species similarities or differences in area requirement, dispersal ability, and other factors producing non-randomly coincident or exclusive distributions without implying direct interactions, as discussed in the second paragraph of this section; on the other hand, direct competitive interactions between species producing non-randomly exclusive distributions in some cases.

Each guild analyzed provided one or two examples of pairs of species with mutually exclusive distributions. These pairs each proved to be the two guild members most similar to each other in morphology and ecology. Often, the two species occupied similar habitats on separate islands less than 1 km apart, and vagrants of one species reached the other island without becoming established. Diamond interpreted these cases of mutually exclusive distributions in terms of interspecific competition. Supporting evidence for a role of competition came from varied other types of evidence, including observations of interspecific fighting, overlap in diet, and niche shifts in habitat and abundance correlated with presence or absence of competitors. The combination of these types of evidence with distributional evidence makes a compelling case for a role of competition. Note, however, that it is only for pairs of especially similar species that the outcome of competition is a mutually exclusive distribution. Most species are sufficiently distinct that they can coexist somehow, on some island(s), with most other species in the guild, by utilizing different spaces, foraging techniques, or food items. Thus, analyses of non-random co-occurrence detect only an infrequent manifestation of competition and greatly underestimate its importance in nature.

THE CONNOR-SIMBERLOFF "NULL HYPOTHESIS"

Connor and Simberloff (1979) raise a number of objections to the analysis of Diamond (1975). Our responses to the less important of these are given in Diamond and Gilpin (1982). We concentrate here on the logic of the method by which they reject Diamond's inference of non-random co-occurrence of species. Connor and Simberloff cannot produce *a priori* predictions for species distributions within archipelagoes in the absence of species interactions. Therefore, they must work *a posteriori* from certain givens of the actual situation. As we note above, as Colwell and Winkler (this volume) conclude, and as Feinsinger *et al.* (1981) have wryly inferred, this practice is fraught with danger.

The idea used by Connor and Simberloff is to attempt to redistribute species randomly on islands and then to compare pairwise or triowise species associations in this randomized simulated data base with those in the observed data base. The dilemma posed at the start of the "Community Assembly and Incidence" section of this chapter immediately arises. For example, a "null hypothesis" with minimal structure, such as a purely random distribution of m species on n islands, would be inappropriate: it would deviate from the observed data base in such gross respects as the S values and would be rejected for reasons unrelated to competition. How can one devise a "null model" that does not implicitly contain effects of competition and that would be rejected as a result of competitive effects in the observed data base but not as a result of other effects? The solution of Connor and Simberloff (1979, p. 1133) is to redistribute the elements of the archipelago presence-absence matrix subject to the following three constraints:

"(i) For each island, there is a fixed number of species, namely, that which is observed.
(ii) For each species, there is a fixed number of occurrences, namely that which is observed.
(iii) Each species is placed only on islands with species numbers in the range for islands which that species is, in fact, observed to inhabit. That is, the 'incidence' range convention is maintained."

With these choices, Connor and Simberloff have undermined the game they propose before it has even begun, for species occurrence frequencies and incidences appear to be strongly influenced by competition. This becomes clear when one compares occurrence frequencies or incidences of a given species in different archipelagoes where it faces different numbers of competitors (Diamond, 1982). Thus, the Connor-Simberloff "null hypothesis" already incorporates competition through two of its three seemingly innocent constraints. The non-randomness that it detects must arise from causes other than competition, plus effects of competition beyond those contained in its constraints. Connor and Simberloff (1979, p. 1133) claim that relaxing their incidence constraint does not affect their results. This claim astonishes us, as we find that a realistic incidence constraint contains most of the observed non-randomness in species co-occurrences (see section on "Analysis of Individual Guilds"). Since Connor and Simberloff present no evidence by which to judge their claim, it is uncertain whether the explanation is that their claim is an inaccurate reflection of their results or that their incidence constraint was too crude to be effective.

One method that Connor and Simberloff use to reshuffle species on islands is repeatedly to find elementary 2×2 submatrices, not necessarily in adjacent rows or columns, of the form

$$\begin{pmatrix} 0 & 1 \\ 1 & 0 \end{pmatrix} \quad \text{or} \quad \begin{pmatrix} 1 & 0 \\ 0 & 1 \end{pmatrix} \qquad \text{(eq. 1)}$$

and then to swap one for the other. Clearly, this does leave column sums (constraint i) and row sums (constraint ii) unaltered. After an unstated number ("several": Connor and Simberloff, 1979, p. 1140) of such swaps are performed (on a digital computer), the resulting "randomized" matrix is analzyed for patterns of association among species pairs or trios, and the average pattern for ten randomized matrices is compared with that for the observed matrix.

RESULTS OF THE CONNOR-SIMBERLOFF TEST

The Connor-Simberloff test works by determining whether the number of species pairs or trios sharing given numbers of islands differs between the randomized and actual data bases more than would be expected by chance. Connor and Simberloff do not restrict their analysis to species within a guild, although this would be the appropriate way to test for competition. Instead, they analyze an entire fauna or, as their sole attempt to choose ecologically reasonable units, sets of confamilial species. The former analysis is inappropriate, while the latter one is little improvement because family boundaries and guild boundaries differ greatly. On the one hand, a given family may contain members of many guilds (*e.g.* New Guinea parrots include nectarivores, frugivores, bark-feeders, and seed-eaters). On the other hand, a given guild may include members of many families (*e.g.* the arboreal frugivore guild in New Guinea is drawn from 14 bird families).

Connor and Simberloff applied their test to three faunas (New Hebrides birds, West Indies birds, West Indies bats). From their stated χ^2 and d.f. values, we recalculate below the confidence with which their "null hypothesis" can be accepted as explaining co-occurrences in these three faunas. We also give the results of our application of their test to a fourth fauna (Bismarck birds), omitting the pointless confamilial tests and the computationally expensive trio tests:

	all species		confamilial species	
	pairs	trios	pairs	trios
New Hebrides birds	$p > 0.90$	$p > 0.95$	$p > 0.99$	$p > 0.99$
West Indies birds	$p < 2 \times 10^{-7}$	—	$p < 10^{-8}$	—
West Indies bats	$p < 0.018$	$p < 10^{-8}$	$p < 10^{-8}$	$p < 10^{-8}$
Bismarck birds	$p < 10^{-8}$	—	—	—

Thus, out of eleven tests, seven reject their "null hypothesis" resoundingly, while the four New Hebrides tests yield very close acceptance.

Why is there such close correspondence between supposedly randomized and actual data for New Hebrides birds and only for this fauna? Connor and Simberloff noticed that all their randomized New Hebrides matrices looked similar to the actual matrix, but they failed to understand what produced this suspicious result, and they proceeded anyway to interpret their New Hebrides results as arguing strongly for their "null hypothesis" and against competition. The reason proves to be twofold: a gross error in performing the New Hebrides tests and a structural flaw in the Connor-Simberloff procedure for constructing supposedly randomized matrices. The error is that, in the last paragraph of a small-print appendix, Connor and Simberloff (1979, p. 1140) mention that they performed only "several" reshufflings by eq. 1 to generate each randomized New Hebrides matrix and that they accepted any randomized matrix whose pattern was "not completely identical" to the observed matrix or to the previous randomized matrix. When one interchanges only several pairs of elements in a 56×28 matrix, the observed and randomized matrices are naturally nearly identical. The result is a spectacular fit of the "null hypothesis" to the jaggedly irregular observed graph for New Hebrides birds (Fig. 1 of Connor and Simberloff, 1979) and acceptance at $p > 0.90$ or $p > 0.99$. This error did not arise in the analyses of West Indies birds and bats, for which Connor and Simberloff (1979, p. 1139, first paragraph of column 2) used a different computer algorithm. When we repeated the Connor-Simberloff test properly on the New Hebrides (all species pairs), executing 500 successful reshufflings instead of several to generate each randomized matrix, and generating 50 instead of 10 such matrices, the confidence of accepting the null hypothesis dropped to $0.10 < p < 0.25$. This is still a much less resounding rejection of the "null hypothesis" than obtained for the other three data sets. This difference is due to a structural flaw in the Connor-Simberloff randomization procedure, which forces the matrix elements back into a pattern very similar to the observed pattern when applied to so-called nested matrices (see paragraph 4 of our section on "Basic Flaws of the Connor-Simberloff Test" and Gilpin and Diamond, 1982, for explanation), of which New Hebrides birds but not the West Indies or Bismarck sets are an example. By our test (section on "Our 'Null Hypothesis'"), which does not suffer from this structural defect, the simulated and actual New Hebrides matrices differ at the level $p < 10^{-8}$.

Thus, of four data bases tested by the Connor-Simberloff procedure, one yields an erroneous result and then a meaningless test when this error is corrected, while three overwhelmingly reject their "null hypothesis," mostly at the $p < 10^{-8}$ level. Remarkably, Connor and Simberloff take these results as substantially confirming their "null hypothesis" and rejecting Diamond's (1975) claim of non-random assembly. They proceed at the end of their paper to moralize on the improper scientific procedures used by students of interspecific competition. (Connor and Simberloff obscured the true conclusion of their analysis by omitting confidence limits from their graphs, interpreting their meaningless New Hebrides results, stating the rejections to be at the $p < 10^{-2}$ level instead of at the actual $p < 10^{-8}$ level, and then attaining better fits between simulated and actual data by a relaxation of constraint (ii) that inevitably yields better fits but makes the simulated matrix a less realistic model of the real matrix.)

At this point one could end by congratulating Connor and Simberloff for having demonstrated, using a new statistical methodology, that our views of non-random community assembly are correct. However, we have found such basic flaws in their approach that we distrust it entirely. By coming to some understanding of these flaws (next section), we have fashioned a better test of our own (following section).

BASIC FLAWS OF THE CONNOR-SIMBERLOFF TEST

Dilution. Competition is possible within guilds of species that utilize shared resources. Among the species of a guild, competition is strongest between those with the greatest resource overlap. The Bismarck archipelago is home to 151 bird species. If we make the extreme assumption that each species competes so strongly with a single other species that their joint biogeographical distribution is affected, then there are 75 pairs for which competition should produce some distributional non-randomness. Yet there are $(151)(150)/2 = 11,325$ total pairwise comparisons to be made for this fauna. Clearly, even a strong signal will be difficult to detect against so much statistical noise. Realizing this, Diamond (1975) explicitly based his analysis on species within the same guild. By expanding the analysis to an entire fauna, Connor and Simberloff submerged instances of competitors with exclusive distributions in an irrelevant mass of data from ecologically remote pairs of species, such as owls and hummingbirds.

Weighting. A different kind of dilution can occur when species are not weighted by their ranges of distribution within the archipelago. A substantial fraction of an archipelago's species may be limited to but a few islands. Many pairs of these may happen by chance to exhibit, for example, "checkerboard" distributions (that is, to be in the zero-islands-shared class of the Connor-Simberloff test). Yet these pairs have little significance compared to a pair such as the Bismarck nectarivores *Nectarinia sericea* and *Myzomela pammelaena*, which also share no islands but occur on 18 and 23 islands, respectively. This latter checkerboard distribution is due to interspecific competition; the probability of its arising by chance is 1 in 200,000,000. The Connor-Simberloff test counts each of their pairs as equal. Analogous problems arise from species sharing any given numbers of islands, not just zero islands. Surely, some method should be developed for weighting observed co-occurrence data by their degree of improbability.

Inability to detect checkerboards. The simplest, clearest, and most non-random pattern caused by interspecific competition is a checkerboard distribution: two or more ecologically similar species having mutually exclusive but geographically irregular distributions on an archipelago's islands. The Connor-Simberloff procedure is structurally incapable of recognizing such a pattern as non-random, for reasons independent of the two above-mentioned deficiencies. To see why this is so, consider the following instance of a checkerboard distribution:

islands

species
$$\begin{matrix} 1 & 0 & 1 & 0 & 1 & 0 & 1 & 0 \\ 0 & 1 & 0 & 1 & 0 & 1 & 0 & 1 \end{matrix}$$

This matrix may be rearranged in numerous ways by swapping the 2×2 submatrices of eq. 1, but any such rearrangement preserves a checkerboard pattern in the resulting "null matrix," which consequently is forced to yield the unreasonable result that any observed checkerboard distribution is consistent with random co-occurrence of species.

Inability to generate meaningful "null matrices" for nested data sets. As mentioned in the section "Results of the Connor-Simberloff Test" and exemplified by New Hebrides bird distributions, the Connor-Simberloff randomization procedure is unable to create meaningful rearrangements of the observed data set, if the set is quite nested. This makes the comparison of the observed matrix and the supposedly randomized matrix an empty one. (By a perfectly nested matrix, we mean one in which each island contains all species on the next less species-rich island, plus additional species, and in which each species occurs on all islands occupied by the next less widely distributed species, plus additional islands.)

Failure to identify non-randomly distributed species combinations. The Connor-Simberloff procedure yields only a "yes/no" answer to the question whether non-randomness is detectable. It does not permit one to attain deeper biological understanding by examining the direction and cause of any non-randomness detected and by identifying the species combinations most responsible.

Competition hidden within the "null hypothesis." Observed sets of island/ species lists are structured by numerous factors other than competition, such as differing island areas, differing dispersal abilities of species, etc. Because Connor and Simberloff had no independent way of incorporating these factors as constraints in their "null hypothesis," they instead took as constraints three given features of the observed data set, tacitly assuming these features to be unaffected by competition. However, two of these three constraints are strongly affected by competition, making their "null hypothesis" inappropriate as a test of competition (section on "The Connor-Simberloff 'Null Hypothesis'"). Unfortunately, this problem is impossible to overcome in any null distribution based in whole or in part on the observed data. We have not been able to correct this problem in our "null hypothesis" either (section on "Our 'Null Hypothesis'"). These obstacles to constructing truly "null" hypotheses in community ecology are in our view the most intractable difficulty in the "null" approach.

Reliance on stochastic simulation. The Connor-Simberloff test relies on Monte Carlo techniques—*i.e.* computer simulations based on pseudo-random numbers. For systems as large as the Bismarcks, which have roughly 10,000 elements in the presence/absence matrix, this approach is computationally expensive. As a result, Connor and Simberloff construct only ten randomized matrices for comparison with each observed matrix. They provide no evidence that this technique adequately covers the sample space. They do not even specify the number of reshuffles by eq. 1 used to create each matrix from a previous matrix except to say that "several of these changes are made" (Connor and Simberloff, 1979, p. 1140); thus, the randomized matrices are surely not independent of each other or of the observed matrix. An analytical method to generate the "null distribution" would clearly be preferable.

OUR "NULL HYPOTHESIS"

We have devised an approach to detecting non-random co-occurrences that cures six of these seven basic flaws in the Connor-Simberloff test (all but the most basic flaw, that of competition hidden within the "null hypothesis"). We recognize fundamental difficulties in any method based on the givens of the field situation, under which both evolution and competition have already been operating for a long time (and probably reached some quasi-steady-state). Nevertheless, it is still possible to devise a test that avoids dilution by focusing on guilds, weights observed species combinations by their deviation from expectations based on randomness, succeeds in detecting checkerboards, succeeds in detecting non-randomness in relatively nested matrices, identifies non-randomly distributed species combinations, and avoids inefficient stochastic simulations.

Our procedure retains the flaw of accepting observed island species numbers and species occurrence frequencies as givens. The Connor-Simberloff approach utilized these constraints by reshuffling 1's and 0's in the presence/absence matrix, then tabulating co-occurrence patterns in the resulting matrix, However, as we saw (paragraph 3 of the preceding section), such reshuffling does not erase checkerboards. Thus, their test is prone to accepting as "null" a data set that actually is highly structured. A better way to incorporate these constraints is to calculate the probabilities p_{ij}, given these constraints, that the ith species will be on the jth island. If such a matrix of incidence probabilities can be obtained, it is straightforward to determine for each pair of species i and k the expected number of islands that they should share, together with a standard deviation, and to compare it to the actual value, thus satisfying our goals stated above.

One way to obtain such probabilities is simply to run the Connor-Simberloff algorithm for reshuffling the presence/absence matrix for a very long time and observe what fraction of the time a species-island element is occupied by a 1. It is immediately apparent that this solves their difficulty with checkerboards, for in the case of the 2-species 8-island checkerboard considered in paragraph 3 of the preceding section the probability elements p_{ij} for occupation of an island by a species uniformly converge to 0.5. That is, the archipelago probability matrix will be:

island

species   .5 .5 .5 .5 .5 .5 .5 .5

.5 .5 .5 .5 .5 .5 .5 .5

whence it can be calculated that the two species should share 2.0 ± 1.5 islands, making the fact that they actually share 0 islands somewhat surprising.

The difficulty with this approach is that it is computationally expensive for such a process to converge to precise estimates of the probability elements. A better method is the standard log linear model of contingency table analysis (Bishop *et al.*, 1975; Fienberg, 1980), which suggests

$$p_{ij} = R_i\, C_j/T,$$

where R_i is the ith row sum (the observed number of islands occupied by the ith species), C_j is the jth column sum (the observed number of species on the jth island), and

$$T = \sum_j C_j = \sum_i R_i$$

(the observed total instances of species on islands). This test is simple and works without problem in many cases, such as the checkerboard above: $R = 4$, $C = 1$, $T = 8$, with $p = 0.5$ for all i and j. In a few cases where it yields values of $p_{ij} > 1$, we have devised three different solutions. We have also used three basically different approaches to estimating p_{ij}'s: the quasi-independent model of contingency table analysis, a recursive approach, and a model in which the p_{ij}'s are transformed logistically. All seven of these methods for estimating p_{ij}'s yield similar results, except for some consequences of structure inherent in four of these methods for detecting positive association.

We have applied this method to the whole Bismarck and New Hebrides avifaunas, as well as to particular Bismarck guilds. Thus, for the Bismarck avifauna, which holds $(151)(150)/2 = 11{,}325$ species pairs, we can obtain the standard deviate (observed minus expected number of islands shared, divided by the standard deviation of the expectation) for each species pair, and we can compare the resulting observed histogram with predictions of the "null model" for a distribution of standard deviates: a normal curve of mean 0 and standard deviation 1 (Figure 17.2). For the Bismarcks ($\chi^2 = 3981$, d.f. $= 66$) as for the New Hebrides ($\chi^2 = 692$, d.f. $= 61$), the deviation of the observed data from randomness is extreme ($p < 10^{-8}$). The Bismarcks exhibit long positive and negative tails (best seen in Figure 17.2 bottom) of species co-occurring more or less often than expected by chance, extending out to $+5.11$ and -5.71 standard deviates, respectively. Even in a fauna containing $(151)(150)/2 = 11{,}325$ species pairs, the chances of finding *any* species pair with such non-random co-occurrences are .002 and .00007, respectively. Furthermore, the two species with the highest negative deviation (most unexpectedly exclusive distribution) are not randomly related ecologically but are the two small nectarivores *Nectarinia sericea* and *Myzomela pammelaena*, which are very close in diet, foraging technique, size, and habitat and hence are inferred on grounds independent of island distributional patterns to be close competitors.

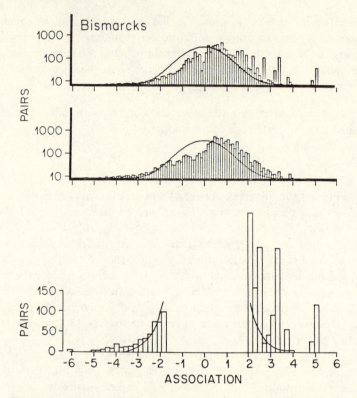

Fig. 17.2. Histogram depicting non-random positive and negative associations in the 11,325 pairs of Bismarck bird species. Abscissa: difference between the observed and expected number of islands shared by a given species pair, expressed in standard deviates of the expected number (this number is that for randomized matrices constructed by our procedure). Ordinate: number of species exhibiting that difference between observed and expected values. Large positive and large negative abscissa values mean that the two species have a much more concordant or much more exclusive distribution, respectively, than expected for random placement on islands. A randomly assembled community would yield the smooth curve shown: a normal distribution with mean zero and standard deviation 1. Top: histogram for all 11,325 species pairs; logarithmic ordinate scale (1 + number of pairs). Middle: histogram for the 3916 species pairs both of whose members occur on three or more Bismarck islands; logarithmic scale. Bottom: like middle graph, but ordinate scale arithmetic and expanded, abscissa intervals wider; only tails of distribution (abscissa values beyond +2 or −2) are shown.

FACTORS CAUSING NON-RANDOM CO-OCCURRENCE

We have scrutinized all Bismarck species pairs exhibiting extremely non-randomly concordant or exclusive distributions (positive and negative associations), in order to understand the biological factors responsible. As discussed in the section "Community Assembly and Incidence," common sense suggests *a priori* many factors other than direct inter-actions of two species that could produce this result. In practice, we interpret the Bismarck data as exhibiting effects of six factors that do not involve direct interactions, of which four produce cases of unex-pectedly coincident distributions and two produce cases of unexpectedly exclusive distributions. We interpret some other cases of unexpectedly exclusive distributions as due to a direct interaction, namely, competition.

Negative: competition. As mentioned above, the most non-randomly exclusive distribution in the Bismarcks is of two convergently similar nectarivores in different families, the sunbirds *Nectarinia sericea* and the honey-eater *Myzomela pammelaena*, occupying 18 and 23 islands respec-tively in a geographically irregular array but sharing no island. The probability that the "null model" would produce such an exclusive distri-bution in *any* Bismarck species pair is only .00007. Similarly, within the three other Bismarck guilds that we analyzed, the two species most similar ecologically and morphologically achieve exclusive or nearly exclusive distributions far beyond expectations for random co-occurrence.

The remaining six factors do not involve direct interactions:

Negative: differing distributional strategies. Within Bismarck guilds, competitively subordinate supertramps are confined to small islands as a result of exclusion by so-called tramp species of the same guild occupying larger islands. Since similar patterns are repeated in various guilds, the supertramps of one guild tend to be negatively associated with the tramps of another guild, even though these ecologically unrelated species pairs do not interact directly. Mostly, such pairs with high negative deviations do not achieve completely exclusive distributions; they share some islands, though many fewer than expected on a random basis. These cases, like the next set of cases, illustrate why tests of co-occurrence data for effects of competition should be performed on a guild rather than on a whole fauna.

Negative: differing geographical origins. Species have invaded the Bis-marcks from three different source areas: New Guinea (to the west), Aus-tralia (to the south), and the Solomon Islands (to the east). Some invaders have not achieved a uniform distribution throughout the Bismarcks. Thus, two species from different source faunas, one species still confined to western islands and the other species still confined to eastern islands, will show an exclusive distribution even though these species are unrelated ecologically.

Positive: shared distributional strategies. This is the converse to the sec-ond point, above. Supertramps of different guilds are similarly confined to small islands and thus occur together more often than expected. Similarly, tramps of different guilds tend to co-occur on large islands.

Positive: shared geographical origins. This is the converse to the third point, above. Species that have reached the Bismarcks from the same source area and that have not yet spread uniformly throughout the Bismarcks tend to co-occur on islands more often than expected for a random placement.

Positive: *shared habitat*. Certain species are restricted to specialized habitats such as marshes, lakes, grasslands, and high mountains, which are present on only a few Bismarck islands. Species restricted to the same habitat tend to co-occur on islands far more often than expected for random placement, and the outcome of competition is differential utilization of habitat patches or food resources on the same island, rather than occupation of different islands. For instance, extensive grasslands occur on only 11 of the 52 surveyed Bismarck islands; hence numerous grassland bird species have similar distributions confined to these islands.

Positive: *single-island endemics*. Of the 25 species confined to single islands in the Bismarcks, most are endemic. But they are not randomly distributed: almost all are on one of the three largest Bismarck islands, with 15 being on the largest island, New Britain. Hence these endemics yield the most extremely non-randomly coincident distributions of the Bismarcks ($p < 1.6 \times 10^{-7}$). The evolutionary explanation is the decrease in population extinction rates and hence increase in endemism with island area (Mayr, 1965). In the middle and lower histograms of Figure 17.2 we have removed these species pairs to expose the rest of the positive tail.

Thus, the analysis of Figure 17.2 yields a detailed appreciation of the dilution effect. When one follows Connor and Simberloff in analyzing all pairwise combinations of species in an archipelago, the instances of non-randomly exclusive distributions due to competition are greatly outweighed by non-randomly exclusive or concordant distributions due to other factors. Terborgh (1981) documents how many instances of exclusive or concordant distributions of West Indian birds arising from these same factors confound the analysis of West Indian birds by Connor and Simberloff.

ANALYSIS OF INDIVIDUAL GUILDS

We have also used our method to analyze pairwise co-occurrence within two Bismarck guilds: the seven species of gleaning flycatchers and the four species of cuckoo-doves discussed by Diamond (1975). As the species in each guild share the same habitat, are geographically widespread through the Bismarcks, and include no single-island endemics, four of the seven factors (nos. 3, 5, 6, and 7) discussed in the preceding section as contributing to non-random co-occurrence in pairwise analysis of the whole avifauna now disappear.

For each guild, the whole pattern of co-occurrence is extremely non-random, and the most deviant single pair is the pair of species most similar ecologically and morphologically, which achieve a perfect checkerboard distribution. For instance, a χ^2 test for the gleaning flycatcher guild rejects the "null hypothesis" at the level of $p < 10^{-4}$; each species coexists on some other island with each other species except for the very similar *Pachycephala pectoralis* and *P. melanura*, which form a perfect checkerboard (the latter on small islands, the former on larger islands); most species pairs co-occur on many more or on many fewer islands than expected; and nine out of 21 pairs are more non-randomly distributed than one would expect of any pair in the guild for a random placement.

To illustrate the importance of incidence constraints, which the Connor-Simberloff test handled crudely and reported to be unimportant, we examined the extent to which the non-random incidence patterns of individual species determine the non-random pattern of species co-occurrence. For each species i we calculated its empirical incidence J_{ij} on the jth island as a function of the island area A_j. In the absence of species interactions beyond those contained in the incidence pattern, the probability that species i and k would co-occur on island j is $J_{ij}J_{kj}$, and the null expectation for number of co-occurrences is $\sum J_{ij}J_{kj}$ summed over all 52 surveyed Bismarck islands. The sets of observed and predicted values for the 21 species pairs of the gleaning flycatcher guild are strikingly close ($p > 0.99$). Thus, a proper incidence constraint accounts for most nonrandomness in co-occurrence of these flycatchers. For the nectarivores *Nectarinia sericea* and *Myzomela pammelaena*, however, their mutually exclusive distribution, which was unexpected by our "null model" without an incidence constraint ($p \sim 6 \times 10^{-9}$), is still unexpected when the incidence constraint is used ($p \sim 4 \times 10^{-3}$). Of course, the "null model" using the incidence constraint is quite meaningless as something against which to test, for it already incorporates manifestations of almost everything of ecological interest, including interspecific competition.

ASSESSMENT OF THE VALUE OF "NULL HYPOTHESES" IN COMMUNITY ECOLOGY

In summary, the "null hypothesis" analysis by Connor and Simberloff (1979) is characterized by hidden structure, inefficiency (reliance on tedious, expensive simulations), lack of common sense (diluting the effect that they seek to demonstrate), imprudence and statistical weakness (failure to note the breakdown of their procedure in two limiting cases: checkerboards and nested distributions), and ultimately by scandalous disregard for the result of their own procedure (claiming acceptance of their "null model" when it was actually rejected at levels as extreme as $p < 10^{-8}$).

As we have shown elsewhere (Diamond and Gilpin, 1982), similar procedural errors vitiate other "null models" by Simberloff and colleagues. These examples include: diluting relevant data in a mass of data that are irrelevant or of varying quality (the models of Simberloff, 1978a, Connor and Simberloff, 1978, Strong et al., 1979, and Connor and McCoy, 1979); using weak and/or inappropriate statistical tests that stack the deck towards acceptance of the "null model" (the models of the same four papers); and hidden incorporation of the posited effect in the "null hypothesis" (the models of Simberloff, 1978a, Connor and Simberloff, 1979, and Strong et al., 1979).

Despite these errors in attempts to date to implement the "null-hypothesis" approach in community ecology, could one claim that the approach is fundamentally sound and merely needs to be done properly? Our experience in devising an improved "null model" for analyzing species co-occurrences is instructive. While our test eliminated procedural flaws of the Connor-Simberloff test and yielded interesting results, it still failed in the extreme to provide a proper distribution against which to test for the single effect of interspecific competition. The proper "null hypothesis" to test effect X consists of "everything-significant-except-X": *all* important biotic and abiotic factors that could structure observed data other than the factor to be tested. Otherwise, rejection of the "null hypothesis" does not actually implicate the factor that one is supposedly testing. On the other hand, the "null hypothesis" must not contain hidden manifestations of the the effect to be tested. Our "null hypothesis," like that of Connor and Simberloff, failed on both counts: it failed to include effects of habitat, geography, and distributional strategies that contributed to rejection of our "null hypothesis" for reasons unrelated to competition, and it still incorporated competition through its constraint of occurrence frequency and, later, of incidence. We see these problems as inevitable in any approach that, like ours and Connor's and Simberloff's, accepts observed data as givens in an attempt to incorporate everything-significant-except-X into the "null model." How can one pretend that one's "null model" is everything-significant-except-X, when it was constructed by rearranging an observed data base that may have been organized by X? Yet we do not know how otherwise to account in a "null model" for effects of species differences in dispersal, habitat preference, persistence ability, and geographical source and for effects of island differences in area, habitat, and location.

The real world is not organized in the simple hierarchical manner that Simberloff and colleagues naively assume, where the effects of obvious factors can be stripped away first, leaving naked and separate for examination all the consequences (if any exist) of the debated factor. The "null models" that these authors offer are misnamed: they are simply one of many unsuccessful competing alternative hypotheses, one that explicitly stresses certain factors and implicitly contains others, including competition. We believe that it is probably inherently impossible to construct a useful everything-except-competition "null hypothesis," and that further efforts in this direction will only sow more confusion.

18.

Neutral Models of Species' Co-occurrence Patterns

EDWARD F. CONNOR

Department of Environmental Sciences, Clark Hall, University of Virginia, Charlottesville, Virginia 22903

DANIEL SIMBERLOFF

Department of Biological Science, Florida State University, Tallahassee, Florida 32306

INTRODUCTION

Oddity in ecological nature is relative to expectation. Whether we respond to facets of nature with wide-eyed amazement or knowing nods depends on what our theories and predispositions have led us to expect. Connor and Simberloff (1979) examined biogeographic patterns of species among islands in this spirit. Can we tell when certain patterns imply a specific cause? Questions of this type require one to attempt to identify the possible configurations of species on islands. In particular, Connor and Simberloff (1979) responded to Diamond's assertion (1975) that biogeographic patterns of species' co-occurrence are shaped largely by competitive interactions between species. We generated "possible worlds"—as biogeographic patterns—in which the sites a species occupied were independent of what other species were present. For if an imaginary world of biogeographic patterns without such interspecific constraints resembles a real one, we need not state that interspecific interactions produced the co-occurrence patterns. Even if the resemblance between real and hypothetical pattern is not perfect, such an exercise can still be useful by telling us to what extent certain patterns might obtain even without interspecific interactions.

Connor and Simberloff (1979) proceeded by abstracting the arrangement of species on islands to form binary 0–1 matrices representing species absences and presences on islands: each column an island and each row a species. The column sums are the number of species on an island, and the row sums are the number of islands on which a species occurs.

To generate a biogeography that is null with respect to species co-occurrences, yet possesses other relevant biologically and environmentally determined structure, Connor and Simberloff (1979) created a similar matrix with identical row and column sums in which presences of species (1's) are independently and randomly determined. This procedure was performed repeatedly to generate an expectation of species arrangement without co-occurrence constraints, and a measure of the variance of this expectation. Summary statistics tallied from these expected biogeographies and the real ones were the numbers of species pairs and trios that co-occur on 0, 1, 2, 3, 4, ... up to all islands in an archipelago. We then compared these results for simulated and real matrices using a χ^2 statistic.

For the three faunas examined by Connor and Simberloff (1979), they found in one instance (New Hebrides birds) a close agreement between observed and expected biogeographies and in two instances (West Indies birds and bats) an excessive number of exclusive arrangements. Even in these two examples a very large number of allopatric arrangements was expected, but the expected was significantly lower than the observed level of allopatry. We suggested several possible reasons for the excessive allopatry observed, among them competitive exclusion, predation, geographical speciation, and unsettled taxonomy. Two quotes summarize our conclusions from these analyses:

> But unless one is willing to ascribe to competition the facts that islands have different numbers of species and that species are found on different numbers of islands, the New Hebrides data still argue heavily against the claim that competition determines most aspects of the distribution of species on islands.

> All this is not to say that species are randomly distributed on islands, or that interspecific competition does not occur. Rather, statistical tests of properly posed null hypotheses will not easily detect such competition, since it must be embedded in a mass of non-competitively produced distributional data.

In sum, Connor and Simberloff (1979) concluded that adducing a role for competition in shaping co-occurrence patterns, over and above whatever role competition may play in determining how many species an island has or how many islands a species occupies, is very difficult since many allopatric arrangements are expected for non-competitive reasons.

The "Seven Flaws" of Connor and Simberloff

Gilpin and Diamond (this volume) raise seven criticisms of our methods and conclusions. Based on these criticisms, they conclude that the Connor and Simberloff (1979) test of the effect of competition on biogeographic co-occurrence patterns is weak, meaningless for the New Hebrides, difficult to interpret biologically, and not truly null and suffers from the inclusion of "irrelevant data."

We argue against these contentions and show that (1) Gilpin and Diamond's (this volume) method is certainly no more powerful than the Connor and Simberloff (1979) procedure, (2) their procedure is apparently impossible to compute as they outline, and (3) their results are consistent with our original interpretation that few biogeographic co-occurrence patterns must be ascribed to competition.

We proceed seriatim with Gilpin and Diamond's (this volume) seven objections.

1. *Dilution.* We analyzed Diamond's "assembly rules," which our first sentence (Connor and Simberloff, 1979) quoted verbatim. We requote them here:

a. If one considers all the combinations that can be formed from a group of related species, only certain ones of these combinations exist in nature.

b. Permissible combinations resist invaders that would transform them into forbidden combinations.

c. A combination that is stable on a large or species-rich island may be unstable on a small or species-poor island.

d. On a small or species-poor island, a combination may resist invaders that would be incorporated on a larger or more species-rich island.

e. Some pairs of species never coexist, either by themselves or as part of a larger combination.

f. Some pairs of species that form an unstable combination by themselves may form part of a stable larger combination.

g. Conversely, some combinations that are composed entirely of stable subcombinations are themselves unstable.

The rules said nothing about guilds or other faunal subsets except for rule (a), which referred only to "related" species. Gilpin and Diamond (this volume) have now changed the original argument and feel that our analysis errs because we treated entire faunas and sought co-occurrences and exclusions among all possible pairs and trios: by analyzing an entire fauna rather than an ecologically defined guild, "Connor and Simberloff submerged instances of competitors with exclusive distributions in an irrelevant mass of data from ecologically remote pairs of species. . . ." Since the rules we were examining did not mention guilds, we did not analyze guilds. We will be the first to agree that effects of competition on biogeographic patterns will be devilishly difficult to extract from a community-wide analysis, since they will be buried in a welter of non-competitively produced data. In fact, we said exactly this in the very paper Gilpin and Diamond (this volume) attack: "statistical tests of properly posed null hypotheses will not easily detect such competition, since it must be embedded in a mass of non-competitively produced distributional data" (Connor and Simberloff, 1979). We are pleased that Gilpin and Diamond (this volume) agree with us.

But confining the analysis to guilds is no trivial matter, for assigning species to guilds requires detailed data on resource use by all species in the community. Root (1967), originator of the guild concept, defines a "guild" as a "group of species that exploit the same class of environmental resource in a similar way. This term groups together species . . . that overlap significantly in their niche requirements." Although this definition does not precisely characterize "class," "same," "similar," and "significantly," it is clear from Root's monograph, which summarizes much of his doctoral research, that an enormous field effort must be expended to delineate even a single small guild. Root's foliage-gleaning bird guild consisted of five species, for each of which he examined numerous stomachs, measuring and classifying all insects contained therein, and studied foraging behavior, determining location, direction, frequency, and manner of attack. Small wonder that "no one has yet been able to analyze all the guilds in a community, and at present we can deal only with a few guilds making up part of a whole community" (Krebs, 1978). Other studies that have delineated guilds have generally rested on massive amounts of field work comparable to Root's original effort and have included the data that rationalize a particular guild assignment. Feinsinger (1976) has thus described a hummingbird guild, and Hespenheide (1971) and Alatalo and Alatalo (1979) have defined flycatcher guilds, while Price (1971) for wasps parasitic on a sawfly, Root (1973) for herbivores on collards, McClure and Price (1975, 1976) for leaf-hoppers on sycamore, Rathcke (1976a) for stemborers of prairie plants, and Faeth and Simberloff (1981b) for leaf-miners on oak trees have all delineated insect guilds.

For no guild, much less for the entire avian community, have Gilpin and Diamond (this volume) provided data to justify guild boundaries. They say that their "guilds are defined in terms of utilization of resources," but these utilization data are not included, nor have they listed all members of even a single guild, much less the partition of the community into guilds. Consequently their analysis and conclusions about guilds cannot be examined.

Even when one has the data required to partition a group of species into guilds, the partition itself is problematic to a degree that renders conclusions vague. For example, Munger and Brown (1981; *cf.* Brown, 1975) have lumped *Dipodomys*, *Perognathus*, and *Peromyscus* into a guild of seed-eating rodents, while Hallett (1982) has separated *Dipodomys* from the other two genera on grounds that, although they exploit the same class of resources (seeds), they do so in different ways. Both partitions are based on exhaustive food and habitat data, suggesting that guild assignment is not automatic. Finally at least certain of these rodents forage differently during different seasons (Brown, 1975; Hallett, pers. comm.), further exacerbating the difficulties of gathering sufficient data to determine guilds.

As Root's original definition (1967) makes clear (see also Krebs, 1978), resource overlap among guild members must be sufficiently great that one might expect competition to occur among its members. Although all guild studies cited above have included some attempt at measuring available resources, two, Rathcke (1976a) and Faeth and Simberloff (1981b), concluded that the resource used to define the guild was not limiting, and thus competition did not appear to be an important force. Alatalo and Alatalo (1979) were unable to determine whether sexual behavior or competition was the more important determinant of foraging behavior, which in turn was a key defining attribute of their flycatcher guild. So it is true that if one actually has the data necessary to justify guild assignment, confining the biogeographic analysis to guilds need not be circular: one need not know in advance that competition is occurring and among which species it is occurring in order to find biogeographic pattern data compatible with a competition hypothesis. But if the members of a particular guild do not compete especially intensely, it is difficult to see how their biogeographic co-occurrence pattern can have been influenced by competition, even if it is compatible with competition. And if the members of a guild *do* compete strongly, and the competition *does* affect the biogeographic pattern in the direction of increasing exclusivity, it is already established (Pielou and Pielou, 1968; Simberloff and Connor, 1981) that non-competitive forces such as differential suitability of different islands for different species could have caused precisely the same type of biogeographic pattern. So one could still not adduce the geographic co-occurrence pattern as strong support for a competition hypothesis.

It is of interest to note here that Whittam and Siegel-Causey (1981) have attempted to subdivide and analyze seabird communities in the manner suggested by Gilpin and Diamond (this volume). Although they do not call their species groups "guilds," group membership was determined by the single criterion of similarity in nesting site preferences (groups 2–5). Whittam and Siegel-Causey (1981) found that the observed pattern of species co-occurrences was non-random, but that most of the non-randomness was due to positive association among species (25 of 29 significant interaction terms were positive in sign).

A final word on guilds: Connor and Simberloff (1979) showed that the existence of some exclusively distributed species pairs is not *prima facie* evidence for competition since independently arranged species would also generate exclusive pairs. We emphasized that one would have to examine the entire matrix of species × sites to see whether there is an extraordinary amount of exclusivity. A similar caution is in order about guilds. Even granted that a single satisfactorily documented guild were available for examination, its geographic pattern could tell us little about competition in general: not only could different guilds be structured by different forces; one would also have to know the null probability that, of *n* examined guilds (all of different size) all of whose members are randomly and independently arranged on an archipelago, one or more guilds would produce a pattern that would have been statistically odd had only that guild been considered. If one or a few guilds are chosen randomly for examination, and not selected *a priori* for likelihood of competitive effect, inferences drawn from them can of course be stronger.

2. *Weighting.* Gilpin and Diamond (this volume) have changed in yet another way their argument from the one that Connor and Simberloff (1979) responded to; Diamond (1975) said nothing about weighting certain species pairs, trios, etc., more than others in assessing biogeographic patterns (see rules above). But Gilpin and Diamond (this volume) inveigh against the Connor-Simberloff test on the grounds that it treats each species pair as equal, rather than weighting each pair by the probability that a random arrangement of that pair would have produced a distribution as exclusive as that observed. They contend that a pair of common allopatric species is less likely to have arisen by independent placement than a pair of rare allopatric species. Since the assembly rules did not address this issue, and said only that pairs of allopatrically arranged species exist, we did not attempt to produce a test that weighted pairs differently. It is certainly true that different pairs of allopatric species isolated *post facto* from a larger matrix have, if viewed alone, different null probabilities for the observed exclusion. But aside from the irrelevance of this observation to the assembly rules, Connor and Simberloff (1979) have shown that the true probability of some observed degree of exclusion for a set of species cannot be determined by focusing on isolated pairs, since some pairs will always be exclusively arranged. Similarly, in a randomly produced matrix with a large enough number of pairs, some pairs will be not only allopatric, but improbably allopatric given their occurrence totals. So the existence of such pairs no more demands a competitive explanation than did the existence of any sort of allopatric pair, for exactly the same reason. In both instances, it is the distribution of exclusivity among the entire set of species, within a pool or only within a guild, that must be examined.

3. *Checkerboard distributions.* Gilpin and Diamond (this volume) aver that our test fails to recognize as non-random "... the simplest, clearest, and most non-random pattern caused by interspecific competition ...," namely, a perfect checkerboard. The one example given, that of two

species each exclusively occupying four of eight sites, constitutes evidence for non-independence of placement only when viewed alone without fixed column sums

$$(Pr = \binom{8}{4}\binom{4}{4}\Big/\binom{8}{4}\binom{8}{4} = 0.0142).$$

When degenerate arrangements are not allowed (that is, arrangements in which species or sites disappear from the matrix because their row or column sums, respectively, are zero), the probability of an exclusive arrangement, in this instance, becomes $Pr = 1.0$ because there is only one matrix (aside from shuffling rows or columns) possible, namely, that depicted by Gilpin and Diamond (this volume)! It is no wonder that our procedure fails to recognize such a pattern as non-random, since we ignore degenerate arrangements. Degenerate arrangements are discussed

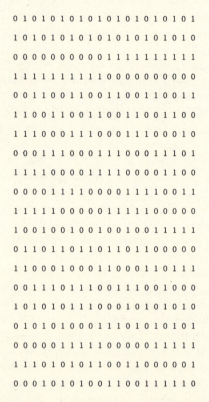

Fig. 18.1. 20 × 20 matrix of species and sites in which all sites contain 10 species and all species are found at 10 sites, but only 10 pairs of species are exclusive in distribution.

further below. Furthermore, consider the perfect checkerboard in Figure 18.1. Twenty species are arranged on 20 islands such that they constitute 10 exclusive pairs, each species is found on 10 islands, and each island contains 10 species. First, we ask if this is an extreme example. Figure 18.2 depicts another perfect checkerboard with the same marginal species and island totals, but with 100 exclusive species pairs. Clearly the checkerboard in Figure 18.1 has a very *low* number of allopatric species pairs relative to the possible number that might obtain; this pattern is exactly opposite to one that we might expect competition to produce. So checkerboardness *per se* need not imply competitive exclusion, and it is not a virtue if an algorithm automatically finds any checkerboard to be improbable.

If one were to ask, as Gilpin and Diamond (this volume) appear to be asking, what is the null probability of exactly 10 or more than 10 pairs of complementary rows (exclusive species) in a 20 × 20 matrix randomly

```
1 1 1 1 1 1 1 1 1 1 0 0 0 0 0 0 0 0 0 0
1 1 1 1 1 1 1 1 1 1 0 0 0 0 0 0 0 0 0 0
1 1 1 1 1 1 1 1 1 1 0 0 0 0 0 0 0 0 0 0
1 1 1 1 1 1 1 1 1 1 0 0 0 0 0 0 0 0 0 0
1 1 1 1 1 1 1 1 1 1 0 0 0 0 0 0 0 0 0 0
1 1 1 1 1 1 1 1 1 1 0 0 0 0 0 0 0 0 0 0
1 1 1 1 1 1 1 1 1 1 0 0 0 0 0 0 0 0 0 0
1 1 1 1 1 1 1 1 1 1 0 0 0 0 0 0 0 0 0 0
1 1 1 1 1 1 1 1 1 1 0 0 0 0 0 0 0 0 0 0
1 1 1 1 1 1 1 1 1 1 0 0 0 0 0 0 0 0 0 0
0 0 0 0 0 0 0 0 0 0 1 1 1 1 1 1 1 1 1 1
0 0 0 0 0 0 0 0 0 0 1 1 1 1 1 1 1 1 1 1
0 0 0 0 0 0 0 0 0 0 1 1 1 1 1 1 1 1 1 1
0 0 0 0 0 0 0 0 0 0 1 1 1 1 1 1 1 1 1 1
0 0 0 0 0 0 0 0 0 0 1 1 1 1 1 1 1 1 1 1
0 0 0 0 0 0 0 0 0 0 1 1 1 1 1 1 1 1 1 1
0 0 0 0 0 0 0 0 0 0 1 1 1 1 1 1 1 1 1 1
0 0 0 0 0 0 0 0 0 0 1 1 1 1 1 1 1 1 1 1
0 0 0 0 0 0 0 0 0 0 1 1 1 1 1 1 1 1 1 1
0 0 0 0 0 0 0 0 0 0 1 1 1 1 1 1 1 1 1 1
```

Fig. 18.2. 20 × 20 matrix of species and sites in which all sites contain 10 species and all species are found at 10 sites, but 100 pairs of species are exclusive in distribution.

filled subject only to the constraints that all row and column sums equal 10, one would have to calculate how many different matrices (*i.e.* matrices that cannot be made identical by a finite sequence of row and column interchanges corresponding to renaming species or sites, respectively) there are with 0, 1, 2, ... pairs of complementary rows. Gilpin and Diamond (this volume) have not done so, and thus cannot state how likely it would be for the checkerboard in Figure 18.1 to occur by chance. Consequently they have no grounds for criticizing our test for failing to recognize this matrix as unlikely to have been randomly produced. In fact, our procedure does recognize such a matrix as unusual ($\chi^2 = 102.35$, $p \ll 0.001$, d.f. = 7). This χ^2 was computed by lumping classes with low expected values (number-of-islands-shared classes 7–20). Degrees of freedom were lowered by 1 because the total number of occurrences is fixed in each matrix (J. Hendrickson, pers. comm.).

Second, suppose one were to posit in advance (hypothesis 1) a collection of 20 species that one *knew* to be competitively structured. This collection consists of 10 pairs such that the members of each pair are competitively prevented from coexisting but are each unaffected by the other 18 species. One could then randomly fill a matrix subject to our original two constraints (fixed row and column totals) plus a third constraint, the 10 disallowed species pairs. The numbers of pairs of complementary rows in any such matrix would exceed ten. In order to claim that our test would fail to detect that there were more exclusively arranged pairs than would be expected if each species were independent of all others (but all had fixed occurrence totals, hypothesis 2), one would first have to generate the expected number of allopatric pairs and its standard deviation for each hypothesis. Then one would have to find such broad overlap between the confidence belts about the two expectations that the test would fail to distinguish matrices generated by the two different algorithms. Gilpin and Diamond (this volume) have done so for neither our test nor theirs, so they have not even shown that our test fails to recognize such a pattern produced by competition, much less that theirs generally outperforms ours in such cases. We have performed such a test, the results of which are presented in Table 18.1 and Figure 18.3. Clearly the Connor and Simberloff (1979) procedure can detect unusual arrangements when they are present. The confidence belts for these two algorithms are such that, for the classes of allopatry encompassing very exclusive arrangements, the observed data are generally acceptable under hypothesis 1 and very suprising according to hypothesis 2. The χ^2 for hypothesis 1 is considerably less than that for hypothesis 2 (respectively, $\chi^2 = 3.36$, $0.020 > p > 0.10$, classes 1–2 and 7–20 lumped, d.f. = 6; and $\chi^2 = 102.35$, $p \ll 0.001$, classes 7–20 lumped, d.f. = 7). Finally, it is worth reiterating that the procedure just outlined would be a valid begin-

Table 18.1. Observed, expected, and standard deviation of the expected number of pairs sharing 0, 1, 2, 3 ... N islands for the 20 × 20 matrix depicted in Figure 18.1. The expected values were computed using the Monte Carlo procedure of Connor and Simberloff (1979) with and without co-occurrence constraints for 10 pairs of species. Number of iterations = 1000. Observed values were those for the matrix in Figure 18.1.

Number of Islands Shared	Actual Number of Pairs	Expected Number of Pairs Without 10 Exclusive Pairs	SD	Expected Number of Pairs With 10 Exclusive Pairs	SD
0	10	1.362	1.799	10.011	0.130
1	2	6.328	3.917	0.266	0.662
2	2	16.097	4.964	3.140	2.281
3	12	31.941	6.174	16.385	4.797
4	42	48.480	6.925	41.845	5.802
5	64	48.029	6.803	56.940	11.469
6	42	27.743	4.999	41.563	5.894
7	12	8.645	2.804	16.321	4.740
8	2	1.298	1.190	3.187	2.274
9	2	0.075	0.278	0.324	0.698
10	0	0.002	0.045	0.016	0.148
11	0	0.000	0.000	0.002	0.045
12	0	0.000	0.000	0.000	0.000
13	0	0.000	0.000	0.000	0.000
14	0	0.000	0.000	0.000	0.000
15	0	0.000	0.000	0.000	0.000
16	0	0.000	0.000	0.000	0.000
17	0	0.000	0.000	0.000	0.000
18	0	0.000	0.000	0.000	0.000
19	0	0.000	0.000	0.000	0.000
20	0	0.000	0.000	0.000	0.000

ning at testing the power of either test only if one knew *in advance*, and not from the biogeographical configurations alone, that there were 10 mutually exclusive pairs. To our knowledge Gilpin and Diamond (this volume) have not presented strong non-biogeographic evidence that even a single pair of species in the Bismarcks or other southwest Pacific archipelagoes cannot coexist for competitive reasons. That the existence of 10 allopatric pairs of species in a 20 × 20 site matrix would not be especially compelling *prima facie* evidence for competition is well demonstrated by the fact that there are $20!/20^{10} \cdot 10! = 654,729,000$ ways of partitioning 20 species into 10 pairs. Once again we must know which species are competing in order for the biogeographical data to show us that competition is occurring.

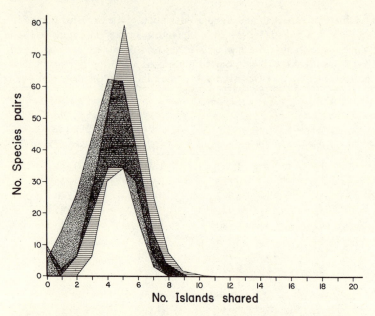

Fig. 18.3. Ninety-five percent confidence belts for hypothesis 1 (lined; includes 10 exclusive pairs) and hypothesis 2 (stippled; all species independent) as explanations of species arrangement depicted in Figure 18.1 and Table 18.1. Note that for very exclusive arrangements (say, 0, 1, or 2 islands shared) the confidence bands overlap only slightly.

4. *Meaningful rearrangements.* For the one instance where Connor and Simberloff's (1979) procedure fails to reject the null hypothesis of independent species arrangements, Gilpin and Diamond (this volume) suggest that the result was "meaningless" and that the outcome of the test was predetermined because the row and column constraints entrained the rearrangements to be mere renamings of species and islands. Hence, the expected and observed levels of co-occurrence should be identical. However, Connor and Simberloff (1979) stated that special care was taken to ensure that all rearrangements were actually "nonequivalent"; that is, each arrangement was *not* derivable from any other by a finite series of row and column interchanges. The matrices do not represent simply renaming of species and/or islands. Further, the fact that the New Hebrides distributions apparently allow fewer distinct matrices than other distributions do will lead to a decrease in the variance of the expected number of pairs at all levels of exclusivity, so that smaller deviations of observed and expected values would constitute evidence for non-random placement. Therefore, the New Hebrides test is particularly conservative. That the expected values of species co-occurrences are so strikingly similar to

those observed thus indicates that the observed levels of allopatry and sympatry of New Hebrides birds are not unusual, given the species' frequencies and island richnesses.

5. *Directionality*. Gilpin and Diamond (this volume) criticize our procedure because it can tell only whether or not non-randomness is present. They feel that a better procedure would also indicate the direction of the non-randomness and which species pairs are "most responsible" for the non-randomness. While we did not dwell upon the direction of the non-randomness, we did point out that it was in the direction of excessive exclusivity. In addition, it is not true that our procedure is unable to determine the direction of non-randomness. It is quite simple to inspect the individual cell contributions to the χ^2 statistic to determine whether the non-randomness is contributed by positive or negative associations. Furthermore, the procedures used by Gilpin and Diamond (this volume) do not produce a test for the "direction" of the non-randomness. Exactly as we did, they compare the observed and expected patterns of species co-occurrences by a χ^2 statistic, which unless partitioned does not indicate the direction of deviation. From a visual inspection of the observed and expected pattern of species co-occurrences in the Bismarck avifauna (their Figure 17.2) it appears that few species pairs lie beyond the lower tail of the expected distribution, but rather that the non-randomness is caused by a disproportionate number of species pairs in the positive half of the distribution. Yet, Gilpin and Diamond (this volume) provide no means to assess the significance of these contributions to the overall pattern of non-randomness.

Regarding Gilpin and Diamond's (this volume) desire to identify the species pairs responsible for the non-randomness, we point out that, without additional evidence of non-biogeographic nature, it is impossible to tell which of the species pairs in the tails of the distribution are *bona fide* competitors or mutualists, which are non-independent for other reasons such as dissimilar habitat requirements, and which are expected by chance alone.

6. *Hidden structure*. Gilpin and Diamond (this volume) contend that Connor and Simberloff's (1979) expected arrangements of species on islands are not truly null with respect to species co-occurrences. They feel that fixing the row and column sums actually builds competition into the expected patterns of co-occurrences, so that our test is biased toward failure to reject the null hypothesis of no effect of species interactions on the pattern of species co-occurrences. However, Gilpin and Diamond (this volume) present no evidence that demonstrates that the row and column sums (and incidence functions) are actually affected by competitions. Nowhere is evidence presented that species actually compete for some limiting resources, much less that these interactions have affected the pattern of species co-occurrences.

It is conceivable that competition does affect the number of species on an island, the number of islands on which a species occurs, and species' "incidence functions," as Connor and Simberloff (1979) clearly stated (see above quote from their paper). But even if competition could be shown to affect row and/or column sums, Gilpin and Diamond (this volume) present no evidence that "hidden structure" in these constraints would necessarily translate into altered co-occurrence patterns.

7. *Monte Carlo techniques.* Gilpin and Diamond's (this volume) final criticism of our procedure is that it is based on a stochastic simulation rather than an explicit computation of the null probabilities of distributional sympatry or allopatry. We agree that an explicit solution to this problem is desirable; it certainly is not necessary. Furthermore, the procedure presented by Gilpin and Diamond (this volume) is at best suspect computationally and probably incorrect (see below). Given fixed marginals (row and column sums), we know of no explicit solution for the expected frequency distribution of number of pairs at each level of exclusivity, nor for the expected degree of sympatry between pairs.

The Gilpin and Diamond Procedure

The major components of Gilpin and Diamond's procedure (this volume) are as follows:

1. Compute the probability that the ith species occurs on the jth island as $p_{ij} = R_i C_j / T$, where R_i is the ith column sum, C_j is the jth row sum, and T is the grand sum (number of 1's in the matrix).
2. Compute the expected overlap between each pair of species and its standard deviation, from the above probabilities (the method is not provided by Gilpin and Diamond (this volume), but must involve a sum of the products of the individual probabilities of occurrence on each island).
3. Standardize the observed values of overlap by subtracting the expected value and dividing by the standard deviation of the expected.
4. Compare the frequency distribution of standardized observations to a (0, 1) standard normal distribution via χ^2, and a graphical representation.

Three major problems beset the Gilpin and Diamond procedure:

1. It counts degenerate rearrangements with species occurring nowhere and/or sites with no species as possible null rearrangements, thereby decreasing the probability of observing any specified level of exclusivity. This practice biases their test in favor of rejecting the null hypothesis.

2. The method of estimating the cell probabilities (p_{ij}'s) produces values that are often greater than 1, and the *ad hoc* procedures used to correct for this problem are inappropriate.

3. Transforming the observed values into standard deviates and comparing the frequency distribution of these values to a (0, 1) normal distribution assumes that these deviates are asymptotically (0, 1) normal when such an assumption is untested and perhaps unwarranted.

1. *Degenerate rearrangements.* Consider the example of two species, each of which occupies two sites out of four (Figure 18.4), each site possessing one species. There are $2^8 = 256$ ways of arranging 1's and 0's in such a matrix subject to neither row- nor column-sum constraints. If we impose row-sum constraints, there are $\binom{4}{2}\binom{4}{2} = 36$ ways of arranging the species presences such that each occupies two sites. Among the totally unconstrained rearrangements, 250 involve the loss of one or more sites and/or species. Of the 36 row-sum-constrained rearrangements, 30 involve the loss of one or more sites, but no species. To ensure that all sites are occupied and all species are represented in the population of possible rearrangements, column-sum constraints must also be placed on the rearranged matrix. When they are, there are 6 possible rearrangements, all of which are perfect checkerboard distributions! Had we considered the population of possible rearrangements to be those without row- or column-sum constraints, the probability of observing an exclusive arrangement would be 6/256, for row-constrained rearrangements 6/36, and for row- and column-sum-constrained rearrangements 6/6. For large matrices of species distributions among sites (*e.g.* New Hebrides 56 × 28, Bismarcks 147 × 50), including degenerate rearrangements in the population of possible rearrangements tremendously increases the number of possible rearrangements. This increase can only result in a lowering of the null expectation of any observed level of exclusivity.

The example presented by Gilpin and Diamond (this volume), of two species on 4 of 8 islands each, is a case in point. There are 2^{16} unconstrained rearrangements, $\binom{8}{4}\binom{8}{4} = 4900$ row-sum-constrained rearrangements, and 70 row- and column-sum-constrained rearrangements. The probabilities of observing an exclusive arrangement are thus 70/65,536, 70/4900, and 70/70, respectively. The results of including these degenerate rearrangements is ultimately to increase the probability of rejecting the null hypothesis that species co-occurrence patterns are random.

1 0 1 0

0 1 0 1

Fig. 18.4. 2 × 4 matrix of species and sites in which all sites contain one species and all species are found at two sites.

2. *Cell probabilities.* The procedure outlined by Gilpin and Diamond (this volume) for computing the probability of the ith species' occurring on the jth island, $p_{ij} = R_i C_j / T$ (where R_i is the ith row sum, C_j the jth column sum, and T the total number of 1's in the matrix), does not compute probabilities; rather it computes expected frequencies. When this procedure is used for the New Hebrides avifauna, fully 235 of 1568 cell "probabilities" are greater than 1. Although Gilpin and Diamond (this volume) suggest that this is the traditional log-linear model of contingency table analysis, this approach is appropriate only for frequency data, not binary presence/absence data. Neither a log-linear nor a logistic model can be fit to these "probabilities" (Bishop *et al.*, 1975). Gilpin and Diamond (this volume) do not fully describe their recursive approach, so it is not possible to determine its suitability for estimating the expected cell probabilities. However, the Monte Carlo procedure outlined but not performed by Gilpin and Diamond (this volume) could be used to estimate cell probabilities as they suggest. In spite of Gilpin and Diamond's (this volume) claim that their procedure is both more powerful and an explicit solution to the problem posed by Connor and Simberloff (1979), it appears that the only means of estimating the cell probabilities is by Monte Carlo techniques.

3. *Asymptotic normality.* The final step in Gilpin and Diamond's (this volume) procedure is to compare the observed frequency distribution of deviates to a $(0, 1)$ standard normal distribution by χ^2. In so doing, they assume that, given no species association, this distribution would be asymptotically $(0, 1)$ normal. However, asymptotic with respect to what? There is no analog to sample size in this analysis, so one could not observe the behavior of the distribution of deviates for larger and larger samples to see if it is, in fact, asymptotically normal. Why assume that the null distribution of these deviates is $(0, 1)$ normal? Could it not be skewed or leptokurtic? Obviously, the outcome of the χ^2 test depends on the shape assumed for the null distribution of deviates. If the true null distribution of deviates is skewed, then the conclusions reached by Gilpin and Diamond (this volume) for the New Hebrides and the Bismarcks avifaunas may be erroneous. However, since there is no way to determine whether or not the null distribution is $(0, 1)$ normal, the test proposed by Gilpin and Diamond (this volume) is suspect.

Which Procedure?

Both the procedure used by Connor and Simberloff (1979) and Simberloff and Connor (1979) and that proposed by Gilpin and Diamond (this volume) are designed to assess the role of competition in shaping patterns of species co-occurrence. However, the problems discussed above lead us to conclude that our procedure is superior. It is not apparent that cell "probabilities" can be explicitly computed as Gilpin and Diamond (this volume) claim. And, if Monte Carlo techniques must be used, then both the statistics reported by Connor and Simberloff (1979) and those reported by Gilpin and Diamond (this volume) could be compiled from the same group of simulated archipelagoes. However, it is difficult to interpret the results of the Gilpin and Diamond (this volume) procedure, since one can only guess what the expected distribution of deviates might be. On the other hand, two problems beset our procedure; both concern the algorithm for constructing expected rearrangements. First, for certain configurations of the marginal totals (few tied rows of columns) the algorithm sometimes fails to complete filling the matrix with 1's subject to the marginal totals. We have tried several other algorithms, but have found none more efficient. Second, since we proceed serially by species in filling the matrix, a bias is introduced in species placement that increases as one proceeds further in the process of filling the matrix. We have tried several algorithms that we know are unbiased, but these fail to complete all but the simplest matrices (2 × 4). Since these algorithms all fail to complete matrices, we have not yet succeeded in determining the effect of this bias. Both of these problems may be serious, but the first can be overcome by alternative procedures, as we did for the New Hebrides birds.

CONCLUSION

We concluded, as we did in 1979, that co-occurrence patterns isolated *post facto* from a multitude of such patterns cannot be considered odd unless shown to be so relative to some expectation of co-occurrence. To posit that a particular co-occurrence pattern arises from competitive exclusion, one must show that whether or not the pattern is improbably exclusive, it is actually caused by competition. Arguing from a co-occurrence pattern alone, one can only show it to be improbably exclusive. Without evidence of a non-biogeographical nature, one cannot specify the exact causes of exclusivity. We reiterate that all of this is not to say that competition does not occur, nor that it has no effect on patterns of co-occurrence, but that to adduce such a role for competition requires evidence beyond the observed pattern of co-occurrence. Finally, for the three biotas we discussed in 1979 we still find no compelling evidence to reject our null hypothesis in one instance or to conclude that competitive interactions are the reasons for rejecting it in the other two instances.

Acknowledgements

We thank R. Dueser, G. Hornberg, J. Rotondo, D. Ramirez, L. G. Abele, and D. Strong for their advice on these problems.

19.

REJOINDERS

MICHAEL E. GILPIN and JARED M. DIAMOND

EDWARD F. CONNOR and DANIEL SIMBERLOFF

BY GILPIN AND DIAMOND

We completed the text of our contribution to this volume in September 1981. In March 1982, shortly before this volume was to go to press, the editors sent us a chapter that one of the editors and a colleague wrote to rebut our chapter (Connor and Simberloff, this volume: abbreviated C and S). In November 1982 the editors sent us a further rebuttal, entitled "Rejoinder by Connor and Simberloff" and repeating some of the same arguments unchanged. These two documents have been helpful to us in understanding the style of reasoning preferred by C and S. We appreciate the editors' special interest in our work, and we are grateful for the opportunity to comment on these critiques.

Three issues are in dispute:

1. Do observed species co-occurrence patterns agree with expectations based on independent (random) placement of species?
2. If not, what biological factors are responsible?
3. What is the best method to test for non-random co-occurrence patterns over broad taxonomic groups?

The comments of C and S on the first two questions can be disposed of briefly.[1]

---

[1] C and S (this volume) repeatedly use a peculiar debating ploy. Our chapter states that it is based on a detailed analysis in two recent papers that were published several months before the C and S critique was sent to us (Diamond and Gilpin (=D and G) 1982, Gilpin and Diamond (=G and D) 1982). Nevertheless, C and S pretend eight times that our conclusions in this volume are unsupported by evidence: "For no guild ... have G and D (this volume) provided data ..."; "G and D (this volume) provide no means to assess ..."; "G and D (this volume) present no evidence ..."; "the method is not provided by G and D (this volume)"; "G and D (this volume) do not fully describe their recursive approach"; etc., etc. (pp. 319, 327, 328, 330). Naturally, the data and full description are not in this volume, as we would thereby have exceeded the editors' page limit; they appear instead in the references in which we cited them as appearing, mainly our two 1982 papers.

As for the first question, it is now clear that the answer is "no": co-occurrence patterns are radically non-random. For the four communities analyzed, both the C and S test and ours reject a "null" model of random co-occurrence at the level of $p < 10^{-8}$, except that C and S accepted "null" models for the New Hebrides at the level of $p > 0.90$ or 0.99. Only one disagreement remains between C and S and us on this first question: C and S continue to ignore the demonstration that their New Hebrides result was partly due to a programming error of theirs (insufficient submatrix reshufflings), whose correction lowers the p level even by their test, and partly to a structural flaw in their test.

Faced with this answer to the first question, C and S (p. 317) now pretend, citing two quotes from C and S (1979), that they never claimed that co-occurrence patterns were random, but merely that the non-randomness was not demonstrably due to competition. Readers can easily satisfy themselves, merely by glancing at C and S (1979), that many other quotes from C and S (1979) actually indicate their belief in substantially random co-occurrence patterns. Just a few examples suffice. They titled their paper "Chance or competition?" rather than "Competition or some other explanation for non-chance patterns?" Their 1979 abstract claimed, "We show that every assembly rule is either tautological, trivial, or *a pattern expected were species distributed at random*" (italics ours). Their 1979 coda claimed, "The remaining three [assembly rules] describe situations which would for the most part be found even if species were randomly distributed on islands." Their p. 1134 claimed, "In a nutshell, there is nothing about the absence of certain species pairs or trios, related or not, in the New Hebrides that would not be expected were the birds randomly distributed over the islands as described above." Their p. 1135 claimed, "The New Hebrides bird distribution fit the random hypothesis even more closely than the West Indies birds or bats." In at least nineteen other passages C and S (1979) presented predictions attributed to a hypothesis of "random" distributions and concluded that this hypothesis fits well to observed patterns. We trust that C and S have now abandoned these erroneous beliefs, although they have yet to admit their error explicitly.

On the second question, we have now discussed many times elsewhere (Diamond, 1975 and many other papers; G and D, 1982; this volume) the multiple biological factors responsible for non-randomness and the types of evidence leading to these conclusions. For example, our 1982 paper (G and D, 1982) was entitled "Factors contributing to non-randomness in species co-occurrences on islands" and proceeded to discuss seven such factors (competition and six others), summarized in this chapter. Diamond (1975) devoted individual sections to the importance of habitat requirements (his pp. 361–364), five area-related factors (pp. 364–371), dispersal (pp. 371–378, 437–438), reproductive rates (pp. 378–380), and transition probabilities (pp. 438–439) for species occurrences and co-occurrences. Nevertheless, C and S pretend to be refuting errors of ours and to be offering corrective insights of their own when they suggest that several factors besides competition could be responsible for non-randomness and that factors other than species presence lists on islands are needed to identify these factors.

We regret having to waste space on the third issue at all. The C and S test has numerous fatal flaws (Wright and Biehl, 1982; D and G, 1982), among which C and S now acknowledge minor ones and continue to ignore the major ones. Our test is not perfect, but it cures most of these flaws. The relative merits of the various tests will be sufficiently clear to anyone using them that it is puzzling why C and S cling to their discredited test. Unfortunately, their chapter (this volume) introduces numerous new errors. We comment on their defenses of their test under seven headings, then on their critique of our test.

1. *Dilution.* As this chapter and D and G (1982) point out, the C and S (1979) analysis of all pairwise species combinations in a fauna obviously buries possible instances of competing species pairs in a huge excess of ecologically related pairs. Hence the proper tests for competition should be done within guilds (for details see D and G, 1982, p. 66; G and D, 1982, p. 76; Colwell and Winkler, this volume). C and S (pp. 318–321) object: a quote from Diamond's (1975) assembly rules says nothing about guilds; guilds are hard to define anyway; our paper in this volume omitted data to delineate guild boundaries; guilds can be structured by forces other than competition; and it is not enough to analyze selected guilds.

The quote from Diamond (1975) omitted the word "guild" because Diamond (1975) had repeatedly made explicit, from his introduction onwards, that his analysis was confined to guilds (*e.g.* second page of introduction: "the various species in a guild can coexist only in certain combinations," Diamond, 1975, p. 346). Each set of assembly rules described by Diamond (1975) was attributed to a specific guild (*e.g.* section heading on p. 393 "Assembly rules for the cuckoo-dove guild" and similarly on pp. 400, 404, 406). Diamond considered it unnecessary to belabor this point in every possible quote about assembly rules, because it was not appreciated that anyone would be so silly as to search for effects of competition in all pairwise species combinations of a fauna, until C and S (1979) did exactly that.

Yes, it is non-trivial to assign species to guilds, just as it is to assign species to genera, art works to periods, and objects of any class to subclasses. We did not waste the entire length allotment of our chapter in defining guilds and listing their species because Diamond (1975) had already done so for each guild discussed, based on the technical literature on New Guinea birds. For example: "All [species of the fruit-pigeon guild] are ecologically similar in being arboreal, living in the crowns rather than in the middle story, being exclusively frugivorous, not taking stones into the gizzard, and hence restricted to eating soft fruits that can be crushed by the gizzard wall" (Diamond, 1975, p. 406); species listed on pp. 407 and 408 and in Rand and Gilliard (1967); utilization functions presented on pp. 407 and 428–433; extensive dietary data published by Crome (1975) and other authors.

Of course factors other than competition contribute to guild structure. The four New Guinea bird guilds that Diamond (1975) analyzed in detail yielded similar assembly rules, including evidence for a role of competition in maintaining exclusive distributions. Of course future analyses of other guilds, or other avifaunas, or other New Guinea taxa, may yield the same or different results. Do C and S really argue that every guild must be analyzed before publication? Why not every fauna? every taxon?

The whole lengthy section by C and S (this volume) entitled "Dilution" is a study in evasion of the obvious point. *Any* attempt to define and utilize guilds would have been preferable to their analysis of all pairwise species combinations in a fauna, which diluted data potentially relevant to competition.

2. *Weighting.* C and S (this volume) again evade the obvious point. Of course, "the true probability of some observed degree of exclusion for a set of species cannot be determined by focusing on individual pairs, since some pairs will always be exclusively arranged" (C and S, p. 321). The correct way to determine the true probability for the whole set is to weight the difference between the observed and predicted exclusivity of each pair in the set by the probability of this difference; our test does weight, and the C and S test does not weight (see D and G, 1982, p. 71, for more details). In addition, weighting reveals that numerous species pairs have a far more exclusive distribution than expected for *any* species pair of the whole fauna on the basis of random co-occurrence (G and D, 1982, p. 79). Focusing on such individual pairs does suffice for demonstrating and discussing significantly exclusive distributions.

3. *Checkerboard distributions.* The section by C and S under this heading is so confused that it requires several pages to straighten it out. C and S begin with a 2-species, 8-island checkerboard and congratulate themselves (p. 322) that their method "fails to recognize such a pattern [a checkerboard] as non-random, since we ignore degenerate [*sic*] arrangements." Yet, a page later C and S are happy to announce that their method does recognize a checkerboard (that of their Figure 18.1) as unusual. How can they have it both ways? For the benefit of readers confused by the sections "Checkerboard distributions" and "Degenerate

rearrangements" and by Table 18.1 and Figures 18.1, 18.3, and 18.4 of C and S (this volume), here is what it is about:

Clearly, if one places two species randomly on islands, it is improbable that their distributions will define a checkerboard (unless this outcome was dictated by hidden constraints used in making a "random" placement). The more islands there are, and the more islands the species occupy, the more improbable is a checkerboard. Any adequate test of non-random co-occurrence must therefore be able to recognize a checkerboard as non-random. Our test does; the C and S test has difficulty in doing so, for the reasons that C and S correctly acknowledge in discussing their 2-species, 8-island case and later their 2-species, 4-island case (their Figure 18.4) and that we have explained further (our section "Inability to detect checkerboards," and D and G, 1982, pp. 70–71). Briefly, the row and column constraints that C and S used to construct a null matrix ensure that, if the observed matrix is a checkerboard, the null matrix will also be a checkerboard, so that it is structurally difficult for their test to recognize a checkerboard as non-random. C and S fail to understand why their test detects the checkerboard of their Figure 18.1 but not other checkerboards, and this inconsistency escaped their notice.[2]

Previously, C and S (1979) showed no awareness of this structural weakness of their test vis-à-vis checkerboards. They now recognize this issue and proceed to befog it by applying the term "degeneracy," with its mathematically and morally pejorative connotations. Suppose one wishes to assess the improbability of a checkerboard by calculating what fraction of all randomly produced matrices are checkerboards. Among the many possible arrangements when one randomly places species on islands are some in which a certain species ends up on no islands or a certain island ends up with no species. C and S invent the term "degenerate" for such matrices, and in their rejoinder (p. 342) they still "felt it was apparent" that such matrices should be deleted from the sample space of all possible rearrangements, thereby increasing the likelihood of acceptance of their

---

[2] After their Figure 18.1, an extreme checkerboard with 10 exclusive species pairs, C and S present their Figure 18.2, an even more extreme checkerboard. As the purpose for presenting Figure 18.2, C and S explain that their Figure 18.1 has fewer exclusive pairs than their Figure 18.2, so they conclude (p. 323), "Clearly the checkerboard in Figure 18.1 has a very *low* number of allopatric species pairs relative to the possible number that might obtain; this pattern is exactly opposite to one that we might expect competition to produce. So ... it is not a virtue if an algorithm automatically finds any checkerboard to be improbable" [*sic*!]. Here is an exact analogue to their reasoning: "The man in Figure 1 weighs 484 lbs., but we have found a man whom we show in Figure 2 and who weighs 740 lbs. Clearly the man in Figure 1 has a very *low* weight relative to the possible weight that a man might have. So ... it is not a virtue if your algorithm automatically identifies the man in Figure 1 as overweight."

"null hypothesis." However, there is nothing immoral about such matrices, and it is a fatal vice rather than a virtue of the C and S procedure to ignore them. The whole question of "degeneracy" is irrelevant to our method, which directly calculates co-occurrence probabilities without having to construct reshuffled matrices. When C and S complain (p. 324) that we have not calculated "how many different matrices ... there are with 0, 1, 2, ... pairs of complementary rows" corresponding to the matrix of their Figure 18.1, they are correct. We do not need to, because our method directly calculates the probability of this checkerboard as approximately 10^{-60}, without constructing the 10^{60} "null" matrices.

C and S conclude this confusion by pretending (p. 325), "To our knowledge Gilpin and Diamond (this volume) have not presented strong non-biogeographic evidence that even a single pair of species in the Bismarcks or other southwest Pacific archipelagoes cannot coexist for competitive reasons." True, not in this volume—because we have extensively presented elsewhere the numerous lines of non-biogeographic evidence demonstrating that exclusive distributions involving many pairs of southwest Pacific species are maintained by competition (G and D, 1982, pp. 82–83; the book Diamond, 1972; monographs Diamond, 1975, and Diamond and LeCroy, 1979; many papers).

4. *Meaningful rearrangements.* C and S (1979) introduced their critique with an analysis of New Hebrides birds, which appeared to be in stunning agreement with their "null hypothesis" ($p > 0.90, 0.95$, or 0.99 depending on the test; remarkably detailed match between observed and predicted distributions). C and S (pp. 326–327) still accept this conclusion and believe their test to be especially careful and conservative. This tenacity is remarkable, when we have twice pointed out (this chapter, section "Results of ...," paragraph 3; D and G, 1982, p. 69) that their result is simply an error due to a programming mistake that they described explicitly in their small-print appendix (performing only "several" reshufflings of 2×2 submatrices to generate only "a number of" randomized New Hebrides matrices that are "not completely identical" to the actual matrix: C and S, 1979, p. 1140).

To make clearer what it is that C and S admit to having done, we have prepared Figure 19.1. The matrix of Figure 19.1a is a 28×14 piece (to save space) of the actual 56×28 New Hebrides matrix. Figure 19.1b is the same matrix with "several" = two 2×2 submatrices reshuffled. Obviously, reshuffling so few pairs of elements in a 28×14 (actually 56×28) matrix guarantees that the reshuffled and actual matrices will be indistinguishable at the $p > 0.90$ or $p > 0.99$ level, regardless of the structure of the actual matrix. The defense of this transparently meaningless procedure by C and S (this volume, their section "Meaningful rearrangements") is another study in evasion of the obvious. They note

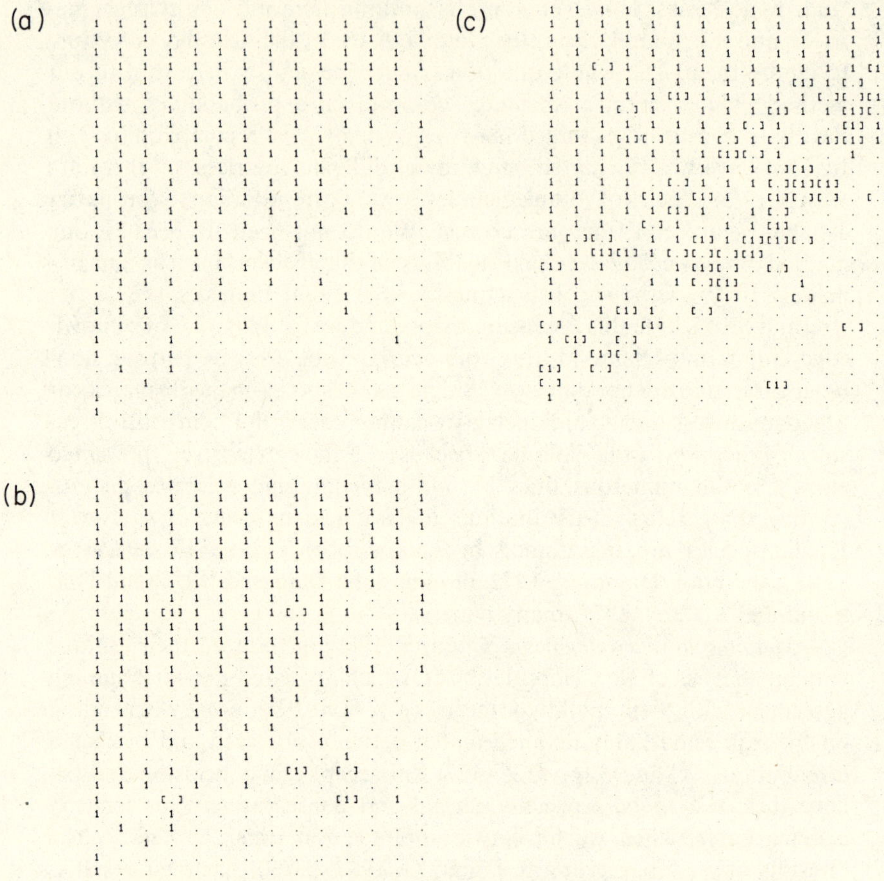

(a)

(b)

(c)

Fig. 19.1. (a) 28 × 14 piece of the New Hebrides bird matrix. Columns are islands, and rows are species. If the element *ij* is indicated as "1", it means that species *i* occurs on island *j*. A blank means that species *i* is absent on island *j*. (b) The same matrix, after two reshufflings of 2 × 2 submatrices. Elements that differ from corresponding elements of Fig. 17.3a are bracketed to call attention to them. The resulting reshuffled matrix is necessarily almost identical to the original matrix. This exemplifies how "null" matrices were constructed for the New Hebrides by C and S, who performed only "several" reshufflings. (They actually placed rows and columns in what they termed canonical form. This alters the matrix's appearance but not the calculated statistics). (c) The matrix of Fig. 17.3a after 500 reshufflings, with changed elements bracketed. The "null" matrix now differs obviously from the original matrix.

that "special care was taken to insure that all rearrangements were actually 'nonequivalent.'" However, they neglect to mention that they had generated only a few such rearranged matrices and that the nonequivalence was in only a few elements. The rejoinder by C and S objects to our describing this procedure as a programming mistake. Specifically, it is an algorithm that is structurally incapable of answering the question for which it was designed—in short, a programming mistake. The rejoinder further states that we have not shown the C and S procedure to be "biased or destined to produce matrices indistinguishable from the original." True, it is not biased, merely guaranteed to produce matrices nearly indistinguishable from the original.

As an illustration of how a proper reshuffling transforms the result, Figure 19.1c is the matrix of Figure 19.1 after 500 reshufflings, yielding a matrix obviously different from Figure 19.1a or 19.1b. When we repeated this procedure to obtain 50 such independently reshuffled matrices, the p level for the New Hebrides dropped from $p > 0.90$ or $p > 0.99$ to $0.1 < p < 0.25$. But why only to this level, rather than to the $p < 10^{-8}$ level obtained by the same test for the three other faunas? Because the adjective "conservative" used by Connor and Simberloff (p. 326) is euphemistic for a further structural flaw: even with adequate reshuffling the C and S test fails for nested matrices such as the New Hebrides matrix, which other tests recognize as radically non-random at the $p < 10^{-8}$ level (see D and G, 1982, pp. 69–70, and G and D, 1982, p. 78, for details).

5. *Directionality*. Our test calculates, for each species pair, the probability that its observed co-occurrence is either more or less exclusive than expected for random placement. Having thus identified the species pairs in the positive and negative tails of the distribution, we were then able to examine at length the biological factors contributing to these patterns of non-random association or exclusion (G and D, 1982, pp. 78–82). C and S (p. 327) protest that their test does have the ability to determine directionality; they neglect to explain that this means only whether there are more or fewer species pairs sharing x islands than expected, not which pairs deviate how much in which direction. They also protest that non-biogeographic evidence is needed to interpret species pairs in the positive and negative tails; this is true, but C and S neglect to add that our test permits the search for such evidence by identifying the pairs in the tails and that their test does not identify the pairs and hence fails to permit such a search.

6. *Hidden structure*. As we pointed out, the C and S test incorporates hidden structure by using row and incidence constraints that are affected by competition. C and S respond (p. 327), "Gilpin and Diamond (this volume) present no evidence that demonstrates that the row and column sums (and incidence functions) are actually affected by competition." Of

course: we stated that the evidence is instead presented in Diamond (1982), which shows that species occurrence frequencies (= the row constraint) and incidence undergo gross shifts·between archipelagoes as a function of the number and identity of competing species. C and S (p. 328) also pretend, "Gilpin and Diamond (this volume) present no evidence that 'hidden structure' in these constraints would necessarily translate into altered co-occurrence patterns." In this case the twofold evidence is actually summarized in our chapter in this volume (sections headed "Inability to detect checkerboards" and "Analysis of Individual Guilds") as well as presented in more detail elsewhere (D and G, 1982, pp. 70–71; G and D, 1982, pp. 81–82). The evidence is that row and column constraints can preserve checkerboard co-occurrence patterns and that incidence constraints preserve much of the non-randomness in co-occurrence patterns.

7. *Monte-Carlo techniques.* The issue is not only that the Monte Carlo approach is inferior in principle, as C and S evidently now recognize, but also that they have implemented it incorrectly in reshuffling far too few times, as they still fail to recognize and as discussed in connection with our Figure 19.1. In addition, we tried the Monte Carlo approach as a method for calculating cell probabilities for our test (see next section) and found it exceedingly slow to converge.

Our procedure. C and S (this volume) criticize our procedure on three substantive grounds, all of which are easily shown to be wrong. In addition, C and S pretend three more times that our chapter omits supporting details for which our chapter actually refers readers to our previous papers. "The method ... not provided by Gilpin and Diamond (this volume)" (C and S, this volume, p. 328) is provided by G and D (1982), p. 77, column 2, equations for E_{ik} and SD_{ik}. "Gilpin and Diamond (this volume) do not fully describe their recursive approach, so it is not possible to determine its suitability ..." (C and S, this volume, p. 330): the recursive approach is described by G and D (1982), p. 84, Method 1. "The Monte Carlo procedure outlined but not performed by Gilpin and Diamond (this volume)" (C and S, this volume, p. 330) was performed by G and D (1982), p. 84, Method 3. Now to the three substantive errors:

First, C and S complain that our method counts "degenerate" arrangements. Here they are completely mistaken: our method does not depend on counting rearranged matrices at all as theirs does, but instead calculates overlap possibilities directly. The only instances where we used a method of counting rearrangements were in discussions of model matrices to illustrate the principles involved (*e.g.* the 2 × 8 checkerboard). We described earlier in this appendix how C and S misunderstood "degeneracy."

Second, C and S question our method for computing cell probabilities. They neglect to mention that we performed a robustness analysis by comparing seven different methods (G and D, 1982, pp. 77 and 84). The remaining comments that C and S make under their heading "Cell probabilities" are simply wrong. The problem of p values greater than 1 that C and S cite arises infrequently, and we found that three different ways to solve it yield essentially the same conclusion (G and D, 1982, p. 77). We have already performed the logistic model fit (G and D, 1982, p. 77) that C and S claim to be impossible. C and S conclude (this volume, p. 330) that "the only means of estimating the cell probabilities is by Monte Carlo techniques." We are gratified that C and S accept the validity of this version of our technique, since it yielded conclusions in agreement with our six other methods. However, the preference of C and S for this method of cell probability determinations is misplaced: of the seven methods, we found it to be the slowest, most expensive of computer time, and least accurate.

Finally, C and S question whether we are correct in assuming a normal distribution for the standard deviates in our "null" model, and they wonder whether it might instead be skewed or leptokurtic. Our method, because it allows the direct calculation of expected co-occurrences, also allows the direct calculation of the variance of this estimate. Since we are summing over multiple islands, the central limit theorem (Feller, 1950) guarantees normality, mean zero and S.D. 1, of the expected distribution of the standard deviates. A glance at our Figure 17.2 shows that its mode is shifted to the right and that in addition it has a long tail to the left. The disagreement with the "null" model is obvious even before one has applied the χ^2 test.

BY CONNOR AND SIMBERLOFF

We are beginning to feel like Br'er Rabbit enmeshed in the Tar Baby: Each of our attempts to elucidate the statistics of species combinations on islands seems only to elicit a mass of obfuscatory goo. With no briar patch in sight, and since our contribution to this volume deals adequately with the issues, our best option seems to be to keep our response very brief and to hope that ecologists read both papers and rejoinders very carefully. We trust that a close reading will vindicate both our substantive ecological claims and our motives. In particular, an examination of Connor and Simberloff (1979) should establish that our contribution to this volume accurately cites and represents our earlier work. We must add, in light of the rejoinder, a few points about detecting whether co-occurrence patterns are surprising with respect to a stated hypothesis.

The dilution effect certainly confounds community-wide analysis of presence/absence matrices, as we observed in this volume and in 1979. Since Gilpin and Diamond view the effect as so debilitating, it is odd that they devote pages 309–311 to analysis of entire avifaunas. Such analyses are unable to indicate biological mechanisms even when they can falsify statistical hypotheses (Pielou and Pielou 1968, Simberloff and Connor 1981), but in this instance we cannot ascertain whether a null statistical model is falsified since the presence/absence data for Bismarck birds are unpublished. Whether Diamond (1975) has precisely listed guilds for the Bismarcks we leave to readers to determine. For the fruit pigeons that Gilpin and Diamond's rejoinder uses as an example, we observe that Diamond (1975) cites no technical literature to justify the guild assignment on pp. 406–411, where the guild is defined, or anywhere else. Crome (1975) is not cited, nor is Crome acknowledged. Crome (1975) discusses only fruit pigeons, so it is impossible based on Crome (1975) to say that the eighteen listed species constitute the guild of exclusively frugivorous arboreal species that do not take stones into the gizzard.

We believe our discussion of checkerboards is sufficiently clear that we need do no more than call attention to our main contribution. Lest this exchange degenerate further, however, we state that "degenerate" has no pejorative connotations when assigned to matrices. It is defined (*Oxford English Dictionary*, Compact Edition, 1971, p. 674) as "d. *Geom.* Of a curve or other locus: To become reduced to a lower order, or altered into a locus of a different or less complex form," and is used in matrix algebra (*e.g.* Kemeny *et al.* 1974, p. 343) without connoting moral or mathematical turpitude. We felt it was apparent that non-existent species (those found on no islands) and empty islands (those with no species) could tell us nothing about forces constraining species coexistence, so that matrix rows or columns representing such species or islands should be omitted. A matrix with such rows or columns thus seemed to us well within the mathematical sense of the term "degenerate." Readers can judge both our logic and our intent in this matter.

There was no programming mistake described in the Appendix of Connor and Simberloff (1979), and Gilpin and Diamond have not shown the size of the universe of matrices for New Hebrides birds or that our sampling of this universe was either biased or destined to produce matrices indistinguishable from the original. We still question whether, if there are very few different, non-degenerate matrices with a given set of row and column sums, there is any statistically powerful way to show that species co-occurrence patterns are surprising or strongly imply any biological force, above and beyond whatever forces determine the island richnesses and species frequencies. In any event, no statistical statement can be made about matrices or overweight people without some notion of the nature of the universe. An extraterrestrial who saw but two humans would have no way of knowing whether a 484-lb. man is heavy or light.

Finally, we hope that readers will pay close attention to those areas, listed in our main contribution, in which Gilpin and Diamond have changed the arguments of Diamond (1975), then inveighed against Connor and Simberloff (1979) for not treating the new arguments. The new arguments deserve to be dealt with on their own merits, and we have begun to do so here, but we could not have anticipated in 1979 how Gilpin and Diamond today would attempt to salvage the approach of Diamond (1975).

20.

A Null Model for Null Models in Biogeography

ROBERT K. COLWELL

Department of Zoology, University of California, Berkeley, California 94720

DAVID W. WINKLER

Museum of Vertebrate Zoology and Department of Zoology, University of California, Berkeley, California 94720

If we could seed a series of virgin, replicate earths with primordial life and set the level of interspecific competition differently in each, could we tell them apart three billion years later by looking at biogeographical patterns? In this paper we present the results of an effort to approximate this experiment by computer simulation, with the purpose of examining the potentials and the pitfalls of several methods of biogeographical analysis. Since we control the intensity of competitive exclusion as a variable in the simulation, biases and limitations in the construction of null models in biogeographical studies can be studied directly. We will show that the effects of competitive exclusion on island biotas are likely to be underestimated or even obscured by several directional biases that are difficult to estimate or avoid in the real world. However, far from counseling despair, we hope that the model we have developed may lead to a better understanding of the complex interactions between evolution, ecology, and chance events in biogeography.

The principal message we hope to convey is that the setting of constraints and assumptions in the development of null models is not an arbitrary matter. The null hypothesis tested in any analysis of biogeographical data (*e.g.* Connor and Simberloff, 1978; Strong *et al.*, 1979) is not that empirical patterns do not differ from random ones, but that they do not differ from patterns generated by a particular model of the world. The design of such a "null model" is critical, but is necessarily based on biological judgments. We will proceed by describing a general model of evolution, biogeography, and ecological interactions within which many null models for biogeographical patterns can be defined.

The Model: Phylogenesis, Morphological Evolution, and Taxonomy

The model consists of two parts. The first, described in this section, builds upon the models of Raup *et al.* (1973) and Raup and Gould (1974) for the stochastic generation of phylogenies, with "phenotypic" characters evolving stochastically at each speciation event. Implemented in a program called "GOD," this part of our model, like that of Raup and Gould (1974), produces "biotas" of any desired size that mimic many of the commonly observed properties of real-world phylogenies (Figure 20.1). The input variables for the program are simply the probabilities of speciation, extinction, and character change and a seed for the random number generator. (The number of characters and the size of the tree must also

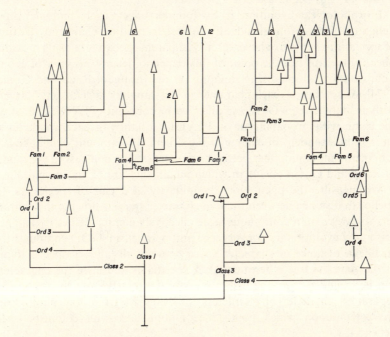

Fig. 20.1. Example of a stochastically generated phylogenetic tree ("Beta"), based entirely on three probabilities: speciation (.20), extinction (.10), and character change (.80). Evolution was stopped at 100 species, 51 of which had become extinct, leaving 49 living species (at the top of the tree) as a "mainland pool" for populating imaginary archipelagoes. The triangles represent the "bills" of some computer creature that partitions a food resource (as every programmer knows, larger bugs mean larger bills). Bill length and bill width are linear transformations of independently evolving Characters 3 and 4, respectively. Each triangle represents a genus, monotypic if no number appears; otherwise, the numbers in or near triangles give the number of species in that genus. The computer-generated taxonomy is based exclusively on the topology of the tree (not on characters) according to a cladistic algorithm.

be set, but they do not affect patterns of evolution.) The tree is initiated with a single species whose "characters" all have initial values of zero. At each iteration of a counter for evolutionary time, this "root" species and each of its living descendant species gives rise to a daughter species with probability "PSPN," goes extinct with probability "PEXT," or continues to exist unchanged with probability (1 − PSPN − PEXT). The process continues until a predefined number of speciation events have occurred, until a predefined number of time units have passed, or until all species are extinct. (When PEXT approximates or exceeds PSPN, many trees die young. Even when PSPN is twice PEXT, some trees become extinct in early stages, but are generally safe from extinction once some "epidemic threshold" has been passed.)

Morphological evolution is by founder effect: each character in a daughter species is independently subject to random change at the time of speciation, but is fixed ever after. The parent species retains its previous phenotype. (It is not the point of this paper to defend this view of evolution; we have adopted it in building on the results of Raup and Gould, 1974). At speciation, each character in the daughter species gets "larger" with probability PEVOL/2, "smaller" with probability PEVOL/2, or stays the same with probability (1 − PEVOL). Change in each character is entirely independent of change in the others. The amount of change is a (uniform) random number between zero and one, in contrast to the model of Raup and Gould (1974), in which all changes are of unit magnitude.

We measure the phenotypic dissimilarity of a pair of species as the euclidean distance between them in character space. Allowing the characters to evolve on a continuous scale effectively eliminates ties in distance arising by evolutionary convergence and parallelism. GOD computes the euclidean distance in character space between all pairs of species living at the time the tree is terminated. A specifiable subset of the characters can be excluded from the distance computation. (We use this option to separate a character or characters for "vagility" from "morphological" characters upon which coexistence among successful colonists is to depend.)

Since one of the techniques we evaluate is sampling within taxonomically constrained biotas, it is necessary to produce a hierarchical classification of the species in each tree. There are, of course, a great number of algorithms available for classifying sets of species on the basis of their phenotypic characters, but these are not appropriate (or necessary) in this case, since the actual phylogeny is completely known. The tree-generating program GOD outputs a taxonomic "name" for each species (as a string of 1's and 2's) that completely specifies a path from that species to the root of the tree, in dichotomous fashion. (Thus, species "2122" is the second species on the tip of the second twig on the first branch of the second trunk of the phylogenetic tree, if all possible branches evolve.) Sister groups at the time of branching end in "1" and "2," respectively, and share all digits to the left. When the tree is complete, some species (including most extinct ones) will have fewer digits in their names than others, depending on the number of nodes between the species in question and the root of the tree. At this point all names shorter than the longest name in the tree are right-filled with 1's, producing monobasic taxa at various hierarchical levels. This step is necessary to preserve the cladistic hierarchy, and to ensure that each species has a unique "name."

The Model: Biogeography, Ecology, Classification, and Sampling

For each evolutionary tree generated, program GOD outputs a list of all species, with information on when each originated; when it went extinct, in the case of "fossil" species; the identity of its immediate ancestor; its score for each character; its dichotomous cladistic "name"; and, for each living species, the euclidean distance from its location in character space to the location of each other living species. This information is used by a separate program, "WALLACE," to carry out a variety of neontological operations. In effect, the living species produced by GOD are assumed to have evolved in a world free of selective forces, including species interactions. Alternatively, one may imagine that selection has operated, but in a way indistinguishable from stochastic character change and stochastic speciation (i.e. that the species in a given tree have all lived their lives allopatrically in some highly dissected continent.)

In this study we designate the set of living species in a large phylogeny produced by GOD to be a "mainland" pool of potential dispersers. From this pool WALLACE assembles sympatric communities. The probability of each species' being chosen from the pool may be a constant, or may be weighted by one or more of its characters representing "vagility." Archipelagoes of island communities can be assembled with a specified number of species per island (sample size) and a specified number of islands (samples)—a great advantage over natural archipelagoes. Interspecific competition can be imposed with adjustable force, either before colonization (in the "mainland" pool) or after assignment of species to islands, by eliminating a specifiable proportion of the species present as victims of competitive elimination, their vulnerabilities based on morphological similarity to other species on the island. Finally, various sampling schemes can be used on archipelagoes with known histories of colonization and competition to investigate the power and biases of each scheme in revealing the forces that produced the biogeographical and morphological patterns in an archipelago.

Before describing the "treatments" applied to communities and archipelagoes, we must explain the algorithms used to eliminate species by "competition" and to classify species hierarchically for taxonomically constrained sampling.

"GAUSE," a subroutine of WALLACE, ranks all the species in any sample sent to it according to their vulnerability to direct and diffuse competition, based on a specifiable subset of the characters of each species. The algorithm used to produce this ranking is as follows. First, the matrix of euclidean distances in morphological space between all pairs of species in the sample is searched for the smallest distance. Then, one or the other of the two species involved is ranked for competitive elimination and struck from the matrix; the smallest distance value in the matrix (and, of course, all other distances involving that particular species as well) is thereby removed from the matrix. Then the next smallest distance is found, one of the two species involved is ranked and eliminated from consideration, and so on, until all species are ranked. (Both members of the last remaining pair are automatically given the last rank.)

A simple criterion determines which of the two species is to be ranked and eliminated from the matrix at each cycle. The species whose removal would produce a greater increase in the mean of all remaining distances in the matrix is chosen—put another way, this is the species whose "diffuse similarity" to the remaining species is greater. This choice is easily accomplished; the program consults (and constantly updates) a vector of marginal totals for the distance matrix, and simply asks which of the two species has the smaller sum of distances to all others. In case of a tie, one of the two species is chosen at random to be ranked for elimination (ties are very scarce when two or more characters are involved and characters evolve on a continuous scale). Figure 20.2 shows examples of the operation of GAUSE on island samples.

"HENNIG," another subroutine of WALLACE, takes the dichotomous code names for all the species in a tree (including the extinct species) and produces a hierarchical classification, according to a specifiable variable "KUT" that tells HENNIG the maximum number of lineal nodes per taxonomic level. For example, if KUT is set at 3, up to 3 levels of branching can be embraced in a single taxonomic level. If all possible speciations have occurred in all lineages, each taxon at this level would include 2-to-the-KUT-power (8) subtaxa. (This is, of course, only rarely the case in all lineages.) Thus KUT sets the maximum number of species per genus, genera per family, and so on. (Subroutine HENNIG appears in WALLACE, rather than in GOD, to permit a series of reclassifications to be performed with different values of KUT, without the need to rerun GOD each time.) Finally, HENNIG produces nomenclature, assigning consecutive numerals to subtaxa within each taxon, as names.

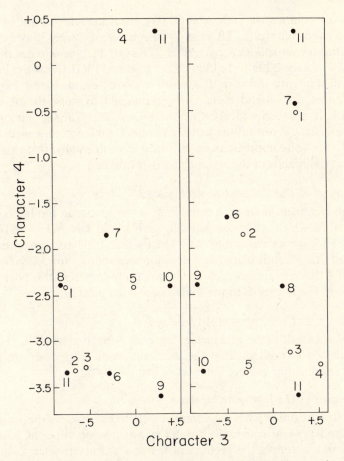

Fig. 20.2. Ranking of island species according to their vulnerability to extinction. The two boxes represent two of the islands of an Icarus-Lack Archipelago, each colonized by twelve living species from tree eta. The numbers are the ranks, from most vulnerable to competitive elimination (1) to least vulnerable (11), within islands. (The ranking algorithm is explained in the text.) If five of the species on each island are eliminated by competition in rank order (PKILL = .42, Figure 20.4, line C), the open circles will be removed, leaving the solid dots. Characters 3 and 4 correspond to the "bill" characters of Figure 20.1, untransformed.

Treatment 1: The Lack Archipelago

From the "mainland pool" (all the living species in a single tree) "K" samples of "N" different species each are drawn at random. (In other words, sampling is without replacement within samples, with replacement between samples.) Each sample is the biota of an island in a K-island archipelago. Next, GAUSE is called to rank the species within each island biota for competitive vulnerability. (We use Characters 3 and 4—"bill length" and "bill width," let us say—to compute distances in two-dimensional character space.) A parameter "PKILL" then determines the proportion of each biota to be eliminated; in effect, PKILL sets the limiting level of similarity scaled to the overall similarity on the island. Next, the portion of each island biota to be eliminated by competition, set by PKILL, is marked for local extinction (Figure 20.2). The surviving species on each island represent the biota of the Lack Archipelago, a place where the ravages of competition can be tuned experimentally. (No migration among the islands of the archipelago is permitted.)

Treatment 2: The Tallahassee Archipelago

This treatment involves a reshuffling of the species in the Lack Archipelago, in which, after competition, each island has $N(1 - \text{PKILL}) =$ "NLEFT" species remaining. Any species that survived competition on at least one island is placed in a pool for resampling. Now, K samples of NLEFT different species each are drawn at random from this pool as the biotas of the K islands in the Tallahassee Archipelago.

Treatment 3: The Kropotkin Archipelago

The K islands of this archipelago are each populated with NLEFT species chosen at random from the mainland pool. All species are mutually congenial; no competition occurs here.

Treatment 4: The Icarus Archipelago

In this treatment, we use Characters 1 and 2 of each species in the mainland pool to compute an index of vagility, or dispersal potential. The index is simply the product of the two characters after a constant is added to each character to bring the minimum character state up to zero (since characters may evolve into negative values). Two characters were combined to reduce the frequency of evolutionary convergence and parallelism. The probability of any particular species' being chosen to populate the Icarus Archipelago is proportional to its value for the vagility index, in contrast to Treatments 1, 2, and 3, in which uniform probabilities are used. The K islands of the Icarus Archipelago are each populated with NLEFT different species chosen from the mainland pool by this method.

Treatment 5: The Icarus-Lack Archipelago

This archipelago is populated in the same way as the Lack Archipelago, except that the colonists are chosen from the mainland pool as for the Icarus Archipelago—by vagility-weighted probabilities. Competitive elimination after colonization is the same as in the Lack Archipelago.

Treatment 6: The Icarus-Tallahassee Archipelago

This archipelago is populated by the same means as the Tallahassee Archipelago, but species are drawn from the repooled survivors of competition in the Icarus-Lack Archipelago. (The resampling is done uniformly, not with vagility-weighted probabilities.)

Treatment 7: The Linnaean Quadrille

This treatment is a resampling of the post-competitive biota of the Lack Archipelago with taxonomic constraints, contrasted with a taxonomically matched sample from the mainland pool. The sampling algorithm begins with the (uniform) random selection of one of the K islands in the Lack Archipelago. Next, a species on that island is chosen at random (the "propositus"), and its taxonomic identity is determined at a designated taxonomic level ("genus," "family," "order," etc.). Now a list is made of all other species on that island that are "contaxonic" (congeneric, confamilial, etc., depending on the designated level) with the propositus species, and one of them is chosen at random. (If there is no other species on that island that is contaxonic with the propositus, a new island and a new propositus are chosen at random.) In this way, a pair of contaxonic, coexisting species is found in the Lack Archipelago; call it the Island Pair. Now a list is made of all species in the mainland pool that are contaxonic with the Island Pair, and a Mainland Pair is chosen from the list. (There must, of course, be at least one such pair, since the archipelago fauna is a subset of the mainland pool.) This process is repeated until 20 matched sets of pairs have been selected. Any redundant sets are then removed before statistical analysis.

RESULTS AND CONCLUSIONS

Several trees were generated by program GOD, some using the same set of input probabilities but with different random number seeds to get an idea of the variability to be expected, some with different values of the input probabilities to explore their effects on the phylogenies and biogeographical analyses.

Tree beta, shown in Figure 20.1, had an extinction probability only half that of the speciation probability, producing rapid radiation and only a few species per lineage in extinct groups. Trees generated with extinction probability equal to speciation probability are less "radiating," with major groups becoming extinct. Tree eta (which produced the raw data for Figures 20.2, 20.3, and 20.4) is such a tree. Raup and Gould (1974) show that there are often strong correlations among characters for the species in a stochastic tree. The branching process itself reproduces certain character state combinations in the descendant species of unusually prolific ancestral species, while other lineages become extinct. Character correlations tend to be higher in "equilibrium" trees (PEXT = PSPN) than in rapidly radiating trees (PSPN > PEXT) because of less homogeneous filling of morphological space.

Even though the amount and direction of character change in our model is a uniform random variate at each speciation event, the frequency distribution of any given character among species in our imaginary biotas typically has a strong mode, usually not located at zero, as also shown by Raup and Gould (1974) for unit character changes. Like many other effects that we will discuss, the existence of such modes is a consequence of the differential propagation, by chance, of certain lineages that represent variations on a theme. Substituting a biologically more realistic rule for character change, such as a normal random distribution, will only emphasize the modality. The tests used by Simberloff and Boecklen (1981) to assess the effects of competitive displacement on characters of coexisting species assume a uniform frequency of character states as the null distribution. Since the average (and minimum) distance between random draws from a uniform distribution is always larger than the distance between random draws from a modal distribution with the same range, their tests are biased to an unknown degree against finding any effects of character displacement, which acts in the same direction on similarity. The uniform random distribution, however appealing as a null hypothesis for character states, cannot be produced by any stochastic evolutionary process known to us, except as a pathological limiting case. Thus any distribution of character states from nature that is uniform (or log-uniform) is, in itself, strong evidence for character displacement.

By comparing the effects of treatments 1–7 on patterns of morphological similarity among "coexisting" species on our imaginary islands and in mainland pools, we can demonstrate three different inherent biases in some techniques of biogeographical analysis designed to test for the effects of interspecific competition. All three biases tend to obscure or even reverse any competitive effects, seriously weakening any inferences drawn using the techniques that produce them.

1. *The Narcissus effect: Sampling from a post-competition pool underestimates the role of competition, since its effect is already reflected in the pool.* When the species of an archipelago are pooled and then sampled at random to test for the role that competition may have played in eliminating closely similar sympatric species, there is a consistent underestimation of the effect of competition, since the most vulnerable species have been eliminated from the entire archipelago (or never evolved in the first place, in the case of a local radiation). In our model, this situation is represented by the comparison of a Lack Archipelago with a Tallahassee

Archipelago. The correct comparison (assuming it to be possible) is between the empirical island pattern (the Lack Archipelago) and a sample from a truly pre-competitive pool—in the model, the Kropotkin Archipelago, sampled at random from the mainland pool, with no competition.

Figure 20.3 shows the Narcissus effect for tree eta. The values plotted are the mean distances in two-character morphology space (Figure 20.2) for first through sixth nearest neighbors, averaged over islands (see Inger and Colwell, 1977). Pooling all pairwise distances obscures the relationship between closely similar species, which is the focus of competition. (Gilpin and Diamond (this volume), in a similar vein, decry the pooling of diverse guilds in sampling for null models.) In Figure 20.3, the dissimilarity of coexisting species in the Lack Archipelago (dashed line) is consistently greater than in the Kropotkin Archipelago (dotted line)—significantly greater for first nearest neighbors. The Tallahassee Archipelago (solid line) differs in the right direction for first nearest neighbors (though not significantly), but quickly unites with the dashed line for more distant neighbors. Pooling and resampling consistently underestimates any role that competition may have played (Grant and Abbott, 1980; Case and Sidell, 1983). It is important to recognize this directional bias, even if it is

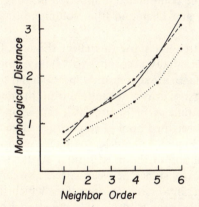

Fig. 20.3. The Narcissus effect for tree eta. Mean euclidean distances to first through sixth nearest neighbor in character space (Characters 3 and 4) for the seven species on each of five islands (each point is the mean of five means of seven scores), contrasted for three treatments. Coexisting species of the Lack Archipelago (42% eliminated by competition—dashed line) are consistently less similar to one another than are the species in the Kropotkin Archipelago (no competition—dotted line), but the effect of competition is only weakly detectable by comparison of the Lack islands with those of the Tallahassee Archipelago (solid line), formed by pooling and resampling the survivors of competition in the Lack Archipelago. Sampling from the mainland pool was uniform, not vagility-weighted, for this analysis.

usually impossible to avoid it. Narcissus could not see the bottom of the pool for his own image, and could not guess its depth.

2. *The Icarus effect*: *Correlations between vagility and morphology can obscure the effects of competition in morphological comparisons of mainland and island biotas.* When comparisons are made between the morphological characteristics of mainland and island biotas, the assumption is often made that the species in the mainland pool do not differ significantly among themselves in vagility (Simberloff, 1978a). This assumption ignores the fact that phylogenies, both real and random (Raup and Gould, 1974), result in character correlations throughout the phenotype. For example, the morphological spectrum of the subset of bird species colonizing any group of islands will include relatively many swallow bills and relatively few tinamou bills, not because of any intrinsic property of their bills, but because swallows have much higher vagility than tinamous, and in the course of their evolution, either because of selection or not, the bills of tinamous have become and have remained invariably different from those of swallows. To the extent that similarities among the high-vagility species colonizing an archipelago *increase* the mean similarity in island faunas over that of mainland pools, the effects of any subsequent or concomitant interspecific interactions that *decrease* the similarity of coexisting species on the island will be confounded. Simberloff (1978a), Grant and Abbott (1980), and Gilpin and Diamond (this volume) discuss this and related problems.

The results for tree eta show this effect (Figure 20.4). Species in the pool selected preferentially according to vagility (line B in Figure 20.4, the Icarus Archipelago) are significantly more similar to one another morphologically than those drawn at random from the mainland pool ignoring their vagilities (line A in Figure 20.4, the Kropotkin Archipelago). When the vagility-weighted pool of colonists is subjected to strong competition (line C in Figure 20.4, an Icarus-Lack Archipelago), the effect of competitive elimination of the most similar species, which increases the mean morphological distance between species, swamps the weaker, opposing effect of vagility-biased colonization, which decreases the mean morphological distance as a result of correlation between morphology and vagility. With competition at this very high level (42% of colonists eliminated), a comparison of samples from the archipelago pool (line D in Figure 20.4, an Icarus-Tallahassee Archipelago) reveals a significant effect of competition. However, when the proportion of species eliminated by competition is reduced to 22% (line E in Figure 20.4)—still a rather high level of competition—the opposing effects are closely balanced, and a comparison with either the mainland pool (line A in Figure 20.4) or with random samples from the archipelago pool (line E in Figure 20.4) reveals no significant difference. The character correlations in our data

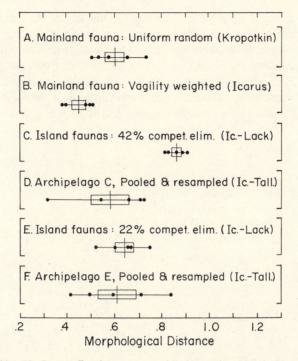

Fig. 20.4. The Icarus effect for tree eta, and its interaction with the Narcissus effect. In each treatment, 7 (NLEFT) species coexist on each island, and there are 5 (*K*) islands per archipelago. Each point is the mean euclidean distance for one of the five islands from each of the species on that island to its nearest neighbor in character space. Boxes represent one standard error above and below the treatment mean. *A priori* comparisons between treatments were made with Mann-Whitney *U*-tests (one-tailed) on the plotted points. When morphology (Characters 3 and 4) is correlated with vagility (Characters 1 and 2), the Icarus effect (A vs. B, $p = .006$) and the effect of competition (B vs. C, $p = .004$; or B vs. E, $p = .004$) are opposed in direction, and may completely cancel each other out (A vs. E, $p = .210$, ns). Under these conditions, competition must be extremely strong to be detected by pooling and resampling (for C vs. D, $p = .004$; but for E vs. F, $p = .345$, ns). For the generation of tree eta, PSPN = PEXT = .30, PEVOL = .70.

are the consequence of the statistical properties of conservative branching processes (the "spreading effect" of Raup and Gould, 1974), and so, presumably, are many of the character correlations in real phylogenies. But to this inevitable background level of intercorrelation are added the potentially powerful effects of selection on functionally integrated phenotypes. Had Icarus had hollow bones and air sacs, the wings he fashioned might have carried him further.

3. *The J. P. Morgan effect*: *The weaker the taxonomic constraints on sampling, the harder it becomes to detect competition.* Results from the Linnaean Quadrille analysis for three trees with identical input probabilities are given in Figure 20.5. The pattern is quite consistent: as sampling is constrained to successively higher taxonomic levels, the difference in morphological distance between contaxonic species in the Lack Archipelago and in the mainland pool decreases monotonically (lower

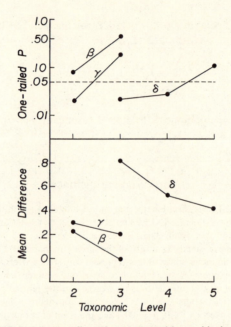

Fig 20.5 The J. P. Morgan effect. If post-competitive archipelago biotas are sampled at random within specified taxonomic limits, then matched with similarly constrained samples from the mainland (source) pool, the effects of competition become less and less detectable as the constraints are weakened (higher taxonomic levels sampled). In the lower graph, we plot the mean difference between the interspecific morphological distance of island species and the interspecific distance of their mainland counterparts (based on Characters 3 and 4). Sample sizes (number of paired pairs) range from 17 to 20. The upper graph shows the significance of one-tailed, paired *t*-tests corresponding to the mean differences in the lower graph. Data are plotted for three trees (beta, delta, and gamma), generated from different random number seeds, but using the same control probabilities (PSPN = .20, PEXT = .10, PEVOL = .80). The island samples are from Lack Archipelagoes (no vagility weighting), with each island losing five of its original twelve species by competition (PKILL = .42), as in Figure 20.2. The computer-generated classification is based on 3 nodes per taxonomic level (KUT = 3). See the text for details.

part of Figure 20.5). The statistical consequences are shown in the upper part of the figure. The null hypothesis of no difference cannot be rejected for higher levels. The cause of this effect is simply that evolution is a conservative branching process: close relatives are more similar than distant ones, and are thus more likely to be incompatible when competition is intense. Including distantly related species in a null sampling pool simply drowns out the signal with noise, progressively weakening the power of the design to detect competition. Others have pointed out this problem (Grant and Abbott, 1980), or related ones (Gilpin and Diamond, this volume), on the basis of biological arguments. We wish to emphasize that it is a general methodological difficulty, demonstrable with imaginary biotas generated under the simplest biological assumptions.

DISCUSSION

The development of neutral models in community ecology has progressed rapidly since the pioneering effort of Caswell (1976). As a result mainly of the efforts of the "Tallahassee School," stochastic models have been proposed as null hypotheses to test for the effect of deterministic features in structuring the composition of isolated biotas (*e.g.* Simberloff, 1970, 1978a, this volume; Connor and Simberloff, 1978, 1979; Strong, this volume; Wilbur and Travis, this volume) and the morphology of their component species (*e.g.* Strong *et al.*, 1979; Ricklefs and Travis, 1980; Schoener, this volume). The development of these models and their application to various aspects of community ecology are forcing a healthy re-examination of the assumptions and conclusions of much of community ecological theory, and the data believed to corroborate it, of the past 20 years. But to insure the most permanent advances, a careful examination of the null hypotheses and their use is as important as a test of their alternatives.

The approach of these studies is that of classical statistical inference. A null hypothesis is chosen, its predictions are compared to the observed data, and the magnitude of the dissimilarity between observed and predicted patterns is evaluated statistically to estimate the probability that the dissimilarity occurred purely by chance. Hypothesis testing at the level of communities differs from more familiar applications only in the difficulty of construction of the null hypothesis.

One of the most important aspects of ecological hypothesis testing and the one that has received the least attention from ecologists is the relevance of considering both Type I and Type II errors in the establishment of statistical criteria for rejection of the null hypothesis. When the investigator has no *a priori* reason to "hedge bets" toward either one of two hypotheses, every effort should be made to equalize alpha and beta and to make them both as small as circumstances allow. Unfortunately, in almost all ecological situations the alternative hypothesis is of the composite type. That is, there is an amorphous class of alternative hypotheses instead of a single, well-defined alternative. When the alternative hypotheses cannot be defined precisely, it is impossible to estimate the probability of a Type II error, beta. In the face of unknown values of beta, establishment of criteria for null hypothesis rejection on the basis of alpha alone becomes totally arbitrary. Furthermore, the custom of reducing the critical (rejection) value of alpha to very low levels unavoidably makes the probability of a Type II error correspondingly higher. In the case of ecological null hypotheses, tests of low power are often used with small sample sizes and low alpha levels (*e.g.* Strong *et al.*, 1979; Connor and McCoy, 1979; Connor and Simberloff, 1978), or probabilities of independent tests of the same (null) hypothesis are not combined (Simberloff and Connor, 1981), producing an unmeasured but potentially overwhelming bias against the rejection of false null hypotheses. Thus, the tests of deterministic community-level effects that have been conducted to date have often been biased against the rejection of the null hypothesis as a result of the way in which the actual statistical comparisons of the null hypotheses with the empirical data have been conducted. These problems are largely a shortcoming of the framework of statistical hypothesis testing, and many of them (*e.g.* the calculation of beta) may not be resolvable. The best that can be done in many situations may be the choice of a test of the highest power available and a setting of liberally high alpha values. (See also Neyman, 1977; Hodges and Lehman, 1970; Grant and Abbott, 1980; Green, 1979; and Hendrickson, 1981, for a discussion of some of these problems.)

We have designed our "null model for null models" to explore the effects of biases of a different kind, namely, the biases inherent in the design of the null hypothesis. Our results demonstrate that construction of an appropriate, unbiased, null hypothesis is very difficult. The counteracting effects of competitive elimination and correlations between vagility and morphology (the Icarus effect) can produce communities that are, with current techniques, indistinguishable from random assemblages of the component species. This problem is exacerbated by the fact that it is impossible to know the evolutionary histories of real communities—even simulated lineages are inconsistent in the degree to which they display this effect. No matter how carefully data from living species are analyzed, ecologists conducting community-level studies will always be plagued by the difficulty of confronting "the ghost of competition past" (Connell, 1980). Our model explicates the danger (the Narcissus effect) pointed out by Grant and Abbott (1980) in using an archipelago biota as the source pool for a null hypothesis implemented in the same archipelago (Strong *et al.*, 1979). The very same danger is associated with the use of real-world mainland biotas as well. In general, the choice of *any* null hypothesis for morphological studies of competition is plagued by this inability to gauge the contribution of current *and* historical competition to morphological differences among community members. Lacking a complete fossil record of the community in question, the ecologist can neither measure the magnitude of this bias nor correct for it in the analysis.

Apart from the problems associated with the construction of null hypotheses *per se*, the general strategy of community-level analysis designed to investigate interspecific processes has problems of its own. If interspecific competition occurs in a community it presumably will be strongest and hence most detectable in a small subset of ecologically similar species. Including other less similar species in an analysis runs the risk of obscuring the effects of competition. Community-level analyses are of great heuristic value in detecting community patterns (Diamond, 1975; Inger and Colwell, 1977). The patterns thus detected, however, must not be overinterpreted as proof of process. The assessment of any interspecific process such as competition will be most profitably pursued and the results most firmly established by an in-depth analysis of small groups of ecologically similar species identified in the community-level analysis (Colwell, 1979).

Acknowledgments

We are grateful to David Raup for sharing the program that does the phylogenetic bookkeeping and tree-plotting in our program GOD. We also wish to thank Ted Case, Hal Caswell, Lloyd Goldwasser, Peter Grant, David Jablonsky, Mary-Claire King, Paul Licht, the students of Zoology 244 (fall, 1979), and the participants in this symposium for discussion and assistance. This work was supported by NSF grant DEB78-12038 (to R.K.C.), and by the U. C. Berkeley Committee on Research.

Note

The programs described in this paper are available upon request from the authors. They are written in FORTRAN, heavily annotated to make modification easy.

21.

The Mechanisms of Species Interactions and Community Organization in Fish

EARL E. WERNER

Kellogg Biological Station and Department of Zoology, Michigan State University, Hickory Corners, Michigan 49060

INTRODUCTION

Ecologists have long been intrigued by patterns in species numbers and relative abundance, and much speculation has been proffered on the extent to which these patterns reveal the underlying processes responsible for community structure. In recent years, Hutchinson's (1957) reformulation of the niche concept and the suggestion of measures relevant to the field (*e.g.* Levins, 1968) have given rise to a burgeoning literature stressing the role of competition in organizing communities and describing patterns in resource partitioning among groups of coexisting species (*e.g.* reviewed by Schoener, 1974b). Despite the popularity of this approach and the great interest it generated, there is no discernible consensus among ecologists on methodology appropriate to the study of ecological communities or the structure of a relevant theory.

Controversy of the sort evidenced in this symposium is bound to develop when inferences concerning the processes underlying community organization are based on descriptive work. Many dynamics can lead to similar equilibria and therefore plausible alternative explanations accounting for any community pattern can always be generated. The major problem, however, is not that we infer various processes from an examination of community structure, but that such investigations are often considered ends in themselves, rather than preludes to studies that test the hypotheses advanced or elucidate the mechanisms postulated to be operating.

Patterns in resource partitioning among coexisting species, for instance, have generally been interpreted as indicating the role of competition in organizing ecological communities. However, the measures traditionally used to document this partitioning (*e.g.* overlap) are difficult to interpret, especially in the absence of data on limiting resources (Colwell and Futuyma, 1971). Alternatively, differences in morphological features or body size between species have been used as indices of differences in resource use. The use of such indices is obviously problematic in the absence of data on the relationship between morphology and resource use as well as resource levels. What clearly is called for is greater resolution of the mechanisms of species interactions so that theory can be elaborated that provides more explicit predictions of expected morphological differences between species or patterns in resource partitioning. If, for instance, body size and morphology were quantitatively related to foraging efficiency and then explicitly coupled to resource use in the theory, one would be able to make much more critical evaluations of the patterns seen in nature. Although some preliminary studies have been made along these lines there has been little systematic attempt to develop such a predictive framework for a given taxon or community.

In this paper, I describe an approach my colleagues and I have adopted in the course of studying fish communities. This approach integrates descriptive investigations of natural lakes with experimental studies in small semi-natural ponds and with foraging studies in the laboratory. We are thus able to take ideas suggested by the descriptive work in the field and to test them under controlled conditions in small ponds. I will first show how we have been able to demonstrate the occurrence of competition and study its action among species of sunfish through this program. I then present laboratory foraging experiments designed to explore the mechanisms underlying these competitive effects and develop a predictive framework to account for such effects. Using optimal foraging theory and estimates of foraging costs and benefits in relation to morphology and body size, accurate predictions of food and habitat use can be made for sunfish in the field. The foraging theory thus provides a descriptor or tool for predicting the behavior of organisms in the absence of information on how decisions are actually made by the animal. To the extent that the theory affords good approximations to what the fish are actually doing, it is useful for studying species interactions; ultimately, of course, these models must explicitly incorporate the mechanisms by which organisms make decisions.

Finally, I show how the above approach can be used to explore interactions between species that have size-structured populations. Using data on the relation between foraging efficiency and body size, I develop a model that enables us to predict ontogenetic niche shifts in fish. Such niche shifts are a universal occurrence in fish (as well as in many other groups) and greatly complicate the study of species interactions. I also show how we can begin to predict behavior when foraging profitability or growth and predation risk are conflicting constraints on the organism. These factors change dramatically as a fish increases in size and greatly influence niche relations and evolution in fish.

SPECIES INTERACTIONS IN THE CENTRARCHIDAE

We have intensively studied interactions among the centrarchid sunfishes, a group of spiny-rayed, warm-water fishes that dominate the fish
faunas of small lakes over much of central North America. Generally 7–10
centrarchids in four genera coexist in the lakes of southwestern Michigan;
five of these species are of the genus *Lepomis*. Adults of these species are
small to moderate in size (10 to 40 cm in length) and exhibit considerable
morphological differentiation (Keast and Webb, 1966). Most members of
the group adapt quite readily to ponds or to the laboratory and so are
excellent experimental material.

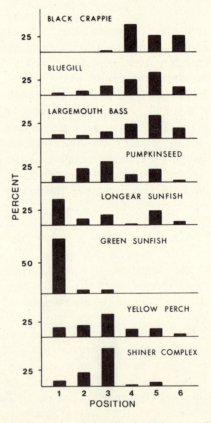

Fig. 21.1. Average midsummer relative abundance (percent of total numbers)
with respect to distance from shore for eight common species in Lawrence
Lake (Barry County), Michigan. Positions are equally spaced in the littoral
zone from <0.5 m deep (position 1) to the deepest extent of the littoral zone
(ca. 5 m, position 6). See Werner *et al.* (1977) for details.

Habitat use and relative abundances of the centrarchids can be readily quantified using SCUBA gear (Werner *et al.*, 1977; Hall and Werner, 1977). Distinct habitat partitioning is evident in the group (Figure 21.1) and food studies indicate that this habitat separation contributes to considerable resource partitioning, at least among the larger size classes (Keast, 1978). The centrarchids separate on all of the traditional axes of niche segregation; species not only separate by distance from shore (depth of water) as indicated in Figure 21.1, but also by stratification in the water column, food size, and time of day in which they feed. The pumpkinseed (*L. gibbosus*) and longear sunfish (*L. megalotis*), for instance, are closely associated with the sediments, whereas the bluegill (*L. macrochirus*) and bass (*Micropterus salmoides*) are found higher in the water column. The bluegill and bass overlap spatially (Figure 21.1) but largely separate by food size (Werner *et al.*, 1977). Further, the black crappie (*Pomoxis nigromaculatus*) migrates into the open water regions at night to feed on vertically migrating plankton, whereas most of the other centrarchids are diurnal. Indeed, upon closer analysis we even find that size classes within a species show spatial and food separation (Figure 21.2; Keast, 1977; Laughlin and Werner, 1980).

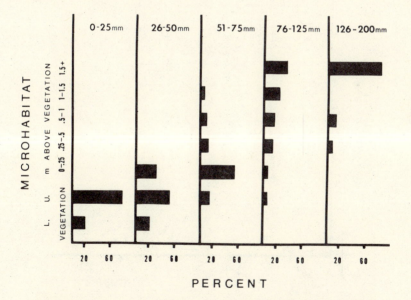

Fig. 21.2. Vertical distribution of five size classes of bluegills in Lawrence Lake, Michigan (percent of total numbers in each size class). L and U refer to the lower and upper one-half of the standing vegetation (ca 0.5 m tall). Data are the average of six transect counts in July and August.

The above patterns in resource partitioning suggest strong competitive interactions among species, and it is relatively straightforward to demonstrate that competition contributes to these patterns. Generally, when fish populations are reduced in small lakes or ponds, the fish remaining realize a dramatic increase in growth rate (*e.g.* Alm, 1946), suggesting some degree of food limitation. It is also clear that growth rates in the field are very much depressed relative to a species' capacity for growth (Figure 21.3).

In order to investigate competition among the centrarchids more directly, we introduced these fish into small experimental ponds where factors such as density, individual size, presence of congeners, and habitat structure could be manipulated. The ponds are large enough (29 m diameter, 1.8 m deep) to contain natural stands of emergent and submersed vegetation and prey populations similar to those of natural littoral zones, yet small enough to afford good experimental control and to permit the manipulation of habitat structure and fish community composition. We performed a large series of experiments with three species of the genus *Lepomis* (bluegill, pumpkinseed, and green sunfish, *L. cyanellus*) that

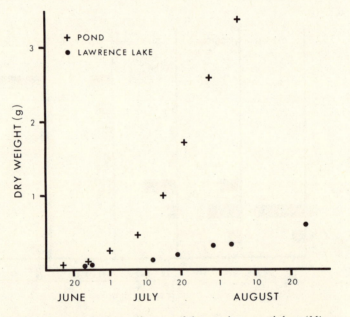

Fig. 21.3. Growth of cohorts of young-of-the-year largemouth bass (*Micropterus salmoides*) in a natural lake (Lawrence Lake, Michigan) and experimental ponds. Densities stocked in the ponds were considerably higher than those extant in the natural lake but no other species were present in the ponds.

clearly demonstrated the competitive effects of these species on each other (Werner and Hall, 1976, 1977, 1979). Each species, when stocked in ponds alone, preferred the vegetation habitat where larger prey are found and evidently higher foraging rates possible. In the presence of congeners, however, the bluegill and pumpkinseed underwent dramatic niche shifts to the plankton and sediment habitats respectively (Werner and Hall, 1976), and all species exhibited reduced growth rates. The green sunfish remained in the vegetation. Habitat use in the presence of congeners thus closely paralleled that actually seen in the field (Werner *et al.*, 1977).

Several subsequent experiments were performed to test the inference that these shifts (or lack thereof in the case of the green sunfish) were due to different abilities to harvest resources from these habitats. By either physical confinement of species to a given habitat alone and in the presence of the congener, or habitat manipulations such as removing all vegetation from a pond, we were able to contrast foraging abilities and growth rates of these species in the different habitats. Using this information, we could then ordinally rank species according to their relative foraging efficiencies in the different habitats (Werner and Hall, 1977, 1979).

These experiments enabled us to predict the temporal order of habitat shifts in a pond as resources declined in the preferred habitat. To test these predictions we chose a pond with abundant vegetation and sediment resources and little plankton of utility to the fish. We predicted that all species would initially feed in the vegetation as foraging rates would be higher there. Because the green sunfish was the most efficient of the three in the vegetation, we predicted that it would not shift habitats as resources declined. We predicted that the pumpkinseed, on the other hand, which is more efficient than the other two species in the sediment habitat, would be the first species to shift habitat use. The bluegill, which was similar in efficiency to the pumpkinseed in the vegetation habitat, but much less efficient in the sediment habitat, should shift later and be effectively "sandwiched" in between the two other species and exhibit reduced relative growth rates. The results of this experiment corroborated our predictions. The pumpkinseed and bluegill both switched habitats in the order predicted (Figure 21.4) and grew at similar rates while in the vegetation. Once the bluegill shifted to feeding predominantly from the sediments, however, its growth rate declined relative to that of the pumpkinseed. Thus the data clearly demonstrated the role of exploitative competition (Werner and Hall, 1979), and, indeed, evidence exists for similar seasonal habitat shifts in natural lakes as peak spring resource levels decline through the summer (Nilsson, 1960; Seaburg and Moyle, 1964; Laughlin, 1979). Though interference competition could be contributing to these patterns, it appears that exploitative interactions are the

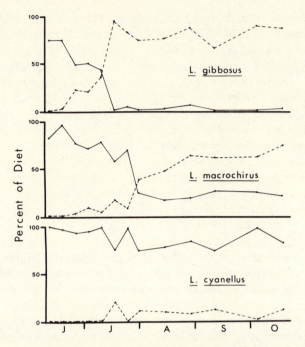

Fig. 21.4. Percent of the diet (by dry weight) comprised by vegetation prey (solid lines) and sediment prey (dashed lines) over the season for the pumpkinseed (*L. gibbosus*), bluegill (*L. macrochirus*), and green sunfish (*L. cyanellus*), respectively. Points are averages for 10 fish on each date. Plots do not total 100% because small amounts of prey were due to species that were not habitat-specific. See Werner and Hall (1979) for details.

dominant force (Werner and Hall, 1977). I will show later that these sorts of shifts can be quantitatively predicted on the basis of resource dynamics.

The pond studies have been extremely useful in demonstrating the consequences of competition among the sunfishes and providing insight into the actual mechanisms operating. Other experimental studies with fish have also shown the strong influence of interspecific competition (Svardson, 1976; Larson, 1980; Hixon, 1980). There is little doubt that the patterns we see in natural lakes are, at least in part, reflecting competition among these species.

Several problems, however, faced us at this point. In order to build mechanistic statements into the theory of species interactions it seemed obvious that we must first be able to predict quantitatively the sorts of diet differences and niche shifts that we saw in the pond experiments. I showed above how different foraging efficiencies among species were responsible for their relative habitat use and success. Thus we needed quantitative relations between morphological features and/or behaviors in the sunfishes and their foraging efficiencies. Second, body size within a species often ranges over orders of magnitude, leading to marked niche shifts during the ontogeny (see below). These shifts have far-reaching and little-understood consequences to species interactions in fish. In our experiments we used a single size class so as not to confound intra- and interspecific size-class interactions. Clearly any greater resolution on niche structure or adaptive radiation in these fish will require that our theory account for how increases in body size through the ontogeny influence success in interactions among species.

SIZE, BODY PLAN, AND FORAGING EFFICIENCY

In order to predict differences in diet and shifts in habitat, the effects of size and morphology must be incorporated in a conceptual framework that accounts for resource choice by the predator. I approached this problem by affixing costs and benefits in the foraging process to body size and morphological differences between species. Using these cost/benefit measures in the context of optimal foraging theory, predictions of predator behavior can be made. Below I first illustrate some of the tradeoffs one sees in the functional morphology of fish and show how these tradeoffs can be quantified. Then I illustrate how we quantified the effects of body size or foraging rates in one of the sunfishes. The following section shows how these data permit predictions of food and habitat use in the field.

Three species, the bluegill, green sunfish, and largemouth bass, are used to illustrate the tradeoffs inherent in their functional morphology (Figure 21.5). These species are representative of the gradient in morphologies of locally coexisting centrarchids. The bluegill's small, protrusible mouth is efficient at sucking in small prey with a current of water; the short, laterally compressed body form provides quickness and economy in maneuvering the body (Alexander, 1967). The bass, on the other hand, has a large mouth, enabling it to catch and consume larger prey but severely compromising its ability to feed on small prey; the more fusiform body gives it the speed to capture large prey. The green sunfish is intermediate in all characters.

One can confirm and measure quantitatively tradeoffs such as those inferred above from gross morphology by bringing fish into the laboratory and measuring costs associated with feeding on different prey types

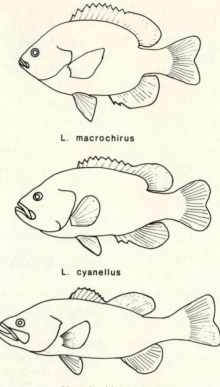

Fig. 21.5. Schematic of the body plan for the bluegill (*L. macrochirus*), the green sunfish (*L. cyanellus*), and the largemouth bass (*M. salmoides*).

(Werner, 1977). As an example, Figure 21.6 presents the time per individual prey captured when these fish were feeding on different densities of small prey (zooplankton—*Daphnia*). Clearly, efficiencies were ordered as the morphology would suggest; the bluegill was twice as efficient as the bass and the green sunfish intermediate between these two species. However, when the prey were larger (small fish—*L. cyanellus*), efficiencies were reversed and the bass was the much more adept predator. One measure of ability on larger prey is the time required to pursue and capture fish at different prey-to-predator size ratios (Figure 21.7); the bass captured prey much more quickly than the other two species in these experiments. Similarly, there were large differences in handling (swallowing) times and the maximum size of prey consumed. More extensive data on these tradeoffs and the aggregate cost curves for each species are provided in Werner (1977).

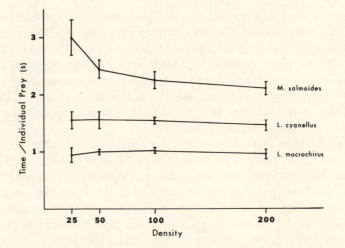

Fig. 21.6. Average time ($\pm$SE) required per capture of individual *Daphnia magna* (mean size of 3.6 $\pm$ 0.05 mm) by the largemouth bass (*M. salmoides*), green sunfish (*L. cyanellus*), and bluegill (*L macrochirus*). Density is number of prey introduced into 110-liter aquaria. See Werner (1977) for details.

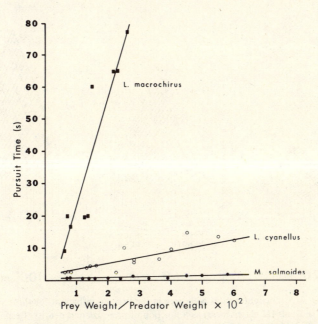

Fig. 21.7. Pursuit time required per capture of small fish by the bluegill (*L. macrochirus*), green sunfish (*L. cyanellus*), and largemouth bass (*M. salmoides*) plotted against the ratio of prey to predator weight. Regressions fit by least squares. See Werner (1977) for details.

Cost in terms of time and energy expenditure can therefore be quantified for fish feeding on different prey types by the above method. Most of these measures should reasonably reflect actual costs in the field. The difficulty with regard to the field, however, is obtaining estimates of searching costs or rates of encounter with prey. Even accurate samples of prey from the environment simply do not represent what is actually available to the fish, *i.e.* what fraction of each prey species is actually encountered by a searching predator.

In order to estimate how morphology and body size affect encounter rates with prey, we set up experiments in the laboratory to measure encounter rates as a function of prey size, prey density, fish size, and habitat type. Since the pond studies indicated that the vegetation, sediments, and open water contained different prey types and were treated as distinct habitats by the sunfish, we set up aquaria simulating the structure of these habitats and introduced prey characteristic of each environment (vegetation, damselfly naiads; sediments, midge larvae; open water, daphnids). Experiments have been performed with both the largemouth bass and the bluegill, but only Mittelbach's (1981) work on the bluegill is presented here.

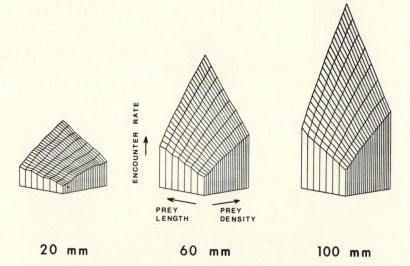

Fig. 21.8. Encounter rates with prey in the vegetation habitat for three sizes of bluegills (*L. macrochirus*) as a function of prey size and density. Response surfaces constructed from multiple regression equations fit to the laboratory data. Prey were damselfly naiads (5–12 mm in length) at densities of 10–45 prey/m³. See Mittelbach (1981) for details.

Individual bluegills were introduced into each habitat and prey encounter rates, handling times, and predator swimming speeds measured. The results of these experiments showed that prey size, prey density, and fish size all significantly affected prey encounter rates in each habitat (Figure 21.8). Using multiple regression techniques we obtained equations predicting encounter rates as a function of these variables (see Mittelbach, 1981, for details). Thus we can estimate the actual numbers and size-frequency distribution of prey encountered in any one of these habitats in the field given the size distribution and density of prey in that habitat.

I have shown then how laboratory experiments can be used to estimate the costs and benefits of foraging associated with differences in body size and species morphology. It remains to incorporate these costs and benefits into an appropriate model that can predict the optimal diet or habitat use for a fish and can be quantitatively tested in the field. In the next section, I present such a model and some data from tests of this model.

FIELD TESTS OF DIET CHOICE AND HABITAT SWITCHING

The following standard foraging model was used to predict the optimal diet. Net energy intake per unit time (E_n/T) from habitat j can be described by

$$E_n/T = \frac{\sum_{i=1}^{n} \lambda_{(ij)} E_{(ij)} - C_s}{1 + \sum_{i=1}^{n} \lambda_{(ij)} H_{(ij)}}, \tag{1}$$

where $E_{(ij)} = Ae_{(ij)} - C_h H_{(ij)}$, and A = assimilable fraction of the energetic content of prey, $e_{(ij)}$ = energetic content of prey size i (cal) found in habitat j, $H_{(ij)}$ = handling time (sec) of prey size i in habitat j, and C_h = energetic costs (cal/sec) of handling prey. C_s = energetic cost (cal/sec) of searching and $\lambda_{(ij)}$ = number of prey size i encountered per second of search in habitat j. The optimal diet is determined by addition of prey to the diet according to their profitability ranking (E_{ij}/H_{ij}) until equation (1) is maximized.

The laboratory experiments with the bluegill enabled us to estimate all the parameters in the model for a given fish size and habitat type. Encounter rates with prey and handling times were estimated directly from regressions fit to the experimental data. Caloric values of prey and assimilation rates were obtained from the literature (Mittelbach, 1981), and the energetic costs of searching for and handling prey estimated from the regressions of Wohlschlag and Juliano (1959), who measured energetics of the bluegill as a function of swimming speed, body size, and temperature.

Thus we had the opportunity to test a foraging model where all parameters were estimated independently of the field situation where the test was conducted.

We have tested predictions of this model in both a natural lake (Mittelbach, 1981) and the experimental ponds (Werner *et al.*, 1983b). In both cases prey were sampled from the three habitats and bluegills sampled from the general area immediately afterward. All samples were taken during or just after the major morning peak in bluegill feeding activity (Sarker, 1977; Wilsmann, 1979). Specifying a fish size, we estimated the parameters for equation (1) and computed the optimal diet for each habitat. The ambient prey distribution and the predicted and actual diets are given in Figure 21.9 for three sizes of bluegills feeding on plankton in the ponds. There is good correspondence between the predicted and actual diets of the fish. Furthermore, the extreme selection of larger prey exhibited by the fish is far greater than that which would result from the bias due to increased visibility of larger prey alone (Mittelbach, 1981). There was also good correspondence for larger fish feeding in the open water and vegetation of a natural lake; examples are provided by Mittelbach (1981) and Werner and Mittelbach (1981).

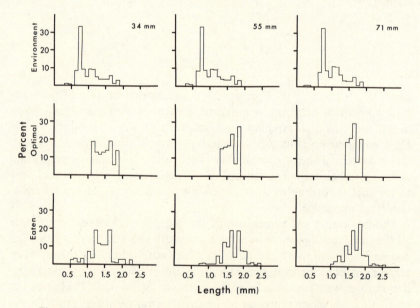

Fig. 21.9. The size-frequency distribution of *Daphnia* in the plankton of a pond, the predicted optimal diet for three size classes of bluegills (*L. macrochirus*), and the actual diet of the bluegills feeding on these *Daphnia* (average of 10 fish/size class).

If the bluegill selects prey within habitats so as to approximate maximal energy return, one might also expect to find individuals choosing to forage in those habitats yielding the maximal net energy gain. Further, because prey dynamics within habitats often have different seasonal patterns (because of extrinsic causes or competitors—see Figure 21.4), we would expect to see fish shift habitats in response to changes in relative habitat profitabilities. If we could predict such shifts on the basis of resource dynamics, it would lend considerable insight into species interactions.

To test the predictions of the model in regard to habitat use I computed the optimal diet and the associated E_n/T (profitability) for each habitat across dates in the pond experiment, yielding the data in Figure 21.9. The predicted return rates for two habitats (open water and sediments—the vegetation habitat was always lower than the open water or sediments in this experiment) are presented in the upper panels of Figure 21.10 for small and large fish, and the fraction of the diet of these fish coming from each habitat is presented in the lower panels. Both size classes of fish exhibit striking habitat shifts when the profitabilities of the open water and sediments cross in late July. Thus the fish also appear to utilize habitats so as to maximize energy return, and we have been able to predict these shifts quite accurately.

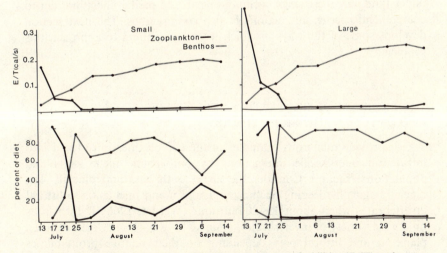

Fig. 21.10. Seasonal pattern of predicted habitat profitabilities (E_n/T) and actual habitat use (percent of diet) of two size classes of bluegills (small = 34–50 mm and large = 71–97 mm) in an experimental pond. Predictions were made for the average fish size on each date. Percent of the diet was calculated on the basis of the dry weight of prey in the diet that were unique to these two habitats. Sample sizes were 10 fish/date for each size class.

The ability to predict energy return rates on a habitat-specific basis further permitted us to discover constraints on the fish that conflict with optimal habitat use. Mittelbach (1981) found in a natural lake that large size classes of bluegills (>100 mm) conformed to the predictions of the model but that smaller fish did not. The small fish spent nearly all their time in the vegetation even when the profitability of the open water was much higher. Since these latter sizes are those vulnerable to the major predator in the system (the largemouth bass, Hall and Werner, 1977), Mittelbach hypothesized that predation risk constrains the smaller size classes to the cover of vegetation.

We subsequently subjected the predation risk hypothesis to a rigorous test in the ponds by monitoring habitat use of size-structured populations of bluegills in the presence and absence of largemouth bass. We found that the smaller size classes of bluegills spent significantly more time in the vegetation in the presence of predators even when foraging rates were as much as fourfold higher in more open habitats (Werner *et al.*, 1983a). But in the absence of predators all size classes foraged in the habitats predicted to be more profitable by the model (see Figure 21.10).

The results integrating the laboratory work with foraging theory to predict diet and habitat use in the field have been very encouraging. As noted earlier, a major incentive for doing this work was the problem of ontogenetic changes in size and related niche shifts. Ultimately we must understand why these shifts occur and be able to predict them in order to comprehend species interactions in fish communities. The next section develops some of the ways in which we are beginning to approach this problem.

THE PROBLEM OF POPULATION STRUCTURE
AND ONTOGENETIC NICHE SHIFTS

Fish provide numerous examples of the complexities introduced into intra- and interspecific interactions by ontogenetic niche shifts and population structure. Consider the largemouth bass and bluegill. The bluegill begins life feeding on the zooplankton and then shifts to littoral invertebrates when about 15–20 mm long (Werner, 1966). It then feeds on a wide variety of invertebrates, often shifting back and forth between plankton and littoral prey depending on their relative profitabilities (Mittelbach, 1981). The largemouth bass typically feeds on zooplankton until it is about 12–20 mm long, then switches to littoral invertebrates, and finally between 30 and 80 mm switches to feeding on fish. Clearly, different competitors and predators will be faced in these various stages; the bass, for instance, overlaps considerably with the bluegill in diet and habitat use during the first months of life. Further, in small ponds and reservoirs there is often an inverse relationship between the abundances of bluegills and bass (Bennett, 1954; Swingle, 1956). The reasons for this relationship are not clear, but it is possible that the bass experiences severe competition with the bluegill during its early life history when it feeds on invertebrates, and that this competition reduces recruitment to the piscivorous sizes. Once the bass negotiates the invertebrate feeding period, however, it can then feed on the bluegill. Complex interactions of this sort are the rule in fish (Larkin, 1978) and in many other communities where species populations are strongly size-structured.

Obviously many of the above problems revolve around the fact that foraging ability, predation risk, etc., are strong functions of body size in animals and that these factors condition the marked ontogenetic niche shifts. The methods we have developed using foraging theory to quantify the consequences of differences in body size and morphology suggest ways in which we might begin to explore these problems using fish. Our studies have shown that fish are extremely flexible ecologically and can switch habitats seasonally as relative resource levels change in these habitats. Because we have been able to predict these shifts, we should also be able to predict ontogenetic niche shifts that occur as relative profit-abilities of different habitats change because of size-related changes in foraging efficiencies.

If we use the laboratory data and estimate the gross energy gained from a habitat with a specified size-frequency distribution of prey for different body sizes of the bluegill, we obtain a curve that initially increases rapidly with body size but gradually begins to level off as body size increases. This curve results from the facts that (1) encounter rates increase with size (Figure 21.8) and that (2) initial increases in size enable the fish to capitalize on larger prey sizes in the habitat and handle larger prey more efficiently, but once the size is reached where all prey in the habitat can be eaten, the latter advantages of increase in size diminish rapidly. The metabolic costs of foraging increase as a power function of body length (Brett and Groves, 1979). The difference between these two curves is the net energy intake available to growth and reproduction (Figure 21.11). This net energy (E_n/T) is the quantity we compute in equation (1) and thus our laboratory work enables us to estimate the habitat-specific gain and cost curves in Figure 21.11.

Consider then a second habitat, which contains larger prey. The gain curve for this habitat will generally be displaced to the right of that for the first habitat as indicated in Figure 21.12 (for simplicity we assume that the energetic costs of foraging are similar for a given size fish in the two habitats). The differences between the gain and cost curves when plotted against body size give the net return curves for each habitat in

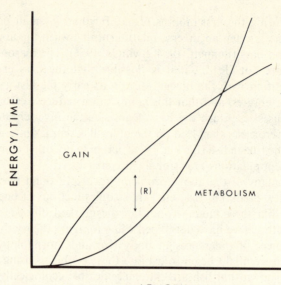

Fig. 21.11. Hypothetical curves for the relation between fish length and metabolic expenditures and foraging gain in a habitat. Curves are patterned after results of the laboratory experiments with bluegills. R is the net energy available for growth and reproduction.

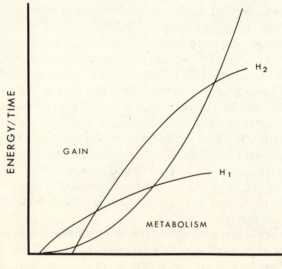

Fig. 21.12. Same as Figure 21.11 except that two habitats are compared. Habitat 2 (H_2) has larger prey than habitat 1 (H_1).

Figure 21.13. With an appropriate scalar these curves could also be converted directly to size-specific growth curves. The predicted size at which the fish should shift from habitat 1 to habitat 2 is the point where the net energy curves cross.

Given that we can define the prey distribution in various habitats, the laboratory work enables us to generate the above curves and predict the size at which ontogenetic habitat shifts should occur. This approach would also be appropriate if the shift is from one prey type to a different one, *e.g.* in the case of the bass where a certain size must be attained before it will switch from feeding on invertebrates to feeding on fish. Obviously these predictions will be affected by seasonal dynamics of prey or by differences among lakes, but the strength of the approach is that predictions can be tied directly to measurable resource dynamics. The morphologies and behaviors of different species will also affect the curves for each habitat and therefore the optimal size to shift habitats. We are currently attempting to test predictions of ontogenetic shifts with both the bluegill and largemouth bass.

The above model allows one to delimit those stages of the life history that will be spent in a given habitat or feeding on a given resource and consequently to predict when different suites of competitors or predators will be important. We hope this formulation will then enable us to identify critical periods or bottlenecks in the life history and to explore the

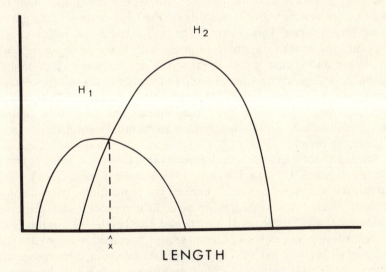

Fig. 21.13. The net return (R) for the two habitats depicted in Figure 21.12. R is calculated as the difference between the gain and metabolism curves. The optimal size at which to shift habitats is denoted by $\hat{x}$.

population-dynamic consequences of novel selection pressures or interactions that occur during a given period. One can assess, for instance, how
the introduction of a competitor that lowers resource levels in one habitat affects a species. The competitor will lower the net energy return curve
for that habitat, obviously moving the optimal switch point to a smaller
or larger size depending on which habitat is affected (*cf.* Figure 21.13).
Furthermore, lowering the net return curve in a habitat also determines
a new growth trajectory through the life history. These changes can have
large effects on life history characteristics and population dynamics. The
time to size of reproductive maturity is lengthened and overall mortality
will be greater as longer times are spent in smaller sizes classes where
the fish are more vulnerable to predators and other mortality factors. In
fact, the major influence of competitors on one another may be through
growth rates, which in turn determine mortality rates by the above mechanism. If the net energy curves are scaled to size-specific growth rates
and we know mortality rates as a function of size, simple models can be
constructed to give the change in mortality on a cohort with changes in
growth rate and to explore changes in population dynamics (Gilliam, in
prep.).

Thus far I have assumed that mortality rates are a monotonically
declining function of size not affected by habitat type. Habitats, however,
generally differ in structure or the types of predators present and therefore
differ in mortality risk as well. If predation risk is much higher in habitat
1 (Figure 21.13) for instance, then the fish should switch into habitat 2
at a smaller size than predicted solely on the basis of the specific growth
or net return curves. This change in the time of the switch will, of course,
also alter the growth trajectory, *i.e.* areas with lower predation risk will
be used at the expense of growth rates. The optimal size to shift habitats
in this case depends on how growth and mortality during these stages
contribute to overall fitness. This growth-mortality tradeoff is very
important and we are now developing models that predict the optimal
switch points based on the contribution to eventual reproductive success
(Gilliam, in prep.).

Since mortality, fecundity, and growth rates are all functions of size in
fish (*e.g.* Bagenal, 1978; Larkin *et al.*, 1956; Shepherd and Cushing, 1980),
we can use the above models to explore the population-dynamic consequences of life history changes or different competitive and predatory
regimes that affect a species at a particular stage in the life history. These
models allow us not only to examine species interactions among the fish,
but also to begin to ask evolutionary questions about convergence and
divergence in body form or morphology when species invade a system. If
we can characterize quantitatively the tradeoffs in morphological features
such as those illustrated earlier with the largemouth bass, green sunfish,

and bluegill, we can then ask how a given change in morphology will affect population growth rates or relative success and postulate what direction evolution of the body plan might take. Analyses along these lines should shed light on the nature of morphological compromises realized in species that utilize very different niches during the life history, and consequently on the overall morphological structure of the communities these species form.

CONCLUSIONS

In the preceding studies I took ideas on species interactions generated from descriptive work in the field and rigorously tested these ideas under controlled conditions in ponds. Results from both the field and ponds suggested laboratory work that, guided by the constructs of foraging theory, enabled us to develop some of the relations underlying the mechanisms of species interactions. In turn, the relationships developed in the laboratory were used to forecast behavior in the field. These predictions were extremely useful in discovering other constraints that the fish were responding to as well as illustrating the potential predictive capabilities of the foraging theory. The interplay of these various approaches has proved very productive and allowed us to capitalize on the strengths of each approach.

I have shown that competition can be an important force in organizing the fish communities of small lakes. Species clearly alter their behavior in the presence of closely related forms such that the utilization of similar resources is reduced. The fish are manifestly capable of assessing the relative profitability of different habitats and behave so as to feed in the more profitable areas. Thus habitat switching is a common and predictable event. Because body size and species' morphologies affect relative foraging efficiency in different habitats, species differentially affect resource levels in different habitats and thus can be expected to behave differently as relative resource levels change. As a consequence, species tend to separate into habitats where they are respectively "better" as resource levels decline, even though all may prefer the same habitat at higher resource levels; to this extent the data conform to the niche compression model.

We have isolated fish in the ponds to study competition experimentally, but it is obvious that competition is not the whole story in the field. Predation, for instance, interacts with competition in a number of important ways. A case in point concerns the spatial distribution of bluegill size classes presented in Figure 21.2. Keast (1977) and others have argued from this pattern that size-class segregation is an evolutionary response to intraspecific competition. The predictions of optimal habitat use in the field, however, indicated that all size classes should be in the same habitats.

These results led to the hypothesis that the spatial segregation of size classes was instead due to a gradient in predation risk. Pond experiments confirmed that in the presence of predators smaller fish remained in the vegetation at the expense of growth rates, which were higher in more open habitats. The resulting *de facto* resource partitioning among size classes has important consequences for species population dynamics (Mittelbach, 1981), but is strictly coincidental to competition. This scenario is a powerful reminder of the problems inherent in drawing inferences from descriptive studies of pattern.

Predation and competition in fish communities also may strongly interact in regulating population sizes. Competition greatly affects growth rates of species and is especially important at smaller sizes (*e.g.* Figure 21.3). Smaller sizes experience greater mortality rates than larger ones, presumably primarily because of predation. Thus reduced growth rates can result in a much greater overall mortality to a cohort as longer time is spent in smaller, more vulnerable stages.

Because predation risk and foraging abilities change with ontogenetic increases in body size in fish, life histories can become very complex, and interactions with other species convoluted. This situation contrasts sharply with that in groups such as birds and mammals, which have served as the basic models for theory in community ecology, and on which most of the data have been collected. Populations of these species can reasonably be considered monomorphic, as juveniles enter the competitive arena at approximately adult sizes. Consequently, little formal attention has been given to interactions among species with strongly size-structured populations. The majority of species, however, more closely approximate the situation with fish, where size-related changes in predation risk and/or foraging abilities can precipitate marked ontogenetic niche shifts, or in extreme cases, metamorphosis. Certainly in fish many of the ontogenetic shifts in resource use are far more dramatic than niche differences that separate different species of birds and mammals.

Our understanding of community structure in most taxa will remain primitive until we characterize the interactions among species with structured populations and complex life histories. Much of what might appear to be a lack of pattern in community structure may simply be alternative configurations resulting from the complex life histories of the interacting species. A major problem with using current theory as a guide is that predictions are so general or qualitative that we know only vaguely how patterns in resource partitioning should differ in different environments or in different taxa.

The point has been made many times before that some of the above problems would be alleviated if more explicit mechanisms of species interactions and relations with resources were embedded in the theory (*e.g.*

Schoener, 1976; Abrams, 1980). The question, of course, is how to proceed. I have attempted to use foraging theory as a framework for developing more mechanistic models of the exploitative interactions among species. This approach has afforded reasonable predictions of food and habitat use under conditions of changing resources. Foraging theory provides a general framework for linking the changes in resource utilization of consumers to the dynamics of multiple resource sets. The lack of such a connection, of course, has long plagued competition studies. Further, the cost/benefit framework enables us to examine quantitatively the trade-offs (or lack thereof) in various behavioral or morphological attributes of species that are considered central to interactions among these species. Bill or body size ratios among coexisting species would obviously mean a great deal more to us if we knew their quantitative import for foraging efficiency. In addition, the flexibility of foraging theory is such that it can be used as a systematic approach to discovering additional constraints under which organisms operate. The above studies are only tentative beginnings and we do not expect the simplistic optimal foraging theory to provide more than an approximation to choice behavior in animals (Werner and Mittelbach, 1981). The predictions of the theory, however appear sufficiently accurate that they can be used to explore certain aspects of the interactions among the sunfish, and as a result will, we hope, provide a framework for incorporating more mechanistic statements into community theory.

Some of the problems inherent in current approaches to community ecology are clearly evident from the proceedings of this symposium. Community ecologists in general tend to be stereotyped in their approach to their systems. Both descriptive/comparative and experimental studies are important and should be exploited for their respective strengths, but both should be directed as part of a general program aimed at contributing to the understanding of mechanisms that will enable us to build greater realism into the theory. That we need to be more critical of the sort of evidence we accept in community ecology is clear; major advances will not be made, however, through arguments as to whether descriptive patterns in community structure fit our expectations based on current competition theory. There is not yet enough biology in competition theory to generate discriminating expectations. A concerted effort to focus our studies on building such a theory would circumvent a lot of the controversy now apparent in community ecology.

Acknowledgments

I wish to thank Patricia Werner, James Gilliam, David Hart, and Deborah Goldberg for critical comments on the manuscript and Laura

Riley for help with the logistics. Much of the work presented here has been a collaborative effort with Donald Hall and, more recently, Gary Mittelbach and James Gilliam. It is a pleasure for me to acknowledge these colleagues. The National Science Foundation has generously supported this work for nearly a decade. This is also contribution number 437 of the W. K. Kellogg Biological Station.

22

Patterns of Flowering Phenologies: Testability and Causal Inference Using a Random Model

BEVERLY J. RATHCKE

Division of Biological Science, University of Michigan, Ann Arbor, Michigan 48109

Kuhn (1970) notes that a concern about philosophy and logic often arises when a science enters a state of controversy over the failing of past theory and the appearance of new approaches. This symposium suggests the time is propitious for such a concern in community ecology.

"To do science is to search for repeated patterns." This quote from MacArthur (1972) captures the essence of a common approach taken in community ecology. By examining patterns such as species distributions, we hope to unveil the processes that cause them. Although unequivocal evidence for causal processes will come only through experimentation, ecologists often examine patterns to guide their research, and in some systems, experimentation may be an impossibility. To gain evidence for the influence of past competition on species adaptations, such as character convergence, we are restricted to interpreting current patterns. The past cause is forever inaccessible to us. But we can ask whether certain patterns do occur, and if they do, their existence will demand explanation and causal inference. Therefore, it is important that we formulate our concepts about patterns into testable hypotheses and that we understand the rules of causal inference in examining our results. In order to discuss these aspects of what I term the pattern-process problem, I present an example of testing and interpreting patterns of flowering phenologies exhibited by plant communities.

In community ecology competition has been invoked as an explanation for patterns to such a degree that it is in danger of becoming a panchreston, a concept that can explain everything (Hardin, 1956). When a theory can explain everything, it loses its power and value because it is removed from the possibility of falsification (Popper, 1959; Brady, 1979). We must submit our theories to tests. Unfortunately, a gap always exists between the language used in discussing theory and the language necessary for use in research (Blalock, 1964). To formulate a hypothesis that can be tested in a research program, we necessarily must operationalize our theory and hence limit it. We can discuss the concept of interspecific competition as a structuring force in communities, but we must test for specific competitive effects. This is true even when we examine current competition through experiments. We cannot see the cause; we can only observe effects and assume a certain cause-effect relationship (Bunge, 1979). The actual existence of competition as a mechanism is often not under test. By specifying an effect that we will accept as evidence for a cause, we limit the possible outcomes for some causal mechanism. We lose generality to gain testability. We may loosen this restriction by considering alternative effects as evidence, but rejection or support of any one specific hypothesis will not usually allow us to reject or accept the theory of competition. Only by continually trying to falsify many hypotheses can we increase or lose confidence in our theory. In community ecology misunderstanding has developed over this limitation. Here I will test a specific hypothesis about how competition might structure flowering phenologies and explicitly discuss the relevance of the results to more general concepts about competition.

The most effective approach for testing hypotheses is to use strong inference, whereby alternative hypotheses are proposed and critical tests are made to distinguish among them (Platt, 1964). Recently, alternative hypotheses to competition have been provided by random models that can generate null hypotheses (*e.g.* Strong, 1980). We can ask, "If no interactions were to occur, what pattern would we observe?" This approach allows us to follow strong inference. Unfortunately, controversy has arisen from the misconception that random modelers are proclaiming that nature is random and only chaos prevails. I hope to diminish this misunderstanding by discussing the application and interpretation of one random model that has been proposed for testing phenological patterns (Poole and Rathcke, 1979).

Flowering phenologies provide a good basis for discussion about causal inference from ecological patterns because competition has often been inferred from the sequential flowering phenologies seen in diverse plant communities (Robertson, 1895; Mosquin, 1971; Frankie *et al.*, 1974; Gentry, 1974; Heithaus, 1974; Reader, 1975; Heinrich, 1976; Stiles, 1977, 1978; Feinsinger, 1978) and recently random models have been applied to phenological patterns (Thomson, 1978a; Parrish and Bazzaz, 1979; Poole and Rathcke, 1979; Anderson and Schelfhout, 1980; Pleasants, 1980; Cole, 1981; Rabinowitz *et al.*, 1981). Here I present a specific example from my own research on the flowering phenologies of shrub species in The Great Swamp in Rhode Island. Because I wish to stress the logical problems inherent in causal inference, I have had to limit discussion of biological details, although these are of great importance and will be reported elsewhere (Rathcke, in prep.). Also, I analyze only a single random model because one can quickly become overwhelmed by the relative merits of methodological variations and mathematical refinements. Specifically, I will address the following questions: (1) How can the predicted and observed patterns be quantified and tested? (2) What is an appropriate null hypothesis? (3) What can and what cannot be inferred from the results of the tests? (4) Where do the results from random models lead us and have they been worthwhile in helping us understand the biology of plant-pollinator systems?

OPERATIONALIZING CONCEPTS

Plants may compete for pollination by attracting pollinators away from one another or through interspecific pollen transfer that results in gametic wastage. In evolutionary time plants might avoid competition through specialization, independence from pollinators, or displacements in flowering times (Levin, 1978). In The Great Swamp shrub species depend upon insects for pollination and commonly share pollinator species (Rathcke, in prep.). Therefore, displacements in flowering time could be a common mode of competitive avoidance among these species and I expected to document the typical staggered flowering pattern that had been interpreted as reflecting the outcome of past competition for pollination. But how should a staggered flowering pattern be defined, and how can this concept be made testable?

A plot of the flowering dates of shrub species in The Great Swamp shows a staggered pattern of flowering throughout the season (Figure 22.1), but what would a pattern of randomly dispersed phenologies look like? A plot of 15 phenologies placed randomly and independently along the time axis also shows this staggered pattern (Figure 22.2). The appearance of organized structure is inevitable given the convention of arranging the species from first to last. Some objective criteria and tests are necessary to distinguish any competitive pattern from one that can be generated by random placement of species. The concept that competition should promote divergence can be operationalized by proposing that the pattern should be non-random and regular (Heithaus, 1974; Schoener, 1974b; Stiles, 1978) and this can be posed as a statistically testable hypothesis: if past competition has promoted divergence, then the flowering times should show a regular dispersion along along the time axis. This hypothesis can be discriminated from an alternative, null hypothesis that the dispersion is not significantly different from a pattern generated by random and independent placement of species. This competition hypothesis specifies and hence limits the possible competitive patterns, but it is testable and subject to falsification.

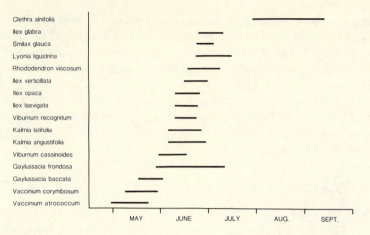

Fig. 22.1. Flowering phenologies in The Great Swamp, Rhode Island, 1977. Dates of first and last flowering are shown.

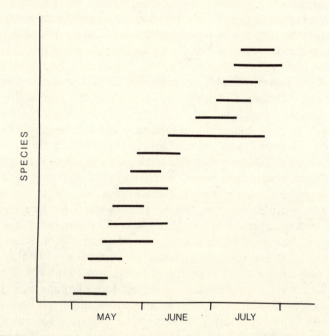

Fig. 22.2. Flowering phenologies in a random swamp. The midpoints of flowering for fifteen shrub species are placed randomly and independently along the time axis.

The degree of dispersion can be statistically tested by comparison of the observed variance of the distances between flowering midpoints of adjacent species with an expected variance based upon randomly and independently placed midpoints (Poole and Rathcke, 1979). If the ratio between the observed and expected variances does not differ significantly from one, the dispersion is not distinguishable from random and the null hypothesis is supported. If the ratio is significantly less than one, the midpoints tend to be regularly dispersed and the competition hypothesis is supported. If the ratio is greater than one, the midpoints tend to be clumped and both the null and competition hypotheses must be rejected. The significance can be tested by chi-square or F distributions and used to judge the fit of the observed pattern to the predicted random pattern.

TESTS AND CAUSAL INFERENCE

A test of the dispersion pattern of the flowering phenologies observed in The Great Swamp shows that the midpoints of the phenologies are significantly clumped or aggregated (Table 22.1); both the competition and the null hypotheses must be rejected, at least in their proposed forms. However, one species, *Clethra alnifolia*, flowered late after the other species and its inclusion increases the degree of statistical clumping. When this last species is excluded and the end of the season set by the penultimate species, the flowering dispersion is not significantly different from random, and both the competition and aggregation hypotheses must be rejected and the null hypothesis is supported.

Given the results from these tests, what can be said and what cannot be said about competition as a causal process? Consider the logical

Table 22.1. Statistical tests of the dispersion of flowering phenologies in The Great Swamp, Rhode Island, 1977.

A. Season: 30 April to 12 September = 135 days
    Species = 16
    Expected (P) = .003
    Observed P = .009
    χ^2 = 47.48, $P < .001$, highly significant

B. Season: 30 April to 6 July = 67 days
    Species = 15
    Expected (P) = .0034
    Observed P = .0030
    χ^2 = 13.24, $P > .50$, not significant

structure of hypotheses: although a true hypothesis must lead to a true prediction, a false hypothesis can lead to either a false prediction or a true prediction. In other words, our observed results may fit our prediction but for the wrong reason (Dayton, 1973). Because many hypotheses can predict the same results, we can never prove hypotheses; we can only support or reject them. Because logically a true hypothesis can never give rise to a false prediction, refutation of a hypothesis is logically more certain than acceptance (Popper, 1959). In reality, because our results are often not clear-cut, we can make errors of either rejecting a true null hypothesis (Type I error) or failing to reject a false null hypothesis (Type II error). Therefore, we must regard either support or rejection as tentative and continue to refine and test further hypotheses generated from our initial premise until we gain or lose confidence in our theory.

Applying these logical rules to the specific example of flowering phenologies, we can reject the competition hypothesis, but we can only conclude that competition has not acted in the manner postulated in this specific hypothesis. We cannot conclude that past or present competition is necessarily absent or unimportant. Competitive selection could have produced other adaptations for avoidance, or it may be constrained by other factors. Competition may be acting in the present, and in the past it may have been too mild, rare, or inconstant and nondirectional to produce competitive divergence. All of these possibilities can be reformulated into testable hypotheses and examined. Such negative findings can force us to reconsider our assumptions, to generate new hypotheses, and to acknowledge the limits of our generalizations. Viewed in this way, falsification is a rewarding process. Rejection of the competition hypothesis can lead to a wealth of new hypotheses and new observations and experiments.

What can be inferred from failure to reject a null hypothesis of a random dispersion of flowering time? The same arguments hold. We cannot conclude that the underlying processes are therefore stochastic (Connor and Simberloff, 1979). Deterministic processes could be acting without generating a pattern distinguishable from random under this model. A single system may require the use of stochastic treatment at one level and deterministic treatment at another (Nagel, 1961). For instance, the action of single gas molecules can be predicted by deterministic Newtonian mechanics, but for predicting gas pressure, this method is hardly feasible and instead a probabilistic average is calculated. A stochastic approach does not deny deterministic processes (Strong et al., 1979). Whether or not stochastic processes are important in nature may always be elusive. Given our errors of measurement in field ecology and the action of so many factors, we may never be able to determine the extent of stochasticity and determinism in natural systems (see Blalock, 1964).

What if the flowering times had shown a regular dispersion pattern? We would accept the competition hypothesis, but we should do so tentatively, for the same cautions apply. Other processes could possibly produce a regular dispersion, and alternative explanations such as selection from fruit dispersers or seed predators must be considered and eliminated in turn. If we truly attempt to falsify a hypothesis and if the hypothesis remains supported under numerous tests, the resulting evidence is the strongest we can garner (Popper, 1959). Unfortunately, a tendency exists to believe in supporting results, to stop searching further for alternative explanations, and to accept our initial assumptions. This tendency can lead to an unjustified sense of security in our hypotheses. On the other hand, we tend to ignore results that do not support our favored prediction. This practice leads to a biased view of how natural systems behave because we must know where certain predictions do not hold. To guard against the tendency to shelter a hypothesis from falsification or to practice advocacy science, whereby one selectively presents only the supporting evidence (Wilson, 1975), alternative hypotheses can be proposed (Chamberlain, 1897). We are thus prevented from having vested interests in single hypotheses and also freed from the single-factor approach, which must be naive for complex ecological systems. Proposal of a null hypothesis from a random model is a significant step in this direction.

CAVEATS FOR THE RANDOM MODELER

Perhaps the greatest difficulty for the interpretation of these random models is that many random models with different predictions can be generated for any problem. This logical dilemma exists for any model or hypothesis or concept. Even a random model is based upon a number of assumptions that may or may not be reasonable, and these assumptions will determine the prediction. It is imperative that we clearly state the assumptions of our model and appreciate their limitations when we apply the model and interpret the results. As an example, I offer the following caveats for the relatively simple random model presented here.

1. The resource axis (time for flowering) was subjectively defined, and different durations yield different results. Obviously, if I had chosen the entire year, I would have found that flowering was aggregated, but this would not have been a particularly interesting result for a temperate plant community. Instead, I asked how the species are dispersed within the observed flowering season, and I tried to make my choices biologically reasonable by using the dates of first and last flowering as the endpoints and by eliminating one species for a second analysis. My doing so was

an attempt to choose the most relevant time period for an examination of temporal displacements that could potentially arise as a result of interactions among species. Statistical problems also arise from these relatively straightforward and simple conventions. Setting the endpoints by the first and last species in the observed system and allowing these to vary in the random model may bias the procedure toward calculating that the observed dispersion is regular (Cole, 1981; Thomson, pers. comm.). Because this bias increases the possibility of rejecting a true null hypothesis (Type I error) and accepting the competition hypothesis, confidence in rejecting the competition hypothesis is increased.

2. A uniform distribution of resources is assumed in this model, *i.e.* every point in time along the axis is considered to be equally favorable for flowering. This is hardly likely to be true for most systems. To make the test more realistic, unfavorable times for flowering, such as dry periods in some tropical habitats, could be excluded from the time axis (Poole and Rathcke, 1979; Stiles, 1979; Cole, 1981), but favorability is also likely to change slowly throughout the season and will be difficult to quantify. One could assume that favorability shows a normal distribution and generate a random model based upon that, but the complexity quickly increases. I have taken the view that if the ultimate factor of competition were strong enough, plants could physiologically shift within the time axis I used and proximate factors determining flowering times could change. Attempts to include pollinator availability in quantifying the resource distribution leave one with the chicken-or-the-egg paradox. For most systems, determining resource availability and favorability may prove to be intractable (Connell, 1980). The uniform distribution chosen here does provide a relatively simple index that can be easily applied and used for comparisons, but the choice may be justified more on pragmatic than on theoretical grounds.

3. Any general test of a community-wide pattern is unlikely to detect specific cases of divergence embedded within it (Strong *et al.*, 1979). Rejection of a regular dispersion pattern would not preclude the possibility that competitive divergence is shown by specific subsets of species. However, for small subsets, random models are not statistically applicable, and each case must be interpreted individually. It should be remembered, however, that the original impetus for such tests was the search for community-wide patterns, and these do demand testing.

4. A critical decision in any community study is how to define the boundaries so as to include the important interacting species and exclude others. What subset of species should be considered? For tests of competitive patterns of flowering, obviously only species that share pollinators should be included, but this choice is problematical because one often tests a pattern postulated to be produced by past competition by examin-

ing pollinator sharing in the current community. If flowering times shifted, pollinator sharing probably changed as well. In The Great Swamp the predominance of generalized pollinators and generalized flowers suggests that most plant species could be potential competitors, and I have used this suggestion as a justification for examining dispersion among this group of species. The increasing number of studies that show competition among unrelated taxa adds to the difficulty of defining the relevant species (*e.g.* Kodric-Brown and Brown, 1979).

Related species may be the strongest competitors and displacements should also be examined within these subsets. However, in this case selection for reproductive isolation may also be influential and other studies indicate that this factor has been more likely to produce divergence than has resource competition (Grant, 1975). Success in separating the contributions of these two mechanisms to displacements among closely related flowering species seems highly improbable.

5. The dispersion of flowering midpoints used in this test does not tell us anything about the avoidance of overlap that is certainly a most relevant parameter for competition. The simple test used here could also be used in conjunction with tests of expected vs. observed overlaps among species (Parrish and Bazzaz, 1979; Thomson, 1978a; Rathcke and Poole, 1979; Cole, 1981; Rabinowitz *et al.*, 1981). However, the use of overlaps increases the number of assumptions used in generating a random model, and interpretations must be carefully considered. For example, if few species were present relative to the time axis, a random dispersion could show no or low overlap. In a diverse community regularly dispersed species could show very high overlaps and little avoidance of competition (Cole, 1981). In fact, if overlap is high, it is quite unclear why species should be regularly dispersed. A species that overlaps completely with one other species and a species that overlaps partially with several species could show the same amount of overlap, but they would certainly experience different ecological consequences, and the same result could demand very different explanations.

6. Is the random and independent dispersion of flowering phenologies the most reasonable non-interactive pattern? One could argue that a non-interactive pattern should be one of coincidence of flowering and that any displacement reflects past competition. Although this pattern seems unlikely for unrelated species, congeners often show the same flowering times, and coincidence may be the more reasonable null hypothesis.

Such a long list of assumptions and their limitations to the model may seem discouraging, but any model has such assumptions and they must be scrutinized. Because of the resistance to random models, these seem to have come under more criticism than other models that perhaps have had more appeal. Perhaps less obvious is that our world view is based upon assumptions that may go unrecognized if we are not forced to confront them and admit to their limitations (Krebs, 1980).

QUO VADIS?

Where do the results from this random model lead us, and has it been valuable in helping us understand the biology of plant-pollinator systems? Despite all the caveats, I think this approach has been valuable in requiring a definition of patterns, by allowing an objective test, and by forcing us to consider the assumptions in the premise. In hindsight, the competition hypothesis predicting regular dispersion of flowering times may seem naive, but it was widely accepted as a reasonable hypothesis until anomalous results appeared. The concept that staggered flowering times demonstrated the importance of past competition was easily supported by the data. Based on the use of tests generated by random models, aggregated flowering has been reported for prairies (Parrish and Bazzaz, 1979; Anderson and Schelfhout, 1980), subalpine meadows (Thomson, pers. comm.) and rain forest *Heliconia* (Poole and Rathcke, 1979); a random pattern has been reported for a Missouri prairie (Rabinowitz *et al.*, 1981); and displaced flowering within plant guilds has been reported for a subalpine meadow (Pleasants, 1980) and in a reanalysis of *Heliconia* (Cole, 1981). These results certainly argue against a monolithic view of repeatable community-wide patterns, and they call for alternative explanations. Finding that flowering times are often aggregated has promoted the consideration of positive interactions among plants (Thomson, 1978b, 1981; Waser and Real, 1979; Rathcke, in prep.) and the possibility that mutualisms may have selected for convergence (Brown and Kodric-Brown, 1979). Consideration of these possibilities has certainly had a positive effect on pollination research.

I propose that the results from tests provided by random models have compelled the development of a more sophisticated view of the complexity and diversity of interactions among plants for pollination and that their use has emphasized the need for stronger evidence. I think the same is true for other studies. In pollination studies emphasis is turning to experimentation (*e.g.* Waser, 1978) and to more detailed investigations of how pollinator visitation may change with neighbors, relative abundances, and other visitors (*e.g.* Inouye, 1978; Thomson, 1978b, 1980; Rathcke, in prep.) and how variable these interactions are year-to-year and site-to-site (*e.g.* Wiens, 1977a).

Although we might have changed our views without random models, I feel their introduction has significantly accelerated this development. The most predictable response to the rejection of this specific competition hypothesis on flowering dispersion in The Great Swamp has been a chorus of alternative explanations. Testing these is proving to be a most interesting and valuable outcome of the application of this random model.

Acknowledgments

I would like to thank Tom Getty, Deborah Rabinowitz, Dick Root, James Thomson, and Don Strong for their stimulating discussions and to express my appreciation and debt to many others who in disagreeing or agreeing have helped me formalize my thoughts. This research was supported by National Science Foundation Grant G-DEB78-24678.

Food Web Design

23.

Food Chains and Return Times

STUART L. PIMM

Graduate Program in Ecology, University of Tennessee, Knoxville, Tennessee 37996

INTRODUCTION

Following MacFadyen's (1963) review of the many possible meanings of the word "community" one might have expected some consensus. Yet, what followed was the addition of a qualifier to the term (*e.g.* bird communities) and a novel meaning. In the words of McNaughton and Wolf (1973): "community is usually applied to organisms with similar life habits," and, I might add, to species that are usually on the same trophic level. Most of the papers in this volume reflect this newer definition; mine does not. I shall consider what might be called a "vertical" property of communities (one that runs across trophic levels) as opposed to a "horizontal" property (one that runs along trophic levels). Beyond this contrast, the philosophies are much the same. I, too, am concerned with the existence of patterns within communities and the processes, if any, that cause them.

The catalogue of vertical community patterns contains a dozen or so food web features (Cohen, 1978; Pimm, 1980a, b; Pimm and Lawton, 1978, 1980; Yodzis, 1980). One process is capable of explaining many of them, and currently, some features are explicable only by this process. This process is the loss of species from systems that lack dynamically stable equilibrium densities; only those systems with stable equilibria are expected to persist long. Now, there is not the space to defend each feature in the catalogue against all rival hypotheses; I do this elsewhere (Pimm, 1982). This paper considers only one feature in detail: the length of food chains. There are many hypotheses for the limited number, usually three or four, of trophic levels observed in real systems (Pimm, 1982). Moreover, the dynamical hypothesis has been around long enough (Pimm and Lawton, 1977) to attract criticism. To reply to these criticisms is the objective of my paper.

THE LENGTH OF FOOD CHAINS

There are four major explanations for the short length of food chains (Pimm, 1982):

Energy flow. Of the energy consumed by a species, only a fraction goes to make new animals. Most of it is dissipated as heat and some of it is not assimilated by the animal. The proportion of the energy consumed that is available to the species' predators is called the ecological efficiency. It is small: on the order of 10%. As energy is transferred along a food chain, that available to any species becomes drastically reduced. Thus, goes the argument, there comes a point where there is insufficient energy to support another trophic level.

Elsewhere, I review the patterns of primary productivity and ecological efficiencies (Pimm, 1982). Primary productivities range over several orders of magnitude. Ecological efficiencies vary from less than 3% in endotherms to nearly 60% in insects. These observations suggest a number of testable hypotheses.

(a) The greater the primary productivity, the longer the food chains could be. Data from tundra and grassland areas show that this is not the case (Pimm, 1982).

(b) There should be some threshold level of primary productivity below which a second and third trophic level cannot be supported. An analysis of the standing crop of the third trophic level against primary productivity shows that at very low levels (< 50 g dry weight m^{-2} yr^{-1}) a third trophic level may be impossible (Pimm, 1982). These levels are at the extreme lower end of primary productivities recorded for natural systems, but the results do confirm that, in the extreme, energy flow imposes limits on the lengths of food chains.

(c) Because, for a given primary productivity, more energy is available to the consumers of insects than of ectothermic vertebrates than of endothermic vertebrates, food chains should be longer in the first of these cases than in the last. I have shown that they are not (Pimm, 1982).

I conclude that energy cannot be the sole factor in limiting the length of food chains.

Size and other design constraints. A Gyrfalcon (*Falco rusticolis*) may be at the end of a food chain because of the physical impossibility of a predator winged and fast enough to catch the bird and large enough to subdue it. Though appealing, such arguments are *ad hoc* and difficult to test. Moreover, the world is full of "impossible" beasts. Examples include pterodactyls nearly an order of magnitude larger than the largest modern birds and small hawks capable of catching such highly maneuverable birds as hummingbirds.

The increased size that accompanies feeding at higher trophic levels means increased energy requirements. These increase approximately as the square of the body length. But increased size also has its benefits. Larger animals feed over larger areas than smaller ones. Energy gain is proportional to an animal's home range, so it, too, increases as the square of the animal's length (Southwood, 1978). The cost to benefit ratio does not inevitably increase with trophic position and body size.

Optimal foraging. Because of the patterns of energy transfer, there is potentially more energy available at lower trophic levels than at higher ones. Food chains may be short because of evolutionary trends for animals to feed at the lowest trophic level consistent with dietary requirements (Hutchinson, 1959; Hastings and Conrad, 1979). Yet, for a particular species, the potentially greater energy available at lower trophic levels may be balanced by the greater number of competitors for that energy as well as increased susceptibility to predators. There are reasons for feeding low in the food chains, but there are also reasons for feeding high in them. Consequently, the limitation in the number of trophic levels may be explained by a balance between these opposing processes.

Dynamics. There are two dynamical arguments. Both involve the idea that, as food chains become longer, the population parameters consistent with stability become more and more restricted. Suppose that population parameters depend on a wide variety of factors that vary over time and space. This dependence implies a limit to how finely the parameters of a system can be "tuned." Thus, as a set of parameters consistent with stability becomes smaller, the system becomes less and less likely to be observed.

For systems best modeled by difference equations—typically short-lived species in seasonal environments—increasing food chain length implies decreasing chances of stability (Beddington and Hammond, 1977). For systems best modeled by differential equations—typically long-lived species in relatively aseasonal environments—the dynamical arguments involve the constraints on a parameter known as the system's return time.

Return time is a measure of how long it takes perturbations in the species' densities to disappear following the disturbance to the species' equilibrium. On average, long food chains have long return times (Pimm and Lawton, 1977). Following May (1973), Lawton and I argued that systems with long return times could not persist in the real world. This limitation imposes a dynamical constraint on return times (discussed in the next section) and, therefore, on the number of trophic levels.

The problem with this argument is that it is not the only argument based on dynamics. There are two ways of modeling stochastic perturbations to species densities. They give diametrically opposite predictions about the lengths of food chains to be expected in the real world, and predict, as a consequence, that either very long or very short food chains are the most likely to persist. Clearly, if the former is true, then dynamics cannot be held responsible for the short food chains actually observed. In this paper I shall concentrate on an examination of these two alternative ways of modeling stochastic perturbations.

Stability in Stochastic Environments

Two Alternatives. The densities of natural populations are constantly buffeted by a variety of factors—including weather. Ecologists seeking to model such stochastic effects face a dilemma: two seemingly reasonable approaches give diametrically opposite results. These are the two arguments:

(A) Alternative 1: The first is that species with densities that return quickly to equilibrium following a perturbation will vary less than those that return to equilibrium slowly (May, 1973). Examples are given in Figure 23.1a, in which two simulations are exhibited of a single-species model. A normally distributed random variable is added to the growth rate each generation. The variance of these stochastic effects is the same in the two simulations. The simulations differ in the rate at which the population returns to equilibrium. In one simulation (closed symbols), there is a long return time, in the other (open symbols) a short one. In the former case, particularly when the population level is small, it cannot easily recover. In the latter case, the population fluctuates less, recovers more quickly, and does not reach the low levels that, in the real world, would mean extinction of the species.

(B) Alternative 2: The second requirement is that the equilibrium density (the carrying capacity) is itself variable. A species with a high rate of return to equilibrium will track the fluctuations in this density. Such a species will have a more variable density than a species that responds more slowly and "sits it out" until more favorable conditions prevail (Turelli, 1978; Whittaker and Goodman, 1979; Luckinbill and Fenton, 1978; May *et al.*, 1978). Examples are given in Figure 23.1b and c where the return times are exactly those of the simulations in 1a. In this alternative, the species with the higher rate of return to equilibrium will fluctuate more and, as a consequence, is more likely to become extinct.

In short, the two alternatives make opposite predictions about how population variability varies with return times.

The definition of stability under stochastic conditions is not simple. There are several criteria that can be applied and some are difficult to apply (Turelli, 1978). One criterion involves the biologically reasonable idea that populations that fluctuate the least are those most likely to per-

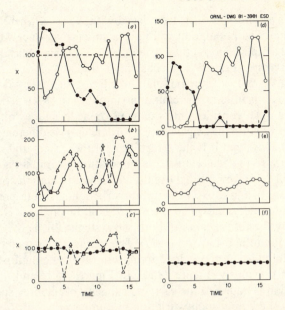

Fig. 23.1. Sample simulations of a discrete version of the logistic growth equation (see text). In (a) the carrying capacity (dashed line) is constant; in one simulation the reproductive rate is high (open symbols), in the other it is low (closed symbols). Randomly distributed terms are added to the species densities each generation. In (b) and (c) the carrying capacity is a random variable (dashed lines with triangles); in (b) the species' reproductive rate is high, in (c) it is low. In (d), (e), and (f) simulations have identical parameters to (a), (b), and (c) respectively, except that an additional trophic level has been added to the systems. The new equilibrium density for the species is 50 in all three simulations. The changes in K for (e) and (f) are exactly as those in (b) and (c) and are, therefore, not shown.

sist. Large fluctuations will eventually bring densities to the low levels from which real populations will stand little chance of recovery.

Though analysis of stochastic models is complex, it is possible to develop the preceding verbal argument analytically. A simple example using a one-species model,

$$\dot{X} = rX(1 - X/K), \tag{1}$$

catches sufficient detail of the argument presented intuitively above. Integrating Eq. (1) over a unit time interval, one obtains Eq. (2):

$$X_{t+1} = (K)[X_t \exp(r)]/[X_t \exp(r) + K - X_t]. \tag{2}$$

Equilibrium at K (the carrying capacity) is approached at a rate dependent on r; $-r$ is the eigenvalue of this system, and $1/r$ its return time.

Under the first alternative, X_t is replaced with $X_t + z_t$, where z_t is a random variable. The population will have a smaller variance when r is infinite (Eq. 2 simplifies to K) than when $r = 0$ (Eq. 2 simplifies to $X_t + z_t$). Under the second alternative stochasticity is added to K; it is replaced by $(K)(1 + z_t)$. When $r = 0$ the population will not fluctuate—it remains at X_t, but when r is infinite, the equation simplifies to $(K)(1 + z_t)$.

These examples consider only a single trophic level: they differ in their return times because the single species parameter (r) is different. However, it is easy to show that it is the return time of the system and not the parameter r *per se* that determines the population's variability. In Figures 23.1d, e, and f, I have repeated exactly the same simulations presented in Figures 23.1a, b, and c, except that another trophic level (a predator) has been added to the systems. The effect of adding the predator is to increase considerably the return times of both the systems: this is the result anticipated by Pimm and Lawton (1977). The effects are as predicted: in (d), where random variation in added to growth rates, fluctuations are much higher than previously (Figure 23.1a). In Figures 23.1e and f, when random variation is added to the carrying capacity of the prey species, both simulations show only slight variation in numbers.

Simply, increasing trophic levels means increased return times for individual species. Certain changes in species parameters will also change return times but, however the changes are brought about, long return times mean less variation in densities if random variation is added to the carrying capacity term and more variation in densities if random variation is added to the growth rates. A comprehensive analysis of this topic is provided by Turelli (1978).

Deciding *a priori* which alternative is likely to be the more important to natural populations is not easy. Neither population density nor carrying capacities are strictly random variables. Death rates are, and so are birth rates, in the sense that animals may fail to achieve their maximum fecundity. This statement says that $\dot{X}$ (birth rate minus death rate) is a random variable without saying precisely how the stochasticity acts. Each hypothesis is little more than a caricature of the ways in which stochasticity varies, and neither hypothesis is likely to be rejected by argument alone. Deciding between the hypotheses *a priori* would require detailed knowledge of the components of birth and death rates and this is not generally available. One must resort to data on the variances of population sizes and their relationship to return times to see which caricature most reasonably captures the essence of what is going on in the real world.

May *et al.* (1978) have suggested that the patterns of variance in yield against harvesting effort for fish populations are in accord with the first alternative. There are experimental data to support alternative 2. Luckinbill and Fenton (1978) experimentally manipulated the carrying capacity of two species of protozoa by altering the supply of bacteria on which they fed. The species with the faster increase from below carrying capacity was also the faster to decline from above it. As predicted by the second alternative, this species varied more than the other and was the first species to become extinct. Such results confirm that alternative 2 can describe a population's behavior, but it is still important to ask which alternative is the better description of the world outside the laboratory.

Bird populations. To decide between the alternatives I have used data from the British Trust for Ornithology's Common Bird Census (CBC) (Battan and Marchant, 1976, 1977; Marchant, 1978). Common farmland and woodland birds have been censused since 1962 and 1965 respectively. The densities are given on a relative scale where the densities of all species are set to 100 in 1966. The available data up to and including 1977 have been analyzed. I calculated the variation in densities as the coefficient of variation (CV). Estimating the rate of return to equilibrium presents more difficulty. As discussed above, a species' rate of return to equilibrium depends on its reproductive parameters but also upon the number of trophic levels in the system in which the species finds itself and the characteristics of the other species in the system. The number of trophic levels is likely to be fairly constant (three or four) and so return times might be closely correlated with reproductive parameters. However, there is no way one can be certain of this. Moreover, clutch sizes (Witherby *et al.*, 1938), though fairly well known, are much bigger than the realized reproductive potential and vary with habitat and population densities. Reproductive parameters may, therefore, seem unsuitable as estimates of return times. However, subsequent to performing the analyses in this paper, I encountered O'Connor's (1981) interesting paper on the ecological characteristics of the CBC species. He shows (his Figure 12) that a measure of reproductive effort (eggs per season) is closely correlated with a measure of resilience (how rapidly a population increases, when rare). His measure of resilience is, in turn, closely related to the estimates of return time I shall now discuss. All these arguments suggests that the results I shall present may not be very sensitive to the exact measure of return time used.

I chose to estimate r from census data using the statistical model,

$$Y_i = \beta_1 X_i + \beta_2 X_i^2 + \varepsilon_i, \tag{3}$$

where the Y_i are the densities in year $n + 1$, X_i the densities in year n, β_j the parameters to be estimated, and the ε_i permit random deviations from the model. Consider the meaning of β_1. First, suppose that each species is adequately described by a single-species population model. Taylor expansion of (2) about K shows that $\beta_1 - 1$ is an estimate of e^r. (Recall that $1/r$ is the return time of this system.) Similarly, systems with more than one species can be linearized about their multispecies equilibrium.

The result is a familiar one (*e.g.* May, 1973) and yields a simple solution to the system of differential equations. For each of the n species, this solution involves n terms, but one term—that involving the largest eigenvalue (λ_{max})—quickly predominates as time progresses. The other $n - 1$ terms can be ignored. The factors in the dominant term include the initial species' density ($X_{i,t}$) and $\exp(\lambda_{max}t)$. Such analyses yield the result that, even for multispecies models, β_1 has a simple relationship to the species' (and the system's) return time.

Now, populations reproducing only during a limited time each year may be modeled by a finite difference equation rather than a differential equation like (1), which has continuous birth and death processes. Reality may be somewhere in between the assumptions of the two models. Nonetheless, for a finite difference analogue of (1),

$$X_{t+1} = X_t + r\,X_t(1 - X_t/K), \tag{4}$$

r is estimated by $\beta_1 - 1$ (Table 23.1). This result also generalizes to multispecies models. Because it is the sign of the relationship between r and CV that is of interest, not the precise functional form, I suggest that β_1 provides a reasonable independent variable to be used in the analysis. Small values of β_1 imply long return times and *vice versa*.

In both (2) and (4) $X_t = 0$ implies $X_{t+1} = 0$, so there is no intercept in the statistical model.

Suppose that single species dynamics were reasonable approximations to the real dynamics. Then one could, by using (4), estimate K as $(\beta_1 - 1)/\beta_2$ (Table 23.1). These estimates were used to evaluate the procedure for each species. The data were collected in two habitats: farmland and woodland. In the farmland data, one species (Carrion Crow) and, in the woodland data, seven species (Greater-spotted Woodpecker, Carrion Crow, Magpie, Blue Tit, Marsh Tit, Song Thrush, and Mistle Thrush) were rejected because estimated values of K were outside the range of observed population densities. In each case the cause was a continuous trend in population densities that makes any model like (1) or (4) inappropriate. Only two other exclusions were made: Spotted Flycatcher had an unusually low β_1 in farmland which reflected its continuous decline there. Whitethroat data were excluded from both habitats *a priori* because of the species' well documented and, so far, persistent crash following the winter of 1968 (Winstanley *et al.*, 1974). Of the remaining species, all estimates of β_1 were reasonable ($\beta_1 > 1$) and all but two (Willow Warbler in woodland, Blackbird in farmland) were <2 (Table 23.1).

These analyses are particularly sensitive to the kind of outlying points that occur during a large drop in density—when X_{t+1} will be "small" and X_t at "normal" levels. The winter of 1962/1963 was a severe one, and

Table 23.1a. Farmland

Species	Data including 1962			Data excluding 1962		
	$\beta_1 - 1$	$(\beta_1 - 1)/\beta_2$	CV	$\beta_1 - 1$	$(\beta_1 - 1)/\beta_2$	CV
Mallard	0.3398	216	36.9	0.4750	212	35.0
Red-legged Partridge	0.4904	114	22.0	0.5095	113	22.2
Partridge	0.2574	104	30.0	0.3632	104	24.3
Pheasant	0.3297	114	18.1	0.3605	114	16.1
Moorhen	0.5160	115	21.3	0.5026	126	22.0
Lapwing	0.9145	135	21.6	0.4902	141	18.6
Turtle Dove	0.4149	124	20.9	0.4369	124	18.5
Cuckoo	—	—	25.5	0.2199	156	25.5
Skylark	0.2950	112	12.1	0.5645	111	11.7
Swallow	0.2353	92	18.2	0.2355	92	18.9
Carrion Crow	0.0800	463	*	0.0972	331	*
Jackdaw	—	—	13.3	0.6392	96	13.3
Magpie	0.5763	97	15.6	0.5494	98	11.0
Great Tit	0.3023	130	23.3	0.3549	129	20.2
Blue Tit	0.3097	134	21.0	0.4028	132	18.0
Long-tailed Tit	—	—	28.0	0.5241	195	28.0
Tree Creeper	—	—	29.9	0.4651	248	29.9
Wren	0.2588	275	52.1	0.4727	291	52.6
Mistle Thrush	0.9280	120	27.2	0.8568	129	28.1
Song Thrush	0.6201	113	19.5	0.8071	119	20.2
Blackbird	0.5071	102	14.7	1.0711	103	12.8
Robin	0.4216	111	20.8	0.6154	111	18.1
Sedge Warbler	0.8509	76	28.0	0.8975	77	28.6
Blackcap	0.5529	142	38.2	0.5334	142	31.5
Garden Warbler	0.5686	82	36.4	0.5534	82	35.9
Whitethroat	0.6076	73	*	0.6011	73	*
Lesser Whitethroat	0.8234	69	24.7	0.9003	71	25.4
Willow Warbler	0.5450	104	19.4	0.5450	104	15.6
Chiffchaff	0.3923	118	36.6	0.3757	118	32.3
Spotted Flycatcher	0.1380	83	*	0.1631	83	*
Dunnock	0.4443	107	19.0	0.6220	107	16.1
Pied Wagtail	0.6996	117	27.4	0.3376	136	26.6
Starling	0.7876	95	10.4	0.5400	96	7.2
Greenfinch	0.4762	106	24.4	0.6389	106	20.9
Goldfinch	0.2264	183	35.7	0.2821	173	34.5
Linnet	0.5958	85	12.9	0.6213	89	13.4
Bullfinch	0.8000	99	22.9	0.9251	100	18.4
Chaffinch	0.4713	101	10.8	0.5915	100	8.0
Yellowhammer	0.5440	98	8.1	0.8532	98	6.6
Corn Bunting	0.4985	91	13.9	0.4198	90	12.2
Reed Bunting	0.2686	161	32.9	0.4330	160	31.5
Tree Sparrow	0.6285	98	17.9	0.8271	99	15.8

* These species were excluded for reasons discussed in the text.

Table 23.1b. Woodland

Species	$\beta_1 - 1$	$(\beta_1 - 1)/\beta_2$	CV
Cuckoo	0.7637	97	11.1
Green Woodpecker	0.8338	147	12.1
Greater-spotted Woodpecker	0.0788	ca. 10,000	*
Carrion Crow	0.1914	197	*
Magpie	0.1961	246	*
Jay	0.7203	90	8.4
Great Tit	0.2898	104	6.0
Blue Tit	−ve	—	*
Coal Tit	0.2879	290	36.5
Marsh Tit	0.1711	74	*
Long-tailed Tit	0.4111	176	30.2
Nuthatch	0.4835	102	18.1
Tree Creeper	0.5558	137	17.6
Wren	0.4403	226	35.6
Mistle Thrush	0.1186	163	*
Song Thrush	−ve	—	*
Blackbird	0.9864	103	5.3
Robin	0.2166	124	8.6
Blackcap	0.7682	104	9.4
Garden Warbler	0.4237	84	18.0
Whitethroat	0.5595	75	49.2
Willow Warbler	1.0230	97	7.0
Chiffchaff	0.3111	83	25.2
Goldcrest	0.5066	305	42.5
Spotted Flycatcher	0.4615	86	16.3
Dunnock	0.6223	111	6.4
Starling	0.5618	—	12.5
Greenfinch	0.4355	105	11.8
Linnet	0.4317	76	12.5
Bullfinch	0.4492	132	12.9
Chaffinch	0.2620	100	6.5
Yellowhammer	0.3140	118	16.1

* These species were excluded for reasons discussed in the text.

population densities following it were strongly depressed. When the pre-crash densities were very different from the long-term average in the years following the species' recovery, they markedly affected the estimates of β_1. The winter of 1961/1962 was also colder than usual, though not exceptional (O'Connor, 1981), and so some species were not at their normal levels during 1962. So, for the farmland data, I estimated β_1 and β_2 excluding 1962. For the woodland data, and for some farmland species, densities were not available for 1962 in any case. Though the estimates

changed dramatically for a few species, the significance of the overall results, discussed in the next paragraph, was changed only trivially, so I shall
simply restrict my discussion to the data including 1962.

The species include two trophic levels: insectivores (almost entirely migratory) and graminivores, with varying tendencies towards omnivory
(largely resident). The migratory species tended to have higher variation
in densities, for a given value of β_1. However, these differences were insufficient to alter the qualitative results I shall now present.

Now, farmland species occur in woodland and *vice versa*, so the two sets
of data cannot be considered independent. Consequently, I performed two
regression analyses of CV versus β_1: farmland species plus additional
woodland species (Figure 23.2a) and woodland species plus additional
farmland species (Figure 23.2b). The slopes of both analyses are negative:
the data show that as some function of the rate of population growth (β_1)

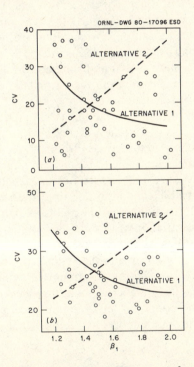

Fig. 23.2. The coefficients of variation (CV) versus a function of how fast
species densities return to equilibrium (β_1) for farmland plus additional woodland species (a), and for woodland plus additional farmland species (b). Two
model alternatives can be fitted to each set of data using Equations (5) and
(6) of the text. The first alternative is that stochastic effects are added to species'
densities; the second is that stochastic effects are added to the species' carrying
capacity.

increases, the variability of the population density decreases. This is the result predicted by alternative 1. Alternative 2 was rejected at levels of 0.02 (Figure 23.2a) and 0.07 (Figure 23.2b). Though the latter level is marginally greater than the normally accepted level of 0.05, I shall show, in the next section, that these levels are conservative. In short, alternative 2 can be rejected with greater certainty than these figures suggest.

Possible biases. A possible criticism of these results stems from the potential interdependence of the dependent (CV) and independent (β_1) variables when both are estimated from the same data. Moreover, there might be violation of the regression model's assumptions when serial data are used. To investigate behavior of parameter estimates obtained from the same data, I performed 10 replicate simulations of Eq. (4) for nine combinations of three values of r (0.1, 0.5, 0.9) and three levels of added variation. I ran each stimulation for 16 generations; this is the same duration as the farmland CBC data. There were two sets of simulations. In the first (alternative 1), variation was added to X_t as randomly distributed terms with an expected mean of zero and standard deviations of 10, 30, and 50. In the second set (alternative 2), I made K a normally distributed variable with a mean of 100 and standard deviations of 5, 10, and 25. The data in Figure 23.1 are from these simulations.

For the first set of simulations, estimates of r were biased, but independent of the level of added variation. As required by the theory, the populations' coefficients of variation decrease with increasing r: there is about a 50% reduction as r increases from 0.1 to 0.9. The reduction, though proportionately the same, is absolutely larger at high levels of variation than at low ones. This feature gives a triangular scatter of data, and leads to a lessened likelihood of rejecting alternative 2 (increased β) because the residual variation about the model at high r values is inflated by the high levels of variation at low r values.

I am indebted to Michael Turelli, who provided a mathematical analysis of this problem. By linearizing the system, an analytical expression is obtained relating the coefficient of variation in population numbers to the variance of the added number, σ^2, and the eigenvalue of the system, $\lambda (= 1 - r)$. This linearization is only valid for small perturbations when the higher order terms in the model are unimportant. Turelli obtains for this alternative:

$$CV^2 = \sigma^2/(1 - \lambda). \tag{5}$$

When $r = 0.1$, $CV/\sigma = 2.29$; when $r = 0.9$, $CV/\sigma = 1$; thus the expected reduction (44%) is close to the value obtained by simulation (50%).

For the second set of simulations, corresponding to alternative 2, coefficients of variation increased with increasing r. Indeed there was about an eightfold increase in CV as r changed from 0.1 to 0.9. The theoretical expectation for this alternative has, again, been provided for me by Michael Turelli. It is:

$$CV^2 = \sigma^2(1 - \lambda)/(1 + \lambda). \tag{6}$$

When $r = 0.1$, $CV/\sigma = 0.23$, and when $r = 0.9$, $CV/\sigma = 0.93$, or about a fourfold increase. The discrepancy between four- and eightfold increases probably reflects the approximation used to obtain the analytical result. This result is valid only for variations about K sufficiently small that linearized equations can be used.

The estimates of r (again biased) did depend on the variance of K in this set of simulations: the values of r were reduced as the variance of K increased. If this effect were strong, then even if alternative 2 were correct, it might be possible to obtain a spurious negative correlation between CV and r. This is not the case. The estimates of r were reduced at most by 18%, as compared with the 400–800% increase in the estimates of CV. In short, when r and CV are calculated from the same data and alternative 2 is true, biases do exist, but they are far too small to negate the conclusion that a negative correlation between CV and r rejects alternative 2.

Finally, using Equations (5) and (6) I estimate the variance of the perturbations (σ^2) under each of the two model alternatives for each species. Then, by taking the mean values, I can fit the expected relationships between CV and r, for each of the two data sets analyzed. I do so in Figure 23.2. While alternative 1 fits the observed data well, alternative 2 is clearly totally inadequate and must be rejected.

CONCLUSIONS

On Modeling Stochastic Effects

I conclude that for birds, stochastic effects are modeled better by perturbing population densities than carrying capacities. This conclusion begs the question: Why? An example is informative. The density of the Song Thrush in farmland (Figure 23. 3) crashed during the unusually cold winter of 1962/1963. Most of the variation in numbers comes from the years 1962 through 1966 as the population climbed back to its carrying capacity. It can be argued that the effect of a hard winter is to cover the ground with snow, reduce the availability of food, and therefore perturb K. That, in general, stochastic effects may be better modeled by perturbations to X, not K, is perhaps explained by the two very different time scales of perturbations to these two variables. The reduction in K lasted, perhaps, two months, but the population breeds only annually and took several years to recover. In the experiments on protozoa described above (Luckinbill and Fenton, 1978) where the species with the larger r became

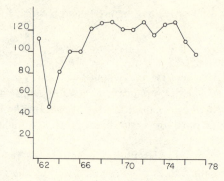

Fig. 23.3 The population density of Song Thrushes in English farmlands in the years 1962–1978, on a relative scale where the density in 1966 was set to 100.

extinct first, the changes to K were as long as, or longer than, the time required for the population to reproduce. I suggest that the apparent conflict between the hypotheses reflects the wide spectrum of frequencies for environmental perturbations and the much smaller range of response time available to natural populations in a seasonal environment. Population models allow continuous population responses and continuous perturbations, and these are unrealistic. Populations that recover from large but transient perturbations will do so with rapidly changing populations recovering the fastest and thus varying the least.

Clearly, if large, infrequent perturbations of short duration (like cold winters) are the norm, then the results of this study on birds are likely to be quite general. There is plenty of anecdotal evidence for this generalization: changes in ocean currents or in planktonic composition ("red tides") and their effects on sea birds are but two examples. However, it is at least possible for perturbations to some systems to be of the other kind: frequent and of long duration. In such circumstances rapidly changing populations would vary the most. There seems no way to be certain of the generality of my conclusion without comparable analyses on many other taxa.

The Consequences of Stochastic Effects

There are two ways of incorporating stochastic effects into population models. One would favor the persistence of systems with long return times and, therefore, long food chains; the other would favor short return times and short food chains. As I have shown for at least bird populations, the latter is a more reasonable description of the real world. The systems in the real world that fluctuate the least (and thus are expected to persist the longest) are those with the fastest returns to equilibrium. Now, I cannot claim to have shown that dynamical limitation is, certainly, the factor that restricts food chain lengths to three or four trophic levels. What I have shown is, merely, that the theoretical basis for this claim is still intact.

What kind of evidence would demonstrate that dynamical constraints are the factor responsible for short food chains? Consider Equation (5): suppose that the CV for all populations has an upper bound—above it, populations will fluctuate so much that extinction is assured. Then, as the environmental uncertainty, σ^2, increases, so the return times of the system must decrease in order to keep the CV in bounds. On average, I would expect shorter food chains in uncertain environments, given comparable communities. Is there any evidence for this prediction? Not a great deal, it seems. But one study on the aquatic systems that accumulate in tree-holes is highly suggestive. Kitching (1983) has studied tree-holes in England and Australia. The two sets of holes are statistically indistinguishable in most of their physical and chemical parameters. Moreover, the mean annual leaf fall—the systems' source of energy—was also the same in the two places. The community in Australia supported three, possibly four, trophic levels, that in England only two. In the former, leaf fall does seem more constant and predictable than in the latter, suggesting that the greater dynamical restrictions shorten the food chains in England.

Interestingly, Kitching's study also throws light on two of the other hypotheses for food chain length given above: size or other design constraints and the balance between evolutionary tendencies to lengthen and shorten food chains. His two communities are very similar taxonomically, with the exception of a Leptodactylid frog at the top of the Australian web. Yet, two insect families common to the two areas contain detritivorous species in England but predatory species in Australia. If design constraints precluded predator habits for one system they should do so for the other. And it is unclear why food chains should shorten to two levels for evolutionary reasons in England but not in Australia—especially since Hastings and Conrad (1979) argue that three trophic levels should be the most likely equilibrium. Kitching concludes, and I agree, that while energetic constraints cannot be totally excluded, the hypothesis of dynamical constraints seems the most attractive alternative to explain his results.

SUMMARY

The causes of the short food chains, so frequently observed in the real world, are far from certain. There are four hypotheses: energetic constraints, size or design restrictions, a balance between evolutionary tendencies to lengthen and shorten chains, and dynamical constraints. The first hypothesis makes several predictions about the relationships between the patterns of energy flow and the number of trophic levels. With one exception, these are not supported. This exception is that extremely low productivities may be unable to support a third trophic level. The second and third hypotheses seem incompatible with the one study that has compared two communities in very different locations: Kitching's (1983) work on tree-hole communities. The fourth hypothesis could be theoretically incorrect. Long food chains are characterized by species that respond only slowly to perturbations. Under one model these species will fluctuate greatly, this fluctuation will cause extinctions, and these, in turn, will lead to shorter food chains. However, under a second model species that respond slowly will fluctuate the least and long food chains would be expected to persist. I have shown that the first of these models more reasonably describes bird populations in English rural areas. How general the result is remains to be seen, though there is anecdotal evidence that the kind of perturbations that lead to it may be general. All this evidence suggests that the hypothesis of dynamical constraints remains a viable one, theoretically. Moreover, Kitching's (1983) results seem to be best explained by this hypothesis.

Acknowledgments

This research was supported jointly by the National Science Foundation's Ecosystem Studies Program under Interagency Agreement No. DEB 77-25781 and the Office of Environmental Research, U.S. Department of Energy, under contract W-7405-eng-326 with Union Carbide Corporation. This is also Publication No. 1793, Environmental Sciences Division, Oak Ridge National Laboratory.

24.

Stability, Probability, and the Topology of Food Webs

MICHAEL J. AUERBACH

Department of Biology, University of North Dakota, Grand Forks,
North Dakota 58202

INTRODUCTION

Analyses of both real and hypothetical ecological communities have demonstrated that any relationships between stability and complexity are more complicated than once presumed and are often contrary to the once fashionable adage that complexity begets stability (*e.g.* May, 1973; Goodman, 1975; Pimm, 1979a). One current approach to stability and complexity in ecological communities is through food-web analysis. Pimm (1979b, 1980b), Pimm and Lawton (1977, 1978), and Rejmánek and Starý (1979) suggest that dynamical constraints are responsible for many hypothesized food-web properties, such as restrictions on the number of trophic levels and interspecific interactions. Here, I examine the underlying assumptions and resultant mathematical properties of the food-web models.

A number of natural assemblages of species that have persisted for many years fail to meet the criterion for stability currently used in food-web analyses (Auerbach, 1979). I will outline some possible reasons for this paradox that will demonstrate how tenuous certain assumptions of the models are. The outline is followed by a detailed examination of several model parameters, in which I suggest that there is a stronger mathematical than biological basis for some current predictions about food webs.

GROUND RULES AND DEFINITIONS

The analyses discussed in this paper follow the cybernetic systems approach that has been widely used since its introduction into ecology by Gardner and Ashby (1970) and May (1972). This technique views a collection of species as a closed system, with each species' abundance affected by interspecific interactions. These interactions are modeled after first-order Lotka-Volterra equations and are presented in matrix form. The percentage of non-zero elements, excluding the principal diagonal, is defined as the connectance of the matrix (C). Intraspecific or outside influence on a species is only incorporated for species deemed to be self-limited, which often include only the autotrophs.

I will denote the total number of species in a food web as "m," and the average strength of interaction between species as "i." I follow Pimm and Lawton's (1978) definition of omnivores as species that feed on more than one trophic level and Cohen's (1978) definitions of sink, source, and community webs. Community webs are defined as sets of organisms selected within a habitat without prior regard to their eating relationships. Sink webs are subunits of community webs. To construct a sink web one chooses one or more species and then traces feeding relationships between these organisms and their prey, their prey's prey, and so on. Source webs are the opposite of sink webs; one starts with an organism and identifies its predators.

The terms "basal species" and "top predators" describe certain species' positions in a food web. Basal species feed on no other species, while top predators occupy the highest trophic level of a web. This differs slightly from Pimm and Lawton's (1980) definition of top predators as species on which nothing else feeds. This change was necessary because, in abstracting food webs from the literature and in constructing model webs, I determined the trophic placement of a species by its food sources. Species were placed one trophic level higher than the highest level occupied by their prey. This method ignores intra-trophic level interactions such as mutualism and interspecific competition.

WHY ARE SOME REAL WEBS UNSTABLE?

The criterion generally employed to assay the stability of a food web is that of Lyapunov or neighborhood stability. "Stable" in this sense is used when the population of any species displaced a small distance from its equilibrium level will return to the equilibrium. Whether natural populations ever exist at equilibrium population values is a question beyond the scope of this paper, but in light of the complexity of predicted behavior among models of population dynamics currently available to the population biologist, the criterion of Lyapunov stability may be too restrictive. Barnett and Storey (1970) have noted that "Many systems that are unstable according to this definition perform quite adequately from a practical point of view. Conversely there are systems which although stable according to the definition would not behave well in practice."

Two other major potential problems exist in analyses of food-web stability. First, these studies assume that ecological communities are definable static systems uninfluenced by migration and extinction. One can argue that present models can be revised to incorporate more complex processes, including stochastic events. However, the obvious requirements remain not only that "communities" exist in the real world but that field biologists be able to delimit them. I will show later that how one goes about identifying a community can profoundly affect stability analyses. Second, these analyses implicitly assume that the modeled interactions, competition or predation, are responsible for both the population dynamics of the species comprising the community and its organization. There would be little justification for a conference on the conceptual issues involving ecological communities and the supporting evidence if these assumptions rested on a well-documented base.

Current techniques of food-web stability analysis highlight several necessary simplifications and generalizations that reduce their predictive value. Foremost is the use of Lotka-Volterra-like equations, which, as noted by Lawton and Pimm (1979), are gross simplifications of biotic interactions. Use of these equations also predisposes stability analyses to a tolerance of the myriad assumptions that underlie Lotka-Volterra population models (Heck, 1976).

The failure of the stability analyses to predict the behavior of some natural systems may stem more from specific characteristics of the models than from validity of the stability criterion, simplifications that current techniques mandate, or lack of realism of certain basic assumptions. Of the three important parameters in the models, the number of species (m) and the connectance (C) are observed quantities in any real food web, while the interaction intensities are either estimated or randomly selected from a predetermined distribution. It is conceivable that m and C may be "wrong" for certain natural webs because of the aforementioned necessity of correctly identifying a true community or web. However, "i" is more likely to be the cause of incorrect predictions.

Problems with measuring species' interaction intensities involve both inter- and intraspecific terms. Altering the number and magnitude of the principal diagonal elements (self-limited species) in a community matrix profoundly influences stability analyses since these terms set the damping time for population oscillations (Saunders, 1978; Yodzis, 1981). These terms have a much greater effect on a stability analysis outcome than the interspecific interaction intensities. In fact Yodzis (1981) found stable solutions for a number of real food webs when all principal diagonal elements were set greater than zero or when only a fraction of these elements were enormous. However, the biological realism of a web with a very high percentage of species experiencing intraspecific interference, or a smaller proportion of species subjected to tremendous intraspecific limitation is questionable. Yodzis's results underscore the paucity of information available for estimating the frequency and strength of interactions that are essential to stability analyses. The possibility also remains that the somewhat standardized distributional ranges of interspecific interaction elements, used primarily by Pimm (1979b) and Pimm and Lawton (1978), are responsible for invalid predictions. Lawton and Pimm (1979) have stressed that the limits they impose on the selection of interaction coefficients are rough approximations of the potential interspecific impact of different types of biotic interactions and are designed to generate qualitative predictions. Currently, we can only speculate that interspecific interaction varies spatially and temporally and at least sometimes includes higher-order (multispecific) effects.

That little is known about the general influence of one species on another, let alone influences in a group or web of interacting species, is not a trivial problem in stability analyses. May (1972) has presented a formula for the transition between stable and unstable regions of parameter space for the criterion of Lyapunov stability that states that the community matrix will almost certainly be stable if

$$i(Cm)^{1/2} < 1$$

and unstable if

$$i(Cm)^{1/2} > 1.$$

Since m and C are fixed for any natural system, with the now familiar proviso that the domain of a web be correctly identified, i becomes the critical parameter if current models are to be viewed as valid representations of the real world.

PREDICTIONS OF FOOD-WEB MODELS

Prediction 1: *Interactions in food webs are organized into blocks.* May's (1972, 1973) analyses of simulated communities suggested a relationship between the size and organization of a food web and its stability: even though webs with few species and a comparatively low connectance exhibit almost no probability of being stable, the probability of a stable solution increased dramatically if interactions between species were arranged into blocks. Pimm (1979b) noted, however, that completely compartmentalized models with the same connectance as unblocked models actually are less likely to be stable. Pimm and Lawton (1980) also compared the degree of blocking in real and simulated webs and concluded that real webs show no greater propensity for compartmentalization than chance alone dictates, except between the major habitat divisions found in a few published webs. The issue was further elaborated by McNaughton (1978) and Rejmánek and Starý (1979), who concluded that several natural systems appear to have discrete subsystems, and Murdoch (1979), who claimed that they do not.

I investigated on a crude scale one specific form of blocking, that is, the possibility that polyphagy at high trophic levels unites food chains in a web with infrequent inter-chain interactions at lower trophic levels. Data for this and all subsequent analyses in this paper were extracted from Cohen's (1978) collection of food webs (although whenever possible original sources were used in trophic placement of species, and Summerhayes and Elton's web was excluded from all analyses) and from several insect-host plant webs (Root, 1973; Force, 1974; Askew, 1975). I calculated connectance of each web, then removed the highest trophic level and re-calculated connectance. This process was repeated until only three trophic levels remained, the minimum number of levels needed to establish the existence of this kind of blocking (Pimm and Lawton, 1980). Figure 24.1 illustrates the result for sink, source, and community webs.

One follows the effect of this "trophic reduction" by starting at the right endpoint of a line and then tracing its path to the left. Connectance generally remains within fairly tight boundaries for most community webs, except some with fifteen or fewer species. Note that this method suffers from the bias that as m decreases there are fewer possible levels of connectance and the deletion of even one link in a web is amplified. These results

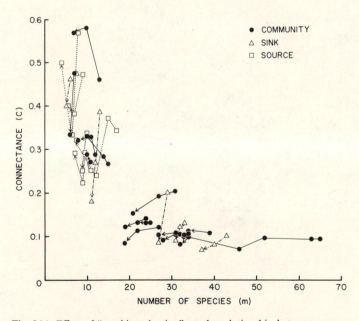

Fig. 24.1. Effect of "trophic reduction" on the relationship between connectance (C) and number of species (m) for community, sink, and source food webs. See text for explanation.

suggest that the connectance of larger community webs is an intrinsic property, indicating that interactions are quite evenly distributed across trophic levels. This is especially true relative to the effect of trophic reduction on sink or source webs.

Figure 24.1 shows my contention that failure to identify all the members of a web can influence stability analyses. For instance, the observed tendency for the connectance of sink webs to decline rapidly as successive trophic levels are removed is probably largely attributable to a greater research effort on the highest trophic level and the prey of these top predators. The converse is probably the case for source webs, except that top predators in these webs are usually parasitoids reared from their prey, so the bias is reduced.

Prediction 2: *There should be an inverse relationship between connectance and number of species among food webs with a comparable average interaction strength.* This prediction originates directly from the influence of these two parameters on a stability analysis as seen in May's formula for qualitative stability. For a fixed value of i, an increase in m must be offset by a decrease in C. Examinations of real webs show the predicted pattern (*e.g.* Rejmánek and Starý, 1979; Yodzis, 1980). Figure 24.2 shows the relationship between C and m for the webs I analyzed. A biological

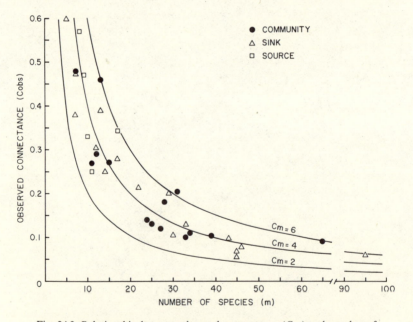

Fig. 24.2. Relationship between observed connectance (C_{obs}) and number of species (m) for community, sink, and source food webs. The three curves represent the hyperbolic functions $Cm = 2$, $Cm = 4$, $Cm = 6$.

interpretation of these results states that as species richness of food webs increases the fraction of actual interspecific interactions of the potential interactions decreases.

Corollary 1: *The product of* C *and* m *among food webs is constant: the relationship between* C *and* m *is hyperbolic.* This hypothesized relationship did not actually originate as a prediction of the stability analyses, but rather was observed in the results of analyses of real and simulated ecological communities. Lawlor (1978) initially documented that constraining the product of C and m could profoundly affect stability analyses. He showed that, in model systems, as m increases the number of biologically feasible configurations rapidly becomes a minuscule fraction of the potential configurations. However, when Lawlor held the product of C and m equal to 10, he found the opposite outcome; that is, as m increased the density of acceptable systems increased and there was a greater probability that a randomly selected system was stable.

Yodzis (1980) examined the relationship between m and C and a theoretical upper bound for C in the different versions of Cohen's (1978) community webs. He found that both sets of connectance values decreased as m increased; however, in neither case did C decrease as fast as m^{-1}, which is the criterion for decreasing connectance to account alone for the stability of increasingly complex systems. But the product of C and m in Rejmánek and Starý's (1979) study of plant-aphid-parasitoid communities yielded a close fit to the relationship $Cm = 3$. They concluded that increased species richness does not cause decreased stability. They also found no evidence that any plant-aphid-parasitoid communities exist outside the range $2 < Cm < 6$.

Rejmánek and Starý's and Yodzis's studies can be viewed slightly differently. Any postulated or observed relationship between C and m can be shown by May's formula to be equivalent to holding the critical value of i constant for different-sized webs. In other words, a plot of $(Cm)^{-1/2}$ against m should have a slope of zero, implying that as complexity increases the average strength of interaction remains constant, but proportionately fewer species interact. Rejmánek and Starý's results appear to fit this pattern, while Yodzis's do not. The slope for the thirteen community webs in the appendix of Cohen's (1978) book, his sink webs, and the additional source webs mentioned earlier does not differ significantly from zero, whether the webs are analyzed together or split into types (Figure 24.3).

The expectation of a constant value for the transition from stability to instability, or the interpretation of results obtained from real or simulated data with respect to this expectation, appears to carry the implicit assumption that ecological communities should meet the criterion of Lyapunov stability. Since several natural communities do not fulfill this

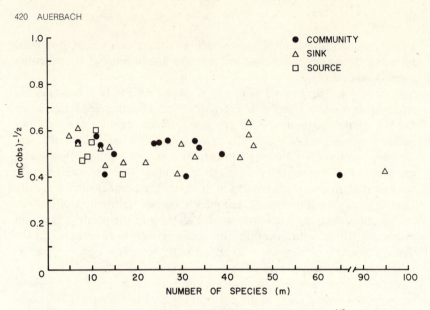

Fig. 24.3. Relationship between the stability threshold $(mC_{obs})^{-1/2}$ and the number of species (m) for community, sink, and source food webs.

expectation (Auerbach, 1979), these second-generation stability studies and layering of hypotheses apparent in many of them seem premature. However, I feel this problem is secondary to a greater methodological flaw. There is currently no way to assess relationships such as the product of C and m because no statistical expectations for the important stability parameters are known. Which patterns have biological relevance and which are simply artifacts of the models? For instance, the products of C and m in Cohen's community webs range from 3 to 6 with a mean of 4.2. Without knowledge of the expected distribution of these products, it is impossible to gauge how constant and significant this relationship is.

I investigated these problems by calculating an expected connectance (C_{exp}) for several real food webs. Minimum possible connectance (C_{min}) for any web is equal to $2/m$. This formula ensures that all species interact with at least one other species and preserves the continuity of the web, precluding isolated chains. Maximum connectance is

$$1 - \frac{\sum_{i=1}^{T} m_i(m_i - 1)}{m^2 - m},$$

where T is the number of trophic levels with more than one species and m_i equals the number of species in the ith trophic level. This formula removes all possible intra-trophic-level interactions. The problem then becomes finding the most likely value within the prescribed range (C_{range}).

Generating expected connectances requires the total number of biologically feasible web configurations for each possible connectance value. The most likely connectance is then assumed to be the value for which there is the greatest number of configurations. I also calculated the mean of the frequency distribution of connectance values, which generally approximates the modal configuration. Figure 24.4 illustrates the calculation procedure. For each level of connectance, simple combinatorics permit determination of the total number of configurations. The illustrated web has a minimum connectance of 0.33, which is equivalent to placing non-zero values in five elements of the matrix. The two darkened

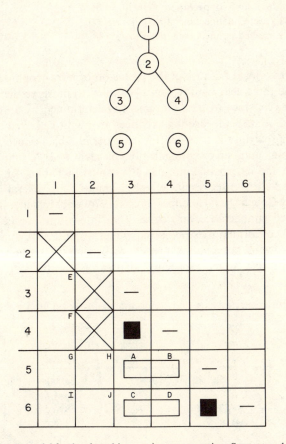

Fig. 24.4. A model food web and its matrix representation. See text and Table 24.1 for details.

Table 24.1 Number of webs for each level of connectance
for the food-web configuration of Figure 24.4.

Number of Links	Connectance	Number of Webs
5	.3333	4
6	.4000	28
7	.4667	85
8	.5333	146
9	.6000	155
10	.6667	104
11	.7333	43
12	.8000	10
13	.8667	1

Minimum connectance = 10/30 = .3333.
Maximum connectance = 26/30 = .8667.
Expected connectance (modal) = .6000.
Expected connectance (mean) = .5778.

cells cannot be filled because of the exclusion of intra-trophic-level interactions. The three cells with X's must be filled to preserve the configuration of the web, since trophic placement is determined by an organism's
prey. For $C = 0.33$ the problem reduces to filling cells A or B and C
or D. This procedure is repeated for each permissible connectance value
until the total number of configurations is identified. A summary table
for the web and matrix of Figure 24.4 is shown in Table 24.1.

Table 24.2 shows the results of this technique for several food webs
compiled by Cohen (1978) and for Force's (1974) web. In general, observed
and expected connectances are close. Two important points should be
emphasized concerning these expected values. First, I have analyzed only
the smallest webs because the procedure becomes extremely time-con-
suming for larger webs, especially identifying feasible configurations.

Table 24.2. Observed and expected connectance of selected food webs.

	Number of Species	Observed Connectance	Expected Connectance (modal)	$\dfrac{C_{exp} - C_{min}}{C_{range}}$
Paine (1966)	5	.6000	.4000	.00
Paine (1966)	7	.3810	.3810	.40
Teal (1962)	7	.4762	.5238	.42
Force (1974)	8	.5714	.4643	.43
Askew (1975)	9	.4722	.5278	.44
Thomas (1962)	17	.2794	.1912	.39

Second, this method assumes all configurations are equally likely. Some predictions concerning web structure, such as Pimm and Lawton's (1978) emphasis on the scarcity of omnivory, imply that some feasible configurations should be weighted more than others. Generally, incorporation of any weighting suggested by current hypotheses either has little effect on or lowers expected connectance.

Although I did not calculate expected connectances for all real webs cited in this paper, I investigated how their size and shape constrained the range of permissible connectance. Figure 24.5 depicts the relationship between minimum feasible connectance (C_{min}) and number of species in these webs. The relationship fits the hyperbolic function $Cm = 2$ almost perfectly, a direct result of the formula for C_{min}. It is no longer surprising that Rejmánek and Starý (1979) could state that no plant-aphid-parasitoid web yet found lies below the $Cm = 2$ line. Their assertion is equivalent to saying that communities with the lowest observed product of C and m simply have the lowest possible connectance. The maximum possible connectance (C_{max}) also declines as m increases (Figure 24.6), but not as rapidly as C_{min} does. The linear fit for C_{max} vs. m is significant for sink webs ($F_{1,14} = 8.70$, $p < .025$) and all webs ($F_{1,32} = 8.99$, $p < .005$), but not for community webs because of the value for the 65-species community.

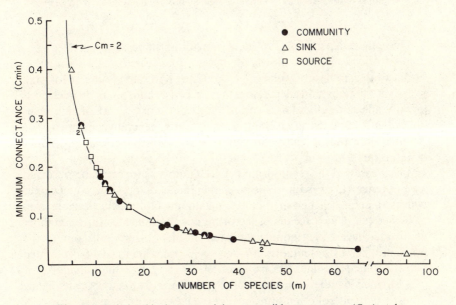

Fig. 24.5. Relationship between minimum possible connectance (C_{min}) and number of species (m) for community, sink, and source food webs. The curve represents the hyperbolic function $Cm = 2$.

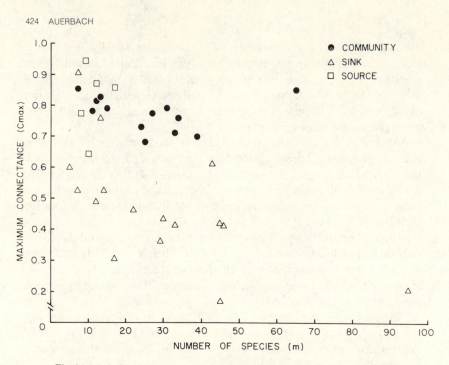

Fig. 24.6. Relationship between the maximum possible connectance (C_{max}) and number of species (m) for community, sink, and source food webs.

One might expect C_{range} to increase with an increase in m, since C_{max} tends to decrease more slowly than C_{min} does. Figure 24.7 shows that it does for community webs ($F_{1,11} = 17.62$, $p < .005$), but not for sink webs or all webs considered together, because number of tropic levels greatly affects C_{max} and as a result C_{range}, but not C_{min}. Most sink webs in Cohen's book have very few trophic levels, generally only two. The number of trophic levels in community webs tends to increase very slowly with an increase in m.

I next assessed the relationship between number of species and observed connectance (C_{obs}), with respect to the range of possible connectance values. I defined a new parameter, maximum biologically feasible connectance (C_{bio}), as a means of factoring out the influence of the number of trophic levels. C_{bio} equals the number of non-zero cells out of the total number of potentially non-zero cells, after removal of prohibited intra-trophic-level interactions. In other words, C_{bio} is the numerator of C_{obs} divided by the numerator of C_{max}. Figure 24.8 shows a tendency toward a hyperolic relationship between C_{bio} and m, just as for C_{min} and C_{obs}. Actual interactions comprise a smaller proportion of biologically permissible interactions as web size increases.

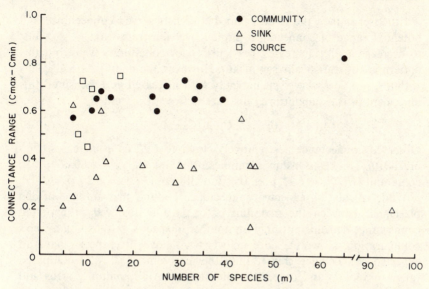

Fig. 24.7. Relationship between the range of connectance (maximum connectance minus minimum connectance) and number of species (m) for community, sink, and source food webs.

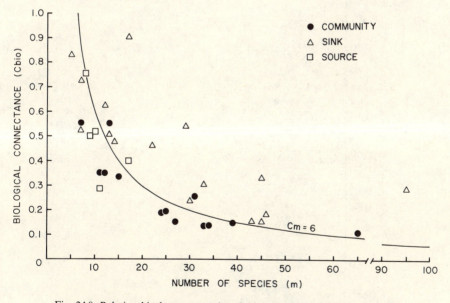

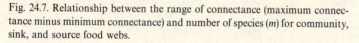

Fig. 24.8. Relationship between maximum biologically feasible connectance (C_{bio}) and number of species (m), for community, sink, and source food webs. The curve represents the hyperbolic function $Cm = 6$.

I then examined the relationship between observed connectance and potential range of connectance. Figure 24.9 illustrates that C_{range} tends to decrease as C_{obs} increases, particularly for community webs, but this pattern is not statistically significant. However, the point where C_{obs} falls within C_{range} does vary systematically with increased m. I measured this placement as the standardized distance between C_{min} and C_{obs}:

$$(C_{obs} - C_{min})/(C_{range}).$$

Observed connectance lies relatively closer to C_{min} as m increases (Figure 24.10). This relationship is significant for sink ($F_{1,14} = 5.80, p < .05$), community ($F_{1,11} = 6.13, p < .05$), and all webs ($F_{1,32} = 9.29, p < .005$).

While calculating expected connectance I found that amount of displacement between the midpoint of C_{range} and observed or expected connectance depends not only on number of species, but also on a web's configuration. However, since no webs with over five species have an expected connectance greater than the midpoint of C_{range}, I compared range midpoints and observed connectances. All community webs and most source (4 of 5) and sink (14 of 16) webs have observed connec-

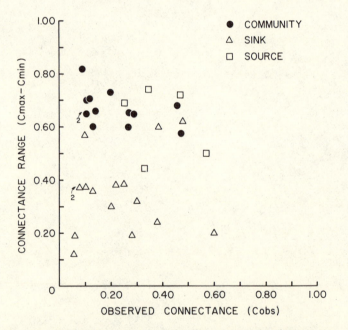

Fig. 24.9. Relationship between the range of connectance (maximum connectance minus minimum connectance) and observed connectance (C_{obs}) for community, sink, and source food webs.

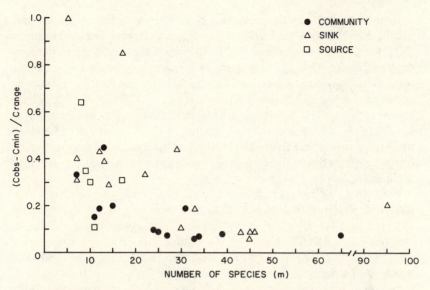

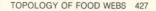

Fig. 24.10. Relationship between the standardized displacement of observed connectance from minimum connectance $[(C_{obs} - C_{min})/(C_{range})]$ and number of species (m), for community, sink, and source food webs.

tances less than the midpoint of C_{range}. In other words, most webs have connectances less than this estimate of expected connectance. Violators of this pattern are usually small webs, such as those of Paine, Force, and Thomas shown in Table 24.2. A cursory examination of large webs indicates that few, if any, have expected connectances that are less than 40% of the range from C_{min} to C_{max}. When C_{obs} are compared to this lower limit of expected connectance, most community (12 of 13), source (4 of 5), and sink (11 of 16, 1 tie) webs still have observed connectances less than expected. This finding implies that real webs are less complex than chance alone dictates, and that this discrepancy increases as web size increases, with the connectance of small webs lying near the expected values.

These results can be interpreted in light of the stability threshold, $(Cm)^{-1/2}$. A constant value of $(Cm)^{-1/2}$ as m increases indicates a hyperbolic relationship between C and m, or in other words, the same threshold value of i for different-sized webs. In simulation studies where the interaction terms are selected from a predetermined distribution, one implicitly assumes that the threshold value of i is constant. There is no *a priori* expectation that as web size increases there should be a reduction in i, which implies that interactions in larger webs are less intense, although this hypothesis has been presented to explain results of some stability studies.

Figure 24.11 shows the relationship between number of species and calculated threshold values for C_{min}, C_{max}, C_{obs}, and the midpoint of C_{range} for Cohen's community webs. As m increases, $(mC_{min})^{-1/2}$ remains constant, as does $(mC_{obs})^{-1/2}$ (Figure 24.3), but values for the expected and maximum connectance decrease. The threshold value for biological connectance (Figure 24.12) also decreases ($F_{1,32} = 17.01$, $p < .001$). These results suggest that there should be a reduction in i as m increases, as May's (1973) "amusing corollary" predicts. As web complexity increases, observed connectance tends to be displaced further away from expected connectance and predicted average interaction strength for stability declines.

Prediction 3: *Food webs have a limited number of trophic levels.* Prediction 4: *Omnivory should be rare in food webs.* Corollary 1: *Omnivores that feed on adjacent trophic levels should be more common than omnivores that feed on separated trophic levels.* I have grouped these hypotheses together because they all pertain to food-web shape. It became clear that topology of a web greatly restricts the range of certain important stability parameters, such as connectance. Although hypotheses have dealt with

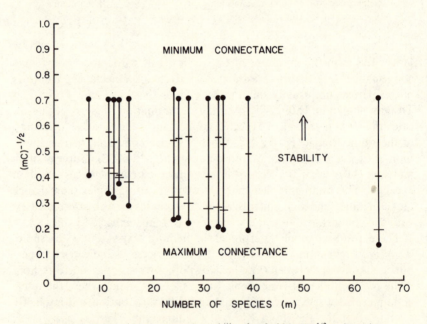

Fig. 24.11. Relationship between the stability threshold $(mC)^{-1/2}$ and number of species (m), for community food webs. Threshold values were calculated for the minimum (upper circles), maximum (lower circles), and observed (short horizontal lines) connectances and the midpoint of the connectance range (long horizontal lines) of each food web.

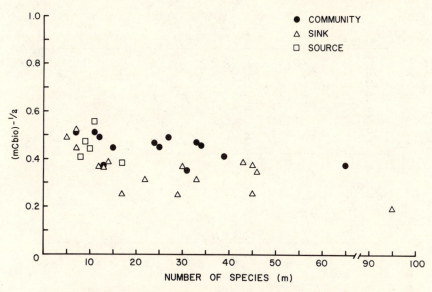

Fig. 24.12. Relationship between the stability threshold based on maximum biologically feasible connectance $(mC_{bio})^{-1/2}$ and number of species (m), for community, sink, and source food webs.

some aspects of food-web shape, such as those listed above, no one has treated the general influences of web topology on stability analyses. I now do so for small (4–6 species) webs.

I have calculated number of configurations and expected connectance for every possible number of trophic levels in four-, five-, and six-species webs. Figure 24.13 illustrates all two- through five-trophic-level configurations for five species. Solid lines represent constrained interactions necessary to maintain the trophic placement of species. Circles with no connections to adjacent levels represent areas where a species could interact with any species in the next lower trophic level and still maintain its web position. The dashed line in the third configuration from the left denotes an "unconstrained" species; that is, one that has no mandatory link with a species in the trophic level above it. The three configurations marked by stars have unillustrated unconstrained analogues. Unconstrained species are always present when there are more than two trophic levels and more than one basal species. The type of web shape represented by the unconstrained configurations is probably rare in nature, since basal species are generally fed upon by at least one member of the next highest trophic level. Table 24.3 lists results of the analysis of these model webs. No values were calculated for expected connectance or number of configurations for unconstrained six-species webs.

5 SPECIES CONFIGURATIONS

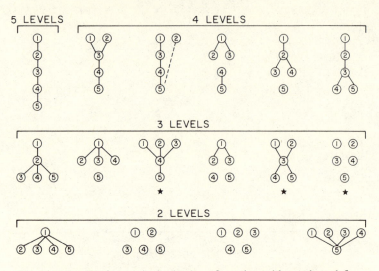

Fig. 24.13. Possible five-species food web configurations with two through five trophic levels. Solid lines represent constrained interactions, dashed line represents an unconstrained interaction. Stars indicate configurations with unconstrained analogues that are not shown.

For more than two trophic levels the most likely configurations within a fixed number of trophic levels tend to have the lowest expected connectances. These webs are the most likely to be stable in a simulation study in which the same range of interaction strength is sampled for different webs. Furthermore, as web size increases, the range of expected connectances observed at a fixed number of trophic levels decreases, the same relationship between C and m predicted by stability analyses and observed among real webs.

The identity between patterns predicted by stability theory and probability theory relates to many hypotheses concerning food webs. For instance, Pimm and Lawton (1977) have argued that the number of trophic levels in food webs should be small. This prediction has been in the ecological literature for a long time, but has been linked to thermodynamics, while Pimm and Lawton's assertion is based on stability theory. It is apparent in Table 24.3 that expected connectance declines as number of trophic levels decreases. The most likely number of trophic levels, defined as the level with the greatest number of possible webs, lies between the minimum and maximum number of levels. For four and five species it is three levels and for six species it is four levels. This hint of a slight increase in the most likely number of trophic levels as m increases is exactly the pattern observed among real community webs analyzed here.

As number of trophic levels decreases for a web with a fixed number of species, the potential number of omnivores decreases. Possibilities of omnivores having prey in non-adjacent trophic levels also decrease as species are restricted to fewer levels. These observations are not directly analogous to Pimm and Lawton's (1978) predictions concerning omnivores because they found that changing the position of omnivores in a web affected stability analyses, even when connectance was fixed. However, the probability approach to web shape suggests alternative explanations for many patterns apparent in real webs, and the scarcity of certain types of interactions may result because there are few possible arrangements with the shape necessary to permit these interactions.

The probability approach provides insight into two other food-web predictions. Pimm and Lawton (1980) analyzed species deletion stability (the stability of a web after a species has been removed) of model webs and found that the position held by the eliminated species influenced whether the reduced web was stable. Removal of a top predator had less effect than removal of a basal species. The results exhibited in Table 24.3 suggest the same conclusion. For a given number of species and trophic levels, webs with the highest expected connectance tend to be those with more top predators than species in any other level. Webs fat in the middle levels (the pattern seen in most real webs analyzed here) or ones fattest at the basal species and second trophic level generally have lowest expected connectances. In these, removal of a top predator usually leads to lower expected connectance than does removal of a basal species. For instance, a six-species web in the 2–2–2 configuration has an expected connectance of .4919. If a top predator is removed the web reduces to a 2–2–1 configuration with C_{exp} of .5421, while the elimination of a basal species leaves a 1–2–2 configuration with C_{exp} equal to .5727.

Pimm (1979b) has also noted that different-chain omnivory will have little impact on stability of a web in comparison to same-chain omnivory. The unconstrained webs in Figure 24.13 are examples of different-chain omnivory. Every unconstrained model I examined has a lower C_{exp} than any configuration with the same number of species and trophic levels. Once again the predictions of stability theory and probability theory are similar.

CONCLUSIONS

I have emphasized two points concerning the potential problems associated with food-web stability analyses, including possible sources of bias or artifactual results. First, current methods suffer from myriad assumptions and inconsistent ground rules, such as differing means of identifying web members. Second, one must examine mathematical properties of the

Table 24.3. Number of webs and minimum, maximum, and expected connectances for different topologies of 4-, 5-, and 6-species food webs.

Number of Species	Number of Trophic Levels	Configuration (Basal Species to Top Predators)	Number of Webs	C_{min}	C_{max}	C_{exp} (mean)	C_{exp}[1] (modal)
4	4	1-1-1-1	8	.5000	1.0000	.7500	.6667 (3), .8333 (3)
4	3	1-1-2	4	.5000	.8333	.6667	.6667 (2)
4	3	1-2-1	6	.5000	.8333	.6389	.6667 (3)
4	3	2-1-1	4	.5000	.8333	.6667	.6667 (2)
4	3	2-1-1*	8	.5000	.8333	.6250	.6667 (4)
4	2	1-3	1	.5000	.5000	.5000	.5000 (1)
4	2	2-2	5	.5000	.6667	.5333	.5000 (4)
4	2	3-1	1	.5000	.5000	.5000	.5000 (1)
5	5	1-1-1-1-1	64	.4000	1.0000	.6844	.7000 (20)
5	4	2-1-1-1	32	.4000	.9000	.6500	.6000 (10), .7000 (10)
5	4	2-1-1-1*	84	.4000	.9000	.6048	.6000 (28)
5	4	1-2-1-1	48	.4000	.9000	.6333	.6000 (16)
5	4	1-1-2-1	48	.4000	.9000	.6333	.6000 (16)

5	4	1-1-1-2	32	.4000	.9000	.6500	.6000 (10), .7000 (10)
5	3	1-1-3	8	.4000	.7000	.5500	.5000 (3), .6000 (3)
5	3	1-3-1	14	.4000	.7000	.5214	.5000 (6)
5	3	3-1-1	8	.4000	.7000	.5500	.5000 (3), .6000 (3)
5	3	3-1-1*	26	.4000	.7000	.5038	.5000 (12)
5	3	1-2-2	44	.4000	.8000	.5727	.6000 (21)
5	3	2-1-2	16	.4000	.8000	.6000	.6000 (16)
5	3	2-1-2*	42	.4000	.8000	.5571	.5000 (16)
5	3	2-2-1	76	.4000	.8000	.5421	.5000 (30)
5	3	2-2-1*	92	.4000	.8000	.5239	.4000 (36)
5	2	1-4	1	.4000	.4000	.4000	.4000 (1)
5	2	2-3	19	.4000	.6000	.4421	.4000 (12)
5	2	3-2	25	.4000	.6000	.4320	.4000 (18)
5	2	4-1	1	.4000	.4000	.4000	.4000 (1)
6	6	1-1-1-1-1	1024	.3333	1.0000	.6667	.6667 (252)
6	5	1-1-1-2	512	.3333	.9333	.6333	.6000 (126), .6667 (126)
6	5	1-1-1-2-1	768	.3333	.9333	.6222	.6000 (196)

(*Continued*)

Table 24.3   (Continued)

Number of Species	Number of Trophic Levels	Configuration (Basal Species to Top Predators)	Number of Webs	C_{min}	C_{max}	C_{exp} (mean)	C_{exp}[1] (modal)
6	5	1-1-2-1-1	768	.3333	.9333	.6222	.6000 (196)
6	5	1-2-1-1-1	768	.3333	.9333	.6222	.6000 (196)
6	5	2-1-1-1-1	512	.3333	.9333	.6333	.6000 (126), .6667 (126)
6	4	1-1-1-3	128	.3333	.8000	.5667	.5333 (35), .6000 (35)
6	4	1-1-3-1	222	.3333	.8000	.5489	.5333 (65)
6	4	1-3-1-1	222	.3333	.8000	.5489	.5333 (65)
6	4	3-1-1-1	128	.3333	.8000	.5667	.5333 (35), .6000 (35)
6	4	1-2-2-1	864	.3333	.8667	.5667	.5333 (231)
6	4	1-1-2-2	576	.3333	.8667	.5778	.6000 (155)
6	4	1-2-1-2	384	.3333	.8667	.5889	.6000 (105)
6	4	2-2-1-1	1720	.3333	.8667	.5346	.5333 (456)
6	4	2-1-2-1	376	.3333	.8667	.5915	.6000 (105)
6	4	2-1-1-2	256	.3333	.8667	.6000	.6000 (70)
6	3	1-2-3	216	.3333	.7333	.5000	.4667 (66)

6	3	1-3-2	160	.3333	.7333	.4867	.4667 (51)
6	3	2-2-2	1288	.3333	.8000	.4919	.4667 (360)
6	3	2-1-3	64	.3333	.7333	.5333	.5333 (20)
6	3	2-3-1	698	.3333	.7333	.4560	.4667 (215)
6	3	3-2-1	882	.3333	.7333	.4573	.4000 (273)
6	3	3-1-2	64	.3333	.7333	.5333	.5333 (20)
6	2	1-5	1	.3333	.3333	.3333	.3333 (1)
6	2	2-4	65	.3333	.5333	.3774	.3333 (32)
6	2	3-3	217	.3333	.6000	.3899	.3333 (90)
6	2	4-2	67	.3333	.5333	.3781	.3333 (32)
6	2	5-1	1	.3333	.3333	.3333	.3333 (1)

[1]Numbers in parentheses indicate number of possible webs with the modal connectance.
*Unconstrained configurations. See text.

stability models to gauge what constraints are placed on important parameters independently of biology and to determine expected values or distributions of these parameters. When these are done, food webs generally appear to have fewer interspecific interactions than chance alone dictates.

Pimm (1980b) recently commented that "Confirmation of all these predictions from stability analyses suggests that system stability places necessary, though not sufficient, limitations on the possible shape of food webs." In conducting analyses of real and hypothetical webs, I have taken the opposite viewpoint. Observed and likely shapes of food webs place constraints on the possible outcome of stability analyses. Interestingly, when Yodzis (1981) rearranged the location of interspecific interactions in real food webs he found real webs had a greater percentage of stable solutions than their permuted analogs. I should emphasize that by using probability theory to generate expected connectances I am not arguing that real webs are organized randomly in nature. I employed this approach simply as one way of examining what intrinsic patterns might lie within the present method of stability analysis. The probability technique demonstrates that there are alternatives to the hypothesis that dynamical constraints determine the organization of food webs.

Community Changes
in Time and Space

25.

On Understanding a Non-Equilibrium World:
Myth and Reality in Community Patterns and Processes

JOHN A. WIENS

Shrubsteppe Habitat Investigation Team, Department of Biology,
University of New Mexico, Albuquerque, New Mexico 87131

The mission of community ecology, as of any scientific endeavor, is to detect the *patterns* of natural systems, to explain them by discerning the causal *processes* that underlie them, and to generalize these explanations as far as possible. We usually detect patterns by making comparisons among features of different natural communities, or between the natural systems and logical or theoretical models of community structuring. Finding patterns, we then often derive a process explanation by inference, by constructing logical and realistic scenarios of how pattern and process might be linked, or by accepting the underlying premises of a theoretical model as causal explanations of the patterns. Manipulative experiments may be performed to define the mechanisms underlying the patterns more rigorously, but this is not a frequent practice, nor is it as impeccable as is often believed (see Schoener, 1974b; Dayton and Oliver, 1980; Underwood and Denley, this volume).

Lying at the foundation of contemporary views of community patterns and the processes producing them is the presumption that these systems are at or close to equilibrium:[1] they are ecologically saturated, resource limited, and governed by biotic interactions, especially competition (Cody, 1974a, b; Schoener, 1974b; Pianka, 1976). This equilibrium need not be static; fluctuations in natural communities are generally acknowledged, often accompanied by the supposition that variations in resource levels

---

[1] "Equilibrium" has been defined in various ways, which primarily separate into considerations of (1) the stability or "steadiness" of community components and (2) the ability of the system to return to some previous state following perturbation, its resilience (Holling, 1973; Leigh, 1975; Harrison, 1979; Smith, 1980). Although meaning (2) is relevant to some of the questions of community ecology (*e.g.* are some species combinations more resistant to invasion than others?), meaning (1) is more appropriate to most community questions. I use "equilibrium" in the sense of meaning (1) here.

are closely tracked by members of the community (Cody, 1981; see also Boyce and Daley, 1980). Ecology has a long history of presuming that natural systems are orderly and equilibrial (the "balance of nature" notion; see Worster, 1977), and the infusion of evolutionary thinking into ecology strengthened this view, providing a mechanism (natural selection) that may lead to the development of optimally structured communities (Cody, 1974b; Cody and Diamond, 1975b; Pianka, 1976). This conception of natural systems stems in part from a worldview derived from Greek metaphysics, which proposes that nature must, ultimately, express an orderly reality (Simberloff, 1980), and in part from our theory, which is largely founded upon equilibrium or near-equilibrium mathematics (Ulanowicz, 1979; Levins, 1979). Rather than simply accepting this assumption as an article of faith (*e.g.* Cody, 1974a, 1981), however, we should investigate the equilibrium status of communities in its own right. Further, even if the patterns predicted from equilibrium-based theory are observed, it does not follow that the community is in fact in equilibrium—one may obtain the "right" patterns for the wrong reasons (Dayton, 1973; Thomson, 1980).

I began my own studies of communities fervently embracing the existing views of competitively structured, equilibrium communities (*e.g.* Wiens, 1969). But I have become skeptical of much of this dogma (*e.g.* Wiens, 1977a), and believe now that we know far less about the patterns and processes of communities than we think we do. Here I will summarize some of the evidence that has fueled my skepticism, and discuss how problems in logic or procedure may cast doubt on many patterns and process explanations in community ecology.

PREDICTIONS AND PATTERNS IN SIMPLE HABITATS

Current ecological theory provides a variety of predictions of the patterns that should be expected of equilibrial, competitive communities. John Rotenberry and I have sought to test this theory by documenting the patterns that characterize the breeding bird communities of structurally simple habitats, the prairies and shrubsteppe of western North America. Most of these habitats are arid or semiarid, with mean annual precipitation ranging from less than 20 cm to almost 100 cm. The avian assemblages breeding in these locations are small, containing 2–7 regular breeding species, with overall densities of less than 100 to somewhat over 600 individuals/km^2 (Cody, 1966; Wiens, 1974a, 1981b; Wiens and Dyer, 1975; Rotenberry and Wiens, 1980a). Our investigations have not included wide-ranging raptors or gallinaceous species; with these forms excluded, the remaining species are all members of a single ground-foraging guild. As these studies are reported in detail elsewhere (Wiens,

1969, 1973, 1974a, b, 1977a, b, 1980, 1981a, b; Wiens and Dyer, 1975; Rotenberry and Wiens, 1978, 1980a, b; Wiens and Rotenberry, 1979, 1980a, b, 1981a, b, c; Rotenberry, 1980a, b), only the basic findings will be summarized here. The critical reader should refer to these publications to evaluate in detail the evidence that has prompted the following statements.

Ecomorphological Patterns

One source of evidence often marshaled to examine the predictions of theories of competitive communities is the patterning of morphology within and among populations (*e.g.* Grant, 1968; Hespenheide, 1973, 1975; Karr and James, 1975; Findley, 1976). Ricklefs and Cox (1977) consider such ecomorphological analysis to be a future focus of studies of community organization, and Maiorana (1978) suggests that morphology may provide "a firm base from which to investigate the nature of competition interactions in natural communities."

Theories of community morphology generate a variety of predictions, such as:

1. In saturated, resource-limited, competitive communities, species should differ in size by some reasonably constant ratio in order to co-exist in equilibrium; a value of approximately 1.3 for the ratio of bill or jaw sizes of adjacent species on the size gradient is widely accepted (Hutchinson, 1959; Sam the Boatman, pers. comm.; but see Horn and May, 1977, for an alternative view of this ratio).

2. Patterns of morphological variation should be closely related to resource use, both within and between species (*e.g.* Schoener, 1965; Hespenheide, 1973; Herrera, 1978). If food is a limiting resource, this relationship should be especially tight between feeding morphology and prey dimensions.

How do our observations relate to these predicted patterns? First, it is apparent that there is little consistency in the patterns of spacing of members of local communities along a morphology spectrum; the ratios of bill size and body weight between adjacent-sized species varied from 1.03 to 3.19 and 1.12 to 3.97, respectively (Wiens and Rotenberry, 1981c). The only obvious regularity among the communities we examined is a gap between a set of small species (10–40 g) and the larger species (100–130 g). This gap is substantially greater than that predicted by theory, however, even if revisions to the basic Hutchinsonian predictions (Maiorana, 1978) are considered.

Variation in morphology within a species should be substantially less than behaviorally mediated variability in resource use. As this variability in resource use may diminish the effectiveness of a given amount of morphological separation between species, the morphological ratios between species should be greater than the ratios of the prey sizes they consume in order to permit coexistence (Maiorana, 1978). Figure 25.1, in addition to showing the scatter of the morphology ratio values, indicates the failure of our observations to meet this expectation for bill-size and prey-size ratios. Moreover, many of the values for comparisons involving small species fall below a prey-size ratio of 1.0; these values indicate situations in which the larger of two adjacent species within a community consumed *smaller* prey, on average, than its counterpart.

This finding casts doubt on the efficacy in our systems of the second of the above predictions, that morphology should be closely linked to resource use. Correlation tests relating bill size or body size to prey size among collections of different species over the areas we sampled (Wiens and Rotenberry, 1980a) indicated significant positive relationships, as expected. This overall pattern, however, is determined by the differences between the set of large species and the small species. If the larger species are deleted from the analysis, and relationships are examined within the set of small species alone, the pattern vanishes. Within single species

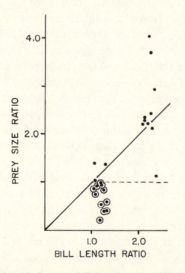

Fig. 25.1. The relationship between bill-length ratios of bird species coexisting in grassland and shrubsteppe locations and the ratios of the mean sizes of prey taken by those species. Each point represents a comparison of the ratios of species collected at a single location in a single year. The solid line indicates equivalence of morphological ratios and prey-size ratios of species pairs (ranked according to their bill sizes). The dashed line represents a prey-size ratio of 1.0; points falling below this line (circled) indicate situations in which the larger member of a coexisting species pair consumed prey of a smaller mean size than the smaller member. After Wiens and Rotenberry (1981c).

populations, we found little substantiation of the prediction that larger individuals should consume larger prey. Of the 23 correlation tests conducted, only six revealed significant correlations between bill lengths and prey sizes, and four of these registered *negative* relationships.

Clear, ecomorphological patterns agreeing with predictions thus emerge in these communities only at the most general level. When the specifics of local populations or communities are considered, they evaporate or are even reversed from those expected from theory.

Diet Niche Relationships

Much current thinking about communities presumes that they are structured by competition related to limitation of food resources (MacArthur, 1972; Schoener, 1974b). Thus the conventional wisdom states that (1) coexisting species should exhibit consistent patterns of divergences in their use of prey types or prey sizes. Further, (2) overlap in diet niche dimensions should be least among those forms that differ the most in body size (as suggested indirectly by Schoener, 1977).

Our studies of the diet composition of breeding birds at four grassland and shrubsteppe locations (Wiens and Rotenberry, 1979; Rotenberry, 1980b) reveal substantial variability in the diets of the species between locations and between years at the same location. Although some coexisting species did differ in dietary composition, dietary niche overlaps were not consistent from year to year at a given location, and many overlap values were quite high. Niche overlap in dietary composition was unrelated to the size difference between coexisting forms (Figure 25.2). A similar lack of overall pattern characterized the utilization of prey sizes by these species. At some locations the prey size utilization curves of co-occurring species diverged sharply, but at other locations or other times the utilization curves of the species were nearly coincident or displayed patterns of overlap or separation opposite to those predicted by theory (see for example, Fig. 5 in Wiens and Rotenberry, 1979). No clear or consistent patterns of overlap or divergence in the use of prey taxa or prey sizes were apparent in these assemblages.

Additional insight into dietary niche dynamics in this system can be derived from Rotenberry's (1980b) more intensive studies at a single shrubsteppe location, where diets were sampled repeatedly through the year. A cluster analysis of these samples based upon similarity in prey taxa composition shows clearly that samples of the same species taken at different times do not cluster together (as would be expected if each species were pursuing a competitively constrained form of dietary specialization). Instead, samples of *different* species taken at the same time cluster together, at rather high similarity levels. Thus, individuals of the different species appear to consume much the same complex of prey items at any

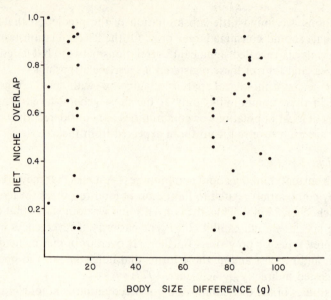

Fig. 25.2. The relationship between niche overlap in diet composition (by prey taxa) and body size difference among coexisting grassland and shrubsteppe birds. Each point represents a species pair coexisting at a given site in a given year. See Wiens and Rotenberry (1979) for additional details.

given time, but their diets change in parallel through time. These dietary features all seem most consistent with the interpretation that the birds are foraging in an independent, opportunistic fashion in a variable but non-limiting environment. We see none of the niche patterns that would be expected in a resource-limited, competitive community at equilibrium.

Patterns of Habitat Occupancy

For many organisms, variation in features of habitat structure represents an important set of niche dimensions that may be partitioned to enhance the coexistence of species (as, for example, in MacArthur's (1958) warbler community). Because habitat (vegetation) structure has been presumed to be especially meaningful to the structuring of bird communities (*e.g.* Cody, 1974a, 1981), several predictions relating habitat equilibrium to community patterns merit examination:

1. Suites of bird species should covary in their distribution and abundance. These represent the sets of ecologically compatible species or "permitted" species combinations that are the outcome of competitive interactions (Diamond, 1975, 1978; Cody, 1981).

2. Sets of bird species and community attributes such as species diversity should exhibit clear patterns in relation to variations in habitat structure, especially vertical layering (MacArthur and MacArthur, 1961; Cody, 1968b).

3. Species within a community that evidence high niche overlap on one habitat niche dimension should have low overlap on another—"niche complementarity" (Schoener, 1974b) should be evident.

4. If habitat structure varies through time, bird species distributions and community composition should change in consistent and related manners, as the birds "track" resource variations (Cody, 1981).

We have conducted our studies of bird/habitat relationships on two geographic scales, and it is instructive to consider how well these predictions are borne out at each scale. On a "continental" scale, spanning the spectrum from tallgrass prairies through Great Basin shrubsteppe (Rotenberry and Wiens 1980a), several clearly defined suites of covarying species were apparent: a set of 4–6 species with distributions centered in the eastern mesic prairies, another group of 3–4 species with affinities for more xeric shortgrass prairies, and a third suite of 3–4 species occupying the western shrubsteppe habitats. This pattern reinforced and extended the findings of an earlier association analysis conducted on a different data set (Wiens, 1973). Whether such suites of species represent limited combinations "permitted" by their ecological complementarity, with all other combinations being "forbidden," is debatable. It is clear, however, that at this scale of observation sets of mutually co-occurring species are arrayed in a definable pattern.

Beyond this finding, most of the species recorded in our continental-scale examination were significantly associated with specific habitat features. By conducting a Principal Component Analysis (PCA) on the data matrix for the 22 habitat features that we measured, we defined three unambiguous, statistically independent dimensions of habitat variation. The first two represented variation in horizontal patchiness and vertical heterogeneity, respectively. Nearly all of the birds belonging to definable suites of species evidenced clear and significant associations with these synthetic dimensions of variation in habitat structuring. Moreover, bird species diversity was significantly correlated with increasing vertical habitat heterogeneity, as expected from conventional theory. Yet, if the two PCA axes are considered to represent independent niche dimensions, we find no evidence of niche complementarity among these bird species— high niche overlap on one dimension is in fact significantly associated with high niche overlap on the other (Figure 25.3).

So at this broad continental scale, some of the patterns anticipated from equilibrium community theory are clearly evident, while others fail to

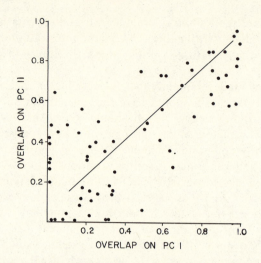

Fig. 25.3. Niche overlap of coexisting grassland and shrubsteppe bird species on the first and second axes of a Principal Components Analysis of features of habitat structure. PC I represents a synthetic gradient in horizontal habitat heterogeneity, PC II a gradient in vertical habitat heterogeneity. The regression is significant ($r = 0.71$, $p < 0.001$). There is no evidence of overall niche complementarity on these two dimensions of habitat variation. See Rotenberry and Wiens (1980a) for further development.

materialize. To seek a finer resolution of these relationships, we conducted a parallel analysis of bird-habitat patterns on a "regional" scale, considering a series of 14 plots located in Great Basin shrubsteppe in Oregon and Nevada (Wiens and Rotenberry, 1980b). On that scale, we were unable to discern any suites of covarying or associated bird species. A PCA conducted using the matrix of bird species abundances yielded six significant components, each of which accounted for roughly 12% of the variation present in the original data matrix. No definable sets of bird species sorted out on any of these components; the species were instead broadly spread over the six dimensions in no visible or statistically detectable patterns. The species thus appear to be distributed quite independently of one another on this scale.

Given this result, it is not too surprising that there were relatively few correlations between distribution and abundance of the bird species and features of habitat structure. The most typical shrubsteppe species were even more weakly related to habitat structuring than were several "grassland" species, for which this habitat might be considered ecologically marginal. The typical shrubsteppe species, however, were more strongly associated with variations in the coverages of several shrub species than

with habitat structure variations; the grassland species evidenced few significant correlations with shrub species coverages. PCA and canonical
correlation tests indicated that less than 20% of the variation in the overall matrix of bird abundances could be related to variation in the habitat
structure data matrix. On this scale of resolution, then, the bird species
appear to vary independently not only of one another, but of habitat
structuring as well.

But what of temporal variation? Perhaps the failure of clear patterns
to emerge on the regional scale is a consequence of our combining temporally separated samples, and thus obscuring the dynamics of resource
tracking that might actually be occurring. Where precipitation varies
greatly from year to year, changes in productivity, and in the resource
base available to birds, should be closely associated with changes in the
bird community (Grant and Boag, 1980; Cody, 1981). Our regional studies
included sampling at each location over three consecutive years. The first
year of our studies followed an extremely dry period (27% of normal precipitation during the October–April "bioyear"), while the second and
third years were unusually wet (136% and 126% of normal, respectively)
(Rotenberry and Wiens, 1980b). Possibly some of the patterns expected
from equilibrium or resource-tracking theory might emerge were we to
consider the years separately.

Habitat structure changed substantially over the sites between the dry
and the following wet year, producing an overall increase in the height
and coverage of vegetation and a decrease in horizontal patchiness. Despite these rather obvious changes in habitat structure between years,
however, the overall relationships of the sites to one another with respect
to their habitat structure remained essentially unchanged—all sites
changed in basically the same way. Shrub species coverages, on the other
hand, exhibited no significant changes between years, which is not unexpected for long-lived woody desert perennials. There was substantial
between-year variability in local population densities of the bird species,
but despite this variability, none of the species demonstrated any tendency
to increase or decrease consistently across all sites between years. Further, the avifaunas of the different sites changed in different ways between
years—the sites retained no statistically discernible pattern of relationship in bird community composition to one another from one year to the
next. Bird populations in this system thus appear to vary in a manner
that is independent of changes in the structural configuration of the habitat, changes in the densities of other coexisting bird species, and changes
in populations of the same species at other sites. We find no evidence of
orderly, patterned tracking of habitat variations during an episode of substantial environmental change by either single bird species or the entire
avian community.

Alternative Explanations

Our search for patterns in these communities, guided by the predictions of equilibrium community theory, has scarcely been a resounding success. We can detect a few general patterns on a broad, biogeographic scale, but these are not replicated on a local scale, which is the arena where the population interactions presumed to produce them should take place. Instead, the bird populations in this system seem not to be in any readily definable equilibrium, vary pretty much independently of one another, do not appear to be resource-limited very often, do not fully saturate the available habitat, and do not provide clear evidence of ongoing competition. We seem to be dealing with non-equilibrium, non-competitive communities, by any conventional definition of those terms.

Previously I have suggested that such loose community structuring should be expected in highly variable or harsh environments (Wiens, 1974a, 1977a). Under such conditions, resource levels and environmental conditions may at times be severely constraining, imposing strong selection on populations of coexisting species and prompting intense competition. Such "ecological crunches" may act in a major way to determine the ecological adaptations of species and the biogeographic patterns of community composition. Much of the time, however, the environment may be more benign, and resource levels essentially non-limiting. Under these conditions, populations and communities may be freed of close, direct biotic or even abiotic control, and vary in manners that erode the clean patterns expected from equilibrium theory. The habitat may not be fully saturated with breeding individuals, and the community not saturated with species. Price (1980) has noted similar attributes of parasite communities that he considers to be non-equilibrial, Dunham (1980) has documented such a scenario in communities of rock-dwelling lizards, and Ehrlich *et al.* (1980) describe the complex effects, ranging from extinction through stability to increase, that a severe drought "crunch" had on local populations of butterflies.

Such an explanation does not set well with advocates of contemporary competitive community theory; Cody (1981), for example, recognizes but summarily dismisses such a possibility. Instead, various *ad hoc* explanations may be offered to reconcile observations such as ours with current views when the *ceteris paribus* assumption apppears not to hold (*e.g.* Hutchinson, 1978:494 ff.). Such *ad-hoc*ing behavior is, of course, characteristic of attempts to preserve a prevailing paradigm or research tradition (see Popper, 1968; Kuhn, 1970; Lakatos, 1970). Thus:

1. Perhaps we have not uncovered the true patterns of our communities because we have failed to measure the proper parameters or niche dimensions (*e.g.* Noon, 1981) or have overlooked varying resource levels that

are actually tracked closely through time by a changing set of consumers (Cody, 1981). Of course, the array of possible niche dimensions that might be considered in searching for the "true" patterns is virtually endless, rendering much of the theory untestable.

2. Ecological systems are often characterized by time lags between events and system responses (May, 1973), so perhaps our failures reflect an inattentiveness to the proper periodicity of time lags. Perhaps if we considered the relations between birds and habitat structure variation using a 2-yr, or 3-yr, or some other time lag, clear patterns might emerge. The array of possible time lags that might possibly be biologically relevant is large, and in general we should attempt to avoid postulating unspecified time lags to explain away anomalous observations (Maynard Smith, 1978). A variation on this *ad-hoc*ing theme is the ressurection of the "ghost of competition past" (Connell, 1980), the suggestion that the contemporary structure of the community has been molded by competition at some time in the past, which, now resolved, is undetectable as a process (see Noon, 1981, for a recent appearance of this "ghost"). Such arguments are not only untestable but logically flawed.

3. Many of the theoretical arguments assume some degree of uniformity or at least continuity in resource distributions. But one may also posit irregular resource distributions to reconcile observations with competitive community theory. Thus, for example, the irregular spacings of species along morphological dimensions that we observe might reflect clear inequities in the availability of different sizes of prey during periods of resource limitation.

4. Perhaps the most malevolent *ad hoc* argument that has been leveled against our findings is that the grassland and shrubsteppe systems we have studied are somehow atypical or weird. These communities may indeed be non-equilibrial, it is argued, but surely most other systems must be "normal," and at or close to equilibrium. As the equilibrium status of these communities has more often been assumed than measured, however, there is little way to determine with any rigor how the grassland/shrubsteppe communities do compare with others. Some preliminary notion may be gained, however, by considering the year-to-year variability in the densities of species populations and total avian density and the yearly turnover in species composition characteristic of a variety of general habitat types (Table 25.1). Myers *et al.* (manuscript) have recently analyzed such measures of equilibrium or stability. They conclude that, although some specific measures of avian community stability do vary significantly among habitat types, most do not. Of the measures shown in Table 25.1, species turnover does vary significantly across the spectrum from forests to deserts, but the two measures of density stability show no habitat-related pattern of variation. In any event, the grassland and

Table 25.1. Between-year variability ($\bar{x} \pm$ 1SD) in population densities and turnover in species composition for several North American habitat types. From Myers *et al.* (manuscript). Variability in population density or total community density is measured as the coefficient of variation (CV) of annual density measures considered over a minimum of three consecutive years at each location. Turnover is measured as the proportion of species present in a censused community in only one of two consecutive years.

Habitat Type	n	CV Total Density	CV Species' Densities*	Species Turnover
Eastern deciduous forest	13	15 ± 13	40 ± 11	14 ± 4
West coast coniferous forest	4	19 ± 9	43 ± 4	12 ± 5
Rocky Mountain coniferous forest	10	26 ± 18	52 ± 18	23 ± 7
Grassland	16	18 ± 9	40 ± 13	16 ± 9
Shrubsteppe	5	18 ± 10	58 ± 22	26 ± 13
Desert	6	25 ± 17	47 ± 15	30 ± 8
Tundra	5	24 ± 3	55 ± 10	22 ± 4

* Averaged over species but weighted by relative abundance.

shrubsteppe avifaunas do not stand out as being notably less stable than those of most other temperate or high-latitude habitat types in North America.

There is another reason to question the view that these systems are atypical. Arid and semiarid grassland, shrubsteppe, and deserts comprise 34% of the total land area of the earth (Whittaker, 1975), and this proportion is increasing with the desertification of many areas. If tundra habitats are included, the percentage is increased to 40%. This seems a rather substantial portion of the earth's land surface to consider as "atypical," especially when compared with the 22% contributed by temperate forests and woodlands, or the 16% contributed by tropical forests.

Finally, Cody's studies (1966, 1968b, 1974a) of bird community organization and niche relationships in grasslands provide what has been considered by some to be fairly strong evidence supporting theories of competitive structuring of equilibrium communities. These findings are developed at length in support of theories of competitive community structuring in some textbooks (*e.g.* MacArthur, 1972:157–159; Collier *et al.*, 1973:323–330; Hutchinson, 1978:186–190; Elseth and Baumgardner, 1981:417–419). Cody's studies and ours have been conducted on the same avian assemblages (and at some of the same locations). One cannot accept one set of data as confirming evidence and discount the other because the system is somehow "weird."

CONCLUSIONS: NON-EQUILIBRIUM, REALITY, AND MYTHS

Community ecology has been dominated too long by the assumption that natural systems are at or close to equilibrium, and that therefore most of the theory that has been developed is directly relevant (see Yodzis, 1981, for a recent effort to legitimize this viewpoint). If one accepts our argument that the bird communities we have studied in grassland and shrubsteppe habitats are non-equilibrial, then the general absence of the patterns predicted by theory is perhaps not surprising. Our work suggests the possibility that other systems might also be non-equilibrial much of the time (see also Sousa, 1979a). I do not suggest, however, that all systems might turn out to be non-equilibrial upon close examination. Natural communities should be viewed as being arrayed along a gradient of states ranging from non-equilibrium to equilibrium (Figure 25.4). At the extremes, some anticipated features of the communities may be defined. Non-equilibrium communities should be characterized by a general "decoupling" of close biotic interactions, and the species should respond to environmental variations largely independently of one another. Habitats may not be fully saturated with individuals, and species may be under-represented. Populations and communities may be more strongly influenced by abiotic agents than by the imposition of ceilings on resource abundance, and population dynamics may be governed by effects that are largely independent of density. Behaviorally, individuals may exploit resources in a more or less opportunistic fashion. Stochastic environmental effects upon the community may be of considerable importance, especially at local scales. Collectively, these factors produce communities that are only loosely structured, and clear and consistent patterns may be largely lacking. At the equilibrium end of the gradient, these features will generally show a tighter organization, leading to the sorts of patterns expected

NONEQUILIBRIUM	EQUILIBRIUM
Biotic decoupling	Biotic coupling
Species independence	Competition
Unsaturated	Saturated
Abiotic limitation	Resource limitation
Density independence	Density dependence
Opportunism	Optimality
Large stochastic effects	Few stochastic effects
Loose patterns	Tight patterns

Fig. 25.4. Natural communities may be arrayed along a spectrum of states from equilibrium to non-equilibrium. At either extreme, several attributes of community structuring or dynamics can be anticipated, as shown.

from theory (Figure 25.4). This conceptualization bears some similarities to the $r-K$ dichotomy (Pianka, 1972) and to Southwood's (1976) "duration stability" spectrum. Finding that a given system is equilibrial or non-equilibrial, of course, does not justify the conclusion that the attributes shown for these states in Figure 25.4 have in fact been demonstrated.

Equilibrium and non-equilibrium states do not represent mutually exclusive states of community (*contra* Thomson, 1980:724), but opposite poles of a spectrum of community states. A major objective of community ecology should be to begin to place various natural systems at positions along this spectrum. Because we have not been attentive to the possibility that natural communities might be spread over such a gradient instead of tightly clustered at the equilibrium end, however, we have little information to guide us in making such placements. Measures of population or community variability, such as those given in Table 25.1, provide a provisional point of departure. It might be more effective, however, to begin establishing the sorts of criteria that would enable us to make *a priori* predictions of the relative position of a given community along the gradient. Thus, we might expect equilibrium communities to develop in situations of moderate to low environmental variation, where abiotic conditions are not extremely unpredictable or harsh, and where there is a moderate to high degree of spatial heterogeneity in the system. Equilibrium also might be likely in assemblages of intermediate biotic complexity, especially if sets of closely linked or coevolved species are included. Domination of trophic web dynamics by intraspecific rather than interspecific interactions may also enhance equilibrium (Yodzis, 1981). Assemblages of populations that are largely resident or sedentary, and that have the potential for relatively rapid population growth following perturbation, may also tend toward equilibrium. Other similar attributes can be imagined.

The prospect that some (perhaps many) communities may be non-equilibrial creates several problems. These affect the degree to which we regard the community patterns that have been detected and their underlying process explanations as reality[2] or as myths. The presumption of equilibrium, for example, affects the methodology that may be employed to gather the observations or evidence that, in turn, reveal patterns (Wiens and Rotenberry, 1981a). Under conditions of equilibrium and saturation, the common practice of measuring only a few resource variables can be justified, as one need consider only those few key niche dimensions that

---

[2] I consider patterns that result from an underlying biological mechanism to be "real." Statistical patterns may emerge upon analysis of observations, and these are "real" in a statistical sense; whether they represent reality or myth, as I use the terms here, depends upon the existence of an associated *biological* reality.

serve to separate the community members. True patterns can be revealed by taking only single samples from a series of locations and then comparing them, as conducting replicate or repeated sampling would presumably produce much the same patterns (Wiens, 1981b). There should be no problems involved in conducting studies at different scales of resolution, because under equilibrium conditions, patterns and the processes that produce them should hold broadly and be independent of scale. If the community studied is *not* in equilibrium, however, the methodology becomes much more involved. Many resource variables must be measured, as many different variables may be important to different species, in different places, at different times. Because the components of the community may vary independently of one another, repeated sampling will be necessary to reveal the dynamics of the system. Studies conducted on different scales of space or time may well produce different results, as the system varies as a result of both deterministic and stochastic events that are not independent of scale (Taylor and Taylor, 1977; Poole, 1978; Sousa, 1979a). What this all means is that any "patterns" that are detected by application of an equilibrium-based methodology to a non-equilibrial community have a strong likelihood of being more myth than reality.

A somewhat more basic problem concerns the reality of the communities that are defined for study. MacArthur's (1971) definition of a community as "any set of organisms currently living near each other and about which it is interesting to talk" may be operationally useful, but it is biologically sterile. Terrestrial animal ecologists, in particular, often tend to define their communities by convenient taxonomic boundaries (*e.g.* lizard communities, bird communities, raptor communities, heteromyid communities), operating on the premise that the competitive processes that structure communities should be most intense among members of a restricted taxon. Such an assemblage, however, may represent only a small portion of the total community that is founded upon utilization of some common resource pool (Reichman, 1979a, and papers reviewed therein; Wiens, 1980). To the degree that the "true" community (as defined by biological relationships) has been incompletely represented in a taxonomically defined community, the "patterns" discerned for the latter assemblage may well be myths.

Bird communities pose additional problems. Many birds are extremely mobile, undertake short- or long-distance migrations, and demonstrate varying degrees of fidelity to previous breeding or wintering locations. These characteristics produce a substantial flux of individuals in local populations, and spread the sources of biological limitation of populations, and the consequent determination of community structuring, over a large and undefined area. Because most local assemblages of birds contain mixtures of species that differ in migratory tendencies and pathways and in longevity and fecundity, the dynamics of any given local assemblage are likely to be driven by an amorphous complex of factors, the effects of which are likely to be different for almost every local assemblage. This uncertainty decreases the probability that any "patterns" that seem apparent in such communities are in fact real.

Permeating all of the above is the additional problem of scale. Community ecologists have generally been insensitive to the effects that conducting studies on different spatial or temporal scales may have on the patterns that are detected and the processes that can be inferred to operate. MacArthur (1972), however, proposed something akin to a "Goldilocks" solution to this problem: we should search for patterns at a scale not too small (because the sample of species and individuals would be inadequate), and not too large (speciation and history will complicate things), but just right. MacArthur suggested that this "just right" scale should be that of relatively small areas of homogeneous habitat containing an adequate number of species. This is in fact precisely the scale on which our own studies of local communities have been conducted.

The problem, really, is that processes do not operate independently of scale in space and time (Taylor and Taylor, 1977; Dayton, 1979a; Simberloff, 1980; Dayton and Oliver, 1980; Wiens, 1981a). If the local populations and communities that we study are non-equilibrial, vary independently of one another, and are governed by a wide array of underlying processes (many of which may have a substantial stochastic element), then the dynamics of the various local communities may well seem chaotic. Nonetheless, the summation of such a heterogeneous collection into larger regional or biogeographic sets may produce apparent patterns in community structuring. These may well be artifacts of the summation procedure, myths that have no biological reality and no unitary process explanation. The hope that lurking "under the colored chaos there rules a more profound unity" (Bronowski, 1977:12) may often be in vain.

We may also question the reality of the process explanations that have been offered of the "patterns" of communities. For example, we have observed that the patterns of species' presence or absence over large areas are relatively stable and predictable (a "global" equilibrium), while the species membership of local assemblages fluctuates considerably (Table 25.1). Sousa (1979a) noted similar evidence of global equilibrium / local non-equilibrium in the species composition of intertidal boulder-field communities. Such observations have led to the suggestion that the species makeup of a community is more important than the numerical dynamics of the component species' populations, as it determines how frequently over evolutionary time a particular species has to contend with a particular competitor (Thomson, 1980). Such arguments, however, assert that process explanations such as competition apply equally well to population and to species patterns. "Species" do not interact competitively as units; individual members of local populations do. There would thus seem to be a logical inconsistency in explanations of global patterns of species' presence or absence that rely upon processes that are carried out among individual members of local populations. Appropriate explanations of broad patterns of species' distributions should be based upon biogeographic and evolutionary processes, not on features of local population behavior or dynamics.

A more serious form of logical flaw in process explanations of community patterns casts greater doubt on their veracity. For example, after describing some niche differences among coexisting species, Diamond (1978:324) states that "these niche differences result from competition: they are the means by which species minimize competition in nature." The cited studies only document ecological differences between species; Diamond relied upon the premises of competitive exclusion theory to infer a process underlying the presumed patterns. No actual evidence of competition was gathered in these studies, and Diamond's "explanation" is an assertion of faith rather than an empirical statement. It is this sort of relaxed logical procedure that allows Diamond to conclude that "ecologists using only the simplest observational methods now routinely document competition in a wide variety of animal and plant groups" (1978: 322).

Such "documentations," unfortunately, are almost always inferential assertions that explain patterns by the unthinking acceptance of the premises (process explanations) contained in a theory whose predicted patterns are at least coarsely matched by the observations—the "fallacy of affirming the consequent" (Northrop, 1959). Alternative hypotheses, some of which may predict much the same patterns from entirely different premises or process explanations (e.g. Wiens and Rotenberry, 1979), are usually not considered. The emphasis is upon verification of existing theory rather than broad and objective testing and refutation among alternative hypotheses. Thus, the process explanations offered following what seem to be accepted logical procedures in much of community ecology may also be more myth than reality. As Dayton and Oliver (1980:115) have observed, "The verification of ideas may be the most treacherous trap in science, as counter-examples are overlooked, alternative hypotheses brushed aside, and existing paradigms manicured."

If indeed many natural communities normally exist in a state some distance away from equilibrium, and the patterns and processes that have been "documented" following equilibrium thinking are more likely to be myth than reality, what do we do now? I would suggest the following:

1. We should question the preconceptions that seem to guide our approach to community studies. These are manifestations of the equilibrium competitive community paradigm, and they constrain the sorts of questions we ask, hypotheses we pose, methods we use, and conclusions we reach (Kuhn, 1970; Dayton and Oliver, 1980). The views of natural communities expressed in textbooks and journal articles should be subjected to critical evaluation of both the likely reality of the patterns that are portrayed and the logical rigor of the process explanations that are inferred.

2. Rather than attempting to test single hypotheses in a piecemeal fashion, we should follow the procedures of strong inference (Platt, 1964) and test suites of alternative hypotheses, structured in such a way that exclusive sets of predictions are available for testing (Wiens and Rotenberry, 1979). The emphasis should be upon refutation of untenable hypotheses, rather than on vain verifications of favored hypotheses.

3. Attention should be devoted to development of a richer body of community theory, less closely tied to equilibrial and near-equilibrial assumptions and more clearly cognizant of the spectrum of non-equilibrium through equilibrium conditions that may occur in nature.

4. In light of the recognition that non-equilibrium and equilibrium are not "either/or" states of communities, but that a gradient between these two extremes exists, greater effort should be devoted to placing various natural systems in their appropriate positions on this gradient.

5. We should acknowledge that local communities may often vary independently of one another, and thus concentrate our efforts on developing a thorough understanding of what the *real* patterns of the local assemblages of populations are, what features characterize their dynamics, and what processes underlie these patterns and dynamics.

6. It seems unlikely that broad-based correlational analyses or models based on little more than logic (*e.g.* Cody 1974b) will be able to reveal the linkages between processes and patterns with any degree of rigor. To do so will require manipulative field experiments. These, however, must be structured with great care, following the tenets of sound experimental design. Such experiments should be coupled with a firm knowledge of the dynamics of the resource bases on which the community is founded.

7. At least some of the problems of contemporary community ecology stem from a tendency to generalize, to divine "rules" for community structuring (*e.g.* Diamond, 1975). The temptation to generalize prematurely should be assiduously avoided.

Following these suggestions may not solve all of the problems of community ecology, but I believe they might redirect us onto a more productive pathway toward understanding how, why, and whether natural communities are structured.

Acknowledgments

Much of this work has been conducted in collaboration with John Rotenberry, who shares credit (and responsibility) for many of the views I have expressed. Many members of the ecology/evolution group at the University of New Mexico had a field day reviewing an early draft of the manuscript; Manuel Molles, Glenn Ford, Fritz Taylor, and Bea Van Horne offered especially helpful direction. Pete Myers generously supplied the unpublished analyses presented in Table 25.1. The research has been supported by the National Science Foundation, most recently through Grant No. DEB-8017445.

26.

Interspecific Morphological Relationships and the Densities of Birds

FRANCES C. JAMES AND WILLIAM J. BOECKLEN[1]

Department of Biological Science, Florida State University, Tallahassee, Florida 32306

INTRODUCTION

Ecologists generally agree that morphological adaptations are evolutionary responses to ecological conditions, and that if species populations remain fairly stable over time, density-dependent factors are probably involved (Krebs, 1972). When these intraspecific concepts are combined and extended to the community level, the question becomes: "Is there a linkage between the densities of individual species and the morphology of other members of the community?" This question is important because several widely held constructs in community ecology have the underlying assumption that community structure is largely attributable to such a linkage. Although the principle of competitive exclusion can be shown to be either trivial or invalid, depending on how it is formulated (Slobodkin, 1961; Den Boer, 1980), its extensions to community ecology, recast as diffuse competition (MacArthur, 1972), limiting similarity (Hutchinson, 1959; MacArthur and Levins, 1967), resource partitioning (Schoener, 1974b), and species packing (Pianka, 1975) are widely accepted as representing the outcome of past or present community-level organizing forces. The evidence for this linkage between the morphology of coexisting species and the structure of communities has been presented either in terms of morphological comparisons among co-occurring species without regard to their commonness or rarity, or in terms of density relationships (density compensation, MacArthur, 1972) without regard to morphology. This separate treatment is unfortunate for three reasons. First, presence/absence patterns in single-sample surveys provide no measure of the processes that underlie the pattern (Haila and Järvinen, 1981; Wiens, 1981b). Second,

---

[1] Present address: Department of Biological Sciences, Northern Arizona University, Flagstaff, Arizona 86001.

they are insensitive to partial effects, such as population fluctuations within species that are consistently present. Estimates of turnover detect only changes in species composition. Third, consideration of density without quantification of morphology leaves the closeness of competitors undefined. We will study this general problem by posing questions based on *a priori* predictions about community organization. Then we will examine seven years of census data for an avian assemblage occupying a deciduous forest in Maryland. We first ask about density relationships *per se*, then whether they are related to morphological shape relationships between pairs of species or among larger sets of species, then whether morphological relationships affect densities within guilds, and finally whether density relationships are related to differences in body size.

If pairwise density correlations were attributable to proximate factors, species pairs that have negative density correlations might be regulated by interspecific interactions, or they might be tracking resources that are negatively correlated. If this effect is large, then the variance of the yearly total densities for the community will be smaller than the sum of the annual variances of the densities for each species. Species pairs having positive density correlations may be tracking the same or covarying resources, or they may depend on one another. If this effect is large, the variance of the yearly total densities will be larger than the sum of the variances for each species.

Seven years may be insufficient for demonstrating statistically significant relationships, but some patterns should begin to emerge in this period if a functional relationship between density and morphology is present. Otherwise, one would have to conclude that the assumptions are incorrect, the data insufficient, the analysis flawed, or the structure of the community is determined by factors that are independent of interspecific relationships.

METHODS

The census data are for spot map breeding bird censuses made by an expert field ornithologist, Chandler Robbins, on a 12 ha plot (approximately 30 acres) of upland hardwood forest in Columbia, Howard Co., Maryland, from 1971 to 1977. The forest is dominated by tulip trees (*Liriodendron tulipifera*), maples (*Acer rubrum*), and oaks (*Quercus alba*, *Q. rubra*), with a rich understory of dogwood (*Cornus florida*). For a full description of the vegetation and the birds see Robbins *et al.* (1971) and subsequent reports in later issues of *American Birds* (Table 26.1).

This area was selected because of its stable vegetation structure relative to other habitats, its predictable climatic regime, and its presumed stable

Table 26.1. Breeding bird community in an upland deciduous forest, Columbia, Maryland, 1971–1977, 12 hectares.*

Species by Guilds	Abbreviation	Range in Number of Territories**
Bark-Probing or -Gleaning		
Red-bellied Woodpecker	RW	1–3
Hairy Woodpecker	HA	.25–1
Downy Woodpecker	DW	2–3.5
Black and White Warbler	BW	0–1
Arboreal Sallying		
Great Crested Flycatcher	CF	0–1
Acadian Flycatcher	AF	6–10
Eastern Wood Pewee	WP	1–3
Ground and Understory-Gleaning		
Carolina Wren	CA	0–3.5
Gray Catbird	GC	1–7
Wood Thrush	WT	4–12
Veery	VE	0–7
White-eyed Vireo	WV	0–2
Worm-eating Warbler	WW	0–1
Ovenbird	OV	0–2
Kentucky Warbler	KW	3–5
Hooded Warbler	HO	3–6.5
Cardinal	CD	6–13
Rufous-sided Towhee	RT	2–8
Arboreal Foliage-Gleaning		
Yellow-billed Cuckoo	YC	0–2
Blue Jay	BJ	.25–1.5
Carolina Chickadee	CC	1.5–3.0
Tufted Titmouse	TT	2.5–4.5
Blue-gray Gnatcatcher	BG	.25–2.5
Yellow-throated Vireo	YV	0–1.5
Red-eyed Vireo	RV	10–19
Parula Warbler	PW	1–2
Cerulean Warbler	CE	1–2.5
American Redstart	AR	0–1.5
Scarlet Tanager	ST	2–5

* Total species = 29; avg. = 25.3/year; standard deviation = 1.6. Average number of territories = 86.2; standard deviation = 5.8; range 78.5–93.5.

** Census results are reported in terms of territories, each of which was being defended by a singing male, assumed to represent the presence of a pair. The number of territories is substantially lower than the actual number of birds present, but it is the standard unit in most breeding bird census work. Data are published in Robbins *et al.* (1971) and *American Birds* 26: 944–945; 27:965; 28:1002; 29:1088–1089; 31:39; 32:25–26.

Table 26.2. Correlations between 6 morphological shape characters and 6 significant canonical discriminant functions of variation in 29 species of birds.

	I	II	III	IV	V	VI
Log bill depth–Log tarsal length	−.49	−.57	−.42	−.38	.33	−.08
Log bill width–Log tarsal length	−.68	−.46	.08	.09	−.26	−.50
Log bill length–Log tarsal length	−.60	.02	.29	−.70	.14	−.21
Log wing length–Log tarsal length	−.66	−.04	.16	.09	.67	−.29
Log tail length–Log tarsal length	−.31	−.41	.51	−.07	.63	−.25
Log toe–Log tarsal length	−.07	−.00	−.07	−.18	.41	−.89
Eigenvalue	57.6	40.5	14.3	10.9	6.0	0.7
Percent of variance accounted for	44.3	31.1	11.0	8.4	4.6	0.6

level of food resources. During the period of study the plot was surrounded by similar vegetation for more than one half mile on all sides. Lynch and Whitcomb (1977) used the data for 1971–1976 as an example of a standard undisturbed eastern hardwood forest. As in their analysis, consideration was restricted to species that normally inhabit the interior of woodland and forest, that had at least one territorial pair in one of the years, and for which the normal territory size is less than half the size of the plot.

Because the census data contain low counts, they are unlikely to be normally distributed. We used either a nonparametric method (Spearman's Rho correlation) or we applied normalizing transformations before comparing means and variances by parametric methods (Box *et al.*, 1978). The number of territories was transformed by taking its square root (Table 26.2), and the variance in number of territories by taking log ($100 \ s^2$). Because these transformations were determined by examining the data, *a priori* hypotheses cannot be formulated, and thus the interpretation of the results will not be stated in terms of significance levels.

Morphological measurements of ≥ 10 study skins of each species consisted of bill length (tip to nasofrontal hinge), bill depth (at the center of the nostril), bill width (at the center of the nostril), tarsal length, wing length (chord), tail length, and the length of the central toe. In cases of sexual dimorphism, 10 birds of each sex were measured. Compared to differences between species, differences between sexes were minor, so sexes were pooled. The specimens measured were birds taken in the breeding season as close to the study plot as possible, but some were from as far away as Florida.

With any multivariate approach to measurements or their logs, most variation is probably attributable to size. In order that this not mask other ecologically important shape relationships, it is usually advisable to remove at least some of the size variation. In our case we subtracted the log of the tarsal length from the log of the other measurements. Using differences between logs of measurements as variables permits the multivariate space to be a shape space, in which subtle but potentially important variation is not masked by overriding size differences among species (Mosimann and James, 1979). We calculated morphological distances among species by canonical discriminant function analysis of the covariance matrix for the log transformed variables, log x–log tarsus (Table 26.2). This procedure establishes a space wherein distances in all directions have the same probabilistic meaning. We used rarefaction curves (Simberloff, 1979; Engstrom and James, 1981; James and Rathbun, 1981; James and Wamer, 1982) to describe community structure in terms of the relationship between species richness and relative abundance.

RESULTS

Of the 29 species of birds recorded in seven years, an average of 25.3 (s.d. 1.6) were present every year (Table 26.1). The total density of birds was also quite stable, averaging 86.2 territorial pairs (range 78.5 to 93.5, s.d. 5.8). A two-factor analysis of variance of the densities showed that there were significant differences in density among species ($F = 29.03$) but not years ($F = .28$). The most common species were the Red-eyed Vireo, Cardinal, Wood Thrush, and Acadian Flycatcher. The least common were the Black and White Warbler, Great Crested Flycatcher, Worm-eating Warbler, Yellow-throated Vireo, and American Redstart. One White-breasted Nuthatch, present only in 1977, was inadvertently omitted from the analysis. Rarefaction curves by years (Figure 26.1) are very similar, indicating that the community structure is stable in relative abundance as well as in total species richness, but note in Table 26.1 that the common species vary substantially in density from year to year (RV 10–19, CD 6–13, WT 4–12, etc.)

Our analysis will be arranged as a discussion of specific questions and a presentation of graphic displays of patterns that might offer evidence of interspecific interactions among the densities of the species of birds in this community.

QUESTION 1. Is there a preponderance of negative associations in density among pairs of species in the total community?

If interspecific interactions determine the density of species in the community, one would expect more negative than positive density correlations. On the other hand, if each species' density is independent of the density of other species in the community, there would be a symmetric distribution of pairwise density correlations. The frequency histogram of the 406 pairwise Spearman's rho (r_s) correlations among the densities

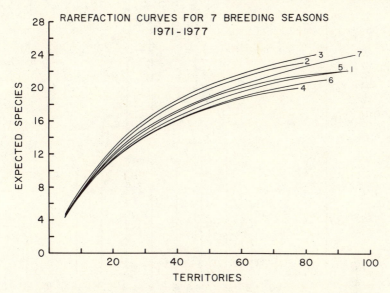

Fig. 26.1. Rarefaction curves for seven breeding seasons, 1971–1977, predict the number of species expected in samples of different numbers of territories. The shape of the rarefaction curve is a function of the relative abundances of the species (Hurlbert, 1971; Simberloff, 1979; James and Rathbun, 1981).

of the 29 species over seven years is nearly symmetrical, and the average (−.008) is very close to zero (Figure 26.2). This symmetric distribution suggests that the density of each species is independent of the densities of other species in the community, but it is not sufficient evidence that no interspecific interactions are present. Some pairs of species could be competing for a mutual resource that is fixed, and others could be responding concordantly to a fluctuating resource.

Järvinen (1979) suggested a quantitative measure of independence for densities in European land bird communities. He calculated the ratio of the sum of the individual species variances to the variance of their sums. If the densities are independent random variables then this ratio will equal 1.0. A ratio less than 1.0 indicates parallel fluctuations, whereas a ratio greater than 1.0 indicates compensatory fluctuations. Because the true variances of the larger population (of which our data are a sample) are unknown, estimates must be used. Some deviations from 1.0 will be expected in these estimates even if the densities are independent. We suggest partitioning the total variance into the sum of the individual species variances and the covariances,

$$\textit{Variance of the sum} = \sum_{N=1}^{N} \sigma_i^2 + \sum_{\substack{i=1 \\ i \neq j}} \sigma_{ij}.$$

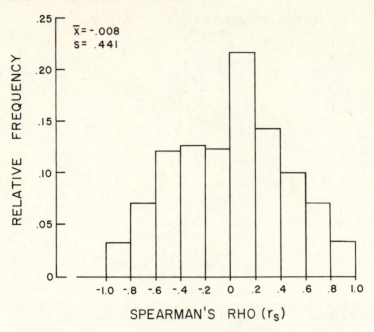

Fig. 26.2. Frequency histogram for 406 Spearman's Rho correlations (r_s) of the square roots of densities between all pairs of species, $\bar{x} = -.008$.

If there is no preponderance of either positive or negative correlations among species (the densities are independent) the covariance term will equal zero, and the variance of the sum will equal the sum of the species variances,

$$\text{Variance of the sum} = \sum_{i=1}^{N} \sigma_i^2.$$

Then we ask whether the observed total variance is greatly different from the observed sum of the individual species variances. Determination of real variances is precluded because data for the different years are not independent. Indeed some of the same individual birds are probably present for more than one year. Because the variance-stabilizing transformation for our data is the square root transformation, we assume that the data follow a Poisson distribution. The variance of the sum of estimated variances for independent Poisson random variables is:

$$\text{Var} \sum_{i=1}^{N} (\sigma_i^2) = \sum_{i=1}^{N} \left(\frac{\lambda_i}{n} + \frac{2\lambda_i^2}{n-1} \right),$$

where λ is the Poisson parameter. We estimate this variance by

$$\widehat{Var} \sum_{i=1}^{N} (\hat{\sigma}_i^2) = \sum_{i=1}^{N} \left(\frac{\bar{x}_i}{n} + \frac{2\bar{x}_i^2}{n-1} \right),$$

where $\bar{x}_i$ is the mean number of territories for the ith species. For the 29 species of birds:

Sum of estimated individual species variances $= \sum_{i=1}^{N} \hat{\sigma}_i^2 = 52.07;$

Estimated variance and standard deviation of the sum of the individual species variances $=$

$$\widehat{Var} \sum_{i=1}^{n} \hat{\sigma}_i^2 = 196.04, \qquad \widehat{S.D.} \sum_{i=1}^{n} \hat{\sigma}_i^2 = 14.00;$$

Sum of estimated covariances $= \sum_{\substack{i=1 \\ i \neq j}}^{N} \hat{\sigma}_{ij} = -18.14;$

Estimated variance of the sum $= 52.07 - 18.14 = 33.93.$

The sum of the estimated individual species variances (52.07) is within 1.3 standard deviations of the estimated total variance (33.93), suggesting that their ratio, 1.53, is not extreme enough to be interpreted as evidence that the species densities are behaving dependently.

Another way to ask the same question, whether species densities are varying independently, is to examine carefully the correlation and covariance matrices. The difference (-18.14) between the sum of the individual species variances (52.07) and the variance of the annual densities (33.93) is twice the sum of the covariances (-9.07). Are there a few large negative covariances, even though the correlations as a set are not skewed? The largest covariance (CD–RV) is positive (7.33) and the next largest are balanced between positive and negative values (CD–GC, -4.41; CD–VE, -4.24; VE–RT, -4.56; CD–RT, 4.23; GC–WT, 4.13; GC–VE, 4.00). It appears that most of the negative covariance term is attributable to an accumulation of extremely small covariances elsewhere in the covariance matrix. Thus, had we calculated Järvinen's ratio (52.07/33.93 = approximately 1.5), it would have been a mistake to have concluded that this positive value indicated compensatory fluctuations. There is no evidence from either the correlations (Figure 26.2) or the covariances that reciprocal interactions are producing community structure. Next we will examine morphological relationships.

QUESTION 2. What are the size and shape relationships of species in this community?

    (a) Are density correlations between pairs of species related to their shape differences?

    (b) Are density correlations between pairs of species related to distances to nearest neighbors in morphological space?

    (c) Do common species influence the densities of their morphological nearest neighbors more than other species do?

    (d) Do species that have closer nearest neighbors in morphological space have greater variation in density?

    (e) Do species on the periphery of the overall morphological space of the community have
    (1) higher or lower densities, or
    (2) greater variation in density?

    (f) Do species with many close morphological neighbors (the 40 pairs within the lower 10th percentile of all 406 pairwise distances) have
    (1) different densities, or
    (2) greater variation in density?

    (g) Do species with a high density of close morphological neighbors (the 40 pairs within the lower 10th percentile of all 406 pairwise distances) have
    (1) different densities or
    (2) greater variation in density?

We calculated canonical discriminant functions from 6 morphological shape variables for 10–20 specimens of each of the 29 species. This procedure produced six significant functions, of which the first three accounted for 86.4% of the variation in the set as a whole (Figure 26.3). The first axis is size relative to tarsal length. One extreme is the woodpeckers with their relatively short tarsi. Moving towards the right in Figure 26.3, the groups of birds are generally flycatchers, vireos, warblers, and mimids. The second axis represents mainly bill shape. The finches have the extreme low values and the woodpeckers, Black and White Warbler, and the Blue-gray Gnatcatcher have high values. The third axis is an interaction in shape between relative bill depth and tail length. These three relationships are displayed in Figure 26.3, but Mahalanobis distances in the full 6-dimensional space, accounting for 100% of the variation in the data set, were used in the analyses that follow.

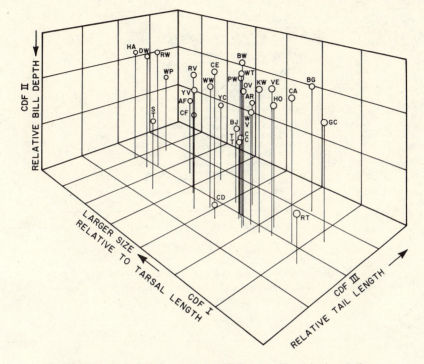

Fig. 26.3. Morphological shape relationships among 29 species in a 3-dimensional space determined by canonical discriminant axes accounting for 86.4% of the variation.

For Question 2a, a graph of correlations among pairwise densities (r_s) for all pairs of species with their distances in 6-dimensional shape space shows what appears to be a completely random pattern (Figure 26.4). If pairs of species with close morphological neighbors had reciprocal density interactions, we would expect a cluster of points in the lower left region of this graphic space. There would be a concentration of negative r_s values associated with smaller distances, but this is not the case.

For Question 2b, a graph of correlations of densities between species and their first nearest neighbor distances in 6-D shape space shows an apparently random overall pattern (Figure 26.5). There is no trend for negative correlations to be associated with smaller distances. Note that the two pairs of species that show extreme reciprocal density patterns, while not close morphologically, are species that have similar feeding habits and occupy similar sections of the habitat. These are the ground-feeding Veery and Ovenbird, and the dense foliage gleaning Gray Catbird

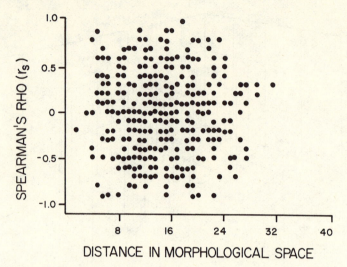

Fig. 26.4. Pairwise density correlations and distances between pairs of species in 6-dimensional morphological shape space are unrelated. Some of the dots in the center of the plot were omitted.

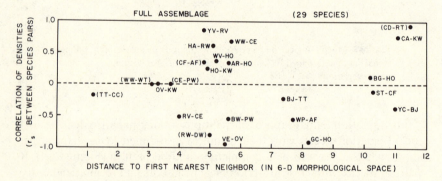

Fig. 26.5. Negative density correlations are not generally associated with closer distances (D^2) in morphological space.

and Hooded Warbler. Also pairs that covary positively have similar habits, the canopy-dwelling Yellow-throated Vireo and Red-eyed Vireo, and the shrub-dwelling Cardinal and Rufous-sided Towhee. To single out special pairs of species in a general analysis and interpret their relationships separately is overinterpreting the data. But certainly some pairs of species could be competing for a mutual resource that is fixed, and other pairs could be responding concordantly to a fluctuating resource.

For Question 2c, if common species influence the density of their morphological nearest neighbors more than rare species do, then we expect an unusual number of cases in which species with high densities do not have close neighbors in morphological space, that is, we expect a positive correlation between average density and first nearest neighbor distance. But the pattern appears to be random (Figure 26.6a).

For Question 2d, species having closer nearest neighbors might vary more in density from year to year, whence we would expect a negative correlation between variance in density and first nearest neighbor distance. In this case there is an apparent relationship but it is in the opposite direction from that predicted (Figure 26.6c). That is, species with close morphological nearest neighbors tend to have less variation in density than others.

For Question 2e, do species of general morphology (near the center of the 6-D space) have lower densities or greater variation in density than other species? Neither relationship (Figures 26.6b and 6d) can be distinguished from a random one.

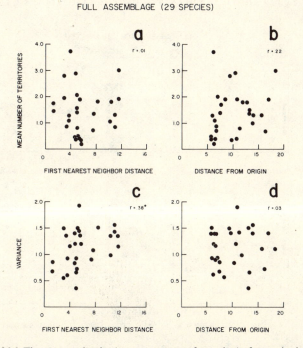

FULL ASSEMBLAGE (29 SPECIES)

Fig. 26.6. The average density (square root transformation) of a species is not related to whether it has a close morphological neighbor (a) or whether it is close to the average morphology of the species in the community (b). The variance in density (log 100 s^2) is positively correlated with morphological distance (c) contrary to the prediction (see text), and it is unrelated to the distance from the origin of the morphological space (d).

For Question 2f, perhaps competition among species is diffuse, such that species having many close morphological neighbors have lower density or greater variation in density from year to year. We then expect a negative correlation in Figure 26.7a and a positive correlation in Figure 26.7c, but neither is apparent.

For Question 2g, do species having many other individuals present of similar morphology have lower density and/or greater variation in density? Again there is no evidence in our data to support either case (Figures 26.7b and 7d).

> QUESTION 3. Are the relationships above different when guilds of species within the community are considered?

We will examine four foraging guilds (Table 26.1) to look for evidence of interspecific interactions among density relationships within guilds. Among the arboreal gleaning birds (Table 26.3), large negative density correlations occur between the Carolina Chickadee and both the Red-eyed and Yellow-throated Vireos, and there is a large positive correlation

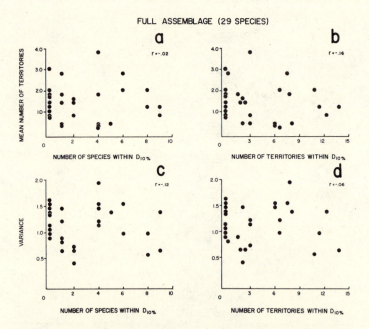

Fig. 26.7. The average density (square root transformation) of a species is not related to the number of species within 10% of the morphological space around it (a), or to the number of individual birds within this space (b). The variance in density ($\log 100 \, s^2$) is not related to the number of species within 10% of the morphological space around it (c), or to the number of individual birds within this space (d).

Table 26.3. Correlation matrix (r_s) of numbers of territories by guilds.

Arboreal Foliage-Gleaning Birds (55 Pairs)

	YC	BJ	CC	TT	BG	YV	RV	PW	CE	AR
BJ	-.37									
CC	.12	.49								
TT	-.25	-.23	-.18							
BG	-.20	.14	.56	-.25						
YV	-.09	-.46	-.83	.09	-.63					
RV	-.08	-.31	-.80	-.25	-.45	.88				
PW	-.73	.17	.71	.24	.00	.00	-.24			
CE	.47	-.53	.08	.24	-.12	-.28	-.51	.00		
AR	.33	-.09	.13	.00	.43	-.61	-.40	-.52	.47	
ST	.11	.23	-.28	-.14	-.20	.48	-.61	-.49	-.75	-.38

Ground and Understory-Gleaning Birds (55 Pairs)

	CW	GC	ST	VE	WV	WW	OV	KW	HO	CD
GC	.21									
WT	.10	.64								
VE	.53	.83	.57							
WV	-.46	-.45	.05	-.52						
WW	.11	-.75	-.36	-.67	.63					
OV	-.51	-.66	-.36	-.94	.39	.56				
KW	.75	-.32	.01	.06	-.19	.43	.00			
HO	-.34	-.91	-.49	-.92	.38	.67	.88	.25		
CD	-.08	-.99	-.45	-.69	.58	.83	.50	.40	.80	
RT	-.34	-.94	-.52	-.85	.70	.82	.66	.11	.85	.94

between the Red-eyed Vireo and the Yellow-throated Vireo. But in such a set of 55 correlations one would expect a few to be significant by chance. These observations may be attributable to real biological responses in the community. In the ground and understory-gleaning guild there are seven positive and six negative correlations. Again the negative correlations may indicate pairwise competition for resources and the positive ones may be evidence of responses to a common fluctuating resource. These sets of species may warrant further research.

We calculated morphological relationships within guilds (Figure 26.8a and b), again determining significant canonical discriminant functions, and we asked the questions (2b, c, d, and e) of the 11-species arboreal foliage-gleaning guild and the 11-species ground and understory-gleaning guild that we asked of the entire association. Again the relationship between r_s and distance to the first nearest neighbor (Figure 26.9) did not have an unusual number of negative values, and correlations between the mean and variance of density were not related to nearest neighbor distance or distance from the origin of the morphological space (Table 26.4).

Of the two small guilds, the bark-probing or -gleaning group contained two strongly negative values of r_s. These were between the Downy and Red-bellied Woodpeckers ($-.82$) and the Black and White Warbler and the Hairy Woodpecker ($-.89$). Although this second pair of species was never represented by more than one pair of each species on the plot, they tended to alternate. Among the 29 species of birds there are three sets of congeners (*Parus*; *Vireo*, 3 species; *Picoides*) and another pair of species that are sometimes interpreted as competitors (WT–VE, Noon, 1981). Among these comparisons the strongest negative correlation is between HA and DW (-0.45), and the strongest positive correlations are RV–YV (0.88), KW–CA (0.75), WV–YV (0.72), WT–VE (0.57).

In all these comparisons we are interested in relationships among the correlations, but not in interpreting significance levels. Examination of sets of pairwise correlations is useful, even when sampling of the populations has not been random and some of the same individual birds are probably present in successive years. To summarize, a few pairs of birds in the 29-species set have strongly negative density correlations. One such pair (RW–DW) are mutual nearest neighbours in morphological shape space, but neither at the community level nor within guilds is there any indication of a tight linkage between density and morphological relationships.

QUESTION 4. Are density relationships among pairs of species in the community related to their differences in size?

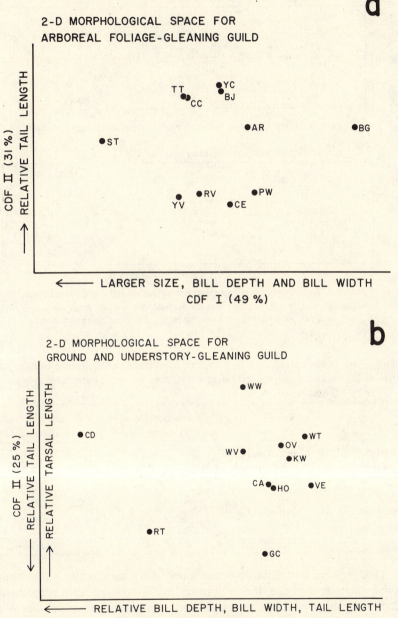

Fig. 26.8. Morphological shape relationships among 11 species in each of two guilds determined by canonical discriminant axes. Size in 8a is relative to tarsal length.

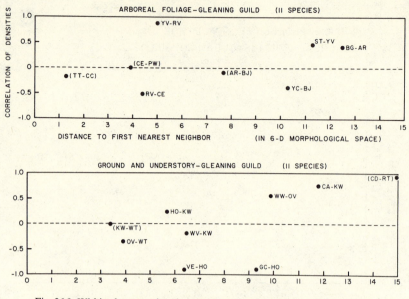

Fig. 26.9. Within the ground and understory-gleaning guild, and the arboreal foliage-gleaning guild, negative density correlations are not generally associated with distances (D^2) to the first nearest neighbor in 6-dimensional morphological space.

Table 26.4. None of the correlations (r) between the average density (mean number of territories) or its variance with morphological distances for two 11-species guilds is significant ($n = 55$) at $a = .05$.

		Arboreal Foliage-Gleaning Guild	Ground and Understory-Gleaning Guild
First nearest neighbor distance	Mean number of territories	−.26	.12
	Variance	.59	.14
Distance from origin	Mean number of territories	.01	.44
	Variance	.26	.15

One explanation for the apparent stability of a community is that birds of similar size compete. If this were so, one would predict a preponderance of negative r_s's to be associated with species pairs of similar size. We examined this possibility by plotting pairwise density correlations against general body size differences (log tarsus$_1$–log tarsus$_2$). Again the prediction was not supported. The variables appear to be independent of one another.

DISCUSSION

The most reasonable explanation for our results is that the species are responding independently to the distribution of resources such as vegetation, food, and nest sites, and that these have a limited range of variation from year to year in this habitat. The extent to which annual abundances are attributable to processes operating at other seasons is unknown but potentially large (Keast and Morton, 1980). The concept of Enemar (1966), Dow (1969), and C. Fowler (1981), that single-species populations in poorer habitats may have greater fluctuations in density, should be explored in detail.

Schluter (manuscript) has surveyed the literature for many groups of animals in many habitats and found that those communities that demonstrate dependence in species populations generally show positive, not negative, density relationships. The finding suggests that the dominant process is generally one of tracking covarying resources (see also Dunning and Brown, 1982).

Thus, we conclude that the apparent stability of this assemblage of breeding birds from year to year, in terms of the number of species and individuals present, and their relative abundance, was not attributable to reciprocal interactions among congeners, species of similar morphology, or members of guilds sharing common resources. In fact, the stability of the community is only apparent. The sum of the variances of the densities of the individual species is not sufficiently larger than the variance of the sums of the yearly totals, and the value of the negative covariance term is too low, to indicate dependence among the species. Thus the structure of the community is epiphenomenal (Simberloff, 1980; Simberloff and Boecklen, 1981) in the sense that its properties are not the result of mechanisms that can be observed at the community level, or even at any level above the species. We gave the community every opportunity to show us that it was organized by interspecific interactions, and it failed it do so.

This conclusion does not require that some species sharing common resources do not affect one another's distribution and abundance. Evidence for such effects has recently been reviewed by Birch (1979; see also Lawton and Strong, 1981). Direct evidence of interspecific interactions among birds, in which space is made unavailable to another species, is well documented for hummingbirds (Pitelka, 1951; Feinsinger, 1976; Kodric-Brown and Brown, 1978; Cody, 1968a; Stiles and Wolf, 1970; Lyon, 1976), Red-winged Blackbirds (Orians and Collier, 1963), vireos (Rice, 1978), wintering finches (Davis, 1973), non-breeding Red-headed Woodpeckers (Williams and Batzli, 1979a, b), and other species (Orians and Willson, 1964; Murray, 1971; Cody, 1974a). Some species of birds are interspecifically territorial at certain times of the year.

But it does not necessarily follow that the overall patterns of community structure within guilds, across habitats, and in similar habitats, are attributable to interspecific interactions. A major experiment that bears on this subject is the famous removal experiment of Stewart and Aldrich (1951; see also Knapton and Krebs, 1974), in which they repeatedly shot all the birds on a 40-acre study plot in Maine. Repopulation occurred overnight, with new males of the same species occupying the same territories. Up to five times as many birds were collected as were originally present. This experiment demonstrated strong population pressure for space (but see Brown, 1969, and Cederholm and Ekman, 1976). More importantly for the argument here, it demonstrated that in spite of overlapping interspecific territories, the appropriate space for each species was unique. This means that interspecific territoriality was not regulating the community.

All of this is contrary to the conclusion of Cody (1974a) that within guilds "a reduced density or absence of one species is most likely to be compensated by an increased density of another species within the group." Competitive exclusion has been overrated as a mechanism driving community organization. The interesting broad-scale community patterns of both animals and plants (Cody, 1974a; Pianka, 1975; Mooney, 1977), even with their apparent convergences in similar environments, should not necessarily be attributed to processes that operate at the scale of the pattern. Indeed, inferences about mechanisms underlying patterns from observations of the patterns are not justified in any case (Brady, 1979). Certainly, the demonstration of non-random patterns of morphological organization of fish communities (*e.g.* Gatz, 1979, 1981) should not be taken as evidence for competitive exclusion. The reason that non-random broad-scale community patterns exist, but are hard to detect at the local level (Wiens, 1981b), is that they are aggregate properties of individual species processes.

Even if we had found evidence of reciprocal density interactions in this study, experiments would be required to test whether the species involved were being favored alternately by different fluctuating resources. Other tests could be designed to look for asymmetrical competition, wherein only one of a pair of species responds (Lawton and Hassell, 1981). It will take more of the cold-hearted approach of Stewart and Aldrich to resolve this question. In the meantime, we doubt that the presence of co-occurring species, regardless of their morphology, is important in maintaining the organization of a community, and we suspect, with Andrewartha and Birch (1954), that the species are generally leading independent lives. In order to formulate hypotheses about the regulation of the distribution and abundance of animals, it would probably be more useful to study single species in detail than to characterize community patterns. While acknowledging that organisms may compete for space, we suspect that a bird community is analogous to Gleason's (1926) plant association, "not an organism, scarcely even a vegetational unit, but merely a coincidence," a collection of species that happen to exist together because of converging accidents of space, time, and environmental needs.

Acknowledgments

We are grateful to C. S. Robbins for conducting the field work, to J. F. Lynch for assistance with the data, to O. Järvinen, C. McCulloch, D. Meeter, J. Mosimann, and J. Travis for advice and ideas, to D. Simberloff, G. Graves, and R. T. Engstrom for comments on the manuscript, to C. NeSmith for measuring the birds, and to J. W. Hardy for lending specimens from the Florida State Museum.

27.

The Structure of Communities of Fish on Coral Reefs and the Merit of a Hypothesis-Testing, Manipulative Approach to Ecology

PETER F. SALE

School of Biological Sciences, The University of Sydney, Sydney, N.S.W. 2006, Australia

INTRODUCTION

Many people would see the aim of science as a search for order or regularities in the universe. MacArthur (1972) put it simply and concisely. "To do science is to search for repeated patterns, not simply to accumulate facts." For him, to do ecology was to search for general patterns in the distribution and abundance of species. While this is an enticingly simple definition of ecology, if taken literally, as it has been by some, as the modus of ecology, it may lead to false conclusions and a lack of the questioning that is one of the hallmarks of science.

These false conclusions arise precisely because vertebrate sensory systems have evolved as devices for detecting patterns. By means of lateral inhibition, temporal adaptations, and related processes, sensory inputs are coded, organized, emphasized, and sometimes ignored. What we see or hear is only an approximation of what is there, an approximation that is simplified, classified, and which can be more ordered than is reality. Given such equipment, and we all have it, the scientist can see faint regularities in his universe. But he can also see regularities that do not exist. In short, if we set out to find patterns we will. We will, whether or not they are there.

The alternative and correct approach is to deny that any patterns exist and to test all apparent regularities against this null hypothesis. Only in this way, by initially refusing to believe our subjective impressions of reality, can we hope to retain the objectivity supposed to characterize science. (A further danger remains even when this approach is adopted. Unless we take great care, our subjective views will influence which of various hypotheses receive serious consideration. Dayton and Oliver,

1980, have considered this problem in some depth and I will not comment further here.)

For ecologists interested in the structure of communities, the hypothesis-testing approach has yielded surprising results. Only very few of the presumed patterns in community structure have been found sufficiently strong to justify their acceptance as true (Simberloff, 1974; Connell and Slatyer, 1977; Wiens, 1977a; Connell, 1978; Connor and Simberloff, 1979).

In the following sections, I will trace some recent discoveries concerning the structure of communities of coral reef fish. The ready visibility of most reef fishes and their relatively sedentary habits following the pelagic larval phase make their communities ideal subjects of study. Ecologists have examined reef fish communities only relatively recently, and initially they brought with them a series of widely agreed on expectations concerning the structure to be found. Those who have used an experimental or hypothesis-testing approach have found that hardly any of these expectations have been fulfilled.

THE EXPECTED PATTERNS

Study of the ecology of reef fishes did not really get started until 1960. Bardach (1958) tagged fish to examine range of movements and (Bardach, 1959) measured standing crops. Stephenson and Searles (1960), and Randall (1961) caged areas to examine the effect of excluding herbivorous fish, and Randall (1963, 1965) subsequently built artificial reefs to examine the relation of the fish community to environmental structure.

By this time there already existed extensive agreement on the patterns that should exist in tropical communities. This agreement derived partly from guesses as to the nature of the tropical environment and partly from a more formal body of theory concerning the nature of competitively organized communities. By 1960, several alternative, but not mutually exclusive hypotheses had been proposed to account for the high density common to tropical communities. These hypotheses held in common that tropical high-diversity systems were equilibrial and competitively organized, and that they achieved their greater species richness by a finer partitioning or a greater degree of overlap in resource use among component species. They differed only in which putative characteristic of the tropics led to this state (see review in Pianka, 1966).

Early studies of reef communities exhibit clear acceptance of this view of their organization. Odum and Odum (1955) in their pioneering energetic study of Enewetak Atoll speak (p. 291) of the reef "unchanged year after year," and suggest that "with such long periods of time adjustments in organismal components have produced a biota with a successful competitive adjustment in a relatively constant environment." They pose the question (p. 291): "How are steady-state equilibria such as the reef ecosystem self-adjusted?" In their summary they conclude that (p. 319) "the reef community is, under present ocean levels, a true ecological climax or open steady-state system." Later Hiatt and Strasburg (1960), in their analysis of the fish of Enewetak Atoll, speak (p. 66) of this "relatively isolated, but complex ecosystem which apparently fluctuates very little, if at all, from year to year, and has over a long period of time acquired a biota successfully adjusted competitively in the relatively constant environment," and they also speak of the reef ecosystem as a "self-adjusted" and "steady-state equilibri(um)." Clearly, man's subjective view of the tropics as idyllic had invaded his perceptions of tropical communities right at the start.

Through the 1960's further hypotheses appeared to account for high tropical diversity, culminating in the stability-time hypothesis (Sanders, 1969; Slobodkin and Sanders, 1969). Among these, only the predation hypothesis (Paine, 1966) did not view the tropical system as competitively structured, although like the others it did hold that it was at equilibrium (Pianka, 1966, 1974). This view of the coral reef system, and therefore of the reef fish community, as equilibrial and stable has persisted almost to the present. Smith and Tyler (1973) stated confidently (p. 2), "There is now abundant evidence that fish communities are more than random assemblages of species whose general habitat requirements happen to be similar. These associations are stable, particularly in the more speciose communities of the tropical littoral environments." Goldman and Talbot (1976), although expressing doubts, stated (p. 126) that "Coral reefs are the epitome of mature, biologically accommodated communities." Many studies of reef fishes have been and are being done under the implicit assumption that resources of one type or another are limiting numbers and determining (a stable) structure in these communities (reviewed in Sale, 1980b).

There are three criteria that must be met if reef fish communities are to conform to this belief. They must show temporal constancy of numbers of individuals, of species, and of species composition. They should show evidence of resource limitation and subtle fine-scale patterns of resource partitioning in response. They should show other evidence of competition, including "assembly rules" (Diamond, 1975) that determine species composition. Evidence concerning each of these follows.

RESOURCE LIMITATION AND PARTITIONING

It is convenient to divide the major resources used by reef fish broadly as food and space. Of these, there has been general agreement that space resources are more likely than food to limit reef fish populations (Smith and Tyler, 1972; Sale, 1977). This assessment is based on several lines of indirect evidence.

1. Newly constructed artificial reefs or denuded natural reefs are rather rapidly colonized by fish (Randall, 1965; Russell *et al.*, 1974; Sale and Dybdahl, 1975, 1978; Molles, 1978; Talbot *et al.*, 1978; Bohnsack and Talbot, 1980).
2. In some highly specialized microhabitats, such as anemones, a relatively constant density of fish is present (Allen, 1972; Moyer and Sawyers, 1973; Ross, 1978).
3. Behavioral observations of many species disclose a persistent and often interspecific defense of living sites or shelter locations (Low, 1971; Reinboth, 1973; Thresher, 1976a, b; Ebersole, 1977; Ehrlich *et al.*, 1977; Robertson *et al.*, 1979).

Corresponding evidence that food supply limits numbers is lacking.

That space resources are more likely than food to limit numbers does not confirm that they are limiting in any given instance. However, this distinction was glossed over by Sale (1977, 1978b), and Smith (1978; Smith and Tyler, 1972), both of whom, while never claiming that fish were always space-limited, implied strongly that most of the time this was the case.

The bulk of data obtained in the 1970's was consistent with space-limitation. Indeed, Robertson *et al.* (1979), in tentatively drawing the conclusion, on largely circumstantial evidence, that three species of surgeonfish were not at carrying capacity, state (p. 31), "We know of no studies to date that show clearly, or even indicate strongly, that the population of any coral reef fish is below the carrying capacity of its habitat." In addition to numerous descriptive studies of space use by reef fishes, long-term monitoring of undisturbed sites, and of sites first denuded of territorial pomacentrid fish, showed a considerable constancy in total area defended as territories (Sale, 1974, 1978b)—which is approximately proportional to numbers of residents—and a recovery of number to pre-denudation levels (Sale, 1976). Artificial reefs filled rapidly (within 3 months) during the summer to what appeared an asymptotic number of individuals (Russell *et al.*, 1974).

Nevertheless, data available quite early should have indicated that numbers of reef fish were not always set by a limit of available space, and further data have appeared recently. Sale (1972) showed that at some sites on Heron Reef the numbers of *Dascyllus aruanus* were strongly correlated with the sizes of coral heads they occupied, indicating an orderly distribution of fish among these habitat patches, and space limitation. But at other sites on the same reef where coral was relatively more abundant, no such correlation existed, despite the fact that few individuals of other species were present. He suggested that in these places numbers of the fish were not limited by the availability of coral. Smith and Tyler (1975) documented a substantial seasonal variation in the numbers of fish supported on a small patch reef (code-named Bimini I in their paper) without concluding that this variation almost certainly meant that space on this patch reef was not limiting throughout the year. Robertson and Sheldon (1979) showed, in a series of experiments in which they manipulated both number of fish and amount of shelter, that despite the fact that *Thalassoma bifasciatum* defends its nocturnal shelter sites from its own and other species, the availability of such shelters was not limiting the population of this species.

Longer-term monitoring of artificial reefs originally described as "filling" with new recruits in 3–4 months (Russell *et al.*, 1974) indicated that an equilibrium number of fish or species was not, in fact, reached. Instead, numbers fluctuated substantially throughout the 32-month study (Figure 27.1) (Talbot *et al.*, 1978). Similar results have been obtained in other studies of artificial reefs (Molles, 1978; Bohnsack and Talbot, 1980) and of

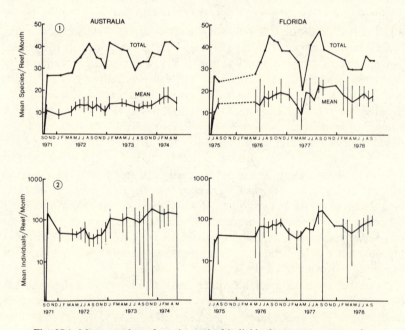

Fig. 27.1. Mean number of species and of individuals present at successive censuses of similar concrete-block artificial reefs built in similar seasons at One Tree Reef (Australia) and Big Pine Key (Florida). Eight replicates were present at each site. Note the variability in mean numbers, some of which is seasonal, and the wide confidence intervals (vertical lines represent 95% confidence limits for numbers on single reefs). Modified from Bohnsack and Talbot (1980).

undisturbed or denuded natural patch reefs (Sale, 1980a; Williams, 1980). Such changes in numbers of fish present indicate that space resources are either not fully used for much of the time or that space requirements of particular species vary dramatically from time to time. No comparably direct experimental evidence is available showing that the amount of structured living space directly limits numbers, although de Boer (1978) was able to increase the density of territorial males of *Chromis cyanea* by adding structure to an area in the form of low vertical opaque screens.

Concomitant with the expectation that resources limit numbers of reef fishes is the belief that species will evolve an efficient partitioning of the resources available. As noted earlier, a large number of descriptive studies have documented differences among species in their use of resources, particularly space resources (reviewed by Sale, 1980b). What has not been demonstrated is that these differences exist because of a competitively directed process of niche diversification. They may exist simply because the species are different in their ecology as well as in their morphology (Sale, 1978b). I have argued (Sale, 1977) that, even though it is difficult to compare degrees of partitioning between communities, there is no indication that reef fishes divide their resources any more finely than do fish in less diverse, temperate communities. Yet finer division is clearly to be expected if the differences in resource use are a consequence of niche diversification in these very diverse tropical communities.

TEMPORAL CHANGE

As noted, far from showing the temporal constancy expected, most studies that have proceeded for longer than one year have demonstrated considerable temporal change in numbers of individuals and of species and in species composition present. Sale and Dybdahl (1975, 1978) showed that isolated coral heads denuded of fish were recolonized within 2–4 months to the extent that they held similar numbers of species and fish as did undisturbed controls. But the numbers held in both experimental and control corals varied seasonally, being greater in the summer, the peak period of recruitment of fish to the reef. Furthermore, species composition varied substantially, both among replicates and between successive collections. Molles (1978), Talbot *et al.* (1978), and Bohnsack and Talbot (1980) have all demonstrated pronounced seasonality in the numbers of fish and species that occupy artificial reefs built of concrete blocks. Talbot *et al.* (1978) stressed that in addition to this overall seasonal fluctuation, there was an aseasonal variation within reefs, and pronounced differences between replicate reefs in the numbers of fish and species present and in species composition. Williams (1980) documented changes in numbers and species composition of pomacentrid fishes (the dominant element of his assemblages) that proceeded in opposite directions on similar patch reefs less than 500 m apart in a shallow lagoon.

Some other authors have provided evidence indicating substantial temporal variability, but without comment. Ogden and Ehrlich (1977) listed numbers of *Haemulon* of several species resident in particular resting schools in the Virgin Islands, censused four times over a 3-year period. Their data exhibit marked differences in numbers (ninefold in one case) between similar seasons, and the variations are not in the same direction. Smith and Tyler (1975) compared censuses of a patch reef made three years apart and argued for a high degree of constancy in the fish fauna present. They obtained a value of 0.88 similarity using a formula that considered only presence or absence of species. Of the 65 species they termed "resident," they listed only 16 as present in roughly equivalent numbers at the two censuses. Twenty-eight of the 65 were present at only one or the other of the censuses. Had they used a similarity coefficient that also considered species abundances, a much lower value of similarity would have resulted. (Their data give similarity of 0.39 using the same formula as in Figure 27.2). As already noted, these authors also documented, without comment, a fourfold increase in numbers of fish from winter to summer on another patch reef.

SPECIES ASSEMBLY RULES

The substantial temporal variability in species composition that has been documented in reef fish assemblages does not, of itself, preclude the existence of strong interspecific interactions determining possible combinations of fish on reefs. But it does suggest that any "stable" or "preferred" species combinations that may exist are subject to frequent disruption.

Sale and Dybdahl (1975, 1978) and Talbot *et al.* (1978) searched for such "preferred" combinations, and thus for species assembly rules demonstrated by their data, but without much success. Sale and Dybdahl sought significant negative associations among pairs of species occupying coral heads of similar type (two types of coral head in each of two habitats, and three types in the reef slope habitat). Their data were such that positive associations between pairs of species might have arisen because larger coral heads held more fish (of all species) than did smaller ones, as well as because of assembly rules. They found no significant negative associations among 105 pairs in live *Acropora pulchra* on the reef flat, 55 pairs in dead coral on the reef flat, 66 pairs in each of live and dead corals in the lagoon, 45 pairs in each of live *Acropora pulchra* and *Acropora hyacinthus* colonies (branched and plate-like forms respectively), and 36 pairs in complex, mixed coral heads all on the reef slope. That is, no

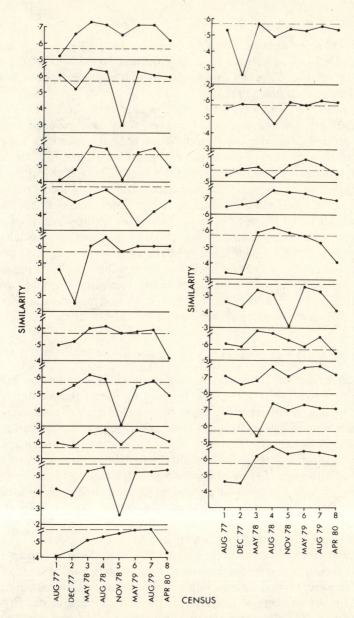

Fig. 27.2. Similarity of the fish fauna present on a patch reef to that present there at all other censuses, for each of eight successive censuses of 20 patch reefs in the One Tree lagoon. Similarity was measured using the formula:

$$C_{ij} = 1 - \tfrac{1}{2} \sum_{k=1}^{s} \left| p_{ik} - p_{jk} \right|,$$

where C_{ij}, percent similarity between censuses i and j, varies from 0.0 to 1.0, and p_{ik} and p_{jk} are the proportions of the fauna present at each census that comprise the kth species. Note the generally low similarity values, and the variability in similarity among censuses of single reefs.

significant negative associations were detected among 418 pairs of species tested. They concluded that their data showed no evidence of the existence of any pair of species with similar habitat requirements but lacking tolerance of each other's presence.

Talbot *et al.* (1978) tested 537 pairs of species of which 42 were significantly negatively associated. However, their data included samples from several different types of reef. When habitat preferences were examined, the members of five of these 42 pairs were found to have significantly different habitat preferences from each other. These might alone be responsible for the negative associations. Only one of the 42 pairs was significantly positively associated in habitat preference—an indication that avoidance of one species by the other was taking place despite the similar habitat requirements. Four pairs were positively associated in occurrence, despite a significant negative association in habitat preferences. The conclusion drawn by Talbot *et al.* is that their data presented only the most tenuous evidence of biotic interactions among species determining distribution of the fish. The conclusion from both studies has to be that assembly rules, if they exist, must be very subtle.

FISH ON NATURAL PATCH REEFS

Since August 1977, I have been monitoring the fish present on each of 20 natural patch reefs in the One Tree lagoon (Great Barrier Reef). The first $2\frac{1}{2}$ years' results amply confirm the conclusions previously reached on variability of fish assemblages (Sale and Douglas, unpublished data).

The reefs were in two groups of 10 located in the northeastern and the southern quadrants respectively of this 2-km diameter lagoon. A series of physical attributes of each reef was measured, including its surface area, volume, percent live coral cover, percent algal-covered rock, substratum diversity, and substratal relief. The two sets of reefs did not differ in any of these characteristics, although the northeastern set rests in marginally deeper water (5.5 m vs. 4 m at low tide).

On the average the reefs held about 128 fish of 21 species, and a total of 143 species was recorded on one or more reefs at one or more of the eight censuses carried out. Step-up multiple linear regressions were used to relate the numbers of fish and of species to aspects of the physical structure of the reefs. Distribution of each of the 10 commonest species (accounting for 75% of individuals seen) was separately analyzed in this way as well. The number of fish present on a reef was significantly related to its surface area ($r = 0.88$), but no other physical attribute significantly improved this relationship. The number of species was best predicted by surface area ($r = 0.82$), and the prediction was significantly, but marginally, improved by the inclusion also of percent live coral ($r = 0.87$).

Both number of fish and number of species varied significantly with season.

Two of the 10 common species had distributions unrelated to any of the physical attributes measured. Five others had distributions related to the size of the reefs (surface area or volume depending on the species), but for one of these, *Chromis nitida*, this relationship only occurred on the northeast reefs. On the south reefs only, this same species was distributed in a way positively associated with degree of substratal relief. Two other species were negatively associated with percent live coral cover, and the final species, *Dascyllus aruanus*, was negatively associated with substratum diversity on south reefs, but not related significantly to any measured attribute on northeast reefs. The clear indication is that reef size, whether measured as surface area or as volume, is a good predictor of numbers of fish or species present, as well as, in many cases, of the numbers of individuals of specific species. However, other attributes of the reef play a minor role, if any at all, in determining the distribution of the fish fauna.

Successive censuses of individual reefs were compared using a simple percent similarity index (*cf.* Sale, 1980a). Pronounced changes occurred in the relative abundance of species present, and many species present at one census were absent at others. Figure 27.2 plots mean similarity of each census to the other seven for each of the 20 reefs. No seasonal trends are apparent in degree of average similarity, and some reefs show greater variability in similarity among censuses than do others. Preliminary analyses carried out to date have failed to relate mean degree of similarity, variance in degree of similarity, or temporal pattern of change in similarity to any physical attribute measured on the reefs or to the size of the assemblage of fish supported. These results confirm and extend the conclusions of Sale (1980a) based on four censuses of six reefs over two years, and those of Talbot *et al.* (1978) and Bohnsack and Talbot (1980) for fish on artificial reefs. The indication is that there is continuous change in the composition of the fish fauna present on small patch reefs, and, presumably, other habitat patches. If identifiable, and stable, combinations of species exist that are more likely than other combinations to occur, they have yet to be found.

RECRUITMENT LIMITATION

It is now apparent that the relatively stable numbers and the evidence of limitation of numbers through availability of space, which Sale (1974, 1975, 1976, 1978a, 1979) found for territorial fish on the Heron Reef slope, is not a situation typical of reef fish communities in general. Even territorial pomacentrids appear not to be space-limited in some other habitats, such as the lagoon of One Tree Reef (Doherty, 1980). Instead, a seasonally and non-seasonally very variable rate of recruitment to reef sites of fish from the plankton appears usually to be responsible for determining numbers of adult fish that are present (Talbot *et al.*, 1978; Bohnsack and Talbot, 1980; Doherty, 1980; Williams, 1980).

The temporal variability in recruitment has been documented in studies that monitored artificial reefs (Russell *et al.*, 1977) or permanent quadrats on a reef face (Luckhurst and Luckhurst, 1977) and in one 13-year study of recruitment of two species of *Siganus* (Kami and Ikehara, 1976) to the reef flat at Guam. More recently, Williams and Sale (1981) have explored spatial and temporal variation in recruitment to identical colonies of branched coral set out on sand patches at four widely separated sites within the One Tree lagoon. In this study, new recruits were removed on a nearly daily basis so that the coral functioned as traps for settling larvae. The study demonstrated many order-of-magnitude differences in recruitment by single, common species both between years and between sites. Disturbingly, even patterns of relative abundance among sites were not always constant between the two years of the study.

Doherty (1980), in a series of manipulative experiments, has examined recruitment and juvenile survivorship of two territorial pomacentrids (*Pomacentrus flavicauda* and *P. wardi*) that occur on patch reefs within One Tree lagoon. He has shown for both species that recruitment and survivorship are independent of adult numbers present (of either species). To determine natural recruitment rates, he monitored arrival of recruits to each of 84 patch reefs of varying size over three successive summers. Numbers of *P. wardi* recruiting to the 2820 m^2 being monitored varied by two orders of magnitude over the three summers, and those of *P. flavicauda*, which only recruited to a subset of 53 small, shallow reefs (185 m^2 total) still varied twenty-five-fold. The third summer was extremely poor for both species.

CONCLUSIONS

That reef fish assemblages are apparently predominantly non-equilibrial and temporally varying not only contradicts our expectations concerning tropical systems. It is also contrary to the view expressed by many workers in the field (*e.g.* Molles, 1978; Smith, 1977, 1978; Gladfelter and Gladfelter, 1978; Anderson *et al.*, 1981). In reaching this conclusion, I have emphasized a number of studies that have in common the use of small assemblages of fish and, usually, a manipulative, experimental approach to understanding them. On a small spatial scale, it is possible to replicate treatments adequately. It is also possible to achieve a high level of accuracy in censusing the fish fauna present (Russell *et al.*, 1978; Sale and Douglas, 1981). Both of these advantages are achieved only with much more difficulty when work is carried out on a larger spatial scale. Workers using larger assemblages usually have fewer replicates, rarely experiment, and often census using abundance classes rather than actual counts of individuals present. With fewer replicates, and without manipulations, evidence of variability is less easy to find, and cruder census methods or emphasis on presence and absence rather than abundance of species inevitably yield higher similarity values, even in small-scale studies (*e.g.* Smith and Tyler, 1975). Among recent workers, it is primarily those who have used larger assemblages (here including Molles, 1978, who pooled small replicates in most of his treatment comparisons) who have espoused the more traditional equilibrial view of community structure. Some among them (Gladfelter and Gladfelter, 1978; Anderson *et al.*, 1981) have argued that the difference in size of the assemblages studied is alone sufficient to account for the difference in conclusions reached.

In fact, if a large reef tract is no more than the sum of its parts, the results of small-scale studies, which have emphasized the apparently extensive temporal and spatial variation in characteristics of the fish fauna, would predict greater temporal constancy than is observed in any one small patch. The small-scale variability arises because of a largely stochastic recruitment from a defined pool of species able to use that habitat. The larger site will normally contain more of the available species pool at any one time, and the assemblage present will thus change less through time. In addition, replicate large sites will contain relatively similar fish assemblages, and for the same reason. This greater spatial and temporal constancy is precisely what is reported by those who have analyzed their data on this broader scale. That they may deal with fewer replicates, seldom experiment, and census more crudely can only enhance this impression of constancy. The importance of the small-scale, manipulative studies is that they caution that there is no basis for interpreting this constancy as the stability of an equilibrial system.

In this paper, I have attempted to show the power of a hypothesis-testing, and where possible, manipulative approach to field studies. In doing this I am no more than restating Connell's (1974) call for field experimentation in ecology. The reef fish story is particularly interesting because the experimental investigations have so largely overturned a pre-existing and remarkably widely agreed on set of expectations. If the results had been to confirm and extend these expectations, the value of the experimental approach could still have been demonstrated, although the tale might have been less interesting in the telling.

There is one additional small message in the story. It illustrates once more the serendipitous nature of much research. Far from conducting a rigorously logical attack on the structure of reef fish assemblages, those of us who have used manipulative approaches have, more often than not, stumbled onto the right questions more by chance than by foresight. The structured research effort in ecology may be as much a myth as the structured reef fish assemblage.

Acknowledgments

This paper is a contribution from the University of Sydney's One Tree Island Field Station, at which much of the experimental work referred to was done. Grants from the Australian Research Grants Committee, the Great Barrier Reef Marine Park Authority, the Great Barrier Reef Committee, the Australian Museum, and the University of Sydney have variously funded my research and that of my (ex-) students. I thank Peter Doherty, Dave Williams, Bill Douglas, and Rand Dybdahl for providing the stimulation of fresh ideas and contradictory results. This manuscript, written in haste for a too-early deadline, did not have the benefit of Tony Underwood's usual critical, nay, caustic review. So I thank him instead just for living next door.

28.

Density Compensation in Vertebrates and Invertebrates: A Review and an Experiment

STANLEY H. FAETH

Department of Zoology, Arizona State University, Tempe, Arizona 85287

INTRODUCTION

Density compensation is the phenomenon by which summed population densities of faunas on islands or insular habitats equal (or, for excess density compensation, exceed) mainland population densities, although number of species is reduced (MacArthur *et al.*, 1972). Density compensation is derived from competition theory. Inherent in most explanations for its occurrence is the idea that for a specified time, community of interacting species, geographic area, and resource base, overall density of individuals is dictated by competition for resources. The species number and actual species composition in the community is deemed relatively unimportant to this optimum density (Crowell, 1962; MacArthur *et al.*, 1972; but see Abbott, 1980, for a differing viewpoint).

Niche and competition theory actually predicts the reverse of density or excess density compensation: overall population densities of faunas on islands with fewer species relative to mainland areas should be lower than mainland densities because remaining species in the "competition community" are less efficacious at "... exploiting resources previously consumed by the absent population(s)" (Case and Gilpin, 1974). When density compensation does occur, it is typically reconciled with competition theory by "niche expansions" or "competitive release" of insular species, which make possible exploitation of a broader resource base (Crowell, 1962; Grant, 1966a; Diamond, 1970; MacArthur *et al.*, 1972; Case and Gilpin, 1974; Case, 1975; Cox and Ricklefs, 1977; Morse, 1977; Kohn, 1978).

Similarly, although niche and competition theory does not explicitly predict occurrence of excess density compensation, modifications of the theory are frequently used to accommodate the phenomenon. In particular, if species that are missing from a "competition community" are interference rather than exploitative competitors, then individuals of the remaining species use unencumbered resources more efficiently (Case and Gilpin, 1974; Case *et al.*, 1979), and the result is an increase in population size.

While many explanations for variation in population sizes on islands are based on competition theory, other mechanisms have been proposed. Predation and parasitism may be reduced on islands, and thus population sizes of prey or host species may increase (Grant, 1966a; MacArthur *et al.*, 1972; Case, 1975). Climate, which tends to be more moderate on islands, could affect population sizes directly through increased survivorship of individuals or indirectly (and in keeping with competition theory) by increasing amounts or decreasing seasonal fluctuations of resources (Case, 1975). Emlen (1978, 1979) suggested that restricted gene flow between island and mainland populations could result in increased levels of adaptation of island species to local conditions, and thus higher population densities. Note that this hypothesis may also involve competition for resources in that highly adapted species would use resources more efficiently. Emlen (1979) and others (Krebs *et al.*, 1969; MacArthur *et al.*, 1972) proposed another explanation, termed the "fence" effect. In this case population densities increase on islands because isolating barriers effectively block escape of individuals that would emigrate or occupy marginal habitats on the mainland.

Williamson (1981) suggests another reason for apparent density compensation on islands. His argument is based on the species-area relationship and is non-biological in origin. Given the ubiquitous species-area relationship of island faunas (Connor and McCoy, 1979) and the lack of *systematic* variation in summed densities of individuals of all species on an archipelago (there can be variation as long as it is random), then average population size per species will decrease as a function of the negative slope of the species-area curve for any geographic region. Williamson (1981) notes therefore that sampling on large areas will give smaller population sizes per species per unit area than sampling on small areas. Thus, comparison of summed population densities on insular areas to those on mainland areas different in size (see Discussion) may result in apparent density compensation simply as a result of sampling effects and the species-area relationship.

Most studies of density compensation have involved vertebrates, including birds (Crowell, 1962; Grant, 1966a; Diamond, 1970; MacArthur *et al.*, 1972; Yeaton and Cody, 1974; Cox and Ricklefs, 1977; Morse, 1977; Nilsson, 1977; Emlen, 1978, 1979; Case *et al.*, 1979; Wright, 1979), lizards (Case, 1975; Wright, 1979), and small mammals (Webb, 1965). Very few studies of density compensation have used invertebrates although there is no strong *a priori* reason to believe density compensation and its causal mechanisms should not be expected. I know of only four: Janzen (1973a), Dean and Ricklefs (1979), and Faeth and Simberloff (1981b) on insects, and Kohn (1978) on gastropods.

The intent of this paper is threefold: (1) to review briefly research on density compensation of vertebrates and invertebrates in insular habitats, (2) to evaluate critically methods used in detecting density compensation and explanatory hypotheses for this phenomenon, and (3) to present an experimental test for density compensation in insects that circumvents methodological problems in prior studies and quantifies the effects of certain causal mechanisms.

EVIDENCE FOR DENSITY COMPENSATION

To demonstrate density or excess density compensation (without regard to *causes*), one must show that mainland and island densities are not significantly different for density compensation or are different for excess density compensation, while accounting for background variation in population densities. While this point may seem obvious, it is rarely evidenced conclusively in studies of density compensation. Because temporal and random variations in most populations can be quite high, discovery of equal or higher densities at a certain point or interval of time does not necessarily mean that population densities are not (or, for excess density compensation, are) significantly different. Analysis-of-variance techniques that account for temporal fluctuations to detect differences in variation between communities seem to be the proper statistics, but are surprisingly absent in density compensation studies. This absence may be in part caused by the frequent difficulty in sampling communities over long periods of time.

Crowell (1962), Grant (1966a), Case (1975), and Morse (1977) suggested that biomass rather than number of individuals should be compared on mainland and islands to detect density (or biomass) compensation. This argument is again based on competition theory. By way of analogy, the resource base on island and mainland could be considered as a resource "urn" and individuals of different species as different-sized marbles of various colors. If resource urns on island and mainland were similar in size, it would take the same volume of marbles (individuals) to fill each urn, but because different-colored marbles are of different size, number of marbles (densities) in island and mainland urns will differ. Unfortunately, the relationship between size of an individual and the proportion of the resource base it encumbers may be complex. Thus, simply comparing biomasses may not reflect true density (biomass) compensation, at least in terms of competitive mechanisms.

Vertebrates

There are numerous studies on density compensation in vertebrates, so I will critique only several exemplary papers, and will refer to other studies with analogous approaches and problems.

Crowell's study (1962) appears to be the first major one on density compensation in birds. He found density of birds (43% greater) and biomass (72% greater standing crop biomass, 49% greater total consuming biomass) higher on Bermuda than on continental North America. Crowell (1962) concluded that ". . . the total population of a fauna does not depend on the number of species present, and . . . the high densities of the Bermuda species can thus largely be accounted for by the presence of few competing species." Apparent niche shifts and expansions of the residual species on Bermuda in the absence of their competitors are cited as evidence supporting this conclusion. Crowell's study suffers from several weaknesses, some of which are also found in later studies on density compensation in birds (Grant, 1966a; Diamond, 1970; MacArthur et al., 1972; Yeaton and Cody, 1974; Cox and Ricklefs, 1977; Morse, 1977; Nilsson, 1977; Emlen, 1978, 1979).

Crowell censused birds by counting singing males. I assume these accurately reflect population sizes of species although there is considerable disagreement among ornithologists regarding methods for estimating density of birds (see Grant, 1966a; Anderson and Ohmart, 1981, and references therein). His censuses were taken on several days over a period of 4 to 10 weeks. These data were compared to densities of birds from every habitat east of 85°W longitude in North America from 13 years of *Audubon Field Notes*. The direct comparability of these data is highly questionable for several reasons, including different observers, greatly different time spans, different years, and vastly different geographic areas. Because only average population densities are reported, it is impossible to determine the amount of seasonal variation, even over the brief censusing period for Crowell's Bermuda censuses. Despite the incompatibility of the two data sets, the proper test hypothesis should have been: Is a difference of 43% between number of individuals on Bermuda and number of individuals on mainland North America significant considering seasonal, yearly, and within-habitat variation within each community? Given the high seasonal and yearly variation in most bird populations, it probably is not. Second, a significantly greater density on Bermuda than on the mainland is expected as a statistical artifact (Williamson, 1981). Average population sizes per unit area are expected to be lower on a large, continental area than on an island simply as a consequence of the species-area relationship. The proper test hypothesis, therefore, is $\mu_{mainland} = \mu_{island}$, but only after the effects of area on densities have been removed by analysis of covariance (Williamson, 1981).

Regardless of whether differences in bird densities are real, is Crowell's interpretation of the causal mechanism based on absence of competitors justified? The only supportive evidence comes from apparent niche expansions, and Crowell himself admits "all observed differences in feeding habits in Bermuda cannot be ascribed to absence of competition *per se*. Local and seasonal variations in type and location of food supply produce appropriate variations in feeding behavior." Moreover, while Crowell claims "Primary productivity (in Bermuda) is about the same as in North America, as estimated from agricultural yields . . . ," it is highly improbable that resource bases in Bermuda and North America are equivalent. Furthermore, one must assume the resource base over a 13-year span in North America is directly comparable to that over the 4–10 week period when Crowell censused in Bermuda. Because resources are dependent on climate, climatic variation must also be assumed to be comparable in these intervals.

The most persuasive evidence for density compensation resulting from competition, given that the assumptions of equivalent areas and resource bases are fulfilled, is evidence of niche expansions by island species. In general, studies on island birds produce conflicting results. Some have no clear evidence of niche expansion (Crowell, 1962; Emlen, 1978; and references in Abbott, 1980) while others do (Grant, 1966a; Keast, 1968; Diamond, 1970; MacArthur *et al.*, 1972; Yeaton and Cody, 1974; Cox and Ricklefs, 1977; Morse, 1977). Abbott (1980) regards niche expansions of birds in insular habitats as exceptions rather than rules. Some claims of apparent niche shifts in bird communities on islands are questioned (Strong *et al.*, 1979). However, even if evidence of niche shifts is accurate, competition need not be the causal force. Case (1975) suggested that predator avoidance could restrict niches of lizards. Predators tend to be underrepresented on islands and niches of prey species could expand into consequently available "predator-free space" (Lawton, 1978a; P. W. Price *et al.*, 1980). Thus niche expansion and numerical response of species could occur on islands merely as a consequence of reduced predation pressure. While reduced predation seems a plausible alternative mechanism to competition for both density compensation and niche shifts for insular faunas, only a few investigators have given it much consideration (Case, 1975; Emlen, 1978). For example, MacArthur *et al.* (1972), in perhaps the most oft cited paper in density compensation, devote only two sentences to predation. Crowell (1962) noted that "In Bermuda, populations are resident rather than migratory, and predators are few." However, he made no statement regarding the possible effects of reduced predation on population sizes or niches. Morse (1977) also noted that predation was less intense on the islands that he studied, but concluded ". . . the competition hypothesis is consistent with the data."

Case (1975) concluded that, for lizards on islands in the Gulf of California, reduced predation, increased insect productivity, and effects of interference competition account for apparent density compensation on at least some small isolated islands. While Case's study is commendable in that he attempted to account for differences in climate, habitat, insect productivity, and predation when comparing mainland and island densities, it suffers from two problems. First, areas of island and mainland study localities are apparently not the same, and at least some observed community differences may be attributable to the probability of lower number of individuals per species per unit sample (Williamson, 1981). Second, baseline variability in density of lizards was not considered. From Case's Table 1, I judge that this variability between years was quite high, both on islands and on mainland areas.

Emlen (1979) also used as insular study sites the Gulf of California islands when comparing bird densities on islands to those on the mainland. Emlen attempted to control for differences in resource distribution and abundance by choosing similar habitats on the mainland and islands, and by crudely estimating foliar insect abundances in all areas. Even if one assumes these variables are properly considered, Emlen's study still has flaws. The first one Emlen himself notes by stating that densities during the short census interval (April 5–28, 1977) could have been atypical, and large differences between mainland and islands may have been only transitory. This difference again can only be deemed significant if background variation is considered. Second, because transect lengths were different on islands and mainlands (Emlen's Table 3) once again a difference in average population size could be simply a statistical consequence of censusing unequally sized areas.

Evidence for difference in predation is conflicting in Emlen's (1979) study: predators of birds appeared uncommon on islands, but lizards preying upon birds' eggs were more abundant. Although no attempt was made to quantify predation pressure, Emlen discounted it as a significant force.

Because Emlen found no evidence of niche expansions or resource difference, he rejected reduced competition as the cause of higher densities. However, increased densities caused by reduced competition need not be consonant with niche shifts (Thomson, 1980) or absence from islands of interference strategists (Case and Gilpin, 1975; Case et al., 1979). Thus, rejecting competition as a cause of density compensation because of lack of niche shifts may be just as invalid as inferring competition because of evidence of niche shifts. Instead Emlen considered higher adaptation to local conditions by island species maintained by restricted gene flow with mainland individuals, and overcrowding caused by the fence effect, to be the primary reasons for higher island densities.

Invertebrates

Studies on the effects of insularity on densities of invertebrates are scarce. Janzen (1973a) examined tropical insect densities by sweep netting in Costa Rica and some Caribbean islands. Numbers of species and individuals were lower on islands than on the mainland during the wet season. Differences in densities were not so pronounced during the dry season. Janzen (1973a) considered this pattern surprising because comparison between mainland sites indicated that those areas with reduced number of species had higher population densities. He attributed this increase to the absence of "potential competitors." Homopteran species on islands, unlike other insects, did show increased population densities. Janzen attributed this difference to their being generalists and feeding on many host plants on the islands (implying niche expansions). The role of predation was not clear in this study; invertebrate predators were more numerous on the islands but vertebrate predators appeared to be fewer. The abundances of parasitoids, which are not accurately censused by sweep netting (Morrison *et al.*, 1979), were not discussed. Since they can have tremendous impacts on populations of insects (Faeth and Simberloff, 1981a), their abundances should be evaluated in all such studies.

Janzen's (1973a) study, at least with regard to the effects of insularity on invertebrate densities, is weakened by the same flaws mentioned for the vertebrate studies. Mainland and island areas are not the same size and do not have the same climate or vegetation composition. The latter is, of course, a critical determinant for both the presence and abundance of phytophagous insects and their predators and parasitoids. Again, no statistics were used to test whether densities on mainland and islands were indeed significantly different.

A second study of invertebrates is that of Kohn (1978), who compared densities of *Conus miliaris* (Gastropoda) on Easter Island to densities of *Conus* species throughout the Indo-West Pacific. He concluded that densities of *C. miliaris* were comparable to or higher than those in other Indo-West Pacific localities harboring more species because of ecological expansion toward resources that usually would be used by the missing congeners. First, Kohn's claim that densities were higher on Easter Island than at other localities is not supported by his data. Using data from Kohn's Table 1 (Kohn did not test statistically for differences), I find that neither mean densities nor maximum densities are significantly greater than at other Indo-West Pacific sites (Wilcoxon Rank Sum Test, average densities, $p > .05$, maximum densities, $p > .15$). In fact, for mean densities, the Eastern Island population is significantly *less* dense than those at other localities (Wilcoxon Rank Sum Test, $p < .01$). Further, all of Kohn's conclusions are questionable because localities compared are of different sizes, and "Major geologic differences between largely basaltic Easter

Island and central IWP coral reef shorelines cause profound differences in habitat structure" (Kohn, 1978). If there is density compensation, is it caused by missing competitors or radically altered habitats and resources?

Dean and Ricklefs (1979) recently examined the possibility that density compensation might have occurred in an insect parasite guild. They found, however, that abundances of a given parasite were positively correlated with the densities and number of co-occurring species. They concluded that parasites did not compete for hosts (see Force, 1980; Bouton et al., 1980, for a differing viewpoint) and therefore density compensation did not occur. Because of the general nature of their study, Dean and Ricklefs (1979) could not demonstrate what other factors might have been operating.

Obviously, there are practical limitations that arise in attempting to quantify parameters that may affect densities in natural situations. Apparently, experiments controlling for various causative factors of community compensation are nonexistent. It is important to emphasize that the three possible outcomes of comparing island to mainland densities—lower island densities, equal densities (density compensation), and higher densities (excess density compensation)—can be explained within the framework of competition theory. Lowered densities are predicted by competition theory because of diminished efficiency of residual species in acquiring resources (Case and Gilpin, 1974). Density compensation is explained by "niche expansions resulting in higher abundances of island species" (e.g. MacArthur et al., 1972). Excess density compensation is interpreted as a result of missing species of interference strategists on islands (Case and Gilpin, 1974). Thus, the simple observation of density differences (or similarities) cannot constitute a falsifiable test (sensu Popper, 1963) that competition is structuring certain island communities. First, to establish that density compensation occurs and, second, to choose among alternative causal mechanisms, it is necessary to control or account for possible sources of variation in the system. It is with this necessity in mind that I describe an experiment that overcomes most of the criticisms I have discussed.

AN EXPERIMENT ON THE EFFECTS OF INSULARITY ON ABUNDANCES OF LEAF MINERS

Methods

To test the effect of insularity of host plants on abundances, species richness, and survivorship of leaf-mining insects, six trees of three oak species (Quercus falcata, Q. nigra, and Q. hemisphaerica) were transplanted into randomly assigned positions (5 m apart) on a grid in an agricultural field. Two individuals of each oak species were transplanted to the edge of the agricultural field to serve as non-isolated but transplanted controls. Isolated (field) and non-isolated (edge) trees were 165 meters apart.

Three trees of each species in the agricultural field were randomly selected and were censused with the control trees over a two-year period (1978 and 1979) for abundances and presence of leaf-mining species.

Survivorship and causes of mortality of leaf miners were documented by inspection or dissection of inactive mines. Sources of mortality can accurately be categorized as (1) predation, (2) parasitism, (3) death for unknown reasons, including weather, nutrition from host leaf, disease, or competitive effects of leaf miners via the host plant (*sensu* Janzen, 1973b).

Details of life histories of the leaf miners, oak host-plant phenology, censusing, and mortality analyses are in Faeth and Simberloff (1981b).

The experiment was designed to minimize weaknesses that pervade previous studies on density compensation. Difficulties that result from comparing densities on areas of unequal size (Williamson, 1981) are avoided because the experimental and control trees have roughly the same surface area (Faeth and Simberloff, 1981b). Entire trees were censused. Absolute leaf-mining densities were used in all analyses; thus any error caused by estimation of densities from a sampling regime was eliminated. Censuses were conducted for leaf-mining densities on control and experimental trees during the growing seasons of 1978 and 1979. Consequently, I was able to account for both within- (seasonal) and between-year variation in densities when determining whether differences in densities between treatment and control were significant.

To determine statistically whether field and edge densities are significantly different while accounting for seasonal and yearly variation, I performed analyses of variance (repeated measures design) on active and total densities for each species (Tables 28.1–6). Sum of squares, mean squares, and F statistics were calculated according to Winer (1971, p. 540).

While subtle chemical and nutritional differences may have existed between experimental trees, trees were at least the same size and apparently healthy. I also randomly selected experimental and control trees from a nearby pool of potentially transplantable trees, again to minimize differences between trees. One would not expect any significant climatic or local weather differences between the edge (non-isolated) and field (isolated) trees (Faeth and Simberloff, 1981b). Yet, the degree of isolation (165 m) was enough to affect at least a portion of potential leaf-mining colonizers (Faeth and Simberloff, 1981b).

Janzen (1973b) suggested that insects that feed on the same host plant may be in intense competition with each other even though direct encounters do not occur and they feed on different plant parts. This competition via the host plant might also occur for leaf miners that feed within

Table 28.1. Summary of analysis of variance for active mines on *Quercus hemisphaerica*.

Source of Variation	Sum of Squares	df	Mean Square	F	Significance
Between Treatments					
A. Treatment	0.821	1	0.821	1.999	$p > .25$
Trees (treatment)	1.232	3	0.411		
Within Treatments					
B. Years	0.021	1	0.021	0.447	$p > .25$
Years-Treatments	0.064	1	0.064	1.362	$p > .25$
Trees (Treatments) × Years	0.140	3	0.047		
C. Dates	2.930	8	0.366	1.718	$p > .10$
Dates-Treatments	2.624	8	0.328	1.540	$p > .10$
Trees (Treatments) × Years	5.112	24	0.213		
D. Years-Dates	1.680	8	0.210	1.008	$p > .25$
Years-Dates-Treatments	2.092	8	0.261	1.253	$p > .25$
Trees (Treatments) × Years × Dates	5.001	24	0.208		

Table 28.2. Summary of analysis of variance for total mines on *Quercus hemisphaerica*.

Source of Variation	Sum of Squares	df	Mean Square	F	Significance
Between Treatments					
A. Treatment	2.027	1	2.072	3.088	$p > .10$
Trees (treatment)	2.684	3	0.671		
Within Treatments					
B. Years	0.302	1	0.302	1.073	$p > .25$
Years-Treatments	0.200	1	0.200	0.711	$p > .25$
Trees (Treatments) × Years	0.844	3	0.281		
C. Dates	3.165	8	0.396	3.328	$p < .05$
Dates-Treatments	0.822	8	0.103	0.866	$p > .25$
Trees (Treatments) × Years	2.866	24	0.119		
D. Years-Dates	1.245	8	0.156	1.386	$p > .10$
Years-Dates-Treatments	0.598	8	0.075	0.666	$p > .25$
Trees (Treatments) × Years × Dates	2.701	24	0.113		

Table 28.3. Summary of analysis of variance for active mines on *Quercus nigra*.

Source of Variation	Sum of Squares	df	Mean Square	F	Significance
Between Treatments					
A. Treatment	0.627	1	0.627	1.507	$p > .25$
Trees (treatment)	1.248	3	0.416		
Within Treatments					
B. Years	0.427		0.427	1.505	$p > .25$
Years-Treatments	0.671	1	0.671	2.365	$p > .10$
Trees (Treatments) × Years	0.851	3	0.284		
C. Dates	4.004	8	0.500	4.100	$p < .01$
Dates-Treatments	1.472	8	0.177	1.451	$p > .10$
Trees (Treatments) × Years	2.972	24	0.122		
D. Years-Dates	1.183	8	0.148	1.359	$p > .10$
Years-Dates-Treatments	1.184	8	0.148	1.359	$p > .10$
Trees (Treatments) × Years × Dates	2.613	24	0.109		

Table 28.4. Summary of analysis of variance for total mines on *Quercus nigra*.

Source of Variation	Sum of Squares	df	Mean Square	F	Significance
Between Treatments					
A. Treatment	0.199	1	0.199	0.397	$p > .25$
Trees (treatment)	1.504	3	1.501		
Within Treatments					
B. Years	1.319	1	1.319	1.506	$p > .25$
Years-Treatments	0.155	1	0.115	0.131	$p > .25$
Trees (Treatments) × Years	2.628	3	0.876		
C. Dates	3.352	8	0.419	6.758	$p > .001$
Dates-Treatments	0.995	8	0.124	2.000	$p > .05$
Trees (Treatments) × Years	1.487	24	0.062		
D. Years-Dates	0.450	8	0.056	1.114	$p > .25$
Years-Dates-Treatments	0.338	8	0.042	0.836	$p > .25$
Trees (Treatments) × Years × Dates	1.206	24	0.050		

Table 28.5. Summary of analysis of variance for active mines on *Quercus falcata*.

Source of Variation	Sum of Squares	df	Mean Square	F	Significance
Between Treatments					
A. Treatment	0.078	1	0.078	0.494	p > .25
Trees (treatment)	0.474	3	0.158		
Within Treatments					
B. Years	0.081	1	0.081	0.890	p > .25
Years-Treatments	0.033	1	0.033	0.363	p > .25
Trees (Treatments) × Years	0.273	3	0.091		
C. Dates	1.655	8	0.207	2.120	p > .05
Dates-Treatments	0.752	8	0.094	0.963	p > .25
Trees (Treatments) × Years	2.343	24	0.098		
D. Years-Dates	1.090	8	0.136	1.541	p > .10
Years-Dates-Treatments	0.780	8	0.097	1.099	p > .25
Trees (Treatments) × Years × Dates	2.118	24	0.088		

Table 28.6. Summary of analysis of variance for total mines on *Quercus falcata*.

Source of Variation	Sum of Squares	df	Mean Square	F	Significance
Between Treatments					
A. Treatment	0.356	1	0.356	0.175	p > .25
Trees (treatment)	6.109	3	2.036		
Within Treatments					
B. Years	0.977	1	0.977	1.597	p > .25
Years-Treatments	0.083	1	0.083	0.136	p > .25
Trees (Treatments) × Years	1.835	3	0.612		
C. Dates	9.023	8	1.128	6.461	p < .001
Dates-Treatments	0.921	8	0.115	0.659	p > .25
Trees (Treatments) × Years	4.190	24	0.175		
D. Years-Dates	0.915	8	0.114	0.704	p > .25
Years-Dates-Treatments	1.393	8	0.174	1.074	p > .25
Trees (Treatments) × Years × Dates	3.876	24	0.162		

leaf tissues, although rarely do they reach high densities, and individual species are at least partially segregated by feeding at specific sites within the leaf (Faeth and Simberloff, 1981a, b). Thus, if interspecific competition is lessened on isolated host plants because of the absence of competitors (in this instance, missing members of the leaf-mining guild), one might expect (1) density compensation or excess density compensation by leaf miners on isolated trees, (2) significant differences in the unknown-cause-of-death category, which includes mortality caused by competition, and (3) evidence for niche expansion of isolated leaf miners in the form of expanded mined regions of leaves on isolated trees.

Results and Discussion

Density Compensation (Species Richness). Previously (Faeth and Simberloff, 1981b), I showed that, for all oak species and trees analyzed together, numbers of species of leaf miners are reduced only by one on isolated trees (12 versus 13 species). This similarity is primarily caused by late leafing and short stature of non-isolated trees of *Quercus falcata;* some early leaf-mining species failed to colonize these trees. Control and experimental trees of *Q. nigra* and *Q. hemisphaerica* were similar in phenology and height. In each of these instances, numbers of leaf-mining species present on field trees were always (except for one census date in one year for *Q. nigra*) lower on isolated field trees in both years (Faeth and Simberloff, 1981b, Figs. 6, 7, 8).

It is interesting to note how slightly altered phenology and plant size in *Q. falcata* affected presence of phytophagous insects. One might speculate that, in mainland-island comparisons such as Janzen's (1973a) where not only phenologies of individual plants are probably altered because of climatic differences, but also abundance and density of the plant species themselves are different, the effect on the phytophagous insect fauna and, subsequently, their predators and parasitoids, would be highly accentuated.

Density Compensation (Abundances). I have shown previously (Faeth and Simberloff, 1981b) that neither active (leaf miners actually feeding at time of censusing) nor total densities (active plus inactive mines, which gives a cumulative index of mining activity) of leaf miners on all species of isolated (field) trees together were significantly different from densities on all non-isolated (edge) trees. I now attempt to determine whether active and total densities on field and edge trees are different for each oak species.

There is considerable seasonal variation and also some yearly variation in both field and edge trees (Figures 28.1–6). From Tables 28.1–6, it is evident there is no treatment (isolation) effect for either active or total densities for any of the three oak species. Thus, even though number of

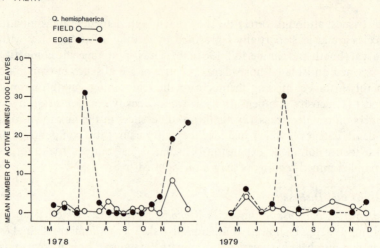

Fig. 28.1. Comparison of mean number of active mines per 1000 leaves on isolated (field) and non-isolated (edge) trees of *Quercus hemisphaerica* for 1978 and 1979.

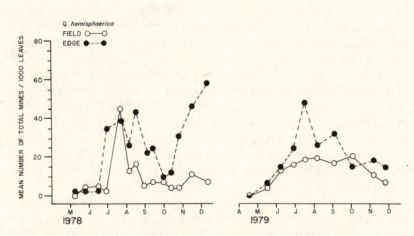

Fig. 28.2 Comparison of mean number of total mines per 1000 leaves on isolated (field) and non-isolated (edge) trees of *Quercus hemisphaerica* for 1978 and 1979.

leaf-mining species was reduced on two of the three oak species, densities are not different and density compensation may have occurred. It is possible that large between-tree variation obscured any effects of isolation. However, only for total densities for *Q. falcata* is there large variation between trees (sum of squares = 6.109) relative to the sum of squares for

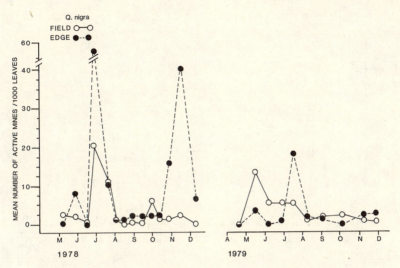

Fig. 28.3. Comparison of mean number of active mines per 1000 leaves on isolated (field) and non-isolated (edge) trees of *Quercus nigra* for 1978 and 1979.

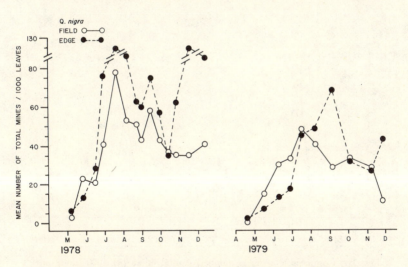

Fig. 28.4. Comparison of mean number of total mines per 1000 leaves on isolated (field) and non-isolated (edge) trees of *Quercus nigra* for 1978 and 1979.

treatment effect. The only significant effect is that of date or seasonal variation in leaf-mining densities, which is found in all cases (Tables 28.1–6) except for active densities on *Q. hemisphaerica*. This result is not surprising when one considers seasonal fluctuations in densities (Figures 28.1–6), but is rarely considered in other studies of density compensation.

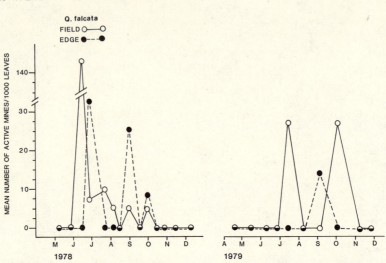

Fig. 28.5. Comparison of mean number of active mines per 1000 leaves on isolated (field) and non-isolated (edge) trees of *Quercus falcata* for 1978 and 1979.

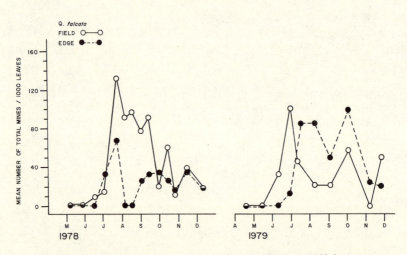

Fig. 28.6. Comparison of mean number of total mines per 1000 leaves on isolated (field) and non-isolated (edge) trees of *Quercus falcata* for 1978 and 1979.

Causes of Mortality

Elsewhere (Faeth and Simberloff, 1981b, Table 2), I determined whether survivorship and various causes of mortality differed significantly (Kolmogorov-Smirnov test) for leaf miners on field and edge trees. Survivorship of larval leaf miners was significantly higher in both years on isolated trees. However, for the three categories of mortality—parasitism, predation, and other causes—only parasitism was significantly different (lower on the isolated trees). Deaths of larval leaf miners on isolated and non-isolated trees from other causes were not significantly different in either year (1978, $p > .05$; 1979, $p > .50$). These results suggest that higher survivorship was caused by lowered parasitism on isolated trees, and that absence of species of leaf miners does not increase survivorship through reduced competition.

The relevance of this experiment to other studies is clear: without actually determining mortality and survivorship, one cannot interpret causes of density compensation even when assumptions of equal area, climate, resources, and true density differences are met. Whether parasitism and/or predation is a significant factor in previous studies is unknown, but certainly must remain a plausible alternative to competitive release.

Niche Shifts

I found no evidence of niche shifts or niche expansions of leaf-mining species on the isolated field trees. In no instance did an individual leaf miner on an isolated tree mine portions of a leaf not consumed by members of the same species on the non-isolated edge trees.

In fact, in another study (Faeth and Simberloff, 1981a) where all but one leaf-mining species were excluded from a tree by a cage, individual leaf miners did not consume areas of leaves left unoccupied by virtue of the missing species of leaf miners. Individuals of the leaf miner within the cage that reached adulthood did not consume more of the leaf (based on length of the mine at emergence) than did individuals of the same species when feeding on leaves of other trees colonized by the full complement of leaf-mining species.

It could be argued that two years is not sufficient for niche shifts to occur or that leaf miners and perhaps invertebrates generally are extraordinarily unplastic with respect to behavioral changes. Obviously, there may be physiological and genetic factors limiting expansion of feeding habits that change only over long (evolutionary) time periods. The important point here is not that niche expansions did not occur but that densities on isolated trees were not different from those on non-isolated trees, and this "compensation" appears to be caused by relaxed parasitism rather than competition.

Niche shifts, if they do occur in the absence of closely related species, need not implicate release from competition solely for a food source. An equally viable hypothesis is expansion into "enemy-free space" (Lawton, 1978a, P. W. Price et al., 1980), prohibited previously by the presence of closely related species. On the other hand, competition may not be involved at all, but rather, when parasites or predators are reduced in insular situations, prey species may expand both in niche parameters and numerically where previously predators and parasites had restricted them.

The latter seems true of leaf miners on oaks: when parasitoids and predators were excluded by cages from a tree harboring a leaf miner, the leaf miner increased dramatically in population size (Faeth and Simberloff, 1981a). This increase suggests that under usual circumstances predators and parasitoids keep leaf-mining populations at levels below those at which both intra- and interspecific competition would occur, a hypothesis certainly not new to ecology (Hairston *et al.*, 1960).

Fence Effects

As Emlen (1978) suggested, barriers to dispersal in insular habitats could result in higher population densities. There is no evidence that the fence effect occurred here: emerging adults appear to fly readily across the isolating distance (165 m), as suggested by the fact that most species did colonize the isolated trees. To the contrary, evidence from this experiment and another (Faeth and Simberloff, 1981b) suggests that emerging adults on small isolated trees probably do not remain where they were born, but rather disperse to nearby stands of oaks to locate mates. Thus, it appears that populations of leaf miners on the isolated trees are maintained by continual immigration from nearby overwintering and pupation sites. Density compensation may result from increased survivorship of each successive cohort of immigrating left miners. Even with increased survivorship of larvae on isolated trees, densities do not increase beyond those on non-isolated trees because no *in situ* reproduction occurs and adults die or disperse from the field before mating.

Summary and Conclusions

I have discussed here major problems and approaches associated with studies of density compensation of insular faunas. In particular, unless one assumes or can demonstrate that (1) densities on island and mainland are accurately assessed, compensating for background variation in population densities, (2) island and mainland areas are equivalent or the effect of area is removed, (3) resource bases are the same on island and mainland, and (4) climatic conditions are alike, the conclusion that density or excess density compensation has occurred is tenuous. All previous studies have failed to demonstrate one or more of these conditions; thus there may be limitations to conclusions drawn from "natural experiments."

If density compensation or excess density compensation is unequivocally shown to occur, then interpretation of causal mechanisms presents further dilemmas. The simple comparison of insular and mainland densities cannot be taken as evidence for competition as a mechanism, even though it often has been, because niche and competition theory can explain all possible outcomes of insular densities relative to mainland abundances (no density compensation, density compensation, or excess density compensation). Evidence for competition as the cause of insular density changes in the form of observed niche shifts is also ambiguous, because niche shifts can be caused by other factors such as release from predation.

Experiments controlling for various external factors that may affect population densities and interpretation of causal mechanisms seem to be more profitable in the study of density compensation than simply observing island and mainland densities and surmising probable causes. Abbott (1980) has described the problems in studies of density compensation of birds where attempts are made to isolate the effects of one factor by comparing island and mainland areas that invariably differ in many factors. Abbott (1980) advocates that "...where possible a truly experimental approach be adopted," and suggests introduction or removal experiments in island/mainland bird studies. It is in this vein that I have presented results from an experiment in which I isolated oak host plants in an agricultural field and monitored species richness, abundances, and causes of mortality of leaf miners. My results showed no significant differences in leaf-mining densities between isolated experimental trees and non-isolated control trees, although number of species was slightly reduced. However, if density compensation occurred, it appears that it was caused by increased survival of larval leaf miners due to decreased parasitism, not by release from competition.

29.

Communities of Specialists: Vacant Niches in Ecological and Evolutionary Time

PETER W. PRICE

Department of Biological Sciences, Northern Arizona University, Flagstaff,
Arizona 86001

The paradigms of ecology concerning community organization cluster around interspecific competition as a central organizing force. Niche occupancy is defined by the presence of competitors on adjacent parts of environmental gradients. Strong competitors have broad niches, weak competitors narrow niches. Similarity between coexisting species is limited through competition, so communities become tightly packed with a regular spacing of Hutchinsonian distances between species. The gradient in species density from high to low latitudes and the changing environmental conditions result in more competition in the tropics, narrower resource utilization and more overlap between species. Other factors such as predation and disturbance moderate the overriding influence of competition in some cases.

The remarkable feature about these paradigms is their general acceptance without rigorous testing except in a very circumscribed set of communities. We have no idea how generally applicable they are, nor even whether they apply to the majority or minority of cases, nor whether they are even worthy of the term paradigm. A second peculiar feature is that general acceptance has even gone beyond the theory, for competition is frequently invoked in high latitudes where theory suggests that disturbance is expected to play a dominant role, while the real part that disturbance plays is seldom evaluated. A third feature is that generalizations about the force of competition in community organization have no limits either geographically, taxonomically, or in terms of resources and how these are exploited. Competition plays a role everywhere no matter what kinds of organisms are involved—small mammals, birds, insects, bacteria, molluscs, algae, angiosperms, or helminths—and no matter what kinds of resources they use—whether they be rapidly renewable like phytoplankton or host tissues, or very seasonal like seeds or flowers, or very patchy like dung or carrion, or very widely dispersed as in tropical forests or coral reefs—and no matter to what degree these resources are exploited by generalists or specialists. And yet, it is most unlikely that the same kinds of organizational forces will play a role in all communities when there can be such a diverse array of geographical locations, taxa, resource states, and modes of exploiting these resources.

One of the most neglected groups of organisms is the specialists. When ecologists have considered specialists these have actually been rather generalized exploiters with simply a narrower choice of food items than an even more generalized organism. But there are whole families of organisms with much more limited choices than this, and the enormous difference between these true specialists and generalists are not generally appreciated (*cf.* Table 2.2 in Price, 1980). Since the role of competition in communities has been studied mostly in communities of generalists, and the paradigms have been developed from such studies, it is valuable to examine the extent to which specialist communities support or contradict the paradigms.

Specialists differ from generalists in several important ways: (1) They are usually small to very small as individuals, taking in a lifetime a minute fraction of the resources available. (2) These resources are narrowly defined in both space and time, with a leaf offering several discrete resources for say, leaf miners, and a resource often being appropriate for attack for a matter of minutes or hours. (3) Many small organisms can normally travel only short distances, so colonization becomes a major obstacle to resource exploitation. (4) The number of resources for specialists in any one habitat, even a very simple one, is extremely large, so that large numbers of species can coexist on discrete resources in one habitat. (5) Time is perhaps the most limiting resource because the resources of substance are ephemeral and colonizing stages are short-lived. These properties of specialists will be illustrated repeatedly in the following discussion and are elaborated upon by Price (1980) (Figure 29.1). Ecology and evolution of specialists operate on a miniature scale hardly appreciated by the student of generalists (Price, 1977, 1980).

These differences between generalists and specialists are fundamental and inescapable. They are sufficiently profound to suggest that communities of specialists may well be organized in ways different from the paradigmatic picture. For example, ecological time may be so limiting that colonization is seldom complete, and communities are usually depauperate, with vacant niches available. Perhaps resources are so diverse that adaptive radiation has not kept pace and vacant niches exist in evolutionary time also. The pool of colonist species is smaller than the pool of resources available in many habitats. Under these circumstances the forces of interspecific competition would be significantly dampened. The question really is how much.

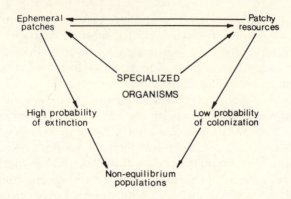

Fig. 29.1. Patch dynamics in populations and species of specialists (after Price, 1980).

Although specialists are so frequently small and inconspicuous, offering problems in their study, they also offer some important advantages: (1) They are intimately associated with their food supply for a large proportion of their lives, so food resources can be accurately described and quantified. This property is important because studies on competition have rarely adequately quantified the resources for which contenders are supposed to be competing, and resource limitation is seldom established. (2) Many habitats are almost complete duplicates of others in terms of structural features, so direct comparisons can be made between communities. For example, a teleost fish, no matter which species, has the same array of organs and juxtaposition of organs available for parasite communities to develop in. Hosts provide homologous habitats. It is therefore possible to evaluate the extent to which resources are fully utilized when vacant niches exist in one habitat and not in another and to estimate the number of species that could potentially coexist in such a habitat. Specialist species can be used to indicate the size of ecological niches and the scale on which we should be looking at niche occupation. (3) Resources for specialists are frequently provided by living organisms themselves, and specificity is limited below the species level by the susceptibility or resistance of individual genotypes. Coevolutionary forces play an important role in dictating presence or absence of resources.

These advantages to the study of specialists make interpretation of relative abundance data for closely related species distributed on a resource gradient simple to gather and interpret. Roskam and van Uffelen's (1981) study on cecidomyiid gall flies on birches in Holland will be used as an example. The two birches *Betula pendula* and *B. pubescens* coexist and hybridize, producing a gradient of genotypes from one species to the other. With a ranking of many characters each birch tree specimen can be placed on this gradient and the abundance of each of two cecidomyiid species, *Semudobia betulae* and *S. tarda*, which form galls on fruits in the catkins, can be estimated. The patterns of abundance of the two species on the gradient are complementary (Figure 29.2). This pattern has a beguiling similarity to expected arrays of species resulting from competition—with competitive displacement and narrowed niche occupation as a result.

Competition cannot be completely discounted in this case, but it is not a parsimonious explanation when the probabilities of encounter between species at a fruit range around 0.003 and 0.007, or on less than 1% of the resources. The more parsimonious explanation is that a speciation event resulted in two *Semudobia* species, one on each *Betula* species, each host-parasite pair coevolving. After the host species became sympatric, hybridization took place, so one galling species can now effectively exploit only

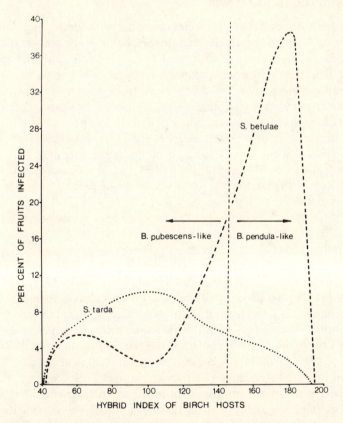

Fig. 29.2. The percent of fruits infected by *Semudobia tarda* and *S. betulae* on a gradient of genotypes from *Betula pubescens*-like hosts on the left to *B. pendula*-like hosts on the right. The lines for each species represent the maximum relative abundances reached on 45 trees sampled (after Roskam and van Uffelen, 1981).

a narrow range of genotypes close to the original host species type. Selective pressures for coevolution to the host, with a 100% encounter rate between host and parasite, will be much stronger than selective pressures from competitors with an encounter rate of less than 1%. Knowing that resources for these cecidomyiids are normally superabundant is an essential part of distinguishing between an explanation depending on competition and one depending upon independent specialized exploitation patterns. This example illustrates the tight link between a specialist and its food resources. Such linkage will be studied in the next section in order to test how geographical patterns in resource exploitation by specialists fit the paradigms on community organization.

GEOGRAPHICAL PATTERNS

Numbers of marine fish species increase with decreasing latitude (Rohde, 1978b). They provide homologous habitats for colonization by gill parasites (Trematoda: Monogenea). In tropical seas many fish species harbor rich monogenean communities: one species is known to support nine monogenean species, and a single host specimen may have seven monogenean species present (Rohde, 1978b); even small fish species a few centimeters long may have a community of five monogenean species (Rohde, 1979). But in spite of a rich monogenean fauna a significant percentage of fish species in the tropics have no monogenean parasites: 17% of 53 species around Lizard Island (15°S), and 31% of 74 species around Heron Island (Rohde, 1978b). Clearly, many vacant niches exist for monogeneans in tropical waters.

In colder waters the majority of fish species have no monogenean parasites: 73% of 33 species in the White Sea (65°N) and 66% of 47 species in the Barents Sea (70°N), and of the remainder a large proportion support only one species of monogenean gill parasite: 21% of 33 species in the White Sea and 30% of 47 species in Barents Sea (Rohde, 1978b). In temperate seas the vast majority of ecological niches remain vacant.

In spite of this latitudinal gradient of species richness in monogenean communities, the niche breadth of parasite species does not change either in site specificity on a host or in the range of hosts utilized (Rohde, 1978a, b, 1979, 1980). Monogenean parasites are just as specific in the absence of potential competitors as they are in their presence (Figure 29.3). This pattern contrasts with that seen in the digenetic trematodes, which show declining specificity with increasing latitude. The combination of the indirect life cycles of digeneans, passive infection by ingestion of an intermediate host by the definitive host, and reduced habitat heterogeneity in more temperate seas may contribute to this difference (see Rohde, 1978a, for further discussion). There is simply a more intimate linkage

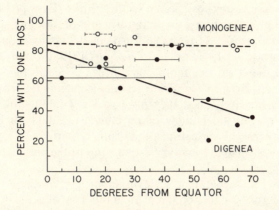

Fig. 29.3. The percent of Monogenea and Digenea parasite species utilizing only one host on fish at various locations on a latitudinal gradient. Open circles Monogenea, closed circles Digenea. Horizontal lines indicate latitudinal range over which sample was taken. Circles placed at mid-point on this range. The regression lines are drawn to clarify the trends (from data in Rohde, 1978a).

between monogenean life cycles and hosts than between digenean life cycles and hosts.

For the monogenean communities Rohde (1978b) concludes that: (1) "Many 'niches' on the gills of non-tropical (and perhaps of tropical) fish are empty." (2) "Many (and perhaps all) habitats can accommodate more species than at present." (3) "Evolution creates niches faster than it can fill them." (4) "Availability of many empty niches reduces the significance of competition" (pp. 416–417).

The paradigms of community ecology are not supported by the evidence that Rohde has gathered on monogenean communities. Species are not more specific in the tropics, niche specificity or niche breadth remains the same independent of the number of species packed into the community, and parasite communities are so undersaturated in terms of numbers of individuals and numbers of species that competition is unlikely to play an important role in community organization.

Other studies on totally different kinds of specialist species support the conclusions reached by Rohde. Specificity of butterfly species to host species appears to be no greater in the tropics than in temperate regions (Price, 1980). (Scriber, 1973, used number of host plant families utilized by papilionid butterflies as an index of specificity and found tropical species to be more specific than temperate species. More data are needed for comparison of butterflies, but analysis of specificity at the familial level of hosts tells us little about community structure.) Bark and ambrosia beetles are less host specific in the tropics than in temperate regions (Beaver,

1979a), and aphids are less specific in the tropics (Eastop, 1972). While the diversity of major hosts, the butterflies and moths, increases toward the tropics, the density of parasitic wasps in the family Ichneumonidae declines (Townes, 1971; Heinrich, 1977; Janzen, 1981), with the obvious consequence that there are more open niches in the tropics and looser species packing. Even in temperate regions ichneumonids could radiate much more extensively than they have so far without utilizing all the resources (Price, 1980). I know of no convincing data on broad geographical patterns of specialist species, other than those for digenetic trematodes illustrated in Figure 29.3, that support the paradigms on community organization.

On a more local geographic scale, in an analysis of species richness of leaf-mining agromyzid flies on the plant family Umbelliferae in the British Isles, the conclusions are consistent with those derived from the larger geographic patterns (Lawton and Price, 1979). Many widely distributed umbellifers have no agromyzids on them; many others have only one species; the maximum number of agromyzid species is five, when other common plants outside the Umbelliferae are known to support nine agromyzid species, and even those hosts are probably well below community saturation for agromyzids and other leaf-mining species (Figure 29.4.). Twelve to fifteen agromyzid species on the most common hosts could be reasonably expected, judging by the richness of species coexisting on some host species on the various plant parts utilized by agromyzids. Agromyzid species richness correlates positively with the species richness of their most likely competitors, the leaf-mining microlepidoptera, and also with the richness of their important enemies, the parasitic braconid wasps. Lawton

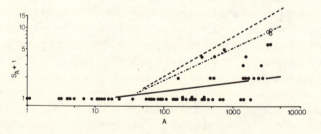

Fig. 29.4. Regression of the number of agromyzid species ($S_A + 1$) on the area occupied (A) by host plants in the family Umbelliferae in Britain (closed circles). The regression (solid line) is significant and accounts for 32% of the variance. Open circles indicate species richness in *Senecio, Taraxacum*, and *Ranunculus* species, not in the Umbelliferae, with the highest number of agromyzids in Britain. The dash-dot line indicates a conservative estimate of the number of species of agromyzids that could be supported by plant species and the dashed line suggests another estimate of the possible richness of communities (see Lawton and Price, 1979, for details).

and Price (1979) concluded that the agromyzids on Umbelliferae do not represent equilibrium assemblages whose richness is limited by competitive interactions, but that many niches remain unexploited, and more evolutionary time may result in richer parasite communities. In addition, it is clear from Spencer's (1972) original data, on which this analysis was based, that niche breadth is no greater when species are alone on a host than when several species coexist.

None of the examples of geographical patterns indicates that competition is playing an important role in organization of communities composed of specialists. Extreme specialization results from forces other than competition, with communities undersaturated and many resources unexploited.

NICHE AND HABITAT OCCUPATION

Niche occupation by specialists frequently shows complementary distributions, which have been interpreted as resulting from interspecific competition (*e.g.* see review by Holmes, 1973). However, an alternative explanation, similar to that relating to Roskam and van Uffelen's data on *Semudobia* species on birches used above, is frequently more parsimonious. For example, Holmes (1971) found the blood flukes *Aporocotyle macfarlani* and *Psettarium sebastodorum* distributed in a complementary way in the gill arches and hearts of rock fishes (Figure 29.5). He argued that their fundamental niches had evolved to occupy adjacent locations in the vascular system through interspecific competition. My alternative explanation (Price, 1980) is that the two species were preadapted to lodge in different parts of the blood system. *Aporocotyle macfarlani* is a squat fluke that naturally lodges in narrowing blood vessels—the gill arches. In contrast, the narrower, looping body of *P. sebastodorum* becomes wedged in heart valves and similar locations well before reaching the gill arches. The original colonization events were probably independent, and current occupation, although complementary, probably remains independent. A subsequent study by Holmes and Price (1980) using more recent information supports the validity of this interpretation. This study reveals the kind of specialization that results from preadaptations independent of any organizing role of competition.

Host-to-host comparisons also suggest that many ecological niches remain vacant, that communities are too loosely packed for competition to play an organizing role, and that niches remain narrow in the absence of competitors. One example concerns metastrongyle nematodes in pulmonary tissues of mammals such as mink, otter, cat, dog, and sheep (Stockdale, 1970). Although six discrete ecological niches are available in pulmonary tissues (Figure 29.6) and all have been colonized in one host

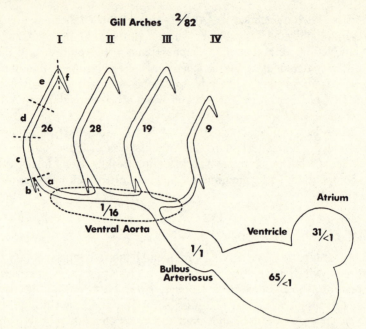

Fig. 29.5. Distribution of *Aporocotyle macfarlani* (in gill arches and below the line, *e. g.* 16% in ventral aorta) and *Psettarium sebastodorum* (above the line, *e.g.* 31% in atrium) expressed as a percentage of the total number of worms per species found in each organ of the host illustrated (from Holmes, 1971).

or another, no metastrongyle community has more than half the niches occupied: mink has three species (Figure 29.6), cat three, dog two, and sheep and otter one. Other examples are analyzed by Price (1980).

When local occupation of habitats and niches is examined in ecological time the same patchiness in occupation prevails. Many patches remain uncolonized and the probability of encounter between potential competitors is low. Examples of this situation include studies by Rathcke (1976a, b), McClure and Price (1976), and Price and Willson (1979).

Low utilization of resources may frequently result from a combination of specialization to a particular resource state and the rapid change in states, resulting in shortage of time when resources can be exploited. For example, the parsnip webworm, *Depressaria pastinacella*, lays eggs only on unopened umbels of wild parsnip, *Pastinaca sativa*, during a period in the spring when plant development from the overwintered rosette to the fully developed flowering stem is very rapid (Thompson and Price, 1977). Any one umbel may remain suitable for oviposition for only 48 hours, meaning for two nights when ovipositing moths fly. Therefore, a small

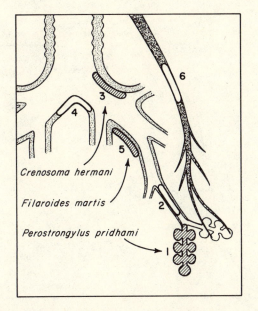

Fig. 29.6. Niche locations of metastrongyle nematodes in pulmonary tissues of
mink (niches 1, 3, and 5) and other niches occupied in other hosts but not in
mink (niches 2, 4, 6) (from descriptions in Stockdale, 1970).

plant with two or three umbels may provide resources for the webworm
moth for only a few days. And, since patches of parsnip are well synchro-
nized, a patch could easily be missed while resources are suitable for attack. Indeed, in a high-density stand, in which resources were available
for about 13 days, less than 50% of umbels were attacked on only 52%
of the plants, leaving about 75% of resources unutilized (Figure 29.7). This
pattern contrasted with utilization of larger plants in uncrowded conditions with more umbels displayed over a longer period of time. Inflorescences were attacked over a period of about 29 days and 90% of plants
and 68% of umbels on these plants were attacked, leaving a little less than
40% of resources unutilized. However, the great majority of umbels in
this old field occurred on plants in dense stands, so resources in general
were superabundant, and this at a time when the webworm had reached
epidemic population levels. In the years 1970–1979, this was one of two
years when webworm numbers were very high.

From these studies and those discussed by Price (1980) I conclude that:
(1) Resources commonly provide more ecological niches than there are
specialist species to fill them. (2) Given more evolutionary time most habitats will contain more specialized species and species packing will become

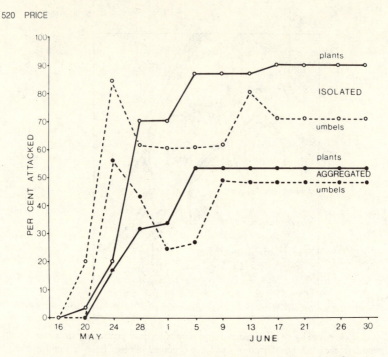

Fig. 29.7. The percent of plants and umbels attacked by the parsnip webworm among large isolated plants and small aggregated plants of wild parsnip (after Thompson and Price, 1977).

tighter. (3) In ecological time many resources remain underexploited even when species exist that are specialized on these resources because of (a) highly patchy habitats, (b) relatively poor colonizing ability, (c) ephemeral resources making time shortage a very significant limiting factor (see Figure 29.1). (4) As a result of factors 1–3, interspecific competition affects a minority of specialist species and does not play an important role in influencing species niches or relative abundance is communities.

ALTERNATIVE PARADIGMS FOR SPECIALIST COMMUNITIES

It would take many pages to review the factors that seem to be important in determining the richness and relative abundance of species in specialist communities. Therefore, only a very general overview will be attempted here in order to summarize my views on assembly of specialists into communities. These will be treated in order of apparent importance based on my own viewpoint. These suggestions are motivated by a need for constructive appraisal of alternatives to the current paradigms, by a feeling that the current paradigms have served an important role in developing our awareness of processes in community organization, and by the hope that these alternatives will not lead to confrontation, for the history of our science shows intermediate positions are usually the most tenable. In fact some points reflect views long held by some ecologists.

The most important factor in assembly of specialists appears to be the individualistic response of each species to the display of resources in a habitat. This view mirrors that held by phytosociologists from Ramensky (1926) and Gleason (1926) to Whittaker (1967). The benefit of this view is that it forces a detailed understanding of the ecology and behavior of each species, a reductionist approach to community ecology that seems to have been strenuously avoided until the recent past. The patterns of presence and abundance of species result from their differential responses to variation in the dispersion of resources, and dominance and diversity will vary as resources vary in quality and quantity. Communities must be studied from the bottom (*i.e.* resources) up, as well as from the top down, although the latter approach has not commonly yielded profitable insights (*e.g.* species per genus ratio and dominance-diversity approaches).

A second pair of important factors concerns the way in which resources are displayed for specialists. They are very patchy and patches are ephemeral. A shortage of time prevails, as Andrewartha and Birch (1954) concluded about 30 years ago. The probability of colonization is therefore low and the probability of extinction is high. Whether these rates are ever similar enough to develop equilibrium communities is doubtful. The mechanisms that are usually invoked (see MacArthur and Wilson, 1967) as stabilizing colonization and extinction rates are unlikely to be present: (1) Populations can be high even in small patches because each individual utilizes such a small quantity of resources. Therefore, extinction rates need not increase rapidly as patches decline in size. (2) Many species closely related in their niche utilization are unlikely to colonize the same patch because few such species exist and colonization probabilities are low for each species. Competitive interactions between species are therefore unlikely to influence extinction rates. (3) It is even more difficult for enemies of the specialists to colonize patches of their prey or hosts since these are more unpredictable and widely dispersed than resources for the specialists. Enemies are therefore unlikely to influence extinction rates. Even in insects, usually regarded as being heavily attacked by parasitic wasps (based on experience largely with pest species), many species can be found with no such parasitoids or with typically less than 1% attacked (*e.g.* see Janzen, 1975b, on parasitoids of bruchid weevils) and some families (*e.g.* Hippoboscidae) have no recorded parasitic wasps in North America (*cf.* Krombein *et al.*, 1979).

The net result of the interaction of the above factors is that communities will not be in an equilibrium state; the hotel of resources will have a low occupancy rate partly because prospective occupants have not arrived, and partly because appropriate occupants do not exist. Figure 29.1 does indeed appear to be appropriate for specialist communities.

These major influences on community structure have little predictive power in terms of the number and kinds of species to be expected in any habitat. But predictive ability is acquired when the individualistic responses of species to resources are understood, and coupled with a probabilistic view of the patches available for colonization.

The final question I wish to address is whether such community organization as envisaged in this paper is common or rare. If it is common in specialists then it is common in nature, since many more specialists occupy this globe than generalists (Price, 1980). In addition, however, claims that patchiness and disturbance dominate community organization have been made repeatedly by ecologists studying disparate groups of organisms exploiting a wide variety of resources (Table 29.1). The examples listed provide a rich source for debate on the validity of the claims made,

Table 29.1. Some real communities that appear to exist in a non-equilibrium state.

Type of resource or substrate	Organisms	References
Plant Communities		
Disturbed but rich soils	Ruderal plants (and similar fungi)	Grime, 1977, 1979
Tropical soils	Trees	Connell, 1979
Tree bark	Epiphytes, especially orchids and bromeliads	Benzing, 1978a, b
Invertebrate Communities		
Intertidal rock surfaces	Barnacles, mussels, and predators (and algae)	Dayton, 1971
Plant stems	Stem-boring insects	Rathcke, 1976a, b
Agricultural plants	Insect herbivores and their enemies	Price, 1976, Levins and Wilson, 1980
Carrion	Flies (Diptera)	Beaver, 1977
Rich cambial layer under bark	Insects	Hamilton, 1978
Plants	Insect herbivores	Strong, 1979
Marine Reefs	Corals	Connell, 1979
Leaf stalks low in nutrients	Bark beetles	Beaver, 1979b
Living hosts	Parasites in general	Price, 1980
Vertebrate Communities		
Grassland	Birds	Wiens, 1973, 1974b
Coral reefs	Fish	Sale, 1977

but the relative recency of the references suggests that the realization that many communities exist in a non-equilibrium state is gathering momentum. We cannot yet deny the probability that such communities, in which competition plays a minor role in organization, are very common in nature.

Acknowledgments

I am grateful to the following colleagues for providing advice and information for this paper: Mark McClure, Beverly Rathcke, Hans Roskam, John Thompson, and Thomas Whitham. Financial support was provided by the National Science Foundation, grant number DEB-8021754.

Literature Cited

Abbott, I. 1973. Birds of Bass Strait. Evolution and ecology of the avifaunas of some Bass Strait Islands, and comparisons with those of Tasmania and Victoria. *Proceedings of the Royal Society of Victoria* 85:197–224.

Abbott, I. 1977. Species richness, turnover, and equilibrium in insular floras near Perth, Western Australia. *Aust. J. Bot.* 25:193–208.

Abbott, I. 1980. Theories dealing with the ecology of landbirds on islands. *Adv. in Ecol. Res.* 11:329–371.

Abbott, I., L. K. Abbott, and P. R. Grant. 1977. Comparative ecology of Galápagos ground finches (*Geospiza* Gould): Evaluation of the importance of floristic diversity and interspecific competition. *Ecol. Monogr.* 47:151–184.

Abele, L. 1979. The community structure of coral-associated decapod crustaceans in variable environments. Pages 165–187 in *Ecological Processes in Coastal and Marine Systems*, R. J. Livingston (ed.). Plenum Press, N.Y.

Abele, L. G. 1982. Biogeography. Pages 241–304 in *Biology of the Crustacea*, Vol. I, L. G. Abele (ed.). Academic Press, N.Y.

Abele, L. G., and E. F. Connor. 1979. Application of island biogeography theory to refuge design: Making the right decision for the wrong reasons. Pages 89–94 in *Proceedings of the First Conference on Scientific Research in the National Parks*, R. M. Linn (ed.). U.S.D.I.

Abele, L. G., and W. K. Patton. 1976. The size of coral heads and the community biology of associated decapod crustaceans. *J. Biogeography* 3:35–47.

Abrams, P. A. 1980. Resource partitioning and interspecific competition in a tropical hermit crab community. *Oecologia* 46:365–379.

Addicott, J. F. 1974. Predation and prey community structure: an experimental study of the effect of mosquito larvae on the protozoan communities of pitcher plants. *Ecology* 55:475–492.

Alatalo, R. V. 1982. Bird species distributions in the Galápagos and other archipelagoes: competition or chance? *Ecology* 63:881–887.

Alatalo, R. V., and R. H. Alatalo. 1979. Resource partitioning among a flycatcher guild in Finland. *Oikos* 33:46–54.

Alexander, R. McN. 1967. *Functional Design in Fishes*. Hutchinson, London.

Allen, E. B., and R. T. Forman. 1976. Plant species removals and old field community structure and stability. *Ecology 57*:1233–1243.

Allen, G. R. 1972. *Anemonefishes*. TFH Publications, Neptune City, N.J.

Alm, G. 1946. Reasons for the occurrence of stunted fish populations with special regard to the perch. *Rep. Inst. Freshw. Res. Drottningholm 25*.

Andersen, R. C., and S. Schelfhout. 1980. Phenological patterns among Tallgrass Prairie plants and their implications for pollinator competition. *Amer. Midl. Natur. 104*:253–263.

Anderson, B. W., and R. D. Ohmart. 1981. Comparisons of avian census results using variable distance transect and variable circular plot techniques. *Studies Avian Biol. 6*:186–192.

Anderson, G.R.V., A. H. Ehrlich, P. R. Ehrlich, J. D. Roughgarden, B. C. Russell, and F. H. Talbot. 1981. The community structure of coral reef fishes. *Amer. Natur. 117*:476–495.

Anderson, J. F. 1970. Metabolic rates of spiders. *Comp. Biochem. Physiol. 33*:51–72.

Anderson, J. F. 1974. Responses to starvation in the spiders *Lycosa lenta* (Hentz) and *Filistata hibernalis* (Hentz). *Ecology 55*:576–585.

Anderson, J. F. 1978. Energy content of spider eggs. *Oecologia 37*:41–57.

Anderson, R. M., and R. M. May. 1979. Population biology of infectious diseases: Part I. *Nature 280*:361–367.

Andrewartha, H. G., and L. C. Birch. 1954. *The Distribution and Abundance of Animals*. The University of Chicago Press.

Andrews, R. H., and T. N. Petney. 1981. Competition for sites of attachment to hosts in three parapatric species of reptile tick. *Oecologia 51*: 227–232.

Anonymous. 1944. British Ecological Society Symposium on "The Ecology of Closely Allied Species." *J. Anim. Ecol. 13*:176–178.

Arntz, W. E. 1980. Predation by demersal fish and its impact on the dynamics of macrobenthos. Pages 121–149 in *Marine Benthic Dynamics*, K. R. Tenore and B. C. Coull (eds.). University of South Carolina Press, Columbia, S.C.

Arp, H., and J. N. Bahcall. 1973. *The Redshift Controversy*. Addison-Wesley, N.Y.

Askenmo, C., A. Bromssen, J. von Eckman, and C. Jansson. 1977. Impact of some wintering birds on spider abundance in spruce. *Oikos 28*: 90–94.

Askew, R. R. 1975. The organization of chalcid-dominated parasitoid communities centered upon endophytic hosts. Pages 130–153 in *Evolutionary Strategies of Parasitic Insects and Mites*, P. W. Price (ed.). Plenum Press, N.Y.

Auerbach, M. A., and D. R. Strong. 1981. Nutritional ecology of *Heliconia* herbivores: experiments with plant fertilization and alternative hosts. *Ecol. Monogr. 51*:63–83.

Auerbach, M. J. 1979. Some real communities are unstable. *Nature 279*: 821.

Auerbach, M. J., and S. D. Hendrix. 1980. Insect-fern interactions: Macrolepidopteran utilisation and species-area association. *Ecol. Ent. 5*: 99–104.

Austin, A. D., S. A. Austin, and P. F. Sale. 1980. Community structure of the fauna associated with the coral *Pocillopora damicornis* (L.) on the Great Barrier Reef. *Aust. J. Mar. Freshw. Res. 31*:163–174.

Ayala, F. J., M. E. Gilpin, and J. R. Ehrenfeld. 1973. Competition between species: theoretical models and experimental tests. *Theor. Pop. Biol. 4*:331–356.

Ayling, A. L. 1978. Population biology and competitive interactions in subtidal sponge-dominated communities of temperate waters. Ph.D. Dissertation, Univ. of Auckland.

Bagenal, T. B. 1978. Aspects of fish fecundity. Pages 75–101 in *Ecology of Freshwater Fish Production*, S. D. Gerking (ed.). John Wiley and Sons, New York and Toronto.

Bakus, G. J. 1981. Chemical defense mechanisms on the Great Barrier Reef, Australia. *Science 211*:497–499.

Balick, M. J., D. G. Furth, and G. Cooper-Driver. 1978. Biochemical and evolutionary aspects of arthropod predation on ferns. *Oecologia 35*: 55–89.

Bardach, J. E. 1958. On the movements of certain Bermuda reef fishes. *Ecology 39*:139–146.

Bardach, J. E. 1959. The summer standing crop of fish on a shallow Bermuda reef. *Limnol. Oceanogr. 4*:77–85.

Barnett, S., and C. Storey. 1970. *Matrix Methods in Stability Theory.* Barnes and Noble, N.Y.

Battan, L. A., and J. H. Marchant. 1976. Bird population changes for the years 1973–74. *Bird Study 23*:11–20.

Battan, L. A., and J. H. Marchant. 1977. Bird population changes for the years 1974–75. *Bird Study 24*:159–164.

Beaver, R. A. 1977. Non-equilibrium "island" communities: Diptera breeding in dead snails. *J. Anim. Ecol. 46*:783–798.

Beaver, R. A. 1979a. Host specificity of temperate and tropical animals. *Nature 281*:139–141.

Beaver, R. A. 1979b. Non-equilibrium "island" communities. A guild of tropical bark beetles. *J. Anim. Ecol. 48*:987–1002.

Beddington, J. R., and P. S. Hammond. 1977. On the dynamics of host-parasitoid-hyperparasitoid interactions. *J. Anim. Ecol. 46*:811–822.

Beebe, W. 1924. *Galápagos: World's End*. Putnams, N.Y.

Bennett, G. W. 1954. Largemouth bass in Ridge Lake, Coles County, Illinois. *Bull. Ill. Nat. Hist. Surv. 26*:219–276.

Benson, R. B. 1968. Hymenoptera from Turkey. Symphyta. *Bull. Brit. Mus. (Nat. Hist.) Entomol. 22*:134.

Benzing, D. H. 1978a. Germination and early establishment of *Tillandsia circinnata* Schlecht. (Bromeliaceae) on some of its hosts and other supports in southern Florida. *Selbyana 5*:95–106.

Benzing, D. H. 1978b. The life history profile of *Tillandsia circinnata* (Bromeliaceae) and the rarity of extreme epiphytism among the angiosperms. *Selbyana 2*:325–337.

Bergquist, P. R. 1978. *Sponges*. University of California Press, Berkeley.

Bernstein, B. B., B. E. Williams, and K. H. Mann. Formation of destructive grazing fronts by the sea urchin *Strongylocentrotus droebachiensis*: the role of predators in modifying aggregation and hiding behavior. *Mar. Biol.* (in press).

Birch, L. C. 1979. The effect of species of animals which share common resources on one another's distribution and abundance. *Fortschr. Zool. 25*:197–221.

Bishop, Y.M.M., S. E. Fienberg, and P. W. Holland. 1975. *Discrete Multivariate Analysis*. MIT Press, Cambridge, Mass.

Blalock, H. M. 1964. *Causal Inferences in Nonexperimental Research*. University of North Carolina Press, Raleigh, N.C.

Blaustein, A. R., and A. C. Risser. 1976. Interspecific interactions between three sympatric series of kangaroo rats (*Dipodomys*). *Anim. Behav. 24*:381–385.

Blegvad, H. 1928. Quantitative investigations of bottom invertebrates in the Limfjord 1910–1927 with special reference to the plaice food. *Rep. Dan. Biol. Stn. 34*:33–52.

Bloom, S. A., J. L. Simon, and V. D. Hunter. 1972. Animal-sediment relations and community analysis of a Florida estuary. *Mar. Biol. 13*:43–56.

Boag, P. T. 1981. Morphological variation in the Darwin's Finches (Geospizinae) of Daphne Major Island, Galápagos. Ph.D. Dissertation, McGill University.

Boag, P. T., and P. R. Grant. 1981. Intense natural selection on a population of Darwin's Finches (Geospizinae) in the Galápagos. *Science 214*:82–85.

Boag, P. T., and P. R. Grant. 1983. The classical case of character release: Darwin's finches (*Geospiza*) on Isla Daphne Major, Galápagos. *Biol. J. Linn. Soc. 15* (in press).

Boer, B. A. de. 1978. Factors influencing the distribution of the damselfish

Chromis cyanea (Poey), Pomacentridae, on a reef at Curacao, Netherlands Antilles. *Bull. Mar. Sci. 28*:550–565.

Bohnsack, J. A., and F. H. Talbot. 1980. Species-packing by reef fishes on Australian and Caribbean reefs: an experimental approach. *Bull. Mar. Sci. 30*:710–723.

Bournier, A. 1977. Grape insects. *Ann. Rev. Ent. 22*:355–376.

Bouton, C. E., B. A. McPheron, and A. E. Weis. 1980. Parasitoids and competition. *Amer. Natur. 116*:876–881.

Bowers, M. A., and J. H. Brown. 1982. Body size and coexistence in desert rodents: chance or community structure? *Ecology 63*:391–400.

Bowman, R. I. 1961. Morphological differentiation and adaptation in the Galápagos finches. *Univ. Calif. Berkeley Publ. Zool. 58*:1–302.

Bowman, R. S., and J. R. Lewis. 1977. Annual fluctuations in the recruitment of *Patella vulgata* L. *J. Mar. Biol. Ass. U.K. 57*:793–815.

Boyce, M. S., and D. J. Daley. 1980. Population tracking of fluctuating environments and natural selection for tracking ability. *Amer. Natur. 115*:480–491.

Box, G.E.P., W. G. Hunter, and J. S. Hunter. 1978. *Statistics for Experimenters.* John Wiley and Sons, N.Y.

Bradley, J. V. 1968. *Distribution-free Statistical Tests.* Prentice-Hall, Englewood Cliffs, N.J.

Brady, R. H. 1979. Natural selection and the criteria by which a theory is judged. *Syst. Zool. 28*:600–621.

Branch, G. M. 1975. Ecology of *Patella* species from the Cape Peninsula, South Africa. 4. Desiccation. *Mar. Biol. 32*:179–188.

Branch, G. M. 1976. Interspecific competition experienced by South African *Patella* species. *J. Anim. Ecol. 45*:507–529.

Branch, G. M., and M. L. Branch. 1980. Competition between *Cellana tramo serica* (Sowerby) (Gastropoda) and *Patireilla exigua* (Lamarch) (Asteroidea) and their influence on algal standing stocks. *J. Exp. Mar. Biol. Ecol. 48*:35–49.

Bray, J. R., and J. T. Curtis, 1957. An ordination of the upland forest communities of southern Wisconsin. *Ecol. Monogr. 27*:325–349.

Brenchley, G. A. 1979. On the regulation of marine infaunal assemblages at the morphological level: a study of the interactions between sediment stabilizers, destabilizers and their sedimentary environment. Ph.D. Dissertation, The Johns Hopkins University.

Brett, J. R., and T.D.D. Groves. 1979. Physiological energetics. Pages 279–352 in *Fish Physiology*, Vol. VIII, W. S. Hoar, D. J. Randall, and J. R. Brett (eds.). Academic Press, N.Y.

Breymeyer, A. 1966. Relations between wandering spiders and other epigeic predatory Arthropoda, *Ekol. Pol. Ser. A. 14*:27–71.

Bronowski, J. 1977. *A Sense of the Future*. MIT Press, Cambridge, Mass.

Brown, J. H. 1971. Mechanisms of competitive exclusion between two species of chipmunks (*Eutamias*). *Ecology 52*:306–311.

Brown, J. H. 1973. Species diversity of seed-eating desert rodents in sand dune habitats. *Ecology 54*:775–787.

Brown, J. H. 1975. Geographical ecology of desert rodents. Pages 315–341 in *Ecology and Evolution of Communities*, M. L. Cody and J. M. Diamond (eds.). Harvard University Press, Cambridge, Mass.

Brown, J. H., and D. W. Davidson. 1977. Competition between seed-eating rodents and ants in desert ecosystems. *Science 196*:880–882.

Brown, J. H., and A. Kodric-Brown. 1977. Turnover rates in insular biogeography: Effects of immigration on extinction. *Ecology 58*:445–449.

Brown, J. H., and A. Kodric-Brown. 1979. Convergence, competition, and mimicry in a temperate community of hummingbird-pollinated flowers. *Ecology 60*:1022–1035.

Brown, J. H., A. Kodric-Brown, T. G. Whitham, and H. W. Bond. 1981. Competition between hummingbirds and insects in the pollination of two species of shrubs. *Southwest. Natur. 26*:133–145.

Brown, J. H., and G. A. Lieberman. 1973. Resource utilization and co-existence of seed-eating rodents in sand dune habitats. *Ecology 54*:788–797.

Brown, J. H., O. J. Reichman, and D. W. Davidson. 1979. Granivory in desert ecosystems. *Ann. Rev. Ecol. Syst. 10*:201–227.

Brown, J. L. 1969. Territorial behavior and regulation in birds. *Wilson Bull. 81*:293–329.

Brown, K. M. 1981. Foraging ecology and niche partitioning in two argiopid spiders: *Argiope trifasciata* Lucas and *Argiope aurantia* (Forskal) (Araneidae, Araneae). *Oecologia 50*:380–385.

Brown, L. H., and D. Amadon. 1968. *Eagles, Hawks, and Falcons of the World*. McGraw-Hill, N.Y.

Brown, W. L., Jr., and E. O. Wilson. 1956. Character displacement. *Syst. Zool. 5*:49–64.

Bunge, M. 1979. *Causality and Modern Science*. Third edition. Dover, N.Y.

Burkenroad, D. D. 1946. Fluctuations in abundance of marine animals. *Science 103*:684–686.

Burton, T. M. 1976. An analysis of the feeding ecology of the salamanders (Amphibia: Urodela) of the Hubbard Brook Experimental Forest, New Hampshire. *J. Herpetol. 10*:187–204.

Carlquist, S. 1974. *Island Biology*. Columbia University Press, N.Y.

Carpenter, F. L. 1978. A spectrum of nectar-eater communities. *Amer. Zool. 18*:809–818.

Case, T. J. 1975. Species numbers, density compensation, and the colonization ability of lizards on islands in the Gulf of California. *Ecology* 56:3–18.

Case, T. J. 1983. Sympatry and size similarity in *Cnemidophorus* lizards. Pages 297–325 in *Lizard Ecology: Studies on a Model Organism*, R. B. Huey, E. R. Pianka, and T. W. Schoener (eds.). Harvard University Press, Cambridge, Mass.

Case, T. J., and M. E. Gilpin. 1974. Interference competition and niche theory. *Proc. Natl. Acad. Sci. USA* 71:3073–3077.

Case, T. J., M. E. Gilpin, and J. M. Diamond. 1979. Overexploitation, interference competition, and excess density compensation in insular faunas. *Amer. Natur.* 113:843–854.

Case, T., and R. Sidell. 1983. Pattern and chance in the structure of model and natural communities. *Evolution* 37:832–849.

Castenholz, R. W. 1963. An experimental study of the vertical distribution of littoral marine diatoms. *Limnol. Oceanogr.* 8:450–462.

Caswell, H. L. 1976. Community structure: a neutral model analysis. *Ecol. Monogr.* 46:327–354.

Caughley, G., and J. H. Lawton. 1981. Plant-herbivore systems. Pages 132–166 in *Theoretical Ecology*, 2nd ed., R. M. May (ed.). Blackwell, Oxford.

Cederholm, G., and J. Ekman. 1976. A removal experiment on Crested Tit *Parus cristatus* and Willow Tit *P. montanus* in the breeding season. *Ornis Scand.* 7:207–213.

Chabreck, R. H., and A. W. Palmisano. 1973. The effects of Hurricane Camille on the marshes of the Mississippi River delta. *Ecology* 54:1118–1123.

Chamberlain, T. C. 1897. The method of multiple working hypotheses. *J. Geol.* 5:837–848.

Chapman, A.R.O. 1979. *Biology of Seaweeds*. University Park Press, Baltimore.

Chappell, M. A. 1978. Behavioral factors in the altitudinal zonation of chipmunks (*Eutamias*). *Ecology* 59:565–579.

Chiang, H. C. 1978. Pest management in corn. *Ann. Rev. Ent.* 23:101–123.

Cody, M. L. 1966. The consistency of intra- and inter-continental grassland bird species counts. *Amer. Natur.* 100:371–376.

Cody, M. L. 1968a. Interspecific territoriality among hummingbird species. *Condor* 70:270–271.

Cody, M. L. 1968b. On the methods of resource division in grassland bird communities. *Amer. Natur.* 102:107–147.

Cody, M. L. 1974a. *Competition and Structure of Bird Communities*. Princeton University Press, Princeton, N.J.

Cody, M. L. 1974b. Optimization in ecology. *Science* 183:1156–1164.

Cody, M. L. 1975. Towards a theory of continental species diversities: bird distributions over Mediterranean habitat gradients. Pages 214–257 in *Ecology and Evolution of Communities*, M. L. Cody and J. M. Diamond (eds.). Harvard University Press, Cambridge, Mass.

Cody, M. L. 1978. Habitat selection and interspecific territoriality among the sylviid warblers of England and Sweden. *Ecol. Monogr. 48*: 351–396.

Cody, M. L. 1981. Habitat selection in birds: the roles of vegetation structure, competitors, and productivity. *BioScience 31*:107–113.

Cody, M. L., and J. M. Diamond (eds.). 1975a. *Ecology and Evolution of Communities*. Harvard University Press, Cambridge, Mass.

Cody, M. L., and J. M. Diamond. 1975b. Introduction. Pages 1–14 in *Ecology and Evolution of Communities*, M. L. Cody and J. M. Diamond (eds.). Harvard University Press, Cambridge, Mass.

Cody, M. L., E. R. Fuentes, W. Glanz, J. H. Hunt, and A. R. Moldenke. 1977. Convergent evolution in the consumer organisms of Mediterranean Chile and California. Pages 144–192 in *Convergent Evolution in Chile and California*, H. A. Mooney (ed.). Dowden, Hutchinson & Ross, Stroudsburg, Pa.

Cody, M. L., and H. A. Mooney. 1978. Convergence versus nonconvergence in Mediterranean-climate ecosystems. *Ann. Rev. Ecol. Syst. 9*:265–321.

Coe, W. E. 1956. Fluctuations in populations of littoral marine invertebrates. *J. Mar. Res. 15*:212–232.

Coen, L. D., K. L. Heck, Jr., and L. G. Abele. 1981. Experiments on competition and predation among shrimps of seagrass meadows. *Ecology 62*:1484–1493.

Cohen, J. E. 1978. *Food Webs and Niche Space*. Princeton University Press, Princeton, N.J.

Cole, B. J. 1981. Overlap, regularity, and flowering phenologies. *Amer. Natur. 117*:993–997.

Cole, L. C. 1957. The measurement of partial interspecific association. *Ecology 38*:226–233.

Colebourn, P. H. 1974. The influence of habitat structure on the distribution of *Araneus diadematus* Clerck. *J. Anim. Ecol. 43*:401–409.

Colinveaux, P. A. 1973. *Introduction to Ecology*. Wiley, N.Y.

Collier, B. D., G. W. Cox, A. W. Johnson, and P. C. Miller, 1973. *Dynamic Ecology*. Prentice-Hall, Englewood Cliffs, N.J.

Colquhoun, M. K. 1941. Visual and auditory conspicuousness in a woodland bird community: a quantitative analysis. *Proc. Zool. Soc. London Ser. A, 110*:129–148.

Colwell, R. K. 1973. Competition and coexistence in a simple tropical community. *Amer. Natur. 107*:737–760.

Colwell, R. K. 1979. Toward a unified approach to the study of species diversity. Pages 75–91 in *Ecological Diversity in Theory and Practice*, J. F. Grassle, G. P. Patil, W. K. Smith, and C. Taillie (eds.). International Cooperative Publishing House, Fairland, Md.

Colwell, R. K., and E. R. Fuentes. 1975. Experimental studies of the niche. *Ann. Rev. Ecol. Syst. 6*:281–310.

Colwell, R. K., and Futuyma, D. J. 1971. On the measurement of niche breadth and overlap. *Ecology 52*:567–576.

Congdon, J. 1974. Effect of habitat quality on distributions of three sympatric species of desert rodents. *J. Mammal. 55*:659–662.

Connell, J. H. 1961a. Effects of competition, predation by *Thais lapillus* and other factors on natural populations of the barnacle *Balanus balanoides. Ecol. Monogr. 31*:61–104.

Connell, J. H. 1961b. The influence of interspecific competition and other factors on the distribution of the barnacle *Chthalamus stellatus. Ecology 42*:710–723.

Connell, J. H. 1970. A predator-prey system in the marine intertidal region. 1. *Balanus glandula* and several predatory species of *Thais. Ecol. Monogr. 40*:49–78.

Connell, J. H. 1971. On the role of natural enemies in preventing competitive exclusion in some marine animals and in rainforest trees. Pages 298–312 in *Dynamics of Populations*, P. J. den Boer and G. R. Gradwell (eds.). Centre for Agricultural Publishing and Documentation, Wageningen, The Netherlands.

Connell, J. H. 1972. Community interactions on marine rocky intertidal shores. *Ann. Rev. Ecol. Syst. 3*:169–192.

Connell, J. H. 1974. Ecology: field experiments in marine ecology. Pages 21–54 in *Experimental Marine Biology*, R. Mariscal (ed.). Academic Press, N.Y.

Connell, J. H. 1975. Some mechanisms producing structure in natural communities: a model and evidence from field experiments. Pages 460–490 in *Ecology and Evolution of Communities*, M. L. Cody and J. M. Diamond (eds.). Harvard University Press, Cambridge, Mass.

Connell, J. H. 1978. Diversity in tropical rain forests and coral reefs. *Science 199*:1302–1310.

Connell, J. H. 1979. Tropical rain forests and coral reefs as open non-equilibrium systems. *Symp. Brit. Ecol. Soc. 20*:141–163.

Connell, J. H. 1980. Diversity and the coevolution of competitors, or the ghost of competition past. *Oikos 35*:131–138.

Connell, J. H., and R. O. Slatyer. 1977. Mechanisms of succession in natural communities and their role in community stability and organisation. *Amer. Natur. 111*:1119–1144.

Connor, E. F., and E. D. McCoy. 1979. The statistics and biology of the species-area relationship. *Amer. Natur. 113*:791–833.

Connor, E. F., and D. Simberloff. 1978. Species number and compositional similarity of the Galápagos flora and avifauna. *Ecol. Monogr. 48*:219–248.

Connor, E. F., and D. Simberloff. 1979. The assembly of species communities: Chance or competition? *Ecology 60*:1132–1140.

Cooley, W. W., and P. R. Lohnes. 1971. *Multivariate Data Analysis.* Wiley and Sons, N.Y.

Coope, G. R. 1978. Constancy of insect species versus inconstancy of Quaternary environments. Pages 176–187 in *Diversity of Insect Faunas*, L. A. Mound and N. Waloff (eds.). *Symp. Roy. Ent. Soc. London 9.*

Coope, G. R. 1979. Late Cenazoic fossil Coleoptera. *Ann. Rev. Ecol. Syst. 10*:247–267.

Corbet, P. S. 1964. Observations on mosquitoes ovipositing in small containers in Zika Forest, Uganda. *J. Anim. Ecol. 33*:141–164.

Cox, G. W., and R. E. Ricklefs. 1977. Species diversity and ecological release in Caribbean land bird faunas. *Oikos 28*:113–122.

Creese, R. G. 1978. Ecology and reproductive biology of intertidal limpets. Ph.D. Dissertation, University of Sydney.

Creese, R. G., and A. J. Underwood. 1982. Analysis of inter- and intraspecific competition amongst intertidal limpets with different methods of feeding. *Oecologia 53*:337–346.

Crisp, D. J. 1974. Factors influencing the settlement of marine invertebrate larvae. Pages 177–265 in *Chemoreception in Marine Organisms*, P. T. Grant and A. N. Mackie (eds.). Academic Press, London.

Crome, F.H.J. 1975. The ecology of fruit pigeons in tropical Northern Queensland. *Aust. Wildl. Res. 2*:155–185.

Crowder, L. B. 1980. Ecological convergence of community structure: a neutral model analysis. *Ecology 61*:194–198.

Crowell, K. L. 1962. Reduced interspecific competition among the birds of Bermuda. *Ecology 43*:75–88.

Culver, D. C. 1973. Competition in spatially heterogeneous systems: an analysis of simple cave communities. *Ecology 54*:102–110.

Culver, D. C. 1976. The evolution of aquatic cave communities. *Amer. Natur. 110*:945–957.

Cushing, D. H., and J. J. Walsh (eds.). 1976. *The Ecology of the Seas.* W. B. Saunders, Philadelphia and Toronto.

Dakin, W. J. 1969. *Australian Seashores*, revised edition. Angus and Robertson, Sydney.

Dakin, W. J., I. Bennett, and E. Pope. 1948. A study of certain aspects of the ecology of the intertidal zone of the New South Wales coast. *Aust. J. Sci. Res. Ser. B 1*:176–230.

Daly, M. A., and A. C. Mathieson. 1977. The effects of sand movement on intertidal seaweeds and selected invertebrates at Bound Rock, New Hampshire, USA. *Mar. Biol. 43*:45–55.

Darwin, C. 1859. *The Origin of Species by Means of Natural Selection.* Murray, London.

Davidson, D. W. 1980. Some consequences of diffuse competition in a desert ant community. *Amer. Natur. 116*:92–105.

Davis, J. 1973. Habitat preferences and competition of wintering juncos and golden-crowned sparrows. *Ecology 54*:174–180.

Davis, P. H., and R. B. Spies. 1980. Infaunal benthos of a natural petroleum seep: study of community structure. *Mar. Biol. 59*:31–41.

Dayton, P. K. 1971. Competition, disturbance, and community organization: the provision and subsequent utilization of space in a rocky intertidal community. *Ecol. Monogr. 41*:351–389.

Dayton, P. K. 1972. Toward an understanding of community resilience and the potential effects of enrichments to the benthos at McMurdo Sound, Antarctica. Pages 81–96 in *Proceedings of the Colloquium on Conservation Problems in Antarctica*, B. C. Parker (ed.). Allen Press.

Dayton, P. K. 1973. Two cases of resource partitioning in an intertidal community: making the right prediction for the wrong reason. *Amer. Natur. 107*:662–670.

Dayton, P. K. 1975a. Experimental evaluation of ecological dominance in a rocky intertidal algal community. *Ecol. Monogr. 45*:137–159.

Dayton, P. K. 1975b. Experimental studies of algal canopy interactions in a sea otter-dominated kelp community at Amchitka Island, Alaska. *Fish. Bull. 73*:230–237.

Dayton, P. K. 1979a. Ecology: A science and a religion. Pages 3–18 in *Ecological Processes in Coastal and Marine Systems*, R. J. Livingston (ed.). Plenum Press, N.Y.

Dayton, P. K. 1979b. Observations of growth, dispersal and population dynamics of some sponges in McMurdo Sound, Antarctica. Pages 271–282 in *Proceedings International Colloquium on Sponge Biology*, Paris, 1978.

Dayton, P. K., and R. R. Hessler. 1972. Role of biological disturbance in maintaining diversity in the deep sea. *Deep-Sea Res. 19*:199–208.

Dayton, P. K., and J. S. Oliver. 1980. An evaluation of experimental analyses of population and community patterns in benthic marine environments. Pages 93–120 in *Marine Benthic Dynamics*, K. R. Tenore and B. C. Coull (eds.). University of South Carolina Press, Columbia, S.C.

Dayton, P. K., G. A. Robilliard, and A. L. DeVries. 1969. Anchor ice formation in McMurdo Sound, Antarctica, and its biological effects. *Science 163*:273–274.

Dayton, P. K., G. A. Robilliard, R. T. Paine, and L. B. Dayton. 1974. Biological accommodation in the benthic community at McMurdo Sound, Antarctica. *Ecol. Monogr. 44*:105–128.

Dean, J. M., and R. E. Ricklefs. 1979. Do parasites of Lepidoptera larvae compete for hosts? No! *Amer. Natur. 113*:302–306.

Dean, T. A., and L. E. Hurd. 1980. Development in an estuarine fouling community: the influence of early colonists on later arrivals. *Oecologia 46*:295–301.

DeAngelis, D. L. 1975. Stability and connectance in food web models. *Ecology 56*:238–243.

DeAngelis, D. L., R. H. Gardner, J. B. Mankin, W. M. Post, and J. H. Carney. 1978. Energy flow and the number of trophic levels in ecological communities. *Nature 273*:406–407.

DeBach, P. 1974. *Biological Control by Natural Enemies.* Cambridge University Press.

Den Boer, P. J. 1980. Exclusion or coexistence and the taxonomic or ecological relationship between species. *Neth. J. Zool. 30*:278–306.

Denley, E. J., and A. J. Underwood. 1979. Experiments on factors influencing settlement, survival and growth of two species of barnacles in New South Wales. *J. Exp. Mar. Biol. Ecol. 36*:269–293.

Diamond, J. M. 1970. Ecological consequences of island colonization by Southwest Pacific birds. I. Types of niche shifts. *Proc. Natl. Acad. Sci. USA 67*:529–536.

Diamond, J. M. 1972. *Avifauna of the Eastern Highlands of New Guinea.* Nuttall Ornithological Club, Cambridge, Mass.

Diamond, J. M. 1975. Assembly of species communities. Pages 342–444 in *Ecology and Evolution of Communities*, M. L. Cody and J. M. Diamond (eds.). Harvard University Press, Cambridge, Mass.

Diamond, J. M. 1978. Niche shifts and the rediscovery of interspecific competition. *Amer. Sci. 66*:322–331.

Diamond, J. M. 1982. Shifts in species incidences and occurrence frequencies due to competition: musical chairs on islands. *Proc. Natl. Acad. Sci. USA 79*:2420–2424.

Diamond, J. M., and M. E. Gilpin. 1982. Examination of the "null" model of Connor and Simberloff for species co-occurrences on islands. *Oecologia (Berl.) 52*:64–74.

Diamond, J. M., and M. LeCroy. 1979. Birds of Karkar and Bagabag Islands, New Guinea. *Bull. Amer. Mus. Nat. Hist. 164*:467–531.

Diamond, J. M., and A. G. Marshall. 1977. Distributional ecology of New Hebridean birds: a species kaleidoscope. *J. Anim. Ecol. 46*:703–727.

Dodson, S. I. 1970. Complementary feeding niches sustained by size-selective predation. *Limnol. Oceanogr. 15*:131–137.

Doherty, P. J. 1980. Biological and physical constraints on the populations of two sympatric territorial damselfishes on the southern Great Barrier Reef. Ph.D. Dissertation, University of Sydney.

Dow, D. D. 1969. Habitat utilization by cardinals in central and peripheral breeding populations. *Can. J. Zool. 47*:409–417.

Dunham, A. E. 1980. An experimental study of interspecific competition between the iguanid lizards *Sceloporus merriami* and *Urosaurus ornatus*. *Ecol. Monogr. 50*:309–330.

Dunham, A. E., D. W. Tinkle, and J. W. Gibbons. 1978. Body size in island lizards: a cautionary tale. *Ecology 59*:1230–1238.

Dunn, E. R. 1926. *The Salamanders of the Family Plethodontidae*. Smith College Anniversary Series, Vol. 7, Northampton, Mass.

Dunning, J. B., and J. H. Brown. 1982. Summer rainfall and winter sparrow densities: a test of the food limitation hypothesis. *Auk 99*: 123–129.

Eastop, V. F. 1972. Deductions from the present day host plants of aphids and related insects. *Symp. Roy. Ent. Soc. London 6*:157–178.

Ebersole, J. P. 1977. The adaptive significance of territoriality in the reef fish *Eupomacentrus leucostictus*. *Ecology 58*:914–920.

Ehrlich, P. R., D. D. Murphy, M. C. Singer, C. B. Sherwood, R. R. White, and I. L. Brown. 1980. Extinction, reduction, stability and increase: the responses of checkerspot butterfly (*Euphydryas*) populations to the California drought. *Oecologia 46*:101–105

Ehrlich, P. R., and P. H. Raven. 1965. Butterflies and plants: a study in co-evolution. *Evolution 18*:586–608.

Ehrlich, P. R., F. H. Talbot, B. C. Russell, and G.R.V. Anderson. 1977. The behaviour of chaetodontid fishes with special reference to Lorenz's "poster colouration" hypothesis. *J. Zool., Lond. 183*:213–228.

Elseth, G. D., and K. D. Baumgardner. 1981. *Population Biology*. D. Van Nostrand. N. Y.

Elton, C. S. 1927. *Animal Ecology*. Sedgwick and Jackson, London.

Elton, C. S. 1946. Competition and the structure of ecological communities. *J. Anim. Ecol. 15*:54–68.

Elton, C. S. 1947. Reviews of the Journal of Animal Ecology, Vols. 14, 15 and 16. *J. Ecol. 35*:266–267.

Elton, C. S. 1958. *The Ecology of Invasions by Animals and Plants*. Methuen, London.

Emlen, J. T. 1978. Density anomalies and regulation mechanisms in land bird populations on the Florida peninsula. *Amer. Natur. 112*:265–286.

Emlen, J. T. 1979. Land bird densities on Baja California islands. *Auk 96*:152–167.

Endean, R. E., and O. A. Jones (eds.). 1973. *Biology and Geology of Coral Reefs*. Academic Press, N.Y.

Enders, F. 1974. Vertical stratification in orb-web spiders and a consideration of other methods of coexistence. *Ecology 55*:317–328.

Enders, F. 1975. The influence of hunting behavior on prey size, particularly in spiders with long attack distances (Araneidea, Linyphiidae, and Salticidae). *Amer. Natur. 109*:737–763.

Enders, F. 1976. Clutch size related to hunting manner of spider species. *Ann. Ent. Soc. Amer. 69*:991–998.

Enemar, A. 1966. A ten-year study on the size and composition of a breeding passerine bird community. *Vär Fägelvärld Suppl. 4*:47–94.

Engstrom, R. T. , and F. C. James. 1981. Plot size as a factor in winter bird-population studies. *Condor 83*:34–41.

Estes, J. A., and J. F. Palmisano. 1974. Sea otters: their role in structuring nearshore communities. *Science 184*:1058–1060.

Ewel, J. J., and A. Madriz. 1968. Zones de Vida de Venezuela. Dirección de Investigacion. Ministerio de Agricultura y Cría, Caracas.

Faegri, K., and L. van der Pijl. 1971. *The Principles of Pollination Ecology*. Pergamon Press, London.

Faeth, S. H., and D. Simberloff. 1981a. Population regulation of a leaf-mining insect, *Cameraria* sp. nov., at increased field densities. *Ecology 62*:620–624.

Faeth, S. H., and D. Simberloff. 1981b. Experimental isolation of oak host plants: effects on mortality, survivorship and abundances of leaf-mining insects *Ecology 62*:624–635.

Fager, E. W. 1964. Marine sediments: effects of a tube-building polychaete. *Science 143*:356–359.

Feare, C. J. 1971. The adaptive significance of aggregation behaviour in the dogwhelk *Nucella lapillus* (L.). *Oecologia 7*:117–126.

Feinsinger, P. 1976. Organization of a tropical guild of nectivorous birds. *Ecol. Monogr. 46*:257–291.

Feinsinger, P. 1978. Ecological interactions between plants and hummingbirds in a successional tropical community. *Ecol. Monogr. 48*:269–287.

Feinsinger, P., R. J. Whelan, and R. A. Kiltie. 1981. Some notes on community composition: assembly by rules or by dartboards? *Bull. Ecol. Soc. Amer. 62*:19–23.

Feller, W. 1950. *An Introduction to Probability Theory and Its Applications*. Wiley and Sons, N.Y.

Fenchel, T. 1975. Character displacement and coexistence in mud snails (Hydrobiidae). *Oecologia 20*:19–32.

Fenchel, T. 1977. Competition, coexistence and character displacement in mud snails (Hydrobiidae). Pages 229–243 in *Ecology of Marine Benthos*, B. C. Coull (ed.). University of South Carolina Press, Columbia, S.C.

Feyerabend, P. 1963. How to be a good empiricist—a plea for tolerance in matters epistemological. Pages 12–39 in *Philosophy of Science. Delaware Seminar*, Vol. 2, B. Baumrin (ed.). Interscience Publishers, N.Y.

Feyerabend, P. 1975. *Against Method*. Humanities Press, Atlantic Highlands, N.J.

Fienberg, S. E. 1980. *The Analysis of Cross-classified Categorical Data*. MIT Press, Cambridge, Mass.

Findley, J. S. 1976. The structure of bat communities. *Amer. Natur.* *110*:129–139.

Fitton, M. G., et al. 1978. *A Check List of British Insects, Part 4: Hymenoptera*. Royal Entomological Society of London.

Force, D. C. 1974. Ecology of insect host-parasitoid communities. *Science* *184*:624–632.

Force, D. C. 1980. Do parasitoids of Lepidoptera larvae compete for hosts? Probably! *Amer. Natur.* *116*:873–875.

Foster, B. A. 1971. On the determinants of the upper limit of barnacles (Crustacea: Cirripedia). *J. Anim. Ecol.* *40*:33–48.

Fowler, C. W. 1981. Density dependence as related to life history strategy. *Ecology* *62*:602—610.

Fowler, N. 1978. Competition and coexistence in an herbaceous plant community. Ph.D. Dissertation, Duke University.

Fowler, N. 1981. Competition and coexistence in a North Carolina grassland. II. The effects of the experimental removal of species. *J. Ecol.* *69*:843–854.

Fowler, N., and J. Antonovics. 1981. Competition and coexistence in a North Carolina grassland. I. Patterns in undisturbed vegetation. *J. Ecol.* *69*:825–841.

Frank, J. H., and G. A. Curtis. 1977. On the bionomics of bromeliad-inhabiting mosquitoes. III. The probable strategy of larval feeding in *Wyeomyia vanduzeei* and *Wy. medioalbipes*. *Mosquito News 37*: 200–206.

Frank, P. W. 1965. The biodemography of an intertidal snail population. *Ecology 46*:831–844.

Frank, P. W. 1968. Life histories and community stability. *Ecology 49*: 355–356.

Frankie, G. W., H. G. Baker, and P. A. Opler. 1974. Comparative phenological studies in tropical wet and dry forests in the lowlands of Costa Rica. *J. Ecol. 62*:881–913.

Fretter, V., and R. Manly. 1977. The settlement and early benthic life of

Littorina neritoides (L.) at Wembury, Devon. *J. Moll. Stud. 43*: 255–262.

Frey, D. G. 1949. Morphometry and hydrography of some natural lakes of the North Carolina coastal Plain: a bay lake as a morphometric type. *J. Elisha Mitchell Sci. Soc. 65*:1–37.

Friedmann, H., L. Griscom, and R. T. Moore. 1950. Distributional checklist of the birds of Mexico. Part I. Pacific Coast Avifauna, No. 29.

Fuentes, E. R. 1976. Ecological convergence of lizard communities in Chile and California. *Ecology 57*:3–17.

Fuentes, E. R. 1980. Convergence of community structure: neutral model vs. field data. *Ecology 61*:198–200.

Gage, D., and D. R. Strong. 1981. The chemistry of *Heliconia imbricata* and *H. latispatha* and the slow growth of a hispine beetle herbivore. *Biochem. Syst. Ecol. 9*:79–81.

Gardner, M. R., and W. R. Ashby. 1970. Connectance of large dynamic (cybernetic) systems: critical values for stability. *Nature 228*:784.

Gatz, A. J., Jr. 1979. Community organization in fishes as indicated by morphological features. *Ecology 60*:711–718.

Gatz, A. J., Jr. 1981. Morphologically inferred niche differentiation in stream fishes. *Amer. Midl. Natur. 106*:10–21.

Gauch, H. G., Jr., R. H. Whittaker, and T. R. Wentworth. 1977. A comparative study of reciprocal averaging and other ordination techniques. *J. Ecol. 65*:157–174.

Gause, G. F. 1934. *The Struggle for Existence*. Williams and Wilkins, Baltimore. Reprinted 1964, Hafner, N.Y.

Gentry, A. H. 1974. Flowering phenology and diversity in tropical Bignoniaceae. *Biotropica 6*:64–68.

Gerdes, D. 1977. The reestablishment of an *Amphiura filliformis* (O. F. Muller) population in the inner part of the German bight. Pages 277–284 in *Biology of Benthic Organisms*, B. F. Keegan, P. O Ceidigh, P.J.S. Boaden (eds.). Pergamon Press, London.

Gertsch, W. J., and S. E. Riechert. 1976. The spatial and temporal partitioning of a desert spider community with descriptions of new species. *Am. Mus. Novitates 2604*:1–25.

Gibson, C.W.D. 1980. Niche use patterns among some Stenodemini (Heteroptera: Miridae) of limestone grassland, and an investigation of the possibility of interspecific competition between *Notostira elongata* Geoffroy and *Megaloceraea recticornis* Geoffroy. *Oecologia 47*:352–364.

Gilbert, F. S. 1980. The equilibrium theory of island biogeography: Fact or fiction? *J. Biogeography 7*:209–235.

Gilbert, L. E. 1977. The role of insect-plant coevolution in the organisation of ecosystems. *Colloq. Int. CNRS 265*:399–413.

Gilpin, M. E., and J. M. Diamond. 1981. Immigration and extinction probabilities for individual species: relation to incidence functions and species colonization curves. *Proc. Natl. Acad. Sci. USA 78*: 393–396.

Gilpin, M. E., and J. M. Diamond. 1982. Factors contributing to non-randomness in species co-occurrences on islands. *Oecologia (Berl.) 52*:75–84.

Gladfelter, W. B., and E. H. Gladfelter. 1978. Fish community structure as a function of habitat structure on West Indian patch reefs. *Rev. Trop. Biol. 26* (Suppl. 1):65–84.

Gleason, H. A. 1926. The individualistic concept of the plant association. *Bull. Torrey Bot. Club 53*:7–26.

Glynn, P. W. 1976. Some physical and biological determinants of coral community structure in the eastern Pacific. *Ecol. Monogr. 46*:431–456.

Glynn, P. W., and R. H. Stewart. 1973. Distribution of coral reefs in the Pearl Islands (Gulf of Panama) in relation to thermal conditions. *Limnol. Oceanogr. 18*:367–379.

Glynn, P. W., R. H. Stewart, and J. E. McCosker. 1972. Pacific coral reefs of Panama: structure, distribution, and predators. *Geol. Rdsch. 61*:483–519.

Goldman, B., and F. H. Talbot. 1976. Aspects of the ecology of coral reef fishes. Pages 125–154 in *Biology and Geology of Coral Reefs*, Vol. 3, Biology 2, O. A. Jones and R. Endean (eds.). Academic Press, N.Y.

Goodall, D. W. 1969. A procedure for recognition of uncommon species combinations in sets of vegetation samples. *Vegetatio 18*:19–35.

Goodman, D. 1975. The theory of diversity-stability relationships in ecology. *Quart. Rev. Biol. 50*:237–266.

Grant, B. R., and P. R. Grant. 1982. Niche shifts and competition in Darwin's Finches: *Geospiza conirostris* and congeners. *Evolution 36*: 637–657.

Grant, K. A., and V. Grant. 1968. *Hummingbirds and Their Flowers*. Columbia University Press, N.Y.

Grant, P. R. 1965a. The adaptive significance of some island size trends in birds. *Evolution 19*:355–367.

Grant, P. R. 1965b. A systematic study of the terrestrial birds of the Tres Marias Islands, Mexico. *Postilla 90*:1–106.

Grant, P. R. 1966a. The density of land birds on the Tres Marias Island in Mexico. I. Numbers and biomass. *Can. J. Zool. 44*:391–400.

Grant, P. R. 1966b.Ecological compatibility of bird species on islands. *Amer. Natur. 100*:451–462.

Grant, P. R. 1967. Bill length variability in birds of the Tres Marias Islands, Mexico. *Can. J. Zool. 45*:805–815.

Grant, P. R. 1968. Bill size, body size, and the ecological adaptations of bird species to competitive situations on islands. *Syst. Zool. 17*: 319–333.

Grant, P. R. 1969. Colonization of islands by ecologically dissimilar species of birds. *Can. J. Zool. 47*:41–43.

Grant, P. R. 1972a. Bill dimensions of the three species of *Zosterops* on Norfolk Island. *Syst. Zool. 21*:289–291.

Grant, P. R. 1972b. Convergent and divergent character displacement. *Biol. J. Linn. Soc. 4*:39–68.

Grant, P. R. 1972c. Interspecific competition among rodents. *Ann. Rev. Ecol. Syst. 3*:79–106.

Grant, P. R. 1975. The classical case of character displacement. *Evol. Biol. 8*:237–337.

Grant, P. R. 1977. Review of D. Lack, 1976, *Island Biology, Illustrated by the Land Birds of Jamaica. Bird-Banding 48*:296–300.

Grant, P. R. 1978. Competition between species of small mammals. Pages 38–51 in *Populations of Small Mammals under Natural Conditions: A Review and Analysis of the Contribution of Long Term Experimental and Descriptive Studies*, D. Snyder (ed.). Special Publ. Ser., Pymatuning Lab. Ecology, Vol. 5. University of Pittsburgh, Pittsburgh, Pa.

Grant, P. R. 1981a. The feeding of Darwin's Finches on *Tribulus cistoides* (L.) seeds. *Anim. Behav. 29*:785–793.

Grant, P. R. 1981b. Speciation and the adaptive radiation of Darwin's Finches. *Amer. Sci. 69*:653–663.

Grant, P. R. 1983. The role of interspecific competition in the adaptive radiation of Darwin's Finches. In *Patterns of Evolution in Galápagos Organisms*, A. Leviton and R. I. Bowman (eds.). Spec. Publ. 1, Amer. Assoc. Adv. Sci., Pacific Division, Washington, D.C. (in press).

Grant, P. R., and I. Abbott. 1980. Interspecific competition, island biogeography and null hypotheses. *Evolution 34*:332–341.

Grant, P. R., and P. T. Boag. 1980. Rainfall on the Galápagos and the demography of Darwin's finches. *Auk 97*:227–244.

Grant, P. R., and B. R. Grant. 1980a. Annual variation in finch numbers, foraging and food supply on Isla Daphne Major, Galápagos. *Oecologia (Berl.) 46*:55–62.

Grant, P. R., and B. R. Grant. 1980b. The breeding and feeding characteristics of Darwin's Finches on Isla Genovesa, Galápagos. *Ecol. Monogr. 50*:381–410.

Grant, P. R., and T. D. Price. 1981. Population variation in continuously varying traits as an ecological genetics problem. *Amer. Zool. 21*: 795–811.

Grant, P. R., T. D. Price, and H. Snell. 1980. The exploration of Isla Daphne Minor. *Noticias de Galápagos No. 31*:22–27.

Grant, P. R., J.N.M. Smith, B. R. Grant, I. Abbott, and L. K. Abbott.

1975. Finch numbers, owl predation and plant dispersal on Isla Daphne Major, Galápagos. *Oecologia (Berl.)* *19*:239–257.

Grassle, J. F. 1977. Slow recolonization of deep-sea sediment. *Nature* *265*:618–619.

Gray, J. S. 1974. Animal-sediment relationships. *Oceanogr. Mar. Biol. Ann. Rev. 12*:223–261.

Green, R. F. 1979. Testing whether island communities are random. Technical Report No. 63. University of California, Riverside, Department of Statistics.

Greenstone, M. H. 1980. Contiguous allotopy of *Pardosa ramulosa* and *Pardosa tuoba* (Araneae, Lycosidae) in the San Francisco Bay region and its implications for the pattern of resource partitioning in the genus. *Amer. Midl. Natur. 104:*305–311.

Grime, J. P. 1977. Evidence for the existence of three primary strategies in plants and its relevance to ecological and evolutionary theory. *Amer. Natur. 111*:1169–1194.

Grime, J. P. 1979. *Plant Strategies and Vegetation Processes*. Wiley, London.

Hagvar, S. 1976. Altitudinal zonation of the invertebrate fauna on branches of birch (*Betula pubsecens* Ehrh.). *Norw. J. Ent. 23*:61–74.

Haila, Y., and O. Järvinen. 1977. Competition and habitat selection in two large woodpeckers. *Ornis Fennica 54*:73–78.

Haila, Y., and O. Järvinen. 1980. Bird communities in a Finnish archipelago 50 years ago and now: a general survey. Pages 151–157 in *Bird Census Work and Nature Conservation*, H. Oelke (ed.). Dachverband Deutscher Avifaunisten, Göttingen.

Haila, Y., and O. Järvinen. 1981. The underexploited potential of bird censuses in insular ecology. *Studies Avian Biol. 6*:559–565.

Haila, Y., and O. Järvinen. 1983. Land bird communities on a Finnish island: species impoverishment and abundance patterns. *Oikos 41* (in press).

Haila, Y., O. Järvinen, and S. Kuusela. Colonization of islands by land birds: prevalence functions in a Finnish archipelago. *J. Biogeogr.* (in press).

Haila, Y., O. Järvinen, and R. A. Väisänen. 1979. Effect of mainland population changes on the terrestrial bird fauna of a northern island. *Ornis Scandinavica 10*:48–55.

Haila, Y., O. Järvinen, and R. A. Väisänen. 1980a. Effects of changing forest structure on long-term trends in bird populations in SW Finland. *Ornis Scandinavica 11*:12–22.

Haila, Y., O. Järvinen, and R. A. Väisänen. 1980b. Habitat distribution and species associations of land bird populations on the Åland Islands, SW Finland. *Annales Zoologici Fennici 17*:87–106.

Hairston, N. G. 1949. The local distribution and ecology of the plethodontid salamanders of the southern Appalachians. *Ecol. Monogr. 19*:47–73.

Hairston, N. G. 1964. Studies on the organization of animal communities. *J. Anim. Ecol. 33* (Suppl.):227–239.

Hairston, N. G. 1973. Ecology, selection, and systematics. *Breviora 414*:1–21.

Hairston, N. G. 1980a. The experimental test of an analysis of field distributions: competition in terrestrial salamanders. *Ecology 61*:817–826.

Hairston, N. G. 1980b. Species packing in the salamander genus *Desmognathus*: what are the interspecific interactions involved? *Amer. Natur. 115*:354–366.

Hairston, N. G. 1981. An experimental test of a guild: salamander competition. *Ecology 62*:65–72.

Hairston, N. G., F. E. Smith, and L. B. Slobodkin. 1960. Community structure, population control, and competition. *Amer. Natur. 94*:421–425.

Hall, D. J., W. E. Cooper, and E. E. Werner. 1970. An experimental approach to the production dynamics and structure of freshwater animal communities. *Limnol. Oceanogr. 15*:839–928.

Hall, D. J., and E. E. Werner. 1977. Seasonal distribution and abundance of fishes in the littoral zone of a Michigan Lake. *Trans. Am. Fish. Soc. 106*:545–555.

Hall, E. R. 1946. *Mammals of Nevada*. University of California Press, Berkeley and Los Angeles.

Hall, E. R. 1981. *The Mammals of North America*. Wiley N.Y.

Hallett, J. G. 1982. Habitat selection and the community matrix of a desert small-mammal fauna. *Ecology 63*:1400–1410.

Hallett, J. G., and S. L. Pimm. 1979. Direct estimation of competition. *Amer. Natur. 113*:593–600.

Hamilton, W. D. 1978. Evolution and diversity under bark. *Symp. Roy. Ent. Soc. London 9*:154–175.

Hardin, G. 1956. Meaninglessness of the word protoplasm. *Scientific Monthly 82*:112–120.

Harger, R. E. 1970. The effect of wave impact on some aspects of the biology of sea mussels. *Veliger 12*:401–414.

Harper, J. L. 1977. *Population Biology of Plants*. Academic Press, London.

Harris, M. P. 1973. The Galápagos avifauna. *Condor 75*:265–278.

Harrison, G. W. 1979. Stability under environmental stress: Resistance, resilience, persistence, and variability. *Amer. Natur. 113*:659–669.

Hastings, H. H., and M. Conrad. 1979. Length and evolutionary stability of food chains. *Nature 282*:838–839.

Hatton, H. 1938. Essais de bionomie explicative sur quelques espèces intercotidales d'algues et d'animaux. *Ann. Inst. Oceanogr. Monaco* *17*:241–348.

Hay, C. 1979. Some factors affecting the upper limit of the southern bell kelp *Durvillaea antarctica* (Chamisso) Hariot on two New Zealand shores. *J. Roy. Soc. N.Z.* *9*:279–289.

Hayward, T. L., and J. A. McGowan. 1979. Pattern and structure in an oceanic zooplankton community. *Amer. Zool.* *19*:1045–1055.

Heck, K. L. 1976. Some critical considerations of the theory of species packing. *Evol. Theory 1*:247–258.

Heck, K. L. 1979. Some determinants of the composition and abundance of motile macroinvertebrate species in tropical and temperate turtle grass (*Thalassia testudinum*) meadows. *J. Biogeography 6*: 183–197.

Heck, K. L., Jr., G. van Belle, and D. Simberloff. 1976. Explicit calculation of the rarefaction diversity measurement and the determination of sufficient sample size. *Ecology 56*:1459–1461.

Heinrich, B. 1976. Flowering phenologies: bog, woodland, and disturbed habitats. *Ecology 57*:890–899.

Heinrich, G. H. 1977. Ichneumoninae of Florida and neighboring states. Arthropods of Florida. Vol. 9. Florida Dept. Agric. Consumer Ser., Gainesville.

Heithaus, E. R. 1974. The role of plant-pollinator interactions in determining community structure. *Ann. Mo. Bot. Gard.* *61*:675–691.

Heller, H. C. 1971. Altitudinal zonation of chipmunks (*Eutamias*): interspecific aggression. *Ecology 52*:312–319.

Hemmingsen, A. M. 1934. A statistical analysis of the differences in body size of related species. *Videnskabelige Meddelelser Dansk naturhistorisk Forening København 98*:125–160.

Hendrickson, J. A. 1979. Analyses of species occurrences in community, continuum, and biomonitoring studies. Pages 361–397 in *Contemporary Quantitative Ecology and Related Econometrics*, G. P. Patil and M. L. Rosenzweig (eds.). International Cooperative Publishing House, Fairland, Md.

Hendrickson, J. A., Jr. 1981. Community-wide character displacement reexamined. *Evolution 35*:794–809.

Hendrix, S. D. 1980. An evolutionary and ecological perspective of the insect fauna of ferns. *Amer. Natur. 115*:171–196.

Herrera, C. M. 1978. Individual dietary differences associated with morphological variation in Robins *Erithacus rubecula*. *Ibis 120*: 542–545.

Hespenheide, H. A. 1971. Food preference and the extent of overlap in some insectivorous birds with special reference to the Tyrannidae. *Ibis 113*:59–72.

Hespenheide, H. A. 1973. Ecological inferences from morphological data. *Ann. Rev. Ecol. Syst. 4*:213–229.

Hespenheide, H. A. 1975. Prey characteristics and predator niche width. Pages 158–180 in *Ecology and Evolution of Communities*, M. L. Cody and J. M. Diamond (eds.). Harvard University Press, Cambridge, Mass.

Hiatt, R. W., and D. W. Strasburg. 1960. Ecological relationships of the fish fauna on coral reefs of the Marshall Islands. *Ecol. Monogr. 30*:65–127.

Hill, M. O. 1973. Reciprocal averaging: an eigenvector method of ordination. *J. Ecol. 61*:237–249.

Hixon, M. A. 1980. Competitive interactions between California reef fishes of the genus *Embiotoca*. *Ecology 61*:918–931.

Hodges, J. L., and E. L. Lehman. 1970. *Basic Concepts of Probability and Statistics*. Holden-Day, San Francisco.

Holdridge, L. R., W. C. Grenke, W. H. Hatheway, T. Laing, and J. A. Tosi, Jr. 1971. *Forest Environments in Tropical Life Zones: A Pilot Study*. Pergamon Press, N.Y.

Holland, A. F., N. K. Mountford, M. H. Hiegel, K. R. Kaumeyer, and J. A. Mihursky. 1980. Influence of predation on infaunal abundance in upper Chesapeake Bay, USA. *Mar. Biol. 57*:221–235.

Holling, C. S. 1973. Resilience and stability of ecological systems. *Ann. Rev. Ecol. Syst. 4*:1–24.

Holmes, J. C. 1971. Habitat segregation in sanguinicolid blood flukes (Digenea) of scorpaenid rockfishes (Perciformes) on the Pacific coast of North America. *J. Fish. Res. Board Can. 28*:903–909.

Holmes, J. C. 1973. Site selection by parasitic helminths: Interspecific interactions, site segregation, and their importance to the development of helminth communities. *Can. J. Zool. 51*:333–347.

Holmes, J. C., and P. W. Price. 1980. Parasite communities: the roles of phylogeny and ecology. *Syst. Zool. 29*:203–213.

Holt, R. D. 1977. Predation, apparent competition and the structure of prey communities. *Theor. Pop. Biol. 12*:197–229.

Hoover, K. D., W. G. Whitford, and P. Flavill. 1977. Factors influencing the distributions of two species of *Perognathus*. *Ecology 58*:877–884.

Horn, H. S., and R. M. May. 1977. Limits to similarity among coexisting competitors. *Nature 270*:660–661.

Horton, C. C., and D. H. Wise. 1983. The experimental analysis of competition between two syntopic species of orb-web spiders (Araneae: Araneidae). *Ecology 64*:929–944.

Hoshiai, T. 1964. Synecological study on intertidal communities. *Bull. Mar. Biol. Station of Asamushi 12*:93–126.

Howard, L.O., and W. F. Fiske. 1911. The importation into the United States of the parasites of the gypsy-moth and the brown-tail moth. U.S. Dept. Agri., Bur. Entomol., Bull. 91.

Huey, R. B. 1979. Parapatry and niche complementarity of Peruvian desert geckos (*Phyllodactylus*): the ambiguous role of competition. *Oecologia 38*:249–259.

Huheey, J. E., and R. A. Brandon. 1973. Rock-face populations of the moutain salamander, *Desmognathus ochrophaeus*, in North Carolina. *Ecol. Monogr. 43*:59–77.

Hull, D. L. 1974. *Philosophy of Biological Science*. Prentice-Hall, Englewood Cliffs, N.J.

Hurlbert, S. H. 1969. A coefficient of interspecific association. *Ecology 50*:1–9.

Hurlbert, S. H. 1971. The nonconcept of species diversity: a critique and alternative parameters. *Ecology 52*:577–586.

Hurlbert, S. H. 1978. The measurement of niche overlap and some relatives. *Ecology 59*:67–77.

Hurlbert, S. H. Pseudoreplication and the design of ecological field experiments. *Ecological Monographs* (in press).

Hutchinson, G. E. 1953. The concept of pattern in ecology. *Proc. Acad. Nat. Sci., Philadelphia, 105*:1–12.

Hutchinson, G. E. 1957. Concluding remarks. *Cold Spring Harbor Symp. Quant. Biol. 22*:415–427.

Hutchinson, G. E. 1959. Homage to Santa Rosalia, *or* Why are there so many kinds of animals? *Amer. Natur. 93*:145–159.

Hutchinson, G. E. 1965. *The Ecological Theater and the Evolutionary Play*. Yale University Press, New Haven, Conn.

Hutchinson, G. E. 1978. *An Introduction to Population Ecology*. Yale University Press, New Haven, Conn.

Hutchinson, G. E., and R. H. MacArthur. 1959. A theoretical model of size distributions among species of animals. *Amer. Natur. 93*:145–149.

Hutto, R. L. 1978. A mechanism for resource allocation among sympatric heteromyid rodent species. *Oecologia 33*:115–126.

Huxley, J. 1942. *Evolution: The Modern Synthesis*. Allen and Unwin.

Inger, R. F., and R. K. Colwell. 1977. Organization of contiguous communities of amphibians and reptiles in Thailand. *Ecol. Monogr. 47*:229–253.

Inger, R. F., and B. Greenberg. 1966. Ecological and competitive relations among three species of frogs (genus *Rana*). *Ecology 47*:746–759.

Inouye, D. W. 1978. Resource partitioning in bumblebees: Experimental studies of foraging behavior. *Ecology 79*:672–678.

Isaacs, J. D. 1973. Potential trophic biomass and trace-substance concentration in unstructured marine food webs. *Mar. Biol. 22*:97–104.

Isaacs, J. D. 1976. Reproductive products in marine food webs. *Bull. S. Calif. Acad. of Sciences 75*:220–223.

Jaccard, P. 1901. Distribution de la flore alpine dans le Bassin des Dranses et dans quelques régions voisines. *Bull. Soc. Vau. Sci. Nat. 37*:241–272

Jackson, J.B.C. 1977. Competition on marine hard substrata: the adaptive significance of solitary and colonial strategies. *Amer. Natur. 111*:743–767.

Jackson, J.B.C. 1979. Morphological strategies of sessile animals. Pages 499–555 in *Systematics Association Special Volume No. 11, Biology and Systematics of Colonial Organisms*, G. Larwood and B. R. Roben (eds.). Academic Press, London and New York.

Jackson, J.B.C. 1981. Gause's principle, Hutchinson and Lack: making natural history theoretically interesting. *Amer. Zool. 21*:889–901.

Jackson, J.B.C., and L. Buss. 1975. Allelopathy and spatial competition among coral reef invertebrates. *Proc. Natl. Acad. Sci. USA 72*:5160–5163.

Jaeger, R. G. 1971. Competitive exclusion as a factor influencing the distributions of two species of terrestrial salamanders. *Ecology 52*:632–637.

James, F. C. 1971. Ordinations of habitat relationships among breeding birds. *Wilson Bull. 83*:215–236.

James, F. C., and S. Rathbun. 1981. Rarefaction, relative abundance and diversity of avian communities. *Auk 98*:785–800.

James, F. C., and N. O. Wamer. 1982. Relationships between temperate forest bird communities and vegetation structure. *Ecology 63*:159–171.

Janzen, D. H. 1966. Coevolution of mutualism between ants and acacias in Central America. *Evolution 20*:249–275.

Janzen, D. H. 1969. Allelopathy of myrecophytes: the ant *Azteca* as an allelopathic agent of *Cecropia. Ecology 50*:147–153.

Janzen, D. H. 1973a. Sweep samples of tropical foliage insects: effects of seasons, vegetation types, elevation, time of day, and insularity. *Ecology 54*:687–708.

Janzen, D. H. 1973b. Host plants as islands. II. Competition in evolutionary and contemporary time. *Amer. Natur. 107*:786–790.

Janzen, D. H. 1975a. *Ecology of Plants in the Tropics*. Edward Arnold, London.

Janzen, D. H. 1975b. Interactions of seeds and their insect predators/parasitoids in a tropical deciduous forest. Pages 154–186 in *Evolu-*

tionary Strategies of Parasitic Insects and Mites, P. W. Price (ed.). Plenum Press, N.Y.

Janzen, D. H. 1981. The peak in North American ichneumonid species richness lies between 38 and 42°N. *Ecology 62*:532–537.

Järvinen, O. 1976. Estimating relative densities of breeding birds by the line transect method. II. Comparison between two methods. *Ornis Scandinavica 7*:43–48.

Järvinen, O. 1979. Geographic gradients of stability in European land bird communities. *Oecologia 38*:51–69.

Järvenin, O., and S. Ulfstrand. 1980. Species turnover of a continental bird fauna: Northern Europe, 1850–1970. *Oecologia (Berl.) 46*: 186–195.

Järvinen, O., and R. A. Väisänen. 1980. Quantitative biogeography of Finnish land birds as compared with regionality in other taxa. *Annales Zoologici Fennici 17*:67–85.

Järvinen, O., and R. A. Väisänen. 1981. Methodology for censusing land bird faunas in large regions. *Studies Avian Biol. 6*:146–151.

Järvinen, O., R. A. Väisänen, and Y. Haila. 1977. Bird census results in different years, stages of the breeding season and times of day. *Ornis Fennica 54*:108–118.

Jewell, P. A. (ed.). 1981. *The Management of Locally Abundant Mammals*. Academic Press, N.Y.

Johnson, N. K. 1978. Review of "Sexual size dimorphism in hawks and owls of North America." *Wilson Bull. 90*:145–147.

Jumars, P. A., and K. Fauchald. 1977. Between community constrasts. Pages 1–20 in *Ecology of Marine Benthos*, B. Coull (ed.). University of South Carolina Press, Columbia, S.C.

Kain, J. M. 1979. A view of the genus *Laminaria*. *Oceanogr. Mar. Biol. Ann. Rev. 17*:101–161.

Kajak, A. 1967. Productivity of some populations of web spiders. Pages 807–820 in *Secondary Productivity of Terrestrial Ecosystems*, K. Petrusewicz (ed.). Pánstwowe Wydawn. Naukowe, Warsaw.

Kajak, A. 1978. Analysis of consumption by spiders under laboratory and field conditions. *Ekol. Pol. Ser. A. 26*:411–428.

Kami, H. T., and I. I. Ikehara. 1976. Notes on the annual juvenile siganid harvest in Guam. *Micronesica 12*:323–325.

Karlson, R. H. 1980. Alternative competitive strategies in a periodic disturbed habitat. *Bull. Mar. Sci. 30*:894–900.

Karr, J. R., and F. C. James. 1975. Ecomorphological configurations and convergent evolution. Pages 258–291 in *Ecology and Evolution of Communities*, M. L. Cody and J. M. Diamond (eds.). Harvard University Press, Cambridge, Mass.

Kastendiek, J. E. 1975. The role of behavior and interspecific interaction in determining the distribution and abundance of *Renilla Kielikeri*, a member of a subtidal sand bottom community. Ph.D. Dissertation, University of California, Los Angeles.

Keast, A. 1968. Competitive interactions and evolution of ecological niches as illustrated by Australian honeyeaters genus *Melithreptus* (Meliphagidae). *Evolution 22*:762–784.

Keast, A. 1977. Mechanisms expanding niche width and minimizing intra-specific competition in two centrarchid fishes. Pages 333–395 in *Evolutionary Biology*, Vol. 10, M. K. Hecht, W. C. Steere, and B. Wallace (eds.). Plenum Press, New York and London.

Keast, A. 1978. Trophic and spatial interrelationships in the fish species of an Ontario temperate lake. *Env. Biol. Fish 3*:7–31.

Keast, A., and E. S. Morton (eds.). 1980. *Migrant Birds in the Neotropics—Ecology, Behavior, Distribution, Conservation*. Smithsonian Institution Press, Washington, D.C.

Keast, A., and D. Webb. 1966. Mouth and body form relative to feeding ecology in the fish fauna of a small lake, Lake Opinicon, Ontario. *J. Fish. Res. Board Can. 23*:1845–1874.

Keegan, B. F., and G. Konnecker. 1973. *In situ* quantitative sampling of benthic organisms. *Helgolander wiss. Meeresunters. 24*:256–263.

Kemeny, J.-G., J. L. Snell, and G. L. Thompson. 1974. *Introduction to Finite Mathematics*, 3rd ed. Prentice-Hall, Englewood Cliffs, N.J.

Keough, M. J., and A. J. Butler. 1979. The role of asteroid predators in the organization of a sessile community on pier pilings. *Mar. Biol. 51*:167–177.

Kerr, R. A. 1980. A new kind of storm beneath the sea. *Science 208*:484–486.

Kessler, A. 1973. A comparative study of the production of eggs in eight *Pardosa* species in the field (Araneae, Lycosidae). *Tijdschr. Entomol. 116*:23–41.

Kirk, A. A. 1977. The insect fauna of the weed *Pteridium aquilinum* (L.) Kuhn (Polypodiaceae) in Papua New Guinea: a potential source of biological control agents. *J. Aust. Ent. Soc. 16*:403–409.

Kirk, A. A. 1980. The insects associated with fern, *Pteridium aquilinum* (L.) Kuhn in Papua New Guinea and their possible use in biological control. M. Phil. Thesis, University of London.

Kitching, J. A., L. Muntz, and F. J. Ebling. 1966. The ecology of Lough Ine. 15. The ecological significance of shell and body forms in *Nucella. J. Anim. Ecol. 35*:113–126.

Kitching, R. L. 1983. Community structure in water-filled tree-holes in Europe and Australia—some comparisons and speculations. In *Phtyotelmata: Terrestrial Plants as the Hosts of Aquatic Insect Com-*

munities, H. Frank and P. Lounibos (eds.). Plexus Press, Marlton, N.J. (in press).

Klopfer, P. H., and R. H. MacArthur. 1961. On the causes of tropical species diversity: Niche overlap. *Amer. Natur. 95*:223–226.

Knapton, R., and J. R. Krebs. 1974. Settlement patterns, territory size, and breeding density in the Song Sparrow (*Melospiza melodia*). *Can. J. Zool. 52*:1413–1420.

Kodric-Brown, A., and J. H. Brown. 1978. Influence of economics, interspecific competition and sexual dimorphism on territoriality of migrant Rufous hummingbirds. *Ecology 59*:285–296.

Kodric-Brown, A., and J. H. Brown. 1979. Competition between distantly related taxa in the coevolution of plants and pollinators. *Amer. Zool. 19*:1115–1127.

Kodric-Brown, A., J. H. Brown, G. S. Byers, and D. F. Gori. Organization of a tropical island community of hummingbirds and flowers. *Ecology* (in press).

Kohn, A. J. 1978. Ecological shift and release in an isolated population: *Conus miliaris* at Easter Island. *Ecol. Monogr. 48*:323–336.

Koponen, S. 1979. Herbivorous insects of birch in northern regions. *Entomol. Tidskr. 100*:231–233.

Krebs, C. J. 1972. *Ecology: The Experimental Analysis of Distribution and Abundance.* Harper and Row, N.Y.

Krebs, C. J. 1978. *Ecology: The Experimental Analysis of Distribution and Abundance*, second edition. Harper and Row, N.Y.

Krebs, C. J., B. L. Keller, and R. H. Tamarin. 1969. Microtus population biology: demographic changes in fluctuating populations of *M. ochrogaster* and *M. pennsylvanicus* in southern Indiana. *Ecology 50*:587–607.

Krebs, J. R. 1980. Ornithologists as unconscious theorists. *Auk 97*: 407–412.

Krombein, K. V., P. D. Hurd, and D. R. Smith. 1979. Catalog of Hymenoptera in America north of Mexico. Vol. 3, pp. 2211–2735. Smithsonian Inst. Press, Washington.

Krzysik, A. J. 1979. Resource allocation, coexistence, and the niche structure of a streamside salamander community. *Ecol. Monogr. 49*:173–194.

Kuhn, G. G., and F. P. Shepard. 1981. Should southern California build defenses against violent storms resulting in lowland flooding as discovered in records of past century? *Shore and Beach 49*:2–10.

Kuhn, T. S. 1970. *The Structure of Scientific Revolutions*, 2nd ed. University of Chicago Press.

Kuris, A. M., A. R. Blaustein, and J. J. Alió. 1980. Hosts as islands. *Amer. Natur. 116*:570–586.

Lack, D. 1940. Evolution of the Galápagos finches. *Nature 146*:324–327.

Lack, D. 1944. Ecological aspects of species formation in passerine birds. *Ibis 86*:260–286.

Lack, D. 1945. The Galápagos finches (Geospizinae): a study in variation. *Occas. Pap. Calif. Acad. Sci. 21*:1–159.

Lack, D. 1946. Competition for food by birds of prey. *J. Anim. Ecol. 15*:123–129.

Lack, D. 1947. *Darwin's Finches.* Cambridge University Press.

Lack, D. 1969a. The number of bird species on islands. *Bird Study 16*:193–209.

Lack, D. 1969b. Subspecies and sympatry in Darwin's Finches. *Evolution 23*:252–263.

Lack, D. 1971. *Ecological Isolation in Birds.* Harvard University Press, Cambridge, Mass.

Lack, D. 1973a. My life as an amateur ornithologist. *Ibis 115*:421–431.

Lack, D. 1973b. The numbers of species of hummingbirds in the West Indies. *Evolution 27*:326–337.

Lack, D. 1976. *Island Biology, Illustrated by the Land Birds of Jamaica.* Blackwell, Oxford.

Laessle, A. M. 1961. A micro-limnological study of Jamaican bromeliads. *Ecology 42*:449–517.

Lakatos, I. 1970. Falsification and the methodology of scientific research programs. Pages 91–96 in *Criticism and the Growth of Knowledge,* I. Lakatos and A. Musgrave (eds.). Cambridge University Press.

Larkin, P. A. 1978. Fisheries management—an essay for ecologists. *Ann. Rev. Ecol. Syst. 9*:57–73.

Larkin, P. A., J. G. Terpenning, and R. R. Parker. 1956. Size as a determinant of growth rate in rainbow trout *Salmo gairdneri. Trans. Am. Fish. Soc. 86*:84–96.

Larson, R. J. 1980. Competition, habitat selection, and the bathymetric segregation of two rockfish (*Sebastes*) species. *Ecol. Monogr. 50*:221–239.

Laughlin, D. R. 1979. Resource and habitat use patterns in two coexisting sunfish species (*Lepomis gibbosus* and *Lepomis megalotis peltastes*). Ph.D. Dissertation, Michigan State University.

Laughlin, D. R., and E. E. Werner. 1980. Resource partitioning in two coexisting sunfish: pumpkinseed (*Lepomis gibbosus*) and northern longear sunfish (*Lepomis megalotis peltastes*). *Can. J. Fish. Aquat. Sci. 37*:1411–1420.

Lawlor, L. R. 1978. A comment on randomly constructed model ecosystems. *Amer. Natur. 112*:445–447.

Lawrence, J. M. 1975. On the relationships between marine plants and sea urchins. *Oceanography and Marine Biology: An Annual Review 13*:213–286.

Lawton, J. H. 1976. The structure of the arthropod community of bracken. *Bot. J. Linn. Soc. 73*:187–216.

Lawton, J. H. 1978a. Host-plant influences on insect diversity: the effects of space and time. Pages 105–125 in *Diversity of Insect Faunas*, L. A. Mound and N. Waloff (eds.). Blackwell, Oxford.

Lawton, J. H. 1978b. *Olethreutes lacunana* (Lepidoptera: Tortricidae) feeding on bracken. *Ent. Gaz. 29*:131–134.

Lawton, J. H., and V. F. Eastop. 1975. A bracken feeding *Macrosiphum* (Hem., Aphididae) new to Britain. *Ent. Gaz. 26*:135–138.

Lawton, J. H., and M. P. Hassell. 1981. Asymmetrical competition in insects. *Nature 289*:793–795.

Lawton, J. H., and S. McNeill. 1979. Between the devil and the deep blue sea: on the problem of being a herbivore. Pages 223–244 in *Population Dynamics*, R. M. Anderson, B. D. Turner, and L. R. Taylor (eds.). *Symp. Brit. Ecol. Soc. 20*:223–244.

Lawton, J. H., and S. L. Pimm. 1978. Population dynamics and the length of food chains. *Nature 272*:190.

Lawton, J. H., and S. L. Pimm. 1979. Reply to Auerbach. *Nature 279*:821–822.

Lawton, J. H., and P. W. Price. 1979. Species richness of parasites on hosts: agromyzid flies on the British Umbelliferae. *J. Anim. Ecol. 48*:619–637.

Lawton, J. H., and D. Schröder. 1977. Effects of plant type, size of geographical range and taxonomic isolation on number of insect species associated with British plants. *Nature 265*:137–140.

Lawton, J. H., and D. R. Strong. 1981. Community patterns and competition in folivorous insects. *Amer. Natur. 118*:317–338

Leigh, E. G., Jr. 1975. Population fluctuations, community stability, and environmental variability. Pages 51–73 in *Ecology and Evolution of Communities*, M. L. Cody and J. M. Diamond (eds.). Harvard University Press, Cambridge, Mass.

Levin, D. 1978. The origins of isolating mechanisms in flowering plants. *Evol. Biol. 11*:185–317.

Levin, S. A., and R. T. Paine. 1974. Disturbance, patch formation, and community structure. *Proc. Natl. Acad. Sci. USA 71*:2744–2747.

Levine, S. H. 1976. Competitive interactions in ecosystems. *Amer. Natur. 110*:903–910.

Levins, R. 1968. *Evolutions in Changing Environments*. Princeton University Press, Princeton, N.J.

Levins, R. 1975. Evolution in communities near equilibrium. Pages 16–50 in *Ecology and Evolution of Communities*, M. L. Cody and J. M. Diamond (eds.). Harvard University Press, Cambridge, Mass.

Levins, R. 1979. Coexistence in a variable environment. *Amer. Natur. 114*:765–783.

Levins, R., and R. Lewontin. 1980. Dialectics and reductionism in ecology. *Synthèse 43*:47–78.

Levins, R., and M. Wilson. 1980. Ecological theory and pest management. *Ann. Rev. Ent. 25*:287–308.

Levinton, J. S. 1972. Stability and trophic structure in deposit-feeding and suspension-feeding communities. *Amer. Natur. 106*:472–486.

Levinton, J. S. 1977. Ecology of shallow water deposit-feeding communities Quisset Harbor, Massachusetts. Pages 191–227 in *Ecology of Marine Benthos*, B. C. Coull (ed.). University of South Carolina Press, Columbia, S.C.

Lewis, J. R. 1964. *The Ecology of Rocky Shores*. English Universities Press, London.

Lewis, J. R. 1977. The role of physical and biological factors in the distribution and stability of rocky shore communities. Pages 417–424 in *Biology of Benthic Organisms*, B. F. Keegan, P. O Ceidigh, and P.J.S. Boaden (eds.). Pergamon Press, London.

Lewontin, R. C. 1974. *The Genetic Basis of Evolutionary Change*. Columbia University Press, N.Y.

Livingston, R. J. 1979. *Ecological Processes in Coastal and Marine Systems*. Plenum Press, New York and London.

Loosanoff, V. L. 1964. Variations in time and intensity of settling of the starfish, *Asterias forbesi*, in Long Island Sound during a twenty-five year period. *Biol. Bull. 126*:423–439.

Lossanoff, V. L. 1966. Time and intensity of settling of the oyster, *Crassostrea virginica*, in Long Island Sound. *Biol. Bull. 130*:211–227.

Lounibos, L. P. 1978. Mosquito breeding and oviposition stimulant in fruit husks. *Ecol. Entomol. 3*:299–304.

Lounibos, L. P. 1981. Habitat segregation among African treehole mosquitoes. *Ecol. Entomol. 6*:129–154.

Low, R. M. 1971. Interspecific territoriality in a pomacentrid reef fish, *Pomacentrus flavicauda* Whitley. *Ecology 52*:648–654.

Lubchenco, J. 1978. Plant species diversity in a marine intertidal community: importance of herbivore food preference and algal competitive abilities. *Amer. Natur. 112*:23–39.

Lubchenco, J. 1980. Algal zonation in the New England rocky intertidal community: an experimental analysis. *Ecology 61*:333–344.

Lubchenco, J., and B. A. Menge. 1978. Community development and persistence in a low rocky intertidal zone. *Ecol. Monogr. 48*:67–94.

Luckens, P. A. 1975a. Competition and intertidal zonation of barnacles at Leigh, New Zealand, *N.Z. J. Mar. Freshwater Res. 9*:355–378.

Luckens, P. A. 1975b. Predation and intertidal zonation of barnacles at Leigh, New Zealand. *N.Z. J. Mar. Freshwater Res. 9*:379–394.

Luckhurst, B. E., and K. Luckhurst. 1977. Recruitment patterns of coral reef fishes on the fringing reef of Curacao, Netherlands Antilles. *Can. J. Zool. 55*:681–689.

Luckinbill, L. S., and M. M. Fenton. 1978. Regulation and environmental variability in experimental populations of protozoa. *Ecology 59*: 1271–1276.

Luczak, J. 1959. The community of spiders of the ground flora of pine forests. *Ekol. Pol. Ser. A. 14*:233–244.

Luczak, J. 1963. Differences in the structure of communities of web spiders in one type of environment (young pine forest). *Ekol. Pol. Ser. A. 11*:159–221.

Luczak, J. 1966. The distribution of wandering spiders in different layers of the environment as a result of interspecies competition. *Ekol. Pol. Ser. A. 14*:233–244.

Lynch, J. F., and N. K. Johnson. 1974. Turnover and equilibria in insular avifaunas, with special reference to the California Channel Islands. *Condor 76*:370–384.

Lynch, J. F., and R. F. Whitcomb. 1977. Effects of the insularization of the eastern deciduous forest on avifaunal diversity and turnover. Pages 461–490 in *Proceedings of a Nat. Symposium on Classification, Inventory and Analysis of Fish and Wildlife Habitat*, A. Marmelstein [Project leader]. U.S. Dept. Int., Phoenix, Ariz.

Lyon, D. L. 1976. A montane hummingbird territorial system in Oaxaca, Mexico. *Wilson Bull. 88*:280–299.

M'Closkey, R. T. 1978. Niche separation and assembly in four species of Sonoran desert rodents. *Amer. Natur. 112*:683–694.

MacArthur, R. H. 1958. Population ecology of some warblers of northeastern coniferous forests. *Ecology 39*:599–619.

MacArthur, R. H. 1971. Patterns of terrestrial bird communities. Pages 189–221 in *Avian Biology*, Vol. 1, D. S. Farner and J. R. King (eds.). Academic Press, N.Y.

MacArthur, R. H. 1972. *Geographical Ecology*. Harper and Row, N.Y.

MacArthur, R. H., J. M. Diamond, and J. R. Karr. 1972. Density compensation in island faunas. *Ecology 53*:330–342.

MacArthur, R. H., and R. Levins. 1967. The limiting similarity, convergence and divergence of coexisting species. *Amer. Natur. 101*:377–385.

MacArthur, R. H. and J. W. MacArthur. 1961. On bird species diversity. *Ecology 42*:594–598.

MacArthur, R. H., and E. O. Wilson. 1963. An equilibrium theory of insular zoogeography. *Evolution 17*:373–387.

MacArthur, R. H., and E. O. Wilson. 1967. *The Theory of Island Biogeography*. Princeton University Press, Princeton, N.J.

McClure, M. S., and P. W. Price. 1975. Competition among sympatric

Erythroneura leaf hoppers (Homoptera: Cicadellidae) on American sycamore. *Ecology 56*:1388–1397.

McClure, M. S., and P. W. Price. 1976. Ecotope characteristics of coexisting *Erythroneura* leaf hoppers (Homoptera: Cicadellidae) on sycamore. *Ecology 57*:928–940.

MacFadyen, A. 1963. *Animal Ecology*. Isaac Pitman and Sons, London.

McGowan, J. A. 1977. What regulates pelagic community structure in the Pacific. Pages 423–444 in *Oceanic Sound Scattering Prediction*, N. R. Anderson and B. J. Zahurenec (eds.). Plenum Press, N.Y.

McGowan, J. A., and P. W. Walker. 1979. Structure in the copepod community of the North Pacific central gyre. *Ecol. Monogr. 49*: 195–226.

McIntosh, R. P. 1980. The background and some current problems of theoretical ecology. *Synthèse 43*:195–256.

McNaughton, S. J. 1978. Stability and diversity of ecological communities. *Nature 274*:251–253.

McNaughton, S. J., and L. L. Wolf. 1973. *General Ecology*. Holt, Rinehart and Winston, N.Y.

Maelfait, J. P., L. Baert, J. Hublé, and A. De Kimpe. 1980. Life cycle timing, microhabitat preference and coexistence of spiders. Pages 69–73 in *Proc. 8th Int. Arachnol. Congr.*, Vienna, 1980.

Maguire, B., Jr., 1963. The passive dispersal of small aquatic organisms and their colonization of isolated bodies of water. *Ecol. Monogr. 43*: 161–185.

Maguire, B., Jr. 1970. Aquatic communities in bromeliad leaf axiles and the influence of radiation. Pages E95–E101 in *A Tropical Rain Forest*, H. T. Odum (ed.). U.S. Atomic Energy Commission, Oak Ridge, Tenn.

Maguire, B., Jr. 1971. Phytotelmata: biota and community structure determination in plant-held waters. *Ann. Rev. Ecol. Syst. 2*:439–466.

Maiorana, V. C. 1978. An explanation for ecological and developmental constants. *Nature 273*:375–377.

Mann, K. H. 1977. Destruction of kelp-beds by sea urchins. A cyclical phenomenon or irreversible degradation? *Helgolander wiss. Meeresunters. 30*:455–467.

Marchant, J. H. 1978. Bird population changes for the years 1975–76. *Bird Study 25*:245–252.

Markham, J. W. 1973. Observations on the ecology of *Laminaria sinclairii* on three northern Oregon beaches. *J. Phycol. 9*:336–341.

Martof, B. S., W. M. Palmer, J. R. Bailey, and J. R. Harrison. 1980. *Amphibians and Reptiles of the Carolinas and Virginia*. University of North Carolina Press, Chapel Hill, N.C.

May, R. M. 1972. Will a large complex system be stable? *Nature 238*: 413–414.

May, R. M. 1973. *Stability and Complexity in Model Ecosystems.* Princeton University Press, Princeton, N.J.

May, R. M. 1975. Patterns of species abundance and diversity. Pages 81–120 in *Ecology and Evolution of Communities*, M. L. Cody and J. M. Diamond (eds.). Harvard University Press, Cambridge, Mass.

May, R. M. 1978. The dynamics and diversity of insect faunas. Pages 188–204 in *Diversity of Insect Faunas*, L. A. Mound and N. Waloff (eds.). Blackwell, Oxford.

May, R. M. 1979. Fluctuations in abundance of tropical insects. *Nature 278*:505–507.

May, R. M., and R. M. Anderson. 1979. Population biology of infectious diseases: Part II. *Nature 280*:455–461.

May, R. M., J. R. Beddington, C. W. Clark, S. J. Holt, and R. M. Laws. 1979. Management of multispecies fisheries. *Science 205*:267–277.

May, R. M., J. R. Beddington, J. W. Horwood, and J. F. Shepherd. 1978. Exploiting natural populations in an uncertain world. *Math. Biosci. 42*:219–252.

May, R. M., and R. H. MacArthur. 1972. Niche overlap as a function of environmental variability. *Proc. Natl. Acad. Sci. USA 69*:1109–1113.

Maynard Smith, J. 1978. Optimization theory in evolution. *Ann. Rev. Ecol. Syst. 9*:31–56.

Mayr, E. 1965. Avifauna: turnover on islands. *Science 150*:1587–1588.

Meadows, P. S., and J. I. Campbell. 1972. Habitat selection by aquatic invertebrates. *Adv. Mar. Biol. 10*:271–382.

Menge, B. A. 1976. Organization of the New England rocky intertidal community: role of predation, competition, and environmental heterogeneity. *Ecol. Monogr. 46*:355–393.

Menge, B. A. 1978a. Predation intensity in a rocky intertidal community. Relation between predator foraging activity and environmental harshness. *Oecologia 34*:1–16.

Menge, B. A. 1978b. Predation intensity in a rocky intertidal community. Effect of an algal canopy, wave action and desiccation on predator feeding rates. *Oecologia 34*:17–35.

Menge, B. A. 1979. Coexistence between the seastars *Asterias vulgaris* and *A. forbesi* in a heterogeneous environment: a non-equilibrium explanation. *Oecologia 41*:245–272.

Menge, B. A., and J. P. Sutherland. 1976. Species diversity gradients: synthesis of the roles of predation, competition, and temporal heterogeneity. *Amer. Natur. 110*:351–369.

Menhinick, E. F. 1967. Structure, stability and energy flow in plants and arthropods in a *Serica lespedeza* stand. *Ecol. Monogr. 37*:255–272.

Merrill, R. J., and E. D. Hobson. 1970. Field observations of *Dendraster excentricus*, a sand dollar of western North America. *Am. Midl. Natur. 83*:595–624.

Miller, A. H., H. Friedmann, L. Griscom, and R. T. Moore. 1957. Distribution checklist of the birds of Mexico. Part II. Pacific Coast Avifauna, No. 33.

Mittelbach, G. G. 1981. Foraging efficiency and body size: a study of optimal diet and habitat use by bluegills. *Ecology 62*:1370–1386.

Miyashita, K. 1968a. Growth and development of *Lycosa T-insignita* BOES et Str. (Araneae: Lycosidae) under different feeding conditions. *Appl. Entomol. Zool. 3*:81–88.

Miyashita, K. 1968b. Quantitative feeding biology of *Lycosa T-insignita* BOES et Str. (Araneae: Lycosidae). *Bull. Nat. Inst. Agr. Sci., Ser. C. 22*:329–344.

Molles, M. C., Jr. 1978. Fish species diversity on model and natural reef patches: experimental insular biogeography. *Ecol. Monogr. 48*:289–305.

Mooney, H. A. (ed.). 1977. *Convergent Evolution in Chile and California*. Dowden, Hutchinson & Ross, Stroudsburg, Pa.

Mooney, H. A., O. T. Solbrig, and M. L. Cody. 1977. Introduction. Pages 1–12 in *Convergent Evolution in Chile and California*, H. A. Mooney (ed.). Dowden, Hutchinson & Ross, Stroudsburg, Pa.

Moore, H. B. 1939. The colonization of a new rocky shore at Plymouth. *J. Anim. Ecol. 8*:29–38.

Moran, M. J. 1980. The ecology, and effects on prey, of the predatory intertidal gastropod, *Morula marginalba*. Ph.D. Dissertation, University of Sydney.

Moreau, R. E. 1966. *The Bird Faunas of Africa and Its Islands*. Academic Press, N.Y.

Morin, P. J. 1981. Predatory salamanders reverse the outcome of competition among three species of larval anurans. *Science 212*:1284–1286.

Morrison, G., M. Auerbach, and E. D. McCoy. 1979. Anomalous diversity of tropical parasitoids: a general phenomenon? *Amer. Natur. 114*:303–307.

Morrison, G., and D. R. Strong. 1980. Spatial variations in egg density and the intensity of parasitism in a neotropical chrysomelid. *Ecol. Entomol. 6*:55–61.

Morse, D. H. 1971. The foraging of warblers isolated on small islands. *Ecology 52*:216–228.

Morse, D. H. 1977. The occupation of small islands by passerine birds. *Condor 79*:399–412.

Mosimann, J. E., and F. C. James. 1979. New statistical methods for allometry with application to Florida red-winged blackbirds. *Evolution 33*:444–459.

Mosimann, J. E., D. F. Sinclair, and D. A. Meeter. 1981. Contiguous size ratios and the randomness of island communities. *Evolution* (submitted).

Mosquin, T. 1971. Competition for pollinators as a stimulus for the evolution of flowering time. *Oikos 22*:398–402.

Moulder, B. C., and D. E. Reichle. 1972. Significance of spider predation in the energy dynamics of forest-floor arthropod communities. *Ecol. Monogr. 42*:473–498.

Moyer, J. T., and C. E. Sawyers. 1973. Territorial behavior of the anemonefish *Amphiprion xanthurus* with notes on the life history. *Jap. J. Ichthyol. 20*:85–93.

Munger, J. C., and J. H. Brown. 1981. Competition in desert rodents: an experiment with semipermeable exclosures. *Science 211*:510–512.

Murdoch, W. W. 1966. Community structure, population control, and competition—a critique. *Amer. Natur. 100*:219–226.

Murdoch, W. W. 1969. Switching in general predators: experiments on predator specificity and stability of prey populations. *Ecol. Monogr. 39*:335–354.

Murdoch, W. W. 1979. Predation and the dynamics of prey populations. *Fortschr. Zool. 25*:295–310.

Murray, B. G. 1971. The ecological consequences of interspecific territorial behavior in birds. *Ecology 52*:414–423.

Myers, A. C. 1977. Sediment processing in a marine subtidal sandy bottom community. I. Physical aspects. *J. Mar. Res. 35*:609–632.

Nagel, E. 1961. *The Structure of Science*. Routledge and Kegan Paul, London.

Neill, W. E. 1974. The matrix and the interdependence of the competition coefficients. *Amer. Natur. 108*:399–408.

Nelson, W. G. 1980. A comparative study of amphipods in seagrasses from Florida to Nova Scotia. *Bull. Mar Sci. 30*:80–89.

Neyman, J. 1977. Frequentist probability and frequentist statistics. *Synthèse 36*:97–131.

Nicholson, S. A., J. T. Scott, and A. R. Breisch. 1979. Structure and succession in the tree stratum at Lake George, New York. *Ecology 60*:1240–1254.

Nilsson, N. 1960. Seasonal fluctuations in the food segregation of trout, char and whitefish in 14 North-Swedish lakes. *Rep. Inst. Freshw. Res. Drottningholm 41*:185–205.

Nilsson, S. G. 1977. Density compensation and competition among birds breeding on small islands in a south Swedish lake. *Oikos 28*:170–176.

Noon, B. R. 1981. The distribution of an avian guild along a temperate elevational gradient: The importance and expression of competition. *Ecol. Monogr. 51*:105–124.

North, W. J. 1971. The biology of giant kelp beds (*Macrocystis*) in California. *Nova Hedwigia Z. Kryptogamented Suppl. 32*.

Northrop, F.S.C. 1959. *The Logic of the Sciences and Humanities*. Meridian Books, N.Y.

Nunney, L. 1980. The stability of complex model ecosystems. *Amer. Natur. 115*:639–649.

O'Connor, R. J. 1981. Comparison between migrant and nonmigrant birds in Britain. Society for Experimental Biology Symposium (in press).

Odum, H. T., and E. P. Odum. 1955. Trophic structure and productivity of a windward coral reef community on Eniwetok Atoll. *Ecol. Monogr. 25*:291–320.

Ogden, J. C., and P. R. Ehrlich. 1977. The behavior of heterotypic resting schools of juvenile grunts (Pomadasyidae). *Mar. Biol. 42*: 273–280.

Olive, C. W. 1980. Foraging specializations in orb-weaving spiders. *Ecology 61*:1133–1144.

Oliver, J. S. 1980. Processes affecting the organization of soft-bottom communities in Monterey Bay, California, and McMurdo Sound, Antarctica. Ph.D. Dissertation, University of California, San Diego.

Oliver, J. S., P. N. Slattery, L. W. Hulberg, and J. W. Nybakken. 1980. Relationships between wave disturbance and zonation of benthic invertebrate communities along a subtidal high-energy beach, Monterey Bay, California, *Fish. Bull. 78*:437–454.

Organ, J. A. 1961. Studies of the local distribution, life history, and population dynamics of the salamander genus *Desmognathus* in Virginia. *Ecol. Monogr. 31*:189–220.

Orians, G. H. 1980. *Some Adaptations of Marsh-nesting Blackbirds*. Princeton University Press, Princeton, N.J.

Orians, G. H., and G. Collier. 1963. Competition and blackbird social systems. *Evolution 17*:449–459.

Orians, G. H., and O. T. Solbrig (eds.). 1977. *Convergent Evolution in Warm Deserts*. Dowden, Hutchinson & Ross, Stroudsburg, Pa.

Orians, G. H., and M. F. Willson. 1964. Interspecific territories of birds. *Ecology 45*:736–745.

Orloci, L. 1967. An agglomerative method for classification of plant communities. *J. Ecol. 55*:193–206.

Osman, R. W. 1977. The establishment and development of a marine epifaunal community. *Ecol. Monogr. 47*:37–63.

Ostmark, E. 1974. Economic insect pests of bananas. *Ann. Rev. Ent. 19*:161–177.

Otte, D., and A. Joern. 1977. On feeding patterns in desert grasshoppers and the evolution of specialized diets. *Proc. Acad. Nat. Sci. Philadelphia 128*:89–126.

Page, C. N. 1976. The taxonomy and phytogeography of bracken—a review. *Bot. J. Linn. Soc. 73*:1–34.

Paine, R. T. 1966. Food web complexity and species diversity. *Amer. Natur. 100*:65–75.

Paine, R. T. 1969. The *Pisaster-Tegula* interaction: prey patches, predator food preference, and intertidal community structure. *Ecology 50*: 950–961.

Paine, R. T. 1971. A short-term experimental investigation of resource partitioning in a New Zealand rocky intertidal habitat. *Ecology 52*: 1096–1106.

Paine, R. T. 1974. Intertidal community structure: Experimental studies on the relationship between a dominant competitor and its principal predator. *Oecologia 15*:93–120.

Paine. R. T. 1977. Controlled manipulations in the marine intertidal zone, and their contributions to ecological theory. Pages 245–270 in *The Changing Scenes in Natural Sciences, 1776–1976*. Academy of Natural Sciences, Philadelphia, Special Publication 12.

Paine, R. T. 1980.. Food webs: linkage, interaction strength and community infrastructure. *J. Anim. Ecol. 49*:667–685.

Palmer, H. D. 1976. Erosion of submarine outcrops, La Jolla submarine canyon, California. *Geol. Soc. Amer. Bull. 87*:427–432.

Palmgren, P. 1930. Quantitative Untersuchungen über die Vogelfauna in den Walden Südfinnlands mit besonderer Berücksichtigung Ålands. *Acta Zool. Fenn. 7*:1–218.

Park, T. 1948. Experimental studies of interspecies competition. I. Competition between populations of the flour beetles, *Tribolium confusum* Duval and *Tribolium castaneum* Herbst. *Ecol. Monogr. 18*:265–308.

Parrish, J.A.D., and F. A. Bazzaz. 1979. Difference in pollination niche relationships in early and late successional plant communities. *Ecology 60*:597–610.

Parsons, T., and M. Takahashi. 1973. *Biological Oceanographic Processes*. Pergamon Press, N.Y.

Patterson, B. D. 1981. Morphological shifts of some isolated populations of *Eutamias* (Rodentia: Sciuridae) in different congeneric assemblages. *Evolution 35*:53–66.

Pearson, T. H., and R. Rosenberg. 1976. A comparative study of the effects on the marine environment of wastes from cellulose industries in Scotland and Sweden. *Ambio 5*:77–79.

Pearson, T. H., and R. Rosenberg. 1978. Macrobenthic succession in relation to organic enrichment and pollution of the marine environment. *Oceanogr. Mar. Biol. Ann. Rev. 16*:229–311.

Peet, R. K. 1978. Ecosystem convergence. *Amer. Natur. 112*:441–444.

Perkins, D. B. 1974. Arthropods that stress waterhyacinth. *PANS 20*:304–314.

Perkins, D. B. 1974. Arthropods that stress waterhyacinth. PANS *20*: 304–314.

Peters, J. L. 1945. *Check-list of Birds of the World*, Vol. V. Harvard University Press, Cambridge, Mass.

Peters, R. H. 1976. Tautology in evolution and ecology. *Amer. Natur. 110*:1–12.

Petersen, C.G.J. 1913. Valuation of the sea. II. The animal communities of the sea bottom and their importance for marine zoogeography. *Rep. Danish Biol. Stat. 21*:1–44.

Peterson, C. H. 1977. Competitive organization of the soft-bottom macrobenthic communities of southern California lagoons. *Mar. Biol. 43*:343–359.

Peterson, C. H. 1979. The important of predation and competition in organizing the intertidal epifaunal communities of Barnegat Inlet, New Jersey. *Oecologia 39*:1–24.

Peterson, C. H. 1980. Approaches to the study of competition in benthic communities in soft sediments. Pages 291–302 in *Estuarine Perspectives*, V. S. Kennedy (ed.). Academic Press, N.Y.

Peterson, C. H., and S. V. Andre. 1980. An experimental analysis of interspecific competition among marine filter feeders in a soft-sediment environment. *Ecology 61*:129–139.

Pianka, E. R. 1966. Latitudinal gradients in species diversity. *Amer. Natur. 100*:33–46.

Pianka, E. R. 1967. On lizard species diversity: North American flatland deserts. *Ecology 48*:333–351.

Pianka, E. R. 1969. Habitat specificity, speciation, and species density in Australian desert lizards. *Ecology 50*:498–502.

Pianka, E. R. 1971. Lizard species density in the Kalahari desert. *Ecology 52*:1024-1029.

Pianka, E. R. 1972. *r*- and *K*-selection or *b*- and *d*- selection? *Amer. Natur. 106*:581–588.

Pianka, E. R. 1974. *Evolutionary Ecology*. Harper and Row, N.Y.

Pianka, E. R. 1975. Niche relations of desert lizards. Pages 292–314 in *Ecology and Evolution of Communities*, M. L. Cody and J. M. Diamond (eds.). Harvard University Press, Cambridge, Mass.

Pianka, E. R. 1976. Competition and niche theory. Pages 114–141 in *Theoretical Ecology*, R. M. May (ed.). W. B. Saunders, Philadelphia.

Pianka, E. R. 1978. *Evolutionary Ecology*. Harper and Row, N.Y.

Piatt, J. 1935. A comparative study of the hyobranchial apparatus and throat musculature in the Plethodontidae. *J. Morphol. 57*:213–251.

Pielou, D. P., and E. C. Pielou. 1968. Association among species of infrequent occurrence: The insect and spider fauna of *Polyporus betulinus* (Bulliard) Fries. *J. Theor. Biol. 21*:202–216.

Pielou, E. C. 1972. 2^k contingency tables in ecology. *J. Theor. Biol.* *34*:337–352.

Pielou, E. C. 1977. *Mathematical Ecology*. Wiley, N.Y.

Pimm, S. L. 1979a. Complexity and stability: another look at MacArthur's original hypothesis. *Oikos 33*:351–357.

Pimm, S. L. 1979b. The structure of food webs. *Theor. Pop. Biol. 16*: 144–158.

Pimm, S. L. 1980a. The bounds on food web connectance. *Nature 285*:511.

Pimm, S. L. 1980b. The properties of food webs. *Ecology 61*:219–225.

Pimm, S. L. 1982. *Food Webs*. Chapman and Hall, London.

Pimm, S. L. 1983. Monte-Carlo analyses in population and community ecology. In *Lizard Ecology: Studies on a Model Organism*, R. B. Huey, E. R. Pianka, and T. W. Schoener (eds.). Harvard University Press, Cambridge, Mass. (in press).

Pimm, S. L., and J. H. Lawton. 1977. Number of trophic levels in ecological communities. *Nature 268*:329–331.

Pimm, S. L., and J. H. Lawton. 1978. On feeding on more than one trophic level. *Nature 275*:542–544.

Pimm, S. L., and J. H. Lawton. 1980. Are food webs divided into compartments? *J. Anim. Ecol. 49*:879–898.

Pitelka, F. A. 1942. Territoriality and related problems in North American hummingbirds. *Condor 44*:189–204.

Pitelka, F. A. 1951. Ecological overlap and interspecific strife in breeding populations of Anna and Allen hummingbirds. *Ecology 32*:641–661.

Platt, J. R. 1964. Strong inference. *Science 146*:347–353.

Pleasants, J. M. 1980. Competition for bumblebee pollinators in Rocky Mountain plant communities. *Ecology 61*:1446–1459.

Poole. R. W. 1978. Some relative characteristics of population fluctuations. Pages 18–33 in *Time Series and Ecological Processes*, H. H. Shugart, Jr. (ed.). SIAM, Philadelphia.

Poole, R. W., and B. J. Rathcke. 1979. Regularity, randomness, and aggregation in flowering phenologies. *Science 203*:470–471.

Popper, K. R. 1959. *The Logic of Scientific Discovery*. Hutchinson, London.

Popper, K. R. 1963. *Conjectures and Refutations: The Growth of Scientific Knowledge*. Harper and Row, N.Y.

Popper, K. R. 1968. *The Logic of Scientific Discovery*. Harper and Row, N.Y.

Post, W. M., III, and S. E. Riechert. 1977. Initial investigation into the structure of spider communities. *J. Anim. Ecol. 46*:729–750.

Powders, V. N., and W. L. Tietjen. 1974. The comparative food habits of sympatric and allopatric salamanders, *Plethodon glutinosus* and

Plethodon jordani in eastern Tennessee and adjacent areas. *Herpeto-logica 30*:167–175.

Power, D. M. 1975. Similarity among avifaunas of the Galápagos islands. *Ecology 56*:616–626.

Preston, F. W. 1962. The canonical distribution of commonness and rarity. *Ecology 43*:185–215; 410–432.

Price, J. H., E.G.G. Irvine, and W. F. Farnham. 1980. *The Shore Environment*. Volume 2: *Ecosystems*. The Systematic Association Special Volume No. 17(a). Academic Press, N.Y.

Price, M. V. 1978. The role of microhabitat in structuring desert rodent communities. *Ecology 59*:910–921.

Price, P. W. 1971. Niche breadth and dominance of parasitic insects sharing the same host species. *Ecology 52*:587–596.

Price, P. W. 1976. Colonization of crops by arthropods: Non-equilibrium communities in soybean fields. *Env. Entomol. 5*:605–611.

Price, P. W. 1977. General concepts on the evolutionary biology of parasites. *Evolution 31*:405–420.

Price, P. W. 1980. *Evolutionary Biology of Parasites*. Princeton University Press, Princeton, N.J.

Price, P. W., C. E. Bouton, P. Grass, B. A. McPheron, J. N. Thompson, and A. E. Weis. 1980. Interactions among three trophic levels: influence of plants on interactions between insect herbivores and natural enemies. *Ann. Rev. Ecol. Syst. 11*:41–65.

Price, P. W., and M. F. Willson. 1979. Abundance of herbivores on six milkweed species in Illinois. *Amer. Midl. Natur. 101*:76–86.

Putwain, P. D., and J. L. Harper. 1970. Studies in the dynamics of plant populations. III. The influence of associated species on populations of *Rumex acetosa* L. and *R. acetosella* L. in grassland. *J. Ecol. 58*: 251–264.

Rabinowitz, D., J. K. Rapp, V. L. Sork, B. J. Rathcke, G. A. Reese, and J. C. Weaver. 1981. Phenological properties of wind- and insect-pollinated plants. *Ecology 62*:49–56.

Ramensky, L. G. 1926. Die Grundgesetzmässigkeiten im Aufbau der Vegetationsdecke. *Bot. Centralblatt. N. F. 7*:453–455.

Rand, A. L., and E. T. Gilliard. 1967. *Handbook of New Guinea Birds*. Weidenfield and Nicolson, London.

Randall, J. E. 1961. Overgrazing of algae by herbivorous marine fishes. *Ecology 42*:812.

Randall, J. E. 1963. An analysis of the fish populations of artificial and natural reefs in the Virgin Islands. *Caribb. J. Sci. 3*:31–47.

Randall, J. E. 1965. Grazing effects on sea grasses by herbivorous reef fishes in the West Indies. *Ecology 46*:255–260.

Ratcliffe, L. M. 1981. Species recognition in Darwin's Ground Finches (*Geospiza* Gould). Ph.D. Dissertation, McGill University.

Rathcke, B. J. 1976a. Competition and coexistence within a guild of herbivorous insects. *Ecology 57*:76–87.

Rathcke, B. J. 1976b. Insect-plant patterns and relationships in the stem-boring guild. *Amer. Midl. Natur. 99*:98–117.

Rathcke, B. J., and R. W. Poole. 1977. Community patterns of flowering phenologies. *Bull. Ecol. Soc. Amer. 58*:15 (abstract).

Raup, D. M., and S. J. Gould. 1974. Stochastic simulation and evolution of morphology—towards a nomothetic paleontology. *Syst. Zool. 23*:305–332.

Raup, D. M., S. J. Gould, T.J.M. Schopf, and D. S. Simberloff. 1973. Stochastic models of phylogeny and the evolution of diversity. *J. Geol. 81*:525–542.

Reader, R. J. 1975. Competitive relationships of some bog ericads for major insect pollinators. *Can. J. Bot. 53*:1300–1305.

Reichman, O. J. 1979a. Concluding remarks. *Amer. Zool. 19*:1173–1175.

Reichman, O. J. 1979b. Factors influencing foraging patterns in desert rodents. Pages 195–213 in *Mechanisms of Optimal Foraging*, A. Kamil and T. Sargent (eds.). Garland, N.Y.

Reinboth, R. 1973. Dualistic reproductive behavior in the protogynous wrasse *Thalassoma bifasciatum* and some observations on its day-night changeover. *Helgolander wiss. Meeresunters. 24*:174–191.

Reiswig, H. M. 1973. Population dynamics of three Jamaican demospongiae. *Bull. Mar. Sci. 23*:191–226.

Rejmánek, M., and P. Starý. 1979. Connectance in real biotic communities and critical values for stability of model ecosystems. *Nature 280*:311–313.

Rey, J. R. 1981. Ecological biogeography of arthropods on *Spartina* islands in northwest Florida. *Ecol. Monogr. 51*:237–265.

Rey, J. R., and E. D. McCoy. 1979. Application of island biogeographic theory to pests of cultivated crops. *Env. Entomol. 8*:577–582.

Rey, J. R., E. D. McCoy, and D. R. Strong. 1981. Herbivore pests, habitat islands, and the species-area relation. *Amer. Natur. 117*:611–622.

Reynoldson, T. B., and L. S. Bellamy. 1971. The establishment of interspecific competition in field populations, with an example of competition in action between *Polycelis nigra* (Mull.) and *P. tenuis* (Ijima) (Turbellaria, Tricladida). Pages 282–297 in *Proc. Adv. St. Inst. Dyn. Nbrs. Pop.* (Oosterbeck, 1970).

Rhoads, D. C. 1974. Organism-sediment relations on the muddy seafloor. *Oceanogr. Mar. Biol. Ann. Rev. 12*:263–300.

Rhoads, D. C., P. L. McCall, and J. Y. Yingst. 1978. Disturbance and production on the estuarine seafloor. *Amer. Sci. 66*:577–586.

Rhoads, D. C., and D. K. Young. 1970. The influence of deposit-feeding organisms on sediment stability and community trophic structure. *J. Mar. Res. 28*:150–178.

Rhoads, D. C., and D. K. Young. 1971. Animal-sediment relations in Cape Cod Bay, Massachusetts. II. Reworking by *Molpadia oolitica* (Holothuroidea). *Mar. Biol. 11*:255–261.

Rice, J. C. 1978. Behavioral interactions of interspecifically territorial vireos. I. Song discrimination and natural interactions. *Anim. Behav. 26:*527–549.

Rice, L. A. 1935. Factors controlling arrangement of barnacle species in tidal communities. *Ecol. Monogr. 5*:293–304.

Richmond, R. C., M. E. Gilpin, S. Perez Salas, and F. J. Ayala. 1975. A search for emergent competitive phenomena: the dynamics of multi-species *Drosophila* systems. *Ecology 56*:709–714.

Ricketts, E. F., J. Calvin, and J. W. Hedgpeth. 1968. *Between Pacific Tides*, 4th ed. Stanford University Press, Stanford, Calif.

Ricklefs, R. E. 1979. *Ecology*. Chiron Press, Newton, Mass.

Ricklefs, R. E., and G. W. Cox. 1977. Morphological similarity and ecological overlap among passerine birds on St. Kitts, British West Indies. *Oikos 29*:60–66.

Ricklefs, R. E., and J. Travis. 1980. A morphological approach to the study of avian community organization. *Auk 97*:321–338.

Ridgway, R., and H. Friedmann. 1901–1950. The Birds of North and Middle America. *Bull. U.S. Nat. Mus. 50* (parts 1–12).

Riechert, S. E. 1974. Thoughts on the ecological significance of spiders. *BioScience 24*: 352—356.

Riechert, S. E. 1977. Games spiders play: behavioral variability in territorial disputes. *Behav. Ecol. Sociobiol. 4*:2–39.

Riechert, S. E. 1978. Energy-based territoriality in populations of the desert spider *Agelenopsis aperta* (Gertsch). *Symp. Zool. Soc. Lond. 42*:211–222.

Riechert, S. E. 1979. Games spiders play. II. Resource assessment strategies. *Behav. Ecol. Sociobiol. 6*:121–128.

Riechert, S. E. 1981. The consequences of being territorial: spiders, a case study. *Amer. Natur. 117*:871–892.

Riechert, S. E., and A. B. Cady. 1983. Patterns of resource use and tests for competitive release in a spider community. *Ecology 64*:899–913.

Rigby, C., and J. H. Lawton. 1981. Species-area relationships of arthropods on host plants: herbivores on bracken. *J. Biogeography 8*: 125–133.

Robbins, C. S., E. F. Connor, D. Buchwald, P. Wagner, and A. Geis. 1971. Upland tulip-tree-maple-oak forest. *American Birds 25*:971.

Robertson, C. 1895. The philosophy of flower seasons, and the phenological relations of the entomophilous flora and the anthophilous insect fauna. *Amer. Natur. 29*:97–117.

Robertson, D. R., N.V.C. Polunin, and K. Leighton. 1979. The behavioural ecology of three Indian Ocean surgeonfishes (*Acanthurus lineatus, A. leucosternon,* and *Zebrasoma scopas*): their feeding strategies, and social and mating systems. *Env. Biol. Fish. 4*:125–170.

Robertson, D. R., and J. M. Sheldon. 1979. Competitive interactions and the availability of sleeping sites for diurnal coral reef fish. *J. Exp. Mar. Biol. Ecol. 40*:285–298.

Rohde, K. 1978a. Latitudinal differences in host-specificity of marine Monogenea and Digenea. *Mar. Biol. 47*:125–134.

Rohde, K. 1978b. Latitudinal gradients in species diversity and their causes. II. Marine parasitological evidence for a time hypothesis. *Biol. Zbl. 97*:405–418.

Rohde, K. 1979. A critical evaluation of intrinsic and extrinsic factors responsible for niche restriction in parasites. *Amer. Natur. 114*:648–671.

Rohde, K. 1980. Comparative studies on microhabitat utilization by ectoparasites of some marine fishes from the North Sea and Papua New Guinea. *Zool. Anz. 204*:27–63.

Root, R. B. 1967. The niche exploitation pattern of the Blue-gray Gnatcatcher. *Ecol. Monogr. 37*:317–350.

Root, R. B. 1973. Organization of a plant-arthropod association in simple and diverse habitats: The fauna of collards (*Brassica oleracea*). *Ecol. Monogr. 43*:95–124.

Rosenzweig, M. L. 1973. Habitat selection experiments with a pair of coexisting heteromyid rodent species. *Ecology 54*:111–117.

Rosenzweig, M. L. 1979. Optimal habitat selection in two-species competitive systems. *Fortschr. Zool. 25*:283–293.

Rosenzweig, M. L., and P. Sterner. 1970. Population ecology of desert rodent communities: body size and seed husking as bases for heteromyid coexistence. *Ecology 51*:217–224.

Rosenzweig, M. L., and J. Winakur. 1969. Population ecology of desert rodent communities: habitats and environmental complexity. *Ecology 50*:558–572.

Roskam, J. C., and G. A. van Uffelen. 1981. Biosystematics of insects living in female birch catkins. III. Plant-insect relation between white birches, *Betula* L., Section *Excelsae* (Koch) and gall midges of the genus *Semudobia* Kieffer (Diptera, Cecidomyiidae). *Neth. J. Zool. 31*:533–553.

Ross, R. M. 1978. Territorial behavior and ecology of the anemonefish *Amphiprion melanopus* on Guam. *Z. Tierpsychol. 46*:71–83.

Rotenberry, J. T. 1980a. Bioenergetics and diet in a simple community of shrubsteppe birds. *Oecologia 46*:7–12.

Rotenberry, J. T. 1980b. Dietary relationships among shrub-steppe passerine birds: competition or opportunism in a variable environment. *Ecol. Monogr. 50*:93–110.

Rotenberry, J. T., and J. A. Wiens. 1978. Nongame bird communities in northwestern rangelands. *Proc. Workshop Nongame Bird Habitat Mgmt. Coniferous Forest Western U.S.* USDA For. Serv. Gen. Tech. Rept. PNW-*64*:32–46.

Rotenberry, J. T., and J. A. Wiens. 1980a. Habitat structure, patchiness, and avian communities in North American steppe vegetation: a multivariate analysis. *Ecology 61*:1228–1250.

Rotenberry, J. T., and J. A. Wiens. 1980b. Temporal variation in habitat structure and shrubsteppe bird dynamics. *Oecologia 47*:1–9.

Roughgarden, J. 1974. Species packing and the competition function with illustrations from coral reef fish. *Theor. Pop. Biol. 5*:163–186.

Roughgarden, J. 1976. Resource partitioning among competing species— a coevolutionary approach. *Theor. Pop. Biol. 9*:388–424.

Roughgarden, J. 1979. A local concept of structural homology for ecological communities with examples from simple communities of West Indian *Anolis* lizards. *Fortschr. Zool. 25*:149–157.

Roughgarden, J. 1979. *Theory of Population Genetics and Evolutionary Ecology: An Introduction.* Macmillan, N.Y.

Ruse, M. 1977. Karl Popper's philosophy of biology. *Phil. Sci. 44*:638–661.

Ruse, M. 1979. Falsifiability, consilience, and systematics. *Syst. Zool. 29*: 530–536.

Russell, B. C., G.R.V. Anderson, and F. H. Talbot. 1977. Seasonality and recruitment of coral reef fishes. *Aust. J. Mar. Freshw. Res. 28*:521–528.

Russell, B. C., F. H. Talbot, G.R.V. Anderson, and B. Goldman. 1978. Collection and sampling of reef fishes. Pages 329–345 in *Monographs on Oceanographic Methodology*, 5, *Coral Reefs: Research Methods*, D. R. Stoddart and R. E. Johannes (eds.). UNESCO, Norwich, U. K.

Russell, B. C., F. H. Talbot, and S. Domm. 1974. Patterns of colonisation of artificial reefs by coral reef fishes. Pages 207–215 in *Proceedings 2nd Intern. Coral Reef Symp.*, Vol. 1, Great Barrier Reef Comm., Brisbane.

Rützler, K. 1970. Spatial competition among porifera: solution by epizoism. *Oecologia 5*:85–95.

Rützler, K. 1975. The role of burrowing sponges in bioerosion. *Oecologia 19*:203–216.

Rymer, L. 1976. The history of ethnobotany of bracken. *Bot. J. Linn. Soc. 73*:151–176.

Rypstra, A. L. 1979. Foraging flocks of spiders. A study of aggregate behavior in *Cyrtophora citricola* Forskal (Araneae: Araneidae) in West Africa. *Behav. Ecol. Sociobiol.* 5:291–300.

Rypstra, A. L. 1981. The effect of kleptoparasites on prey consumption and web relocation in a Peruvian population of the spider, *Nephila clavipes* (L.) (Araneae: Araneidae). *Oikos 37*:179–182.

Sale, P. F. 1972. Influence of corals in the dispersion of the pomacentrid fish, *Dascyllus aruanus. Ecology 53*:741–744.

Sale, P. F. 1974. Mechanisms of coexistence in a guild of territorial fishes at Heron Island. Pages 193–206 in *Proceedings 2nd Intern. Coral Reef Symp.*, Vol. 1, Great Barrier Reef Comm., Brisbane.

Sale, P. F. 1975. Patterns of use of space in a guild of territorial reef fishes. *Mar. Biol. 29*:89–97.

Sale, P. F. 1976. The effect of territorial adult pomacentrid fishes on the recruitment and survivorship of juveniles on patches of coral rubble. *J. Exp. Mar. Biol. Ecol. 24*:297–306.

Sale, P. F. 1977. Maintenance of high diversity in coral reef fish communities. *Amer. Natur. 111*:337–359.

Sale, P. F. 1978a. Chance patterns of demographic change in populations of territorial fish in coral rubble patches at Heron Reef. *J. Exp. Mar. Biol. Ecol. 34*:233–243.

Sale, P. F. 1978b. Coexistence of coral reef fishes—a lottery for living space. *Env. Biol. Fish. 3*:85–102.

Sale, P. F. 1979. Recruitment, loss and coexistence in a guild of territorial coral reef fishes. *Oecologia 42*:159–177.

Sale, P. F. 1980a. Assemblages of fish on patch reefs—predictable or unpredictable? *Env. Biol. Fish. 5*:243–249.

Sale, P. F. 1980b. The ecology of fishes on coral reefs. *Oceanogr. Mar. Biol. Ann. Rev. 18*:367–421.

Sale, P. F., and W. A. Douglas. 1981. Precision and accuracy of visual census technique for fish assemblages on coral patch reefs. *Env. Biol. Fish. 6*:333–339.

Sale, P. F., and R. Dybdahl. 1975. Determinants of community structure for coral reef fishes in an experimental habitat. *Ecology 56*:1343–1355.

Sale, P. F., and R. Dybdahl. 1978. Determinants of community structure for coral reef fishes in isolated coral heads at lagoonal and reef slope sites. *Oecologia 34*:57–74.

Sanders, H. L. 1958. Benthic studies in Buzzards Bay. I. Animal-sediment relationships. *Limnol. Oceanogr. 3*:245–258.

Sanders, H. L. 1969. Benthic marine diversity and the stability time hypothesis. *Brookhaven Symp. Biol. 22*:71–81.

Santelices, B., J. C. Castilla, J. Cancino, and P. Schmiede. 1980. Comparative ecology of *Lessonia Nigrescens* and *Durvillaea antarctica* (Phaeophyta) in central Chile. *Mar. Biol. 59*:119–132.

Santos, S. L., and J. L. Simon. 1980. Response of soft-bottom benthos to annual catastrophic disturbance in a south Florida estuary. *Mar. Ecol. Prog. Ser. 3*:347–355.

Sarker, A. L. 1977. Feeding ecology of the bluegill, *Lepomis macrochirus*, in two heated reservoirs of Texas. III. Time of day and patterns of feeding. *Trans. Am. Fish. Soc. 106*:596–601.

Saunders, P. T. 1978. Population dynamics and the length of food chains. *Nature 272*:189.

Schaefer, M. 1975. Experimental studies on the importance of interspecies competition for the lycosid spiders in a salt marsh. Pages 86–90 in *Proc. 6th Int. Arachnol. Congr.*, Amsterdam, 1974.

Schaefer, M. 1978. Some experiments on the regulation of population density in the spider *Floronia bucculenta* (Araneida: Linyphiidae). *Symp. Zool. Soc. Lond. 42*:203–210.

Schluter, D. 1982a. Distributions of Galápagos ground finches along an altitudinal gradient: the importance of food supply. *Ecology 63*: 1504–1517.

Schluter, D. 1982b. Seed and patch selection by Galápagos ground finches: relation to foraging efficiency and food supply. *Ecology 63*: 1106–1120.

Schluter, D. A variance ratio test for detecting species associations, with some example applications. Manuscript.

Schluter, D., and P. R. Grant. 1982. The distribution of *Geospiza difficilis* on Galápagos islands: tests of three hypotheses. *Evolution 36*:1213–1226.

Schoener, A., and T. W. Schoener. 1981. The dynamics of the species-area relation in marine fouling systems: 1. Biological correlates of changes in the species-area slope. *Amer. Natur. 118*:339–360.

Schoener, T. W. 1965. The evolution of bill size differences among sympatric species of birds. *Evolution 19*:189–213.

Schoener, T. W. 1969a. Models of optimal size for solitary predators. *Amer. Natur. 103*:277–313.

Schoener, T. W. 1969b. Size patterns in West Indian *Anolis* lizards: I. Size and species diversity. *Syst. Zool. 18*:386–401.

Schoener, T. W. 1970. Size patterns in West Indian *Anolis* lizards: II. Correlation with the sizes of particular sympatric species—displacement and convergence. *Amer. Natur. 104*:155–174.

Schoener, T. W. 1974a. Competition and the form of habitat shift. *Theor. Pop. Biol. 6*:265–307.

Schoener, T. W. 1974b. Resource partitioning in ecological communities. *Science 185*:27–39.

Schoener, T. W. 1975. Presence and absence of habitat shift in some widespread lizard species. *Ecol. Monogr. 45*:233–258.

Schoener, T. W. 1976. Alternatives to Lotka-Volterra competition: models of intermediate complexity. *Theor. Pop. Biol. 10*:309–333.

Schoener, T. W. 1977. Competition and the niche. Pages 35–136 in *Biology of the Reptilia*, Vol. 7, C. Gans and D. W. Tinkle (eds). Academic Press, N.Y.

Schoener, T. W., and D. H. Janzen. 1968. Notes on environmental determinants of tropical versus temperate insect size patterns. *Amer. Natur. 102*:207–224.

Schonbeck, M., and T. A. Norton. 1978. Factors controlling the upper limits of fucoid algae on the shore. *J. Exp. Mar. Biol. Ecol. 31*:303–314.

Schroder, G. D., and M. L. Rosenzweig. 1975. Perturbation analysis of competition and overlap in habitat utilization between *Dipodomys ordii* and *Dipodomys merriami. Oecologia 19*:9–28.

Schultz, J. C., D. Otte, and F. Enders. 1977. *Larrea* as a habitat component for desert arthropods. Pages 176–207 in *Creosote Bush*, T. J. Mabry, J. H. Hunziker, and D. R. Difeo, Jr. (eds.). Dowden, Hutchinson & Ross, Stroudsburg, Pa.

Scott, H. 1914. The fauna of "reservoir-plants." *Zoologist 18*:183–195.

Scriber, J. M. 1973. Latitudinal gradients in larval feeding specialization of the World Papilionidae (Lepidoptera). *Psyche 80*:355–373.

Seaburg, K. G., and J. B. Moyle. 1964. Feeding habits, digestion rates, and growth of some Minnesota warm water fishes. *Trans. Am. Fish. Soc. 93*:269–285.

Seaward, M.R.D. 1976. Observations on the bracken component of the pre-Hadrianic deposits at Vindolanda, Northumberland. *Bot. J. Linn. Soc. 73*:177–185.

Seed, R. 1969. The ecology of *Mytilus edulis* L. (Lamellibranchiata) on exposed rocky shores. 1. Breeding and settlement. *Oecologia 3*:277–316.

Seifert, R. P. 1975. Clumps of *Heliconia* inflorescences as ecological islands. *Ecology 56*:1416–1422.

Seifert, R. P. 1980. Mosquito fauna of *Heliconia aurea. J. Anim. Ecol. 49*:687–697.

Seifert, R. P. 1981. Principal components analysis of biogeographic patterns among *Heliconia* insect communities. *J. N.Y. Entomol. Soc. 89*:109–122.

Seifert, R. P. Neotropical *Heliconia* insect communities. *Quart. Rev. Biol.* (in press).

Seifert, R. P., and R. Barrera R. 1981. Cohort studies on mosquito (Diptera: Culicidae) larvae living in the water-filled floral bracts of *Heliconia aurea* (Zingiberales: Musaceae). *Ecol. Entomol. 6*:191–197.

Seifert, R. P., and F. H. Seifert. 1976a. A community matrix analysis of *Heliconia* insect communities. *Amer. Natur. 110*:461–483.

Seifert, R. P., and F. H. Seifert. 1976b. Natural history of insects living on inflorescences of two species of *Heliconia. J. N.Y. Entomol. Soc. 84*:233–242.

Seifert, R. P., and F. H. Seifert. 1979a. A *Heliconia* insect community in a Venezuelan cloud forest. *Ecology 60*:462–467.

Seifert, R. P., and F. H. Seifert. 1979b. Utilization of *Heliconia* (Musaceae) by the beetle *Xenarescus monocerus* (Oliver) (Chrysomelidae: Hispinae) in a Venezuelan forest. *Biotropica 11*:51–59.

Selvin, H. C., and A. Stuart. 1966. Data-dredging procedures in survey analysis. *Amer. Statist. 20*:20–23.

Shepherd, J. G., and D. H. Cushing. 1980. A mechanism for density-dependent survival of larval fish as the basis of a stock-recruitment relationship. *J. Con. Int. Explor. Mer. 39*:160–167.

Sheppard, D. H. 1971. Competition between two chipmunk species (*Eutamias*). *Ecology 52*:320–329.

Simberloff, D. 1970. Taxanomic diversity of island biotas. *Evolution 24*: 23–47.

Simberloff, D. 1974. Equilibrium theory of island biogeography and ecology. *Ann. Rev. Ecol. Syst. 5*:161–182.

Simberloff, D. 1976a. Experimental zoogeography of islands: Effects of island size. *Ecology 57*:629–648.

Simberloff, 1976b. Species turnover and equilibrium island biogeography. *Science 194*:572–578.

Simberloff, D. 1976c. Trophic structure determination and equilibrium in an arthropod community. *Ecology 57*:395–398.

Simberloff, D. 1978a. Using island biogeographic distributions to determine if colonization is stochastic. *Amer. Natur. 112*:713–726.

Simberloff, D. 1978b. Use of rarefaction and related methods in ecology. Pages 150–165 in *Biological Data in Water Pollution Assessment: Quantitative and Statistical Analyses*, K. L. Dickson, J. Cairns, Jr., and R. J. Livingston (eds.). ASTM, Philadelphia.

Simberloff, D. 1979. Rarefaction as a distribution-free method of expressing and estimating diversity. Pages 159–176 in *Ecological Diversity in Theory and Practice*, J. F. Grassle, G. P. Patil, W. K. Smith, and C. Taillie (eds.). International Cooperative Publishing House, Fairland, Md.

Simberloff, D. 1980. A succession of paradigms in ecology: essentialism to materialism and probabilism. *Synthèse 43*:3–39.

Simberloff, D. 1981. Community effects of introduced species. Pages 53–83 in *Biotic Crises in Ecological and Evolutionary Time*, M. H. Nitecki (ed.). Academic Press, N.Y.

Simberloff, D., and W. Boecklen. 1981. Santa Rosalia reconsidered: Size ratios and competition. *Evolution 35*:1206–1228.

Simberloff, D., and E. F. Connor. 1979. Q-mode and R-mode analyses of biogeographic distributions: Null hypotheses based on random colonization. Pages 123–138 in *Contemporary Quantitative Ecology and Related Ecometrics*, G. P. Patil and M. L. Rosenzweig (eds.). International Cooperative Publishing House, Fairland, Md.

Simberloff, D., and E. F. Connor. 1981. Missing species combinations. *Amer. Natur. 118*:215–239.

Simberloff, D., K. L. Heck, E. D. McCoy, and E. F. Connor. 1981. There have been no statistical tests of cladistic biogeographic hypotheses. Pages 40–63 in *Vicariance Biogeography: A Critique*, G. Nelson and D. E. Rosen (eds.). Columbia University Press, N.Y.

Simberloff, D., and E. O. Wilson. 1969. Experimental zoogeography of islands: The colonization of empty islands. *Ecology 50*:278–296.

Simberloff, D., and E. O. Wilson. 1970. Experimental zoogeography of islands: A two-year record of colonization. *Ecology 51*:934–937.

Slobodkin, L. B. 1961. *Growth and Regulation of Animal Populations*. Holt, Rinehart, and Winston, N.Y.

Slobodkin, L. B., and H. L. Sanders. 1969. On the contribution of environmental predictability to species diversity. *Brookhaven Symp. Biol. 22*: 82–93.

Slobodkin, L. B., and F. E. Smith, and N. G. Hairston. 1967. Regulation in terrestrial ecosystems, and the implied balance of nature. *Amer. Natur. 101*:109–124.

Smith, C. L. 1977. Coral reef fish communities—order and chaos. Pages xxi–xxii in *Proceedings 3rd Intern. Coral Reef Symp.*, University of Miami, Miami, Fla.

Smith, C. L. 1978. Coral reef fish communities: A compromise view. *Env. Biol. Fish 3*:109–128.

Smith, C. L., and J. C. Tyler. 1972. Space resource sharing in a coral reef fish community. *Nat. Hist. Mus. Los Angeles County Sci. Bull. 14*: 125–170.

Smith, C. L., and J. C. Tyler. 1973. Population ecology of a Bahamian suprabenthic shorefish assemblage. *Amer. Mus. Novitates 2528*: 1–38.

Smith, C. L., and J. C. Tyler. 1975. Succession and stability in fish communities of dome-shaped patch reefs in the West Indies. *Amer. Mus. Novitates 2572*:1–18.

Smith, D. R., and J. H. Lawton. 1980. Review of the sawfly genus *Eriocampidea* (Hymenoptera: Tenthredinidae). *Proc. Ent. Soc. Wash. 82*:447–453.

Smith, J.N.M., P. R. Grant, B. R. Grant, I. Abbott, and L. K. Abbott.

1978. Seasonal variation in feeding habits of Darwin's ground finches. *Ecology 59*:1137–1150.

Smith, O. L. 1980. The influence of environmental gradients on ecosystem stability. *Amer. Natur. 116*:1–24.

Smith-Gill, S. J., and D. E. Gill. 1978. Curvilinearities in the competition equations: an experiment with ranid tadpoles. *Amer. Natur. 112*: 557–570.

Snedecor, G. W., and W. G. Cochran. 1967. *Statistical Methods*, 6th ed. Iowa State University Press, Ames, Iowa.

Snow, B. K., and D. W. Snow. 1972. Feeding niches of hummingbirds in a Trinidad valley. *J. Anim. Ecol. 41*:471–485.

Snyder, N.F.R., and J. W. Wiley. 1976. Sexual size dimorphism in hawks and owls of North America. *Ornithol. Monogr. 20.*

Soulé, M. E., and B. A. Wilcox (eds.). 1980. *Conservation Biology: An Evolutionary-Ecological Perspective.* Blackwell, Oxford, and Sinauer, Sunderland, Mass.

Sousa, W. P. 1979a. Disturbance in marine intertidal boulder fields: the nonequilibrium maintenance of species diversity. *Ecology 60*:1225–1239.

Sousa, W. P. 1979b. Experimental investigations of disturbance and ecological succession in a rocky intertidal algal community. *Ecol. Monogr. 49*:227–254.

Soutar, A., and J. D. Isaacs. 1974. Abundance of pelagic fish during the 19th and 20th centuries as recorded in anaerobic sediments off the Californias. *Fish. Bull. 72*:257–273.

Southwood, T.R.E. 1976. Bionomic strategies and population parameters. Pages 26–48 in *Theoretical Ecology*, R. M. May (ed.). W. B. Saunders, Philadelphia.

Southwood, T.R.E. 1978. The components of diversity. Pages 19–40 in *Diversity of Insect Faunas*, L. A. Mound and N. Waloff (eds.). *Symp. Roy. Ent. Soc. London* 9.

Southwood, T.R.E. 1980. Ecology—a mixture of pattern and probabilism. *Synthèse 43*:111–122.

Spencer, K. A. 1972. Handbooks for the identification of British insects. Vol. 10. Part 5(g). Diptera, Agromyzidae. Roy. Ent. Soc. London.

Spight, T. M. 1975. Factors extending gastropod embryonic development and their selective cost. *Oecologia 21*:1–16.

Springett, B. P. 1968. Aspects of the relationship between burying beetles, *Necrophorus* spp., and the mite *Poecilochus necrophori* Vitz. *J. Anim. Ecol. 37*:417–424.

States, J. B. 1976. Local adaptations in chipmunk (*Eutamias amoenus*) populations and evolutionary potential at species borders. *Ecol. Monogr. 46*:221–256.

Steele, G. D., and J. H. Torre. 1960. *Principles and Procedures of Statistics*. McGraw Hill, N.Y.

Steele, J. H. 1974. *The Structure of Marine Ecosystems*. Harvard University Press, Cambridge, Mass.

Stephenson, T. A., and A. Stephenson. 1972. *Life between Tidemarks on Rocky Shores*. Freeman and Co., San Francisco, Calif.

Stephenson, W., and R. B. Searles. 1960. Experimental studies on the ecology of intertidal environments at Heron Island. I. Exclusion of fish from beach rock. *Aust. J. Mar. Freshw. Res. 11*:241–267.

Stewart, J. G. 1982. Anchor species and epiphytes in intertidal algal turf. *Pacific Science 36*:45–59.

Stewart, R. E., and J. W. Aldrich. 1951. Removal and repopulation of breeding birds in a spruce-fir forest community. *Auk 68*:471–482.

Stiles, F. G. 1977. Coadapted competitors: the flowering seasons of hummingbird pollinated plants in a tropical forest. *Science 198*:1177–1178.

Stiles, F. G. 1978. Temporal organization of flowering among the hummingbird foodplants of a tropical wet forest. *Biotropica 10*:194–210.

Stiles, F. G. 1979. Regularity, randomness, and aggregation in flowering phenologies. Reply to Poole and Rathcke. *Science 203*:471.

Stiles, F. G., and L. L. Wolf. 1970. Hummingbird territoriality at a tropical flowering tree. *Auk 87*:467–491.

Stockdale, P.H.G. 1970. Pulmonary lesions in mink with a mixed infection of *Filaroides martis and Perostrongylus pridhami. Can. J. Zool. 48*: 757–759.

Stockton, W. L., and T. E. DeLaca. 1982. Food falls in the deep sea: occurrence, quality and significance. *Deep-Sea Res. 29*:157–169.

Storer, R. W. 1966. Sexual dimorphism and food habits in three North American accipiters. *Auk 83*:423–436.

Strong, D. R. 1974. Rapid asymptotic species accumulation in phytophagous communities: The pests of cacao. *Science 185*:1064–1066.

Strong, D. R. 1977a. Insect species richness: hispine beetles of *Heliconia latispatha. Ecology 58*:573–582.

Strong, D. R. 1977b. Rolled-leaf hispine beetles (Chrysomelidae) and their Zingiberales host plants in Middle America. *Biotropica 9*:156–169.

Strong, D. R. 1979. Biogeographical dynamics of insect-host plant communities. *Ann. Rev. Ent. 24*:89–119.

Strong, D. R. 1980. Null hypotheses in ecology. *Synthèse 43*:271–285.

Strong, D. R. 1982a. Harmonious coexistence of hispine beetles on *Heliconia* in experimental and natural communities. *Ecology 63*:1039–1049.

Strong, D. R. 1982b. Potential interspecific competition and host specificity: hispine beetles on *Heliconia. Ecol. Entomol. 7*:217–220.

Strong, D. R., and D. A. Levin. 1979. Growth form of host plants and species richness of their parasites. *Amer Natur. 114*:1–22.

Strong, D. R., E. D. McCoy, and J. R. Rey. 1977. Time and the number of herbivore species: The pests of sugarcane. *Ecology 58*:167–175.

Strong, D. R., and J. R. Rey. 1982. Testing for MacArthur-Wilson equilibrium with the arthropods of the miniature *Spartina* archipelago at Oyster Bay, Florida. *Amer. Zool. 22*:355–360.

Strong, D. R., and D. Simberloff. 1981. Straining at gnats and swallowing ratios: Character displacement. *Evolution 35*:810–812.

Strong, D. R., L. A. Szyska, and D. Simberloff. 1979. Tests of community-wide character displacement against null hypotheses. *Evolution 33*: 897–913.

Sugihara, G. 1980. Minimal community structure: an explanation of species abundance patterns. *Amer. Natur. 116*:770–787.

Sugihara, G. 1981. $S = CA^z$, $z \simeq 1/4$: a reply to Connor and McCoy. *Amer. Natur. 117*:790–793.

Sulloway, F. J. 1982. Darwin and his finches: the evolution of a legend. *J. Hist. Biol. 15*:1–53.

Suppe, F. 1977. *The Structure of Scientific Theories*. University of Illinois Press, Chicago.

Sutherland, J. P. 1970. Dynamics of high and low populations of the limpet *Acmaea scabra* (Gould). *Ecol. Monogr. 40*:169–188.

Sutherland, J. P. 1978. Functional roles of *Schizoporella* and *Styela* in the fouling community at Beaufort, North Carolina. *Ecology 59*:257–264.

Sutherland, J. P. 1980. Dynamics of the epibenthic community on roots of the mangrove *Rhizophora mangle*, at Bahia de Buche, Venezuela. *Mar. Biol. 58*:75–85.

Svärdson, G. 1949. Competition and habitat selection in birds. *Oikos 1*: 157–174.

Svärdson, G. 1976. Interspecific population dominance in fish communities of Scandinavian lakes. *Rep. Inst. Freshw. Res. Drottningholm 55*: 145–171.

Svensson, S. 1978. Territorial exclusion of *Acrocephalus schoenobaenus* by *A. scirpaceus* in reedbeds. *Oikos 30*:467–474.

Swingle, H. S. 1956. Appraisal of methods of fish population study, Part IV. Determination of balance in farm fish ponds. *Trans. N. Am. Wildl. Conf. 21*:298–322.

Talbot, F. H., B. C. Russell, and G.R.V. Anderson. 1978. Coral reef fish communities: unstable high-diversity systems? *Ecol. Monogr. 49*:425–440.

Taub, M. L. 1977. Differences facilitating the coexistence of two sympatric, orb-weaving spiders, *Argiope aurantia* Lucas and *Argiope trifasciata* (Forskal) (Araneidae, Araneae). Master's Thesis, University of Maryland, College Park.

Taylor, L. R., and R.A.J. Taylor. 1977. Aggregation, migration and population mechanics. *Nature 265*:415–421.

Teal, J. M. 1962. Energy flow in the salt marsh ecosystem of Georgia. *Ecology 43*:614–624.

Teesdale, C. 1941. Pineapple and banana plants as sources of *Aedes* mosquitoes. *East African Medical Journal 18*:260–267.

Tegner, M. J., and P. K. Dayton. 1977. Sea urchin recruitment patterns and implications of commercial fishing. *Science 196*:324–326.

Tegner, M. J., and P. K. Dayton. 1981. Population structure, recruitment and mortality of *Strongylocentrotus franciscanus* and *S. purpuratus* in a kelp forest near San Diego, California. *Mar. Ecol. Prog. Series 2*: 255–268.

Tenore, K. R., and C. B. Coull (eds.). 1980. *Marine Benthic Dynamics*, University of South Carolina Press, Columbia, S.C.

Terborgh, J. 1981. Discussion of a paper by Simberloff *et al.* Pages 64–68 in *Vicariance Biogeography: A Critique*, G. Nelson and D. E. Rosen (eds.). Columbia University Press, N.Y.

Thomas, J. D. 1962. The food and growth of brown trout (*Salmo trutta* L.) and its feeding relationships with the salmon parr (*Salmon salar* L.) and the eel (*Anguilla anguilla* L.) in the River Teify, West Wales. *J. Anim. Ecol. 31*:175–205.

Thompson, J. N., and P. W. Price. 1977. Plant plasticity, phenology, and herbivore dispersion: Wild parsnip and the parsnip webworm. *Ecology 58*:1112–1119.

Thomson, J. D. 1978a. Competition and cooperation in plant pollinator systems. Ph.D. dissertation. University of Wisconsin, Madison.

Thomson, J. D. 1978b. Effects of stand composition in two-species mixtures of *Hieracium*. *Amer. Midl. Natur. 100*:431–440.

Thomson, J. D. 1980. Implications of different sorts of evidence for competition. *Amer. Natur. 116*:719–726.

Thomson, J. D. 1981. Spatial and temporal components of resource assessment by flower-feeding insects. *J. Anim. Ecol. 50*:49–59.

Thorson, G. 1950. Reproductive and larval ecology of marine bottom invertebrates. *Biol. Rev. 25*:1–45.

Thorson, G. 1966. Some factors influencing the recruitment and establishment of marine benthic communities. *Neth. J. Sea Res. 3*:267–293.

Thresher, R. E. 1976a. Field analysis of the territoriality of the threespot damselfish, *Eupomacentrus planifrons* (Pomacentridae). *Copeia 1976*: 266–276.

Thresher, R. E. 1976b. Field experiments on species recognition by the threespot damselfish, *Eupomacentrus planifrons* (Pisces: Pomacentridae). *Anim. Behav. 24*:562–569.

Tilley, S. G. 1968. Size-fecundity relationships and their evolutionary implications in five desmognathine salamanders. *Evolution 22*:806–816.

Tilley, S. G. 1980. Life histories and comparative demography of two salamander populations. *Copeia 1980*:806–821.

Tilley, S. G. 1981. A new species of *Desmognathus* (Amphibia: Caudata: Plethodontidae) from the southern Appalachian Mountains. *Occasional Papers of the Museum of Zoology, University of Michigan 695*: 1–23.

Timko, P. 1975. High density aggregation in *Dendraster excentricus* (Eshescholtz): analysis of strategies and benefits concerning growth, age structure, feeding, hydrodynamics, and reproduction. Ph.D. Dissertation, University of California, Los Angeles.

Tinkle, D. W. 1982. Results of experimental density manipulation in an Arizona lizard community. *Ecology 63*:57–65.

Townes, H. 1971. Ichneumonidae as biological control agents. *Proc. Tall Timbers Conf. Ecol. Anim. Control by Habitat Manage. 3*:235–248.

Tryon, R. M., Jr. 1941. A revision of the genus *Pteridium. Rhodora 43*: 1–31, 37–67.

Turelli, M. 1978. A reexamination of stability in randomly varying versus deterministic environments with comments on the stochastic theory of limiting similarity. *Theor. Pop. Biol. 13*:244–267.

Turnbull, A. L. 1973. Ecology of the true spiders. (Araneomorphae). *Ann. Rev. Ent. 18*:305–348.

Turner, M., and G. A. Polis. 1979. Patterns of co-existence in a guild of raptorial spiders. *J. Anim. Ecol. 48*:509–520.

Turner, R. D. 1973. Wood-boring bivalves, opportunistic species in the deep sea. *Science 180*:1377–1379.

Turnipseed, S. G., and M. Kogan. 1976. Soybean entomology. *Ann. Rev. Ent. 21*:247–282.

Uetz, G. W. 1977. Coexistence in a guild of wandering spiders. *J. Anim. Ecol. 46*:531–542.

Uetz, G. W. Web building and prey capture in communal orb weavers. In *Spider Webs and Spider Behavior*, W. A. Shear (ed.). Stanford University Press, Stanford, Calif. (in press).

Uetz, G. W., A. D. Johnson, and D. W. Schemske. 1978. Web placement, web structure, and prey capture in orb-weaving spiders. *Bull. Brit. Arachnol. Soc. 4*:141–148.

Ulanowicz, R. E. 1979. Prediction, chaos, and ecological perspective. Pages 107–117 in *Theoretical Systems Ecology*, E. Halfon (ed.). Academic Press, N.Y.

Underwood, A. J. 1975. Intertidal zonation of prosobranch gastropods; analysis of densities of four coexisting species. *J. Exp. Mar. Biol. Ecol. 19*:197–216.

Underwood, A. J. 1978a. An experimental evaluation of competition between three species of intertidal prosobranch gastropods. *Oecologia 33*:185–202.

Underwood, A. J. 1978b. A refutation of critical tidal levels as determinants of the structure of intertidal communities on British shores. *J. Exp. Mar. Biol. Ecol. 33*:261–276.

Underwood, A. J. 1979. The ecology of intertidal gastropods. *Adv. Mar. Biol. 16*:111–210.

Underwood, A. J. 1980. The effects of grazing by gastropods and physical factors on the upper limits of distribution of intertidal macroalgae. *Oecologia 46*:201–213.

Underwood, A. J. 1981. Structure of a rocky intertidal community in New South Wales: Patterns of vertical distribution and seasonal changes. *J. Exp. Mar. Biol. Ecol. 51*:57–85.

Underwood, A. J., E. J. Denley, and M. J. Moran. 1983. Experimental analyses of the structure and dynamics of mid-shore rocky intertidal communities in New South Wales. *Oecologia 56*:202–219.

Underwood, A. J., and P. Jernakoff. 1981. Interactions between algae and grazing gastropods in the structure of a low-shore intertidal algal community. *Oecologia 48*:221–233.

Van Hook, R. I. 1971. Energy and nutrient dynamics of spider and orthopteran populations in a grassland ecosystem. *Ecol. Monogr. 41*:1–26.

VanBlaricom, G. R. 1978. Disturbance, predation and resource allocation in a high-energy sublittoral sand-bottom ecosystem: experimental analysis of critical structuring processes for the infaunal community. Ph.D. Dissertation, University of California, San Diego.

Vandermeer, J. H. 1969. The competitive structure of communities: an experimental approach with protozoa. *Ecology 50*:362–371.

Vandermeer, J. H., and D. Boucher. 1978. The varieties of mutualistic interactions in population models. *J. Theor. Biol. 74*:549–558.

Virnstein, R. W. 1977a. The importance of predation by crabs and fishes on benthic infauna in Chesapeake Bay. *Ecology 58*:1199–1217.

Virnstein, R. W. 1977b. Predator caging experiments in soft sediments: caution advised. Pages 261–273 in *Estuarine Interactions*, M. L. Wiley (ed.). Academic Press, N.Y.

Virnstein, R. W. 1980. Measuring effects of predation on benthic communities in soft sediments. Pages 281–290 in *Estuarine Perspectives*, V. S. Kennedy (ed.). Academic Press, N.Y.

Vollrath, F. 1976. Konkurrenzvermeidung bei tropischen Kleptoparasitischen Haubennetzspinnen der Gattung *Argyrodes* (Arachnida: Araneae: Theridiidae). *Ent. Germ. 3*:104–108.

Vollrath, F. 1979. Behavior of the kleptoparasitic spider *Argyrodes elevatus* (Araneae; Theridiidae). *Anim. Behav. 27*:515–521.

Wake, D. B. 1966. Comparative osteology and evolution of the lungless salamanders, family Plethodontidae. *Memoirs of the Southern California Academy of Sciences*, Vol. 4.

Warme, J. E., and N. F. Marshall. 1969. Marine borers in calcareous rocks of the Pacific coast. *Amer. Zool. 9*:763–774.

Warme, J. E., T. B. Scanland, and N. F. Marshall. 1971. Submarine canyon erosion: contribution of marine rock burrowers. *Science 173*: 1127–1129.

Waser, N. M. 1978. Competition for hummingbird pollination and sequential flowering in two Colorado wildflowers. *Ecology 59*:934–944.

Waser, N. M., and L. A. Real. 1979. Effective mutualism between sequentially flowering plant species. *Nature 281*:670–673.

Wattel, J. 1973. Geographical differentiation in the genus *Accipiter*. Publication of Nuttall Ornithological Club, No. 13.

Webb, W. L. 1965. Small mammal populations on islands. *Ecology 46*: 479–498.

Weiler, C. S. 1980. Population structure and *in situ* division rates of *Ceratium* in oligotrophic waters of the North Pacific central gyre. *Limnol. Oceanogr. 25*:610–619.

Wells, B. W. 1932. *The Natural Gardens of North Carolina*. University of North Carolina Press, Chapel Hill, N.C.

Werner, E. E. 1977. Species packing and niche complementarity in three sunfishes. *Amer. Natur. 111*:553–578.

Werner, E. E., and D. J. Hall. 1976. Niche shifts in sunfishes: experimental evidence and significance. *Science 191*:404–406.

Werner, E. E., and D. J. Hall. 1977. Competition and habitat shift in two sunfishes (Centrarchidae). *Ecology 58*:869–876.

Werner, E. E., and D. J. Hall. 1979. Foraging efficiency and habitat switching in competing sunfishes. *Ecology 60*:256–264.

Werner, E. E., D. J. Hall, D. R. Laughlin, D. J. Wagner, L. A. Wilsmann, and F. C. Funk. 1977. Habitat partitioning in a freshwater fish community. *J. Fish. Res. Board Can. 34*:360–370.

Werner, E. E., J. F. Gilliam, D. J. Hall, and G. G. Mittelbach. 1983a. An experimental test of the effects of predation risk on habitat use in fish. *Ecology 64* (in press).

Werner, E. E., and G. G. Mittelbach. 1981. Optimal foraging: field tests of diet choice and habitat switching. *Amer. Zool. 21*:813–829.

Werner, E. E., G. G. Mittelbach, D. J. Hall, and J. F. Gilliam. 1983b. Experimental tests of optimal habitat use in fish: The role of relative habitat profitability. *Ecology 64* (in press).

Werner, R. G. 1966. Ecology and movements of bluegill sunfish fry in a small northern Indiana lake. Ph.D. Dissertation, Indiana University.

Whitaker, J. O., Jr., and D. C. Rubin. 1971. Food habits of *Plethodon jordani metcalfi* and *Plethodon jordani shermani* from North Carolina. *Herpetologica 27*:81–86.

Whittaker, R. H. 1956. Vegetation of the Great Smoky Mountains. *Ecol. Monogr. 26*:1–80.

Whittaker, R. H. 1967. Gradient analysis of vegetation. *Biol. Rev. 42*:207–264.

Whittaker, R. H. 1975. *Communities and Ecosystems*, 2nd ed. Macmillan, N.Y.

Whittaker, R. H., and D. Goodman. 1979. Classifying species according to their demographic strategy I: population fluctuations and environmental heterogeneity. *Amer. Natur. 113*:185–200.

Whittam, T. S., and D. Siegel-Causey. 1981. Species interactions and community structure in Alaskan seabird colonies. *Ecology 62*:1515–1524.

Wiebe, P. H., E. M. Hulburt, E. J. Carpenter, A. E. Jahn, G. P. Knapp, S. H. Boyde, P. B. Ortner, and J. L. Cox. 1976. Gulf stream cold core rings: large-scale interaction sites for open ocean plankton communities. *Deep-Sea Res. 34*:695–710.

Wiens, J. A. 1969. An approach to the study of ecological relationships among grassland birds. *Ornith. Monogr. 8*:1–93.

Wiens, J. A. 1973. Pattern and process in grassland bird communities. *Ecol. Monogr. 43*:237–270.

Wiens, J. A. 1974a. Climatic instability and the "ecological saturation" of bird communities in North American grasslands. *Condor 76*:385–400.

Wiens, J. A. 1974b. Habitat heterogeneity and avian community structure in North American grasslands. *Amer. Midl. Natur. 91*:195–213.

Wiens, J. A. 1977a. On competition and variable environments. *Amer. Sci. 65*:590–597.

Wiens, J. A. 1977b. Model estimation of energy flow in North American grassland bird communities. *Oecologia 31*:135–151.

Wiens, J. A. 1980. Concluding comments: are bird communities real? Pages 1088–1089 in *XVII Congr. Intern. Ornithol., Berlin*.

Wiens, J. A. 1981a. Scale problems in avian censusing. *Studies Avian Biol.* *6*:513–521.

Wiens, J. A. 1981b. Single-sample surveys of communities: are the revealed patterns real. *Amer. Natur. 117*:90–98.

Wiens, J. A., and M. I. Dyer. 1975. Rangeland avifaunas: their composition, energetics, and role in the ecosystem. *Proc. Symp. Mgmt. Forest Range Habitats for Nongame Birds.* USDA For. Serv. Gen. Tech. Rept. WO-1:146–182.

Wiens, J. A., and J. T. Rotenberry. 1979. Diet niche relationships among North American grassland and shrubsteppe birds. *Oecologia 42*:253–292.

Wiens, J. A., and J. T. Rotenberry. 1980a. Patterns of morphology and ecology in grassland and shrubsteppe bird populations. *Ecol. Monogr. 50*:287–308.

Wiens, J. A., and J. T. Rotenberry. 1980b. Bird community structure in cold shrub deserts: competition or chaos? Pages 1063–1070 in *Acta XVII Congr. Intern. Ornithol., Berlin.*

Wiens, J. A., and J. T. Rotenberry. 1981a. Censusing and the evaluation of avian habitat occupancy. *Studies Avian Biol. 6*:522–532.

Wiens, J. A., and J. T. Rotenberry. 1981b. Habitat associations and community structure of birds in shrubsteppe environments. *Ecol. Monogr. 51*:21–41.

Wiens, J. A., and J. T. Rotenberry. 1981c. Morphological size ratios and competition in ecological communities. *Amer. Natur. 117*:592–599.

Wilbur, H. M. 1972. Competition, predation, and the structure of the *Ambystoma-Rana sylvatica* community. *Ecology 53*:3–21.

Williams, C. B. 1964. *Patterns in the Balance of Nature.* Academic Press, London.

Williams, D. McB. 1980. The dynamics of the pomacentrid community on small patch reefs in One Three Lagoon (Great Barrier Reef). *Bull. Mar. Sci. 30*:159–170.

Williams, D. McB., and P. F. Sale. 1981. Spatial and temporal patterns of recruitment of juvenile coral reef fishes to coral habitats within One Tree Lagoon, G.B.R. *Mar. Biol. 65*:245–253.

Williams, G. B. 1964. The effects of extracts of *Fucus serratus* in promoting the settlement of larvae of *Spirorbis borealis* (Polychaeta). *J. Mar. Biol. Ass. U.K. 44*:397–414.

Williams, J. B. and G. O. Batzli. 1979a. Interference competition and niche shifts in the bark-foraging guild in central Illinois. *Wilson Bull. 91*:400–411.

Williams, J. B., and G. O. Batzli. 1979b. Competition among bark-foraging birds in central Illinois: experimental evidence. *Condor 81*:122–132.

Williamson, M. 1981. *Island Populations.* Oxford University Press, Oxford.

Wilsmann, L. A. 1979. Resource partitioning and mechanisms of coexistence of blackchin and blacknose shiners (*Notropis*: Cyprinidae). Ph.D. Dissertation, Michigan State University.

Wilson, D. S. 1980. *The Natural Selection of Populations and Communities.* Benjamin/Cummings Publishing Company, Menlo Park, Calif.

Wilson, E. O. 1965. The challenge from related species. Pages 7-24 in *The Genetics of Colonizing Species*, H. G. Baker and G. L. Stebbins (eds.). Academic Press, N.Y.

Wilson, E. O. 1975. *Sociobiology: The New Synthesis.* Harvard University Press, Cambridge, Mass.

Wimsatt, W. C. 1980. Randomness and perceived-randomness in evolutionary biology. *Synthèse 43*:287–331.

Winer, B. J. 1971. *Statistical Principles in Experimental Design.* McGraw-Hill, N.Y.

Wingerden, W.K.R.E. van. 1975. Population dynamics of *Erigone arctica* (White) (Araneae: Linyphiidae). Pages 71–76 in *Proc. 6th Int. Arachnol. Congr.*, Amsterdam, 1974.

Wingerden, W.K.R.E. van. 1978. Population dynamics of *Erigone arctica* (White) (Araneae: Linyphiidae). II. *Symp. Zool. Soc. Lond. 42*:195–202.

Winslow, E.E.A. 1943. *The Conquest of Epidemic Disease.* Princeton University Press, Princeton, N.J.

Winstanley, D., R. Spencer, and K. Williamson. 1974. Where have all the Whitethroats gone? *Bird Study 21*:1–14.

Wise, D. H. 1975. Food limitation of the spider *Linyphia marginata*: experimental field studies. *Ecology 56*:637–646.

Wise, D. H. 1979. Effects of an experimental increase in prey abundance upon the reproductive rates of two orb-weaving spider species (Araneae: Araneidae). *Oecologia 41*:289–300.

Wise, D. H. 1981a. Inter- and intraspecific effects of density manipulatioms upon females of two orb-weaving spiders (Araneae: Araneidae). *Oecologia 48*:252–256.

Wise, D. H. 1981b. A removal experiment with darkling beetles: lack of evidence for interspecific competition. *Ecology 62*:727–738.

Wise, D. H. 1982. Predation by the commensal spider, *Argyrodes trigonum*, upon its host: an experimental study. *J. Arachnol. 10*:111–116.

Wise, D. H. 1983. Competitive mechanisms in a food-limited species: Relative importance of interference and exploitative interactions among labyrinth spiders (Araneae: Araneidae). *Oecologia 58*:1–9.

Wise, D. H., and J. L. Barata. 1983. Prey of two syntopic spiders with different web structures. *J. Arachnol. 11*:271–282.

Witherby, H. F., F.C.R. Jourdain, N.F. Ticehurst, and B. W. Tucker. 1938. *The Handbook of British Birds.* Witherby London.

Wolcott, T. G. 1973. Physiological ecology and intertidal zonation in limpets (*Acmaea*): a critical look at "limiting factors." *Biol. Bull. 145*:389–422.

Wolda, H. 1978. Fluctuations in abundance of tropical insects. *Amer. Natur. 112*:1017–1045.

Wolschlag, D. E., and R. O. Juliano. 1959. Seasonal changes in bluegill metabolism. *Limnol. Oceanogr. 4*:195–209.

Woodin, S. A. 1974. Polychaete abundance patterns in a marine soft-sediment environment: the importance of biological interactions. *Ecol. Monogr. 44*:171–187.

Woodin, S. A. 1976. Adult-larval interactions in dense infaunal assemblages: patterns of abundance. *J. Mar. Res. 34*:25–41.

Woodin, S. A. 1977. Algal "gardening" behavior by Nereid polychaetes: effects on soft-bottom community structure. *Mar. Biol. 44*:39–42.

Woodin, S. A. 1978. Refuges, disturbance, and community structure: a marine soft-bottom example. *Ecology 59*:274–284.

Worster, D. 1977. *Nature's Economy.* Sierra Club Books, San Francisco, Calif.

Wright, S. J. 1979. Competition between insectivorous lizards and birds in Central Panama. *Amer. Zool. 19*:1145–1156.

Wright, S. J. 1981. Extinction-mediated competition: the *Anolis* lizards and insectivorous birds of the West Indies. *Amer. Natur. 117*:181–192.

Wright, S. J., and C. C. Biehl. 1982. Island biogeographic distributions: testing for random, regular, and aggregated patterns of species occurrence. *Amer. Natur. 119*:345–357.

Yeaton, R. I., and M. L. Cody. 1974. Competitive release in island song sparrow populations. *Theor. Pop. Biol. 5*:42–58.

Yodzis, P. 1978. *Competition for Space and the Structure of Ecological Communities.* Springer-Verlag, Berlin and New York.

Yodzis, P. 1980. The connectance of real ecosystems. *Nature 284*:544–545.

Yodzis, P. 1981. The stability of real ecosystems. *Nature 289*:674–676.

Zimmerman, E. C. 1948. *Insects of Hawaii.* Volume I. *Introduction.* University of Hawaii Press, Honolulu.

Author Index

Abbott, 95, 138, 204–205, 207, 210, 217,
    220, 223, 235, 243–244, 246, 248–249,
    252–253, 491, 495, 509
Abele, 101, 123–125, 128, 135, 184
Abrams, 381
Addicott, 55, 113
Alatalo, 211, 319–320
Alexander, 367
Allen, E. B., 113
Allen, G. R., 481
Alm, 364
Anderson, B. W., 494
Anderson, D. T., 159
Anderson, G.R.V., 13, 488–489
Anderson, J. F., 43, 45
Anderson, R. C., 384, 392
Anderson, R. M., 15
Andrewartha, 476, 521
Andrews, 41
Anonymous, 3
Arntz, 189
Arp, 4
Askenmo, 52
Askew, 417, 422
Auerbach, M. A., 38
Auerbach, M. J., 7, 72, 413, 420
Austin, 126–128
Ayala, 55
Ayling, 184

Bagenal, 376
Bakus, 184
Balick, 72
Bardach, 479
Barnett, 414
Battan, 403
Beaver, 515, 522
Beddington, 399
Beebe, 218

Bennett, 375
Benson, 97
Benzing, 522
Bergquist, 184, 187
Bernstein, 187–188
Birch, 16, 40, 475
Bishop, 309, 330
Blalock, 383, 388
Blaustein, 291–292
Blegvad, 189
Bloom, 113
Boag, 218, 220
Boatman, Sam the, 441
Bohnsack, 134, 481–483, 487–488
Bournier, 93
Bouton, 498
Bowers, 291
Bowman, R. I., 204–205, 217–218, 222, 245
Bowman, R. S., 169
Box, 461
Boyce, 440
Bradley, 243
Brady, 383, 476
Branch, 160, 175, 179, 183
Bray, 114
Brenchley, 189
Brett, 375
Breymeyer, 44
Bronowski, 454
Brown, J. H., 15, 55, 101, 230, 282,
    284–286, 288–289, 291, 292, 295,
    319–320, 392
Brown, J. L., 476
Brown, K. M., 44, 46
Brown, L. H., 257–259, 271
Brown, W. L., 217
Bunge, 384
Burkenroad, 169
Burton, 19

Taxonomic Index

Subject Index

LIBRARY OF CONGRESS CATALOGING IN PUBLICATION DATA
Main entry under title:

Ecological communities.

  Bibliography: p.
  Includes indexes.
  1. Biotic communities—Addresses, essays, lectures.
I. Strong, Donald R., 1944–
QH541.145.E258 1984   574.5′247   83–43093
ISBN 0-691-08340-1
ISBN 0-691-08341-X (pbk.)